# Design for Additive Manufacturing: Methods and Tools

# Design for Additive Manufacturing: Methods and Tools

Editors

**Marco Mandolini**
**Patrick Pradel**
**Paolo Cicconi**

MDPI • Basel • Beijing • Wuhan • Barcelona • Belgrade • Manchester • Tokyo • Cluj • Tianjin

*Editors*

Marco Mandolini
Department of Industrial Engineering and Mathematical Sciences, Università Politecnica delle Marche, Via Brecce Bianche 12, 60131 Ancona, Italy

Patrick Pradel
Design School, Loughborough University, Loughborough LE11 3TU, UK

Paolo Cicconi
Department of Industrial, Electronic and Mechanical Engineering, Università degli Studi Roma Tre, Via della Vasca Navale, 79, Roma, Italy

*Editorial Office*
MDPI
St. Alban-Anlage 66
4052 Basel, Switzerland

This is a reprint of articles from the Special Issue published online in the open access journal *Applied Sciences* (ISSN 2076-3417) (available at: https://www.mdpi.com/journal/applsci/special_issues/Design_Additive_Manufacturing).

For citation purposes, cite each article independently as indicated on the article page online and as indicated below:

LastName, A.A.; LastName, B.B.; LastName, C.C. Article Title. *Journal Name* **Year**, *Volume Number*, Page Range.

**ISBN 978-3-0365-4925-5 (Hbk)**
**ISBN 978-3-0365-4926-2 (PDF)**

# Contents

# About the Editors

**Marco Mandolini**

Marco Mandolini is Assistant Professor in "Design and Methods of Industrial Engineering" at the Department of Industrial Engineering and Mathematical Sciences of the Faculty of Engineering of Università Politecnica delle Marche (UNIVPM). His research activities concern methods and tools for design (manufacturing, additive, assembly, disassembly, cost, and environment considerations). He is the author of approximately 130 scientific publications in international journals and conference proceedings. He is the Editor-in-Chief for the "Additive Manufacturing: Design, Opportunities, and Applications" topic of MDPI. He is the deputy head of the B+ laboratory (https://bplus.univpm.eu/) at UNIVPM concerning additive manufacturing. He teaches the "Product Engineering" and "Methods and tools for biomedical products design" courses, respectively, for the master's degrees in Mechanical Engineering and Biomedical Engineering.

**Patrick Pradel**

Patrick Pradel is a lecturer in Product Industrial User-Centered Design and a member of the Digital Design and Fabrication research group (https://tinyurl.com/2fwwsfj4) at Loughborough University, School of Design and Creative Arts. Since 2008, he has been interested in the intricate relationship between digital fabrication and product design, and he has collaborated on several national and international projects in this area. Currently, Patrick is the Principal Investigator for the UK Design for Additive Manufacturing Network (www.designforam.ac.uk) funded by the UK Engineering Physical Sciences Research Council and the "When Design Never Ends" project (www.design-never-ends.ac.uk), funded by the UK Arts and Humanities Research Council. He is the module leader for the "Live Projects" module for the BSc in Product Design and Technology and the "Digital Fabrication Skills" module for the MSc in Integrated Industrial Design.

**Paolo Cicconi**

Paolo Cicconi is an Assistant Professor in Design and Methods of Industrial Engineering at Università degli Studi Roma Tre (Roma Tre), Department of Industrial, Electronic and Mechanical Engineering. In 2011, he obtained a Ph.D. in Mechanical Engineering at Università Politecnica delle Marche, where he continued with a post-doctoral position. From 2018 to 2021, he was a contract professor at the Università degli Studi eCampus for the courses of Computer Graphics and BIM and Design Tools for UAV. Since 2019, he has been teaching Mechanical Drawing at Roma Tre for the degrees in Ocean Engineering and Mechanical Engineering. His research is focused on tools and methods for engineering design. He is the Topic Editor for "Additive Manufacturing: Design, Opportunities, and Applications" at MDPI.

# Preface to "Design for Additive Manufacturing: Methods and Tools"

Additive manufacturing (AM), one of the nine enabling technologies of Industry 4.0, is experiencing rapid growth. Nowadays, there are no industrial sectors that have not employed AM (e.g., automotive, energy, medicine, electronics, cultural heritage). Nevertheless, the implementation of technologies by industries is still limited compared with their intrinsic potential. The main challenges that limit the adoption of such technologies are as follows: the lack of skills—it is important to train those engineers who are capable of designing and managing these new technologies; the sustainability of new processes—we must develop cost models and environmental models which are capable of considering the economic and environmental sustainability of AM processes and their related supply chains; design—there is a need for innovative design paradigms and design for additive manufacturing (DfAM) software tools.

ASTM ISO/ASTM52910 gives requirements, guidelines, and recommendations for using AM in product design. This regulation helps determine which design considerations can be utilised in a design project and permits practitioners take advantage of the capabilities an AM processes. The overall DfAM strategy presented in ASTM ISO/ASTM52910 acts as a lighthouse for other methods which can be conceived for specific problems and products.

In recent years, DfAM software tools have advanced quickly, allowing for the prediction and thus optimization of designs for manufacture. In addition to traditional CAD tools, design freedoms have pushed software houses toward the development of systems which are able to model organic shapes (a result of topology optimisation and generative design approaches). Simulation systems increasingly support designers and production technologists in identifying problems prior to 3D printing (e.g., deformation, accumulation of tension, and porosity).

This Special Issue book highlights the current design methodologies, methods, and tools conceived for supporting product development (from conceptualisation to detail design). The text refers to the Special Issue "Design for Additive Manufacturing: Methods and Tools" of *Applied Sciences* (MDPI). This Special Issue book presents tools for better evaluating the benefits of the technologies and developments which are available in the scientific literature.

**Marco Mandolini, Patrick Pradel, and Paolo Cicconi**
*Editors*

applied sciences

MDPI

*Editorial*

# Design for Additive Manufacturing: Methods and Tools

**Marco Mandolini [1,*], Patrick Pradel [2] and Paolo Cicconi [3]**

1 Department of Industrial Engineering and Mathematical Sciences, Università Politecnica delle Marche, Via Brecce Bianche, 12, 60131 Ancona, Italy
2 School of Design and Creative Arts, Loughborough University, Loughborough LE11 3TU, UK; p.pradel@lboro.ac.uk
3 Department of Industrial, Electronic and Mechanical Engineering, Università degli Studi Roma Tre, Via della Vasca Navale, 79, 00146 Roma, Italy; paolo.cicconi@uniroma3.it
* Correspondence: m.mandolini@staff.univpm.it

**Citation:** Mandolini, M.; Pradel, P.; Cicconi, P. Design for Additive Manufacturing: Methods and Tools. *Appl. Sci.* **2022**, *12*, 6548. https://doi.org/10.3390/app12136548

Received: 22 June 2022
Accepted: 23 June 2022
Published: 28 June 2022



## 1. Introduction

Additive Manufacturing (AM), one of the nine enabling technologies of Industry 4.0, is experiencing rapid growth. Nowadays, there are no industrial sectors that have not employed AM (e.g., automotive [1], energy [2], medicine [3,4], electronics [5], cultural heritage [6]). Nevertheless, the implementation of technologies by industries is still limited compared to their intrinsic potential. The main challenges that limit the adoption of such technologies are lack of skills (need to train engineers capable of designing and managing these new technologies), sustainability of new processes (need to develop cost and environmental models capable of considering economic and environmental sustainability of AM processes and related supply chain) and design (need of innovative design paradigms and Design for Additive Manufacturing–DfAM–software tools).

ASTM ISO/ASTM52910 [7] gives requirements, guidelines, and recommendations for using AM in product design. This regulation helps determine which design considerations can be utilised in a design project or to take advantage of the capabilities of an AM process. The overall DfAM strategy presented in Figure **??** is the lighthouse for other methods conceived for specific problems and products [1,7–11].

In recent years, DfAM software tools quickly advanced, allowing for the prediction and thus optimising designs for manufacture [12]. In addition to the traditional CAD tools, the design freedom pushed software houses toward developing systems able to model organic shapes (a result of topology optimisation or generative design approaches). Simulation systems are increasingly supporting designers and production technologists in identifying problems before 3D printing (e.g., deformation, accumulation of tension, porosity) [13–15].

This editorial aims to highlight the current design methods and tools conceived for supporting the product development (from conceptualisation to detail design). The editorial refers to the Special Issue "Design for Additive Manufacturing: Methods and Tools" of *Applied Sciences* (MDPI). The paper presents some applications to evaluate better the benefits of what is available in the scientific literature. Future trends on DfAM methods and tools conclude the editorial.

## 2. DfAM Methodologies and Methods

The special issue covers different studies on DfAM methodologies and methods, with the majority focusing on developing and implementing Topology Optimisation (TO) for capitalising on AM opportunities. Ahmad, Bici and Campana [6] investigated how TO can be implemented as a concept design tool to conceive the inner supporting frame of an ancient statue. In this application, TO can close the gap between experts and young designers, supporting preliminary target evaluations and topology conception. Dalpadulo et al. [1] proposed a design methodology to optimise components by integrating TO into the redesign process for an AM Assistive Device. Their paper shows how TO can be integrated into a redesign process to achieve mechanical and AM Key Performance Indicators. The DfAM workflow proposed by Sotomayor, Caiazzo and Alfieri [10] integrates optimisation, design, and simulation tools to minimise the number of iterative design evaluations. The authors proposed introducing TO, lattice infill optimisation, and generative design in earlier phases of the design process to maximise AM capabilities. Rosso et al. [9] presented a DfAM workflow for embodiment design that combines Computer-Aided Design tools for the geometric modelling of the part and Computer-Aided Engineering tools for the optimisation and simulation phases. Their workflow considers the possibility of using size optimisation to obtain lattice structures with optimised beams and TO to obtain optimised organic shapes. Fu et al. [16] proposed a new element-based TO method based on Langelaar's AM filter to generate mechanically optimised and self-supporting designs for AM. The approach integrates restrictive DfAM constraints such as overhang angle and length to create support free designs and reduce post-processing costs. Finally, in their review of TO in the design of steel structures and components for AM, Ribeiro et al. [17] examined the methods, applications and research trends of TO in civil and structural engineering over the past five years.

Alongside TO, the remaining authors studied a range of DfAM methodologies and methods. Raffaeli et al. [8] proposed a systematic approach to support designers in understanding the opportunities offered by AM in the design of machined parts. The approach implements multi-criteria decision making to systematically assess the suitability of AM for a given set of design requirements. The study demonstrates that this approach can be advantageous, especially in the early stages of the design process. Rolinck et al. [11] presented a holistic design and development methodology for hydraulic manifolds and fluid components with improved product performance and optimised for Laser Powder Bed Fusion (LPBF). Lastly, in a comparative study, Ikeuchi et al. [14] developed a new artificial neural network model to create a data-efficient modelling approach for simulating cold spray additive manufacturing. The model can predict an as-fabricated product allowing designers to understand the economic and environmental impact of design decisions on material waste after post-processing.

## 3. DfAM Tools

AM process allows realising parts with highly complex shapes that are impossible to produce by any other technologies. The complexity and variability of the additive processes require advanced design tools to optimise the product geometry and the settings of the machine parameters.

The completed design workflow of AM parts requires different kinds of software and the employment of highly skilled users. The specific design tools concern the CAD modelling of the geometrical details, the Finite Element Method (FEM) analysis to simulate the mechanical behaviour of the part after printing, and the FEM simulation of the phases of the 3D printing process, etc. The literature already shows several DfAM tools; however, the 3D printing process's complexity and variability limit the widespread of general purposes tools.

A completed review of Computer-Aided Engineering (CAE) tools for AM is proposed by Nieto and Sanchez [12]. Their work aims to maximise the vantages of AM technology by presenting the use of specific tools for product development. They analysed engineering and CAD platform tools such as mesostructured design, optimisation software, process management, and simulation solutions. Here, the DfAM tools are classified into seven categories: Design, Topology Optimisation, 3D mesh, Process Simulation, In-service Simulation, and Preprinting Slicing. These seven design tools are employed in five design AM phases: Design, Verification, Slicing, Printing, and Finishing. The most ambitious category concerns the simulation of the 3D printing phases because of the novelty of this manufacturing process. These simulations mostly regard the powder bed fusion processes, allowing geometrical deformations and residual stresses to be calculated. Therefore, the process simulations allow optimising the process parameters to minimise the accumulated tensions and the deformations of the pieces during and after printing. The designer can also use the simulation report to analyse the printing position, select the material, and optimise the support structures.

While some DfAM tools are well known in the scientific literature and community, others are new and exclusive for specific additive manufacturing phases and processes. For example, Kim and Kang developed a specific DfAM tool for Carbon-Fibers-Reinforced Polymer filament in Fusion Deposition Modeling (FDM) [13]. They implemented a tool-path algorithm to optimise the orientation of the fibres in the FDM process. The main idea was to reinforce fibres aligned with the profile of the high-stress region, which was previously calculated by FEM analysis. This DfAM tool can be classified in Preprinting Slicing, according to the classification described in [12]. Moreover, it can be employed in the phases of Design and Slicing.

Zouaoui et al. [15] described an example of a design tool for In-service Simulation. They developed a simplified Finite Element Method (FEM) to predict the mechanical behaviour of parts manufactured by fused filament fabrication (FFF) of polymeric pre-structured materials. This paper considers the filament orientation along with the principal stress directions. They also considered the anisotropy of the material related to complex deposition trajectories by introducing specific constants. The simulation confirms that the longitudinal loading direction has the highest yield strength than the transversal one for the FFF process.

The study of DfAM tools is still an open topic in research. They can be classified by objectives, design phase, materials, or processes; however, they are still so specific that they cannot be applied as general-purpose solutions in AM design. In this context, the literature shows relevant cases of these tools for enhancing their replicability in different applications.

## 4. Applications of Methods and Tools for Additive Manufacturing

AM, intended to integrate design software and 3D printing machines to complete the product's manufacturing [18], guarantees multiple benefits in Industry 4.0. The most significant are: customisation, design and development, prototyping, virtual inventory, reduced wastages, speed, flexibility, risk reduction, customer satisfaction, accuracy, productivity, improved profitability, improved supply chain performance, and cost reduction [18].

The variety of materials employed in AM processes (i.e., metals and alloys, polymers, composites, ceramics, and concrete) makes this technology applicable to several industrial sectors. The most trending could be summarised as follows [19]:

- *Biomedical and Biofabrication*: the biomedical market represents around 16% of the total AM market share (Wohlers' report). AM can guarantee the realisation of high-complex shapes [4,20] for innovative biomedical implants, organs, and controlled drug delivery systems. Open-source design methodologies can be used to quickly share 3D AM CAD files among researchers [3].
- *Aerospace*: this sector represents around 18% of the total AM market share. AM may allow the production of complex geometries resulting from parts consolidation, fluidodynamics simulation (to maximise efficiency [2]), and topology optimisation (to maximise the performance to weight ratio). Furthermore, functional electronics (e.g., resistive sensors) can be easily 3D printed [5].
- *Buildings, architecture, and cultural heritage*: this is an infancy sector expected to grow. AM can reduce construction time, increase geometry complexity, and reduce the consumption of heart resources. Reconstruction, documentation, and preservation are the typical AM purposes for cultural heritage [6].
- *Transport and automotive*: AM can produce lightweight components in this sector, which are increasingly requested to comply with energy consumption regulations. AM can also speed up the product development phase. For example, AM is currently adopted

to realise soft assembly tools to increase productivity and manage customised parts without increasing production complexity and cost [21].

## 5. Future Trends in DfAM Methods and Tools

DfAM methods and tools continuously evolve to manage new production technologies and materials and support design engineers adopting AM. Among the future trends for this topic, hereunder are the most relevant to be addressed in future and specific research:

- Methodologies to support designers in disruptive/innovative rather than incremental design [22].
- AM for improving environmental sustainability. Methods to evaluate the environmental sustainability for multi-criteria decision making, product remanufacturing, etc. [22].
- Adopt Life Cycle Costing approaches to evaluate the sustainability of AM [22].
- Extend and update DfAM rules related to the emerging manufacturing technologies.
- Characterisation of complex lattice structures [23].
- Improved simulation software tools for accurately predicting product performances [23].
- Software tools to manage multiple materials, meta material design, multiple function design, and multi-scale [22].
- Knowledge-based 3D CAD systems to automatically model optimised AM shapes [24].
- Increase the adoption of AM and definition of DfAM rules for thermofluid, optical and electronic applications [22].

## 6. Conclusions

As emerged in this editorial, Additive Manufacturing is one of the most exciting research topics in the industry. New design methods and innovative software tools aim to support designers in considering this production edge-breaking technology during product development. The papers published in the Special Issue "Design for Additive Manufacturing: Methods and Tools" of *Applied Sciences* Journal entirely focus on this goal. Nevertheless, since the rapid growth of AM technologies and materials, design methods and tools should follow this trend to continuously support designers of the involved industrial sectors for applying AM and exploiting related unique benefits.

**Author Contributions:** Conceptualization, M.M., P.P. and P.C.; methodology, M.M., P.P. and P.C.; formal analysis, M.M., P.P. and P.C.; resources, M.M., P.P. and P.C.; writing—original draft preparation, M.M., P.P. and P.C.; writing—review and editing, M.M., P.P. and P.C.; supervision, M.M., P.P. and P.C.; All authors have read and agreed to the published version of the manuscript.

**Funding:** This research received no external funding.

**Acknowledgments:** This issue would be impossible without the contributions of various talented authors, hardworking and professional reviewers, and a dedicated editorial team of *Applied Sciences*. Congratulations to all authors—no matter the final decisions of the submitted manuscripts, the reviewers' and editors' feedback, comments, and suggestions helped the authors improve their papers. We want to take this opportunity to record our sincere gratefulness to all reviewers. Finally, we place on record our gratitude to the editorial team of *Applied Sciences*.

**Conflicts of Interest:** The authors declare no conflict of interest.

## References

1. Dalpadulo, E.; Gherardini, F.; Pini, F.; Leali, F. Integration of topology optimisation and design variants selection for additive manufacturing-based systematic product redesign. *Appl. Sci.* **2020**, *10*, 7841. [CrossRef]
2. Murdy, P.; Dolson, J.; Miller, D.; Hughes, S.; Beach, R. Leveraging the advantages of additive manufacturing to produce advanced hybrid composite structures for marine energy systems. *Appl. Sci.* **2021**, *11*, 1336. [CrossRef]
3. Kukko, K.; Akmal, J.; Kangas, A.; Salmi, M.; Björkstrand, R.; Viitanen, A.; Partanen, J.; Pearce, J. Additively manufactured parametric universal clip-system: An open source approach for aiding personal exposure measurement in the breathing zone. *Appl. Sci.* **2020**, *10*, 6671. [CrossRef]

4. Safonov, A.; Maltsev, E.; Chugunov, S.; Tikhonov, A.; Konev, S.; Evlashin, S.; Popov, D.; Pasko, A.; Akhatov, I. Design and fabrication of complex-shaped ceramic bone implants via 3D printing based on laser stereolithography. *Appl. Sci.* **2020**, *10*, 7138. [CrossRef]
5. Watschke, H.; Goutier, M.; Heubach, J.; Vietor, T.; Leichsenring, K.; Böl, M. Novel resistive sensor design utilizing the geometric freedom of additive manufacturing. *Appl. Sci.* **2021**, *11*, 113. [CrossRef]
6. Ahmad, A.; Bici, M.; Campana, F. Guidelines for topology optimization as concept design tool and their application for the mechanical design of the inner frame to support an ancient bronze statue. *Appl. Sci.* **2021**, *11*, 7834. [CrossRef]
7. *ISO/ASTM 52910*; Additive manufacturing—Design—Requirements, Guidelines and Recommendations. International Organization for Standardization: Geneva, Switzerland, 2018.
8. Raffaeli, R.; Lettori, J.; Schmidt, J.; Peruzzini, M.; Pellicciari, M. A systematic approach for evaluating the adoption of additive manufacturing in the product design process. *Appl. Sci.* **2021**, *11*, 1210. [CrossRef]
9. Rosso, S.; Uriati, F.; Grigolato, L.; Meneghello, R.; Concheri, G.; Savio, G. An optimization workflow in design for additive manufacturing. *Appl. Sci.* **2021**, *11*, 2572. [CrossRef]
10. Sbrugnera Sotomayor, N.; Caiazzo, F.; Alfieri, V. Enhancing design for additive manufacturing workflow: Optimisation, design and simulation tools. *Appl. Sci.* **2021**, *11*, 6628. [CrossRef]
11. Rolinck, N.; Schmitt, M.; Schneck, M.; Schlick, G.; Schilp, J. Development workflow for manifolds and fluid components based on laser powder bed fusion. *Appl. Sci.* **2021**, *11*, 7335. [CrossRef]
12. Moreno Nieto, D.; Moreno Sánchez, D. Design for additive manufacturing: Tool review and a case study. *Appl. Sci.* **2021**, *11*, 1571. [CrossRef]
13. Kim, J.; Kang, B. Enhancing structural performance of short fiber reinforced objects through customized tool-path. *Appl. Sci.* **2020**, *10*, 8168. [CrossRef]
14. Ikeuchi, D.; Vargas-Uscategui, A.; Wu, X.; King, P. Data-efficient neural network for track profile modelling in cold spray additive manufacturing. *Appl. Sci.* **2021**, *11*, 1654. [CrossRef]
15. Zouaoui, M.; Gardan, J.; Lafon, P.; Makke, A.; Labergere, C.; Recho, N. A finite element method to predict the mechanical behavior of a pre-structured material manufactured by fused filament fabrication in 3D printing. *Appl. Sci.* **2021**, *11*, 5075. [CrossRef]
16. Fu, Y.; Ghabraie, K.; Rolfe, B.; Wang, Y.; Chiu, L. Smooth design of 3D self-supporting topologies using additive manufacturing filter and SEMDOT. *Appl. Sci.* **2021**, *11*, 238. [CrossRef]
17. Ribeiro, T.; Bernardo, L.; Andrade, J. Topology optimisation in structural steel design for additive manufacturing. *Appl. Sci.* **2021**, *11*, 2112. [CrossRef]
18. Haleem, A.; Javaid, M. Additive manufacturing applications in industry 4.0: A review. *J. Ind. Inf. Integr.* **2019**, *4*, 1930001. [CrossRef]
19. Ngo, T.D.; Kashani, A.; Imbalzano, G.; Nguyen, K.T.Q.; Hui, D. Additive manufacturing (3D printing): A review of materials, methods, applications and challenges. *Compos. Part B Eng.* **2018**, *143*, 172–196. [CrossRef]
20. Nsiempba, K.; Wang, M.; Vlasea, M. Geometrical degrees of freedom for cellular structures generation: A new classification paradigm. *Appl. Sci.* **2021**, *11*, 3845. [CrossRef]
21. Vasco, J.C. Chapter 16—Additive manufacturing for the automotive industry. In *Handbooks in Advanced Manufacturing, Additive Manufacturing*; Elsevier: Amsterdam, The Netherlands, 2021; pp. 505–530. [CrossRef]
22. Lopez Taborda, L.L.; Maury, H.; Pacheco, J. Design for additive manufacturing: A comprehensive review of the tendencies and limitations of methodologies. *Rapid Prototyp. J.* **2021**, *27*, 918–966. [CrossRef]
23. Zhu, J.; Zhou, H.; Wang, C.; Zhou, L.; Yuan, S.; Zhang, W. A review of topology optimisation for additive manufacturing: Status and challenges. *Chin. J. Aeronaut.* **2021**, *34*, 91–110. [CrossRef]
24. Biedermann, M.; Beutler, P.; Meboldt, M. Automated knowledge-based design for additive manufacturing: A case study with flow manifolds. *Chem. Ing. Tech.* **2022**, *94*, 1–10. [CrossRef]

applied sciences

MDPI

*Article*

# A Systematic Approach for Evaluating the Adoption of Additive Manufacturing in the Product Design Process

**Roberto Raffaeli [1,*], Jacopo Lettori [1], Juliana Schmidt [2], Margherita Peruzzini [2] and Marcello Pellicciari [1]**

[1] Department of Sciences and Methods for Engineering, Università degli Studi di Modena e Reggio Emilia, via Amendola 2, 42122 Reggio Emilia, Italy; jacopo.lettori@unimore.it (J.L.); marcello.pellicciari@unimore.it (M.P.)
[2] Department of Engineering "Enzo Ferrari", Università degli Studi di Modena e Reggio Emilia, via Vivarelli 10, 41125 Modena, Italy; juliana.schmidt@unimore.it (J.S.); margherita.peruzzini@unimore.it (M.P.)
* Correspondence: roberto.raffaeli@unimore.it

**Abstract:** Additive Manufacturing (AM) technologies have expanded the possibility of producing unconventional geometries, also increasing the freedom of design. However, in the designer's everyday work, the decision regarding the adoption of AM for the production of a component is not straightforward. In fact, it is necessary to process much information regarding multiple fields to exploit the maximum potential of additive production. For example, there is a need to evaluate the properties of the printable materials, their compatibility with the specific application, redesign shapes accordingly to AM limits, and conceive unique and complex products. Additionally, procurement and logistics evaluations, as well as overall costs possibly extending to the entire life cycle, are necessary to come to a decision for a new and radical solution. In this context, this paper investigates the complex set of information involved in this process. Indeed, it proposes a framework to support and guide a designer by means of a structured and algorithmic procedure to evaluate the opportunity for the adoption of AM and come to an optimal design. A case study related to an ultralight aircraft part is reported to demonstrate the proposed decision process.

**Keywords:** additive manufacturing; design for additive manufacturing; product design; multi criteria decision-making

**Citation:** Raffaeli, R.; Lettori, J.; Schmidt, J.; Peruzzini, M.; Pellicciari, M. A Systematic Approach for Evaluating the Adoption of Additive Manufacturing in the Product Design Process. *Appl. Sci.* **2021**, *11*, 1210. https://doi.org/10.3390/app11031210

Academic Editor: Giangiacomo Minak

Received: 28 December 2020
Accepted: 25 January 2021
Published: 28 January 2021



## 1. Introduction

In the last few years, Additive Manufacturing (AM) has become an actual manufacturing option and many industrial applications clearly show it is able to overcome and substitute traditional processes such as machining or various forms of forming technologies [1]. Even if the AM process is usually more expensive compared to other traditional technologies and suffers from thermal stresses and deformations, it can be advantageous when the design of the components can benefit when highly optimized performances are required, such as light weight, complex shapes, parts number reduction, first-class structural or thermo-fluid dynamic performances [2].

Traditional technologies, i.e., milling, casting and forging, are employed following consolidated and well-known Design for Manufacture and Assemblies (DFMA) principles aimed at guiding engineers and designers to develop product shapes that facilitate the manufacturing and assembly processes, thus reducing production costs and times [3]. Focusing on products' geometry and characteristics, AM allows a strong customization of shapes at different measurement scales (i.e., lattice structures) and the possibility to obtain custom graded and composite materials by depositing different type of materials in different patterns, thus optimizing the part characteristics. Design objectives stress the minimization of the overall product volume, thus reducing material consumption, production times and costs. Additional issues are related to the part orientation in order to minimize the support volumes. Additional problems that must be faced are the development of optimal designs combining some novel CAD modelling functionalities with process simulation tools [4], the application of proper post treatments, and the estimation of realization times and costs. Therefore, the designer must cope with a wide set of information and knowledge to approach AM in an optimal way, considering that the full leverage of given possibilities is often reached with a complete rethinking of the existing solution, possibly starting from its functional structure [5,6].

The evaluation of the opportunity to design a part for AM could be time consuming and finally lead to the conclusion that it is impractical. Conversely, some opportunities could be missed since the advantages of a different manufacturing process are not considered. In this context, the main contribution of this paper is to provide a holistic approach composed of four main evaluation phases of increasing complexity: Preliminary Analysis, Potential Analysis, In-depth Analysis and Cost Assessment. These phases are aimed to support the designer throughout the product development in a structured and efficient way, starting from selecting parts which are suitable to this production type, required material, up to the steps to obtain a final geometry compliant with the chosen technology.

The paper is structured as follows: Section 2 presents a revision of constraints and opportunities provided by AM in the literature for the early assessment of related design opportunities, as well as related background; Section 3 presents an approach to support decision-making for the development of new products and for the redesign of existing parts; in Section 4 the approach is then applied on a test case; finally, conclusions and future works are outlined in Section 5.

## 2. Background Work

### 2.1. Adoption of AM in the Industry

The adoption of AM in the industry is becoming significant and it is pushed by the Industry 4.0 digitalization trend [1,7,8]. The benefits offered by this technology will continue to grow with the efficient processing of digital data for process and product improvements, especially when low production quantities or sophisticated geometric shapes are required [9]. The fast evolution of technical solutions, as well as the expiration of principal patents, has enormously expanded the choice on the market of systems for each process type.

A few different knowledge related spaces need to be considered and faced for an effective application in companies. The first area concerns technology awareness. AM techniques have been classified by the type of process, by how the final construct is obtained and by the source of energy which is used for the phase change of the raw material. Seven categories have been defined so far according to the deposition method used to build the parts, i.e., VAT Polymerization, Fused Deposition Modelling, Powder Bed Fusion (PBF), Sheet Lamination, Material Jetting, Direct Energy Deposition and Binder Jetting [10].

Secondly, the possible application fields to be identified in the specific business being run can be recognised in:

1. Structural components, whose goal is to maximize stiffness while reducing weight;
2. Products that need a complex and specific shape, aimed at increasing fluid dynamic efficiency or temperature distribution, such as in conformal cooling;
3. Rapid tooling, using additive techniques to produce tools for other parts, for instance moulds;
4. Parts where high customization is required, such as in fashion or biomedical prostheses and braces requiring lattice structures and specific superficial texturing [1];
5. Prototypes, which are used for physical design evaluations, even if the production of the final product uses other techniques.

Even though AM machines are already used on the shopfloor to produce prototypes and complex parts, they are still not applied systematically and extensively. Like any other production system, AM presents strengths and limitations [11], as summarized in the following Table 1.

**Table 1.** Strengths and limitations of Additive Manufacturing (AM), on a general basis.

| Strengths | Limitations |
|---|---|
| Possibility of freeform parts with limited additional cost | Parts need surface finishing in coupling faces |
| Possibility of hierarchical complexity (i.e., lattice structure) | Low dimensional and geometrical tolerances |
| Customized performance by multi-material structures | Low building volume |
| Material usage only where it is needed | Long processing time |
| Possibility of light weight parts | Need of support structures for overhanging portions |
| Reduction in logistic costs | High energy consumption |
| Reduction in lead time | Lower mechanical properties compared to other processes |
| Consolidation of assemblies | Residual thermal tensions and distortions in the parts |
| | Bounding surface weakness between two different materials [12] |
| | Intrinsic process anisotropy |

Finally, some design organization issues must be faced. In fact, the wide and mostly unexplored design space that AM offers is a very attractive landscape for the industry. However, the complexity and novelty of reasoning patterns could result in significant factors to impede the implementation of AM techniques in practice.

Leaving the design teams to evaluate the possibilities of adopting AM technology based on subjective and unstructured considerations may lead to skewed considerations and a loss of opportunities. Designers have reported that, in many cases, extensive evaluations and design attempts have concluded without feasible solutions. Such failures have had a negative impact on the management due to the poor rate of success in AM adoption. In particular, design leaders complain about [13,14]:

1. The effort required to identify the right components to be redesigned;
2. The effort required to obtain new designs which need to face novel geometric shapes and careful structural and thermal simulations;
3. The variability of metallurgical properties of 3D printed materials makes the quality and reliability of new products uncertain;
4. Necessity of dealing with a new supply chain and the relative logistic aspects.

The method presented in the paper aims to provide a framework to systematically operate in such a complex environment.

### 2.2. Review from the Literature of Decision Methods in AM Adoption

A design process targeting AM consists of a digital dataflow that precedes the transformation of the raw materials into final parts [1]. The development of complex and organic shapes is based on traditional geometric modelling in CAD systems, even if new emerging tools and Topology Optimization (TO) systems are also used for this purpose [15]. The 3D CAD models are exchanged with other tools using STL, or better, by AMF, 3MF, STEP and STEP-NC formats, which are all geometric exchange file formats which provide appropriate set of data [16]. Simulations with specific Computer Aided Design (CAE) software are usually necessary to foresee manufacturing distortions due to thermal effects and improve the surface quality by operating on part orientation and production parameters. Then, slices and the tool paths (G-CODE) are generated and sent to the machine. At this stage, it is possible to manage further parameters to optimize the geometric, finishing, and mechanical characteristics of the final piece. Post processing activities, such as post-curing, cleaning, successive machining, and treatments, must be carefully assessed since they strongly influence the resources needed to obtain the parts.

The depicted workflow reveals that, at an early stage, there are many heterogeneous aspects a designer must consider and anticipate when deciding to adopt AM. Before starting such activities and investing considerable time, the design requirements must be carefully assessed against the possibilities and constraints given by the manufacturing technology to finally succeed in the AM adoption. In this context, Design for Additive Manufacturing (DFAM) guidelines [17] and some related standards [18–20] are just at the beginning and more studies are needed to obtain a more complete view [21]. Design methods and tools are needed whereby functional performance and/or other key product life-cycle considerations such as producibility, reliability, validation and costs can be optimized and subjected to the capabilities of AM technologies [17]. In the industrial practice, the absence of clear DFAM patterns and guidelines often still results in occasional and unstructured usage of AM at the design stage. Additionally, there are only a few software programs that fully support the specific AM digital process, since most tools cover just limited aspects of the whole process, such as geometric modelling or thermal distortion simulations.

Moreover, there are a few procedures to establish that a product is manufacturable with additive techniques. As a matter of fact, there are numerous product development approaches in the literature for various design contexts. Additionally, tools have been developed to support the designer to select the most suitable approach according to requirements [22]. However, there is a need to develop specific tools for AM product design in order to guide engineers and designers to understand this novel production method. For example, Kumke et al. [23] have developed a modular framework on the basis of VDI 2221 aimed at guiding designers throughout product development in an AM context. Lindemann et al. in [24] have developed a Trade-Off Methodology (TOM) matrix that guides designers and engineers to understand which products should be redesigned to exploit the advantages of AM. Additionally, Klahn et al. [25] have developed a tool to evaluate the suitability of products to be produced with additive technologies by using four criteria: integrated design, individualization, lightweight and efficient design. However, according to [2], a holistic framework in an AM perspective aimed at guiding the engineer or designer throughout product development is still missing, guaranteeing process optimization, thus minimizing material waste and costs. Furthermore, the theme of sustainable design [26,27], a topic of high interest nowadays, needs to be embraced.

### 2.3. *Multi Criteria Decision Methods as a Formal Support to an Objective Design Process*

In the context of fostering the spread of AM applications, it is beneficial to help designers understand when and how to adopt AM technologies to improve designs with better performing and cost-saving solutions. There is a need for a broader and holistic assessment that includes diverse aspects, as well as the possibility of adopting different and new design principles. The discipline of DFAM targets the geometric shape definition phase, while additional efforts are needed to fully understand the potential of AM in an earlier stage, when multiple, conflicting, and often fuzzy requirements must be evaluated. Therefore, to define a structured evaluation workflow to assess the design opportunities, decision-making approaches must be considered.

Multi Criteria Decision Analysis (MCDA) [28] is the discipline aimed at supporting the decision-maker while operating with numerous and conflicting evaluations, allowing them to obtain a compromise solution in a structured way [29,30]. Contrary to numeric simulation tools, which rely on precise quantities, MCDA is better suited to weight subjected, discrete or fuzzy evaluations.

Scarce applications of such methods in early AM design phases are found in the literature. Among them, in [31] the authors have used the Analytical Hierarchical Process (AHP) procedure to rank the most appropriate technologies and machines. Another study in this sense was developed by Zaman et al. in [32], where the authors have introduced a generic decision methodology to provide a set of compromised AM materials, processes, and machines and to guide designers to achieve feasible material–machine combinations from a current database of AM vendors in the world.

Weighted Sum Model (WSM) and AHP appear to be able to support the qualitative and quantitative analyses required by the scope of this paper. The two approaches are briefly recalled here. The AHP method is based on a series of pairwise comparisons between a set of criteria, giving them a score of relative importance and assigning percentage weights [33]. An example of the values for the pair evaluation is reported in Table 2 below.

**Table 2.** Analytical Hierarchical Process (AHP) method: scale of relative importance.

| Driver Dij | Meaning |
|---|---|
| 1 | i and j are equally important |
| 3 | i is slightly more important than j |
| 5 | i is quite a bit more important than j |
| 7 | i is much more important than j |
| 9 | i is largely more important than j |
| 1/3 | i is slightly less important than j |
| 1/5 | i is quite a bit less important than j |
| 1/7 | i is much less important than j |
| 1/9 | i is largely less important than j |

Subsequently, a matrix *nxn* is constructed, where $n$ is equal to the number of criteria to be compared. Supposing that $n$ = 3 (D1, D2 and D3), the matrix has the following construction (Table 3):

**Table 3.** Example of pairwise matrix.

| | D1 | D2 | D3 |
|---|---|---|---|
| **D1** | 1 | X | Y |
| **D2** | 1/X | 1 | Z |
| **D3** | 1/Y | 1/Z | 1 |

In Table 3 X is the relative importance value of D1 compared to D2, Y is the relative importance value of D1 compared to D3 and Z is the relative importance value of D2 compared to D3. To calculate the vector of the percentage weights to assign to each driver, it is enough to determine the maximum eigenvalue $\lambda$ and the relative eigenvector $V_\lambda$ of the matrix [32]. Normalizing the eigenvector $V_\lambda$ so that the sum of its elements is equal to 1, the vector of the percentage weights or priorities relating to the driver Di is obtained. Once the priority vector has been determined, it is fundamental to understand the quality of the pairwise matrix, measuring the consistency of the decision-maker's subjective judgments. To this aim, the Consistency Index (*CI*) is defined as (Equation (1)):

$$CI = \frac{\lambda - n}{n - 1}, \tag{1}$$

Finally, the Consistency Ratio (*CR*) is defined as follows (Equation (2)):

$$CR = \frac{CI}{RI}, \tag{2}$$

where *RI* is the Random Index, which is a value that depends on the number of criteria as explained in [33]. For $n > 4$, the pairwise matrix is consistent if *CR* is lower than 0.1.

On the other hand, the WSM is used to find the best solution among those available, thus solving a decision-making problem. Here, the criteria values (*aij*) are multiplied by the weights calculated with AHP (*wj*), and finally the sum is performed for each alternative (*Ai*), as in Equation (3). The criteria values should be normalized to have the same unit [34].

$$Ai = \sum_{j=1}^{n} wj \times aij, \tag{3}$$

## 3. Methodology and Guidelines

As anticipated above, it is desirable to guide the designer to select suitable candidates for AM efficiently, reducing wasted time during more complex phases for unattractive products. Therefore, the main goal of this paper is to provide guidelines which are framed in a general framework to support designers, even the less experienced, to better understand the potential offered by additive technologies in a given design context. The framework acts as a foundation of specific step guidelines, which are defined on an algorithmic basis to also be implementable in a software supporting tool.

The proposed framework is presented in Figure 1; it is framed in a systematic process [35], even if it presents its specificities. It can be applied both in the case of new design or redesign of an already existing product. A decision-making process has been defined to select feasible candidates and then progressively move on to further analysis, only if the previous phase has succeeded. Specifically, the process is composed of four main phases which are explored in the following sections:

1. Preliminary Analysis;
2. Potential Analysis;
3. In-depth Analysis;
4. Cost assessment.

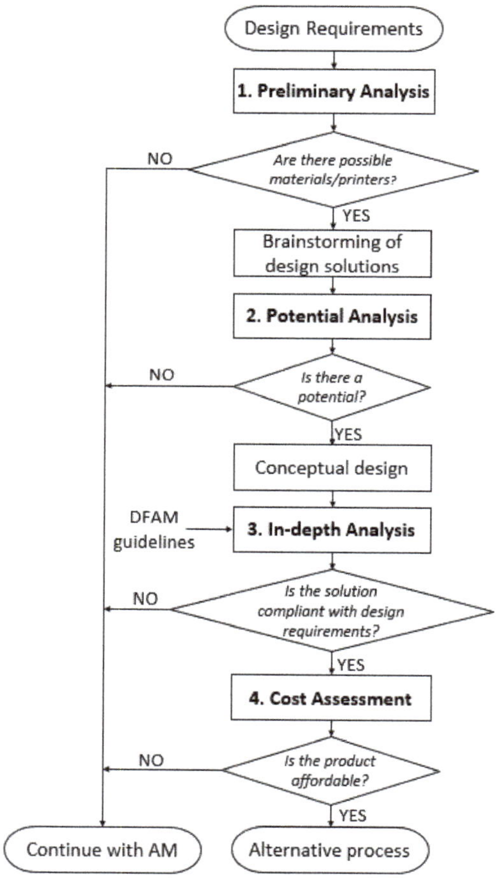


**Figure 1.** General framework of the design approach for AM parts.

Figure 2 reports an example of the timeline required to accomplish each step. The times have been collected by the authors based on some informal interviews with designers who have had experience in designing parts for AM in their companies in the field of machined parts of limited size. The observation of the timeline, even if referring to a particular product category, highlights the necessity of tackling the design activities in a systematic manner, quitting the process as soon as the solution spaces become empty.

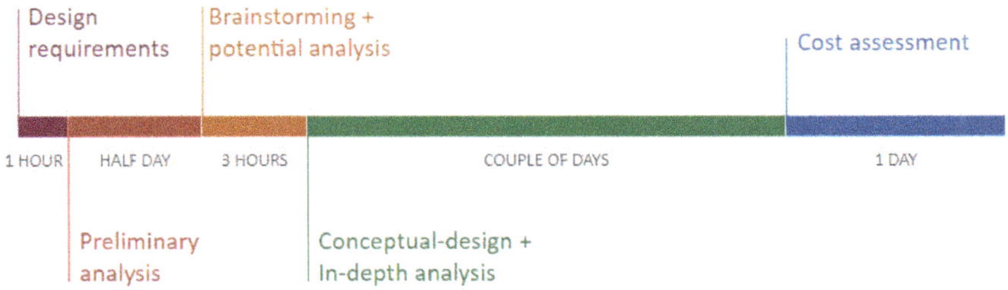


**Figure 2.** Example of timeline for a common design process involving additive technologies in the field of machined parts.

In the following sections, each step of the proposed approach is presented in detail, providing specific application guidelines.

### 3.1. Design Requirements

The process moves from a collection of basic facts and main design requirements of the product under analysis. Table 4 reports a selection of possible data to be gathered, many of them being optional depending on the application.

**Table 4.** Example of initial requirements of a product to be evaluated.

| Requirement | Value |
|---|---|
| Application | Structural, prototype, complex shape, tooling, customized |
| Expected overall size | SizeX × SizeY × SizeZ |
| Load | Rough description of expected loads |
| Constrains | Rough description of expected constraints |
| Target properties of the material | Current material, in case of redesign process<br>Density [kg/m$^3$]<br>Stiffness [GPa]<br>Max working T [°C]<br>Conductivity E<br>Conductivity T<br>Hardness [Rock]<br>Wear resistance<br>Corrosion resistance<br>Tensile strength [N/mm$^2$]<br>Fatigue strength [N/mm$^2$]<br>Yield strength [N/mm$^2$] |
| Production method | Report current production process, in case of redesign process |
| Buy-to-fly ratio | Report current ratio, in case of redesign process |
| Batch size | Expected parts/year |
| Time-to-Market | Required time-to-market |
| Production cost | Report current production cost, in case of redesign process |
| Other requirements | … |

Generic numbers and information are sufficient at this stage, also in the form of expected values or possible ranges. More precise parameters and requirements will be defined or refined in successive evaluation steps. It is important to point out that the table can be adapted, expanding or reducing the requirements, according to the specific application field.

### 3.2. Preliminary Analysis of AM Constraints

Following the flow in Figure 1, the next step of the process is the Preliminary Analysis, which is a rough examination of main manufacturing constraints to exclude unfeasible applications. Three aspects have been identified as overwhelming limitations that could impede the usage of AM. Even if the three evaluations can be regarded as subjective and depending on the evolution of the technologies, it is reasonable to evaluate them early and to skip the process from the beginning if one of them cannot be surely met by the application:

- Lack of a material with some mandatory properties;
- Excessively large or small product dimensions for currently available technologies;
- Excessively large production batches.

In Figure 3, the sub-steps of the Preliminary Analysis are presented:

1. Screening of materials compatible with the given constraints (i.e., minimum stiffness, minimum yield strength, electric conductibility, etc.);
2. Computation of a set of indices according to Ashby [36] for each compatible material;
3. Definition of index weights using AHP and computation of an overall Compliant Index (*CI*) for each material through WSM normalization;
4. Materials ranking;
5. Selection of possible printers proceeding in the sorted list of materials;
6. Check of printer capabilities according to required volume and batch size.

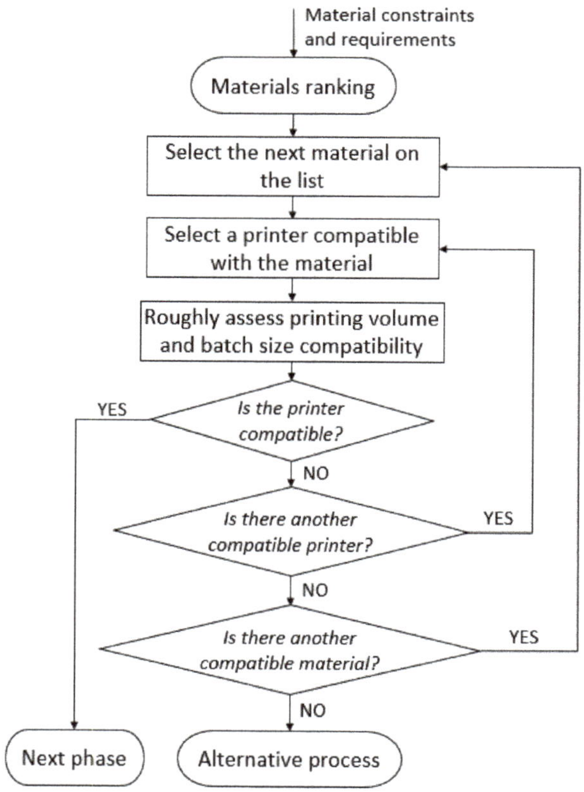


**Figure 3.** Workflow of the preliminary analysis of AM constraints.

Therefore, the first step is to identify suitable materials out of a list. The materials database is built from data provided by technology suppliers, mostly available on the web [36], where the value range of properties of available materials is reported. It is possible that only a few or no materials are compatible if numerous and strict constraints are imposed, making AM an impractical solution for the imposed requirements.

A list of seven indices, drawn from Ashby [37], are suggested to evaluate the materials. The selected indices are:

1. I1 = σr/ρ
2. I2 = σy/ρ
3. I3 = σf/ρ
4. I4 = E/ρ
5. I5 = E^(1/2)/ρ
6. I6 = σy/(ρ*C)
7. I7 = E/(ρ*C)

where σr = ultimate tensile strength ($N/mm^2$), σy = yield strength ($N/mm^2$), σf = fatigue strength ($N/mm^2$), E= yield modulus (GPa), ρ = density ($kg/cm^3$), C = raw material cost (EUR/kg).

The first four indices are related to the mechanical performance of the material, the fifth index refers to the vibration behaviour, while the last two indices consider the cost. The relative weight of the indices must be assigned by the user according to company production range and company strategy. Such indices are normalized to calculate a *CI* for each material through the WSM. The index weights are determined by the AHP method according to design strategies. Then, the materials are ranked from best (high *CI*) to worst (low *CI*). This process is summarized in Figure 4.

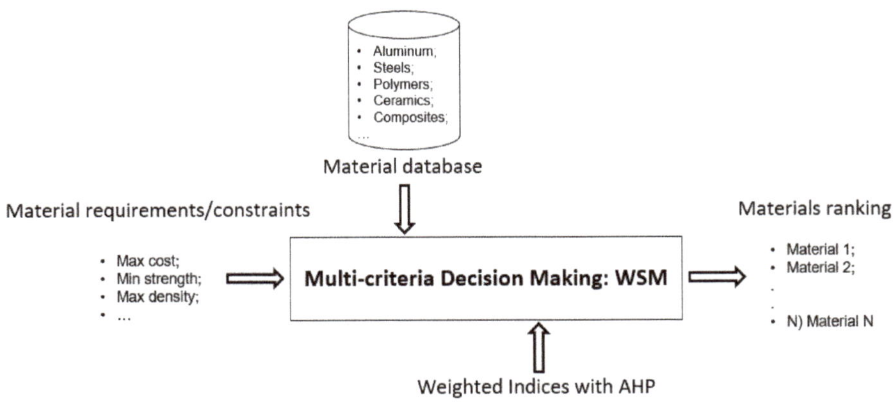


**Figure 4.** Materials ranking process.

Printers compatible with the selected material are then listed, verifying the production requirements relating to the volume of the product and the batch size. Additionally, the designer must consider the intrinsic material anisotropy of AM processes, the bounding surface weakness between two different materials [12] and the technology deposition maturity, thus selecting the most suitable printer. If no printer is compatible with the production requirements, the next material in the ranking is selected, repeating the check of the compatibility of available printers with the imposed constraints. A printer database is built for this purpose from data available on the web [36] or from supplier's data. If a valid combination of material and printer is found that meets the defined requirements, Potential Analysis is performed.

### 3.3. Potential Analysis

The aim of this phase is to guide brainstorming around the product considering a list of design drivers. Each driver is given a score which is combined to compute an overall potentiality index. Fourteen drivers, D1 to D14, have been identified to evaluate the potentiality of an AM employment [2]. Table 5 presents the drivers and list guidelines, mainly considerations and clues, which have been gathered in the literature and from experts to foster the decision on a score to assign to each of them.

**Table 5.** Description of the drivers to evaluate the potentiality of AM in a specific application.

| Driver | Guidelines to Help the Designer to Decide a Score |
|---|---|
| (D1) Does the use of AM lead to potential cost savings considering the entire life cycle? | AM is generally more expensive than a forming or machining technique due to long cycle times and high energy consumption. However, such a gap may reduce if considering the entire product lifecycle. It is the case of complex shapes optimized for highly efficient fluid-dynamic performances, devices such as blowers, diffusers, etc. Indeed, in some applications, AM allows us to reduce the mass, which is beneficial for product operating costs. For example, airplane mass reduction leads to fuel saving. |
| (D2) Can assembly, logistic or procurement cost be reduced? | Forming technologies allow the realization of complex shapes but require investments for tooling that are recovered only if many parts are produced. Conversely, AM does not require fixed costs (apart from the printer itself) and it allows us to consolidate parts in less components of more complex components. A reduced number of parts is beneficial in terms of logistics, assembly times and supply chain management. |
| (D3) Can the product benefit from a reduced time-to-market? | AM can be set up rapidly since it is based on a digital toolchain. Even if the realization time is high and some post processing activities can be required (support removal, post treatments, additional machining, surface coatings, … ), the whole process can be generally accomplished in few days, which is competitive with processes which require the elaboration of CNC toolpaths or the design and manufacturing of tooling and several sequential separated machines. Thus, the overall time-to-market could be significantly reduced. |
| (D4) Is the production batch size limited? | As any other technology requiring limited fixed costs but longer manufacturing cost for each part, AM is more suited to small batch quantities. |
| (D5) Is the post processing phase limited? | AM aims to produce near net shape parts. However, depending on the specific technology, post processing to consolidate the material, annealing residual stresses or improving surface quality may be required. The quantity of such activities can strongly reduce the competitiveness of additive solutions. |
| (D6) Can material waste be reduced? | Nowadays, many materials are available to be processed with AM, some of them being expensive advanced metals. The advantage of adding material only where it is needed, conceiving complex organic shapes, and the possibility of recovering unused powders can lead to a reduction in material waste and improved sustainability. |
| (D7) Would the product benefit from a mass reduction? | Mass reduction is one of the major strengths of AM parts, which has strongly pushed its diffusion. |
| (D8) It is beneficial to increase the complexity of the part? | In some applications, such as fluid dynamics, biomedical applications (e.g., increase the osteointegration of the prostheses thanks to a porous layer), multiple functions integration, etc., key benefits arise from the freedom in the complexity of the shapes. |
| (D9) It is feasible to consolidate parts? | Considerable advantages in terms of cost are gained if the number of parts can be reduced. AM can leverage the consolidation of parts if assembly and functional constraints are respected, also benefiting from grading material properties which can be conferred thanks to inner cavities or lattice structures. |
| (D10) Is customization required? | The limited investment costs of AM pave the way to the paradigm of Mass Customization. Specific parts could be produced according to specific customer requirements, controlling the increase in production costs. |
| (D11) Would the product benefit from areas with differentiated material properties? | One of the features of AM refers to the capability of calibrating the properties of the material along the part, recurring to modulation of micro-structures up to vary the distribution and parameters of lattice cells. For example, material properties can be varied accordingly to applied force, leading to uniform stress level and deflection behaviour that cannot be obtained with uniform material properties. If combining regions with different materials, the compatibility and weakness at the interfaces must be carefully assessed [12]. |
| (D12) Would the product benefit from varied density along the volume? | The volumetric mass of a product can be modulated according to specific design objectives, thanks to lattice structures, inner voids or honeycomb structures. |
| (D13) Is a volumetric lattice or a superficial texture required? | Superficial or volumetric structures can be obtained, enhancing weight reduction, or increasing osteointegration as well as grip. |
| (D14) Is it advantageous to add functionalities in the parts? | The extended design freedom of AM allows multiple functionalities to be incorporated in a part, such as the possibility of integrating electric circuits. |

A value from 1 to 10 is required for each driver, where 1 corresponds to an irrelevant driver and 10 corresponds to a major driver for the product under development. The Potential Index (*PI*) is then calculated according to Equation (4):

$$PI = \frac{\sum_{i=1}^{14} Di \times Qi}{10} * 100, \quad (4)$$

where $Di$ = weight of $i$-th driver, $Qi$ = value assigned by the development team to the $i$-th driver.

The weights of drivers should be calculated using the AHP method, incorporating strategic views of the company. High index values correspond to a great potential of the part to be produced through AM. The *PI* threshold can be fixed by the user based on a trade-off analysis between the risk of losing opportunities and the need to develop deeper and costly analyses with the possibility of having poor results.

### 3.4. In-Depth Analysis

After the Potential Analysis, a preliminary design solution is required. In fact, deeper assessments on the manufacturability by AM and its convenience against other production systems follow a conceptual design of the product itself, at least at a sufficient level of geometrical detail. Generally speaking, the conceptualization phase is highly onerous in terms of time and work and impacts on the design department. So, before developing a conceptual design, it is advisable to discard unpromising ideas following the steps described above, thus focusing efforts only on the most attractive candidates. Therefore, the In-depth Analysis stage has been conceived to include the following evaluations (Figure 5):

1. Develop a conceptual design and its preliminary geometry;
2. Evaluate the geometry according to three indices: Customization Index, Complexity Index, Printability Index;
3. If necessary, refine the geometry or change the intended orientation of the part in the printer;
4. Accurately evaluate the manufacturing process on the selected printer by suitable numeric tools.

The conceived geometry, either by direct CAD modelling or TO tools, is then evaluated according to three indices that are strictly related to the designed shapes:

1. Customization Index: the term customization means the product development according to individual/personal specifications or preference. This index refers to the level of customization that the product must comply to satisfy the customer's requirements, ranging from no customized products to truly unique parts [38,39].
2. Complexity Index: it assesses the complexity of geometries in relation to the difficulty of achieving the shape with AM competing techniques. For instance, in [40], the authors have developed such an index to calculate the shape complexity of cast parts.
3. Printability Index: it is composed of indices related to the limits of AM in obtaining certain geometric features [41]. For example, excessively bulk geometries can lead to distortions; too thin walls and too small holes are difficult to produce; cantilevered surfaces require supports which must then be removed. In Table 6, indices to be considered are summarized. The weights to balance the indices should be assigned accordingly to the specific needs and application context. Indeed, the Printability Index is closely related to the part orientation in the printer.

The first two indices are used to estimate the level of customization and complexity of the geometry according to standard technologies. The Printability Index aims to highlight the criticalities in the conceived design according to the main DFAM guidelines and common geometric limits of AM. Such an index allows the designer to identify changes in the geometry as well as more convenient orientations of the part.

At this stage, numeric simulations are required to verify the stress level in the part according to operating conditions and the peculiar characteristics of the metallurgic microstructure or the anisotropy induced by the layered deposition. Moreover, the thermal gradients usually arising in the process lead to residual stress, which must be carefully assessed with dedicated software tools which are consolidating in the market.

This process leads to a detailed and definitive geometry which allows more precise evaluations on the printing process. Parts nesting can be computed, and selected machine volume optimally occupied. Indeed, if more printers are available and compatible with the chosen material and part size, the optimal one can be selected by iteratively simulating the printing process and expected parts quality against the fixed requirements.

Ultimately, if there is at least one printer compatible with the production requirements, it is possible to move on to the last step of the workflow.

### 3.5. Cost Assessment

The final step of the proposed approach is an assessment of the reached design in terms of production and life cycle costs, possibly including additional considerations such as logistical aspects and production lead time. So, it is necessary to evaluate whether the use of AM leads to an overall economic advantage compared to other production methods. In Figure 6, the flow is represented. Here, the economic aspects are highlighted to understand if there is a saving in terms of costs and/or time-to-market. Strategic decisions must be taken by the company management in case of contrasting results.

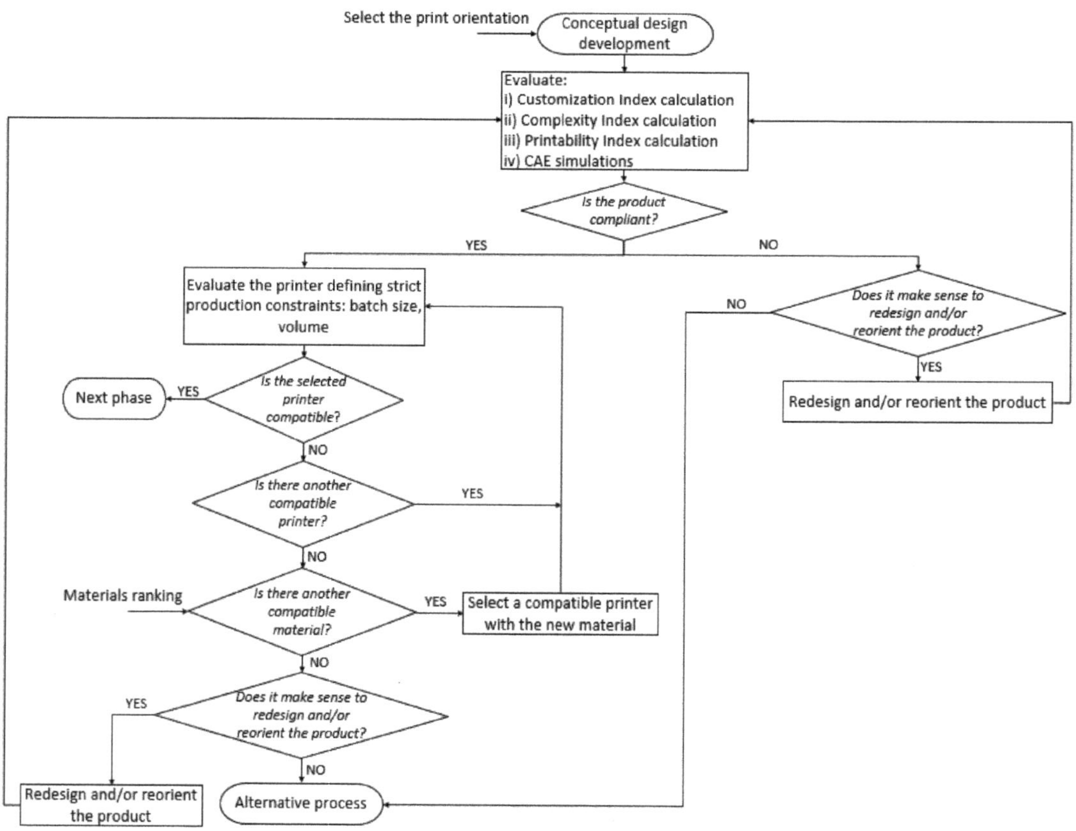


**Figure 5.** In-depth Analysis.

**Table 6.** Indices contributing to the Printability Index.

| Name | Equation | Description |
|---|---|---|
| BuyToFly index | $Cv = \frac{Vtot}{Vbb}$ | Where *Vtot* is the product volume and *Vbb* is the volume of the bounding box. |
| Functional surface index | $Cf = \frac{\sum Sf}{Stot}$ | Where $\sum Sf$ is the sum of functional surface and *Stot* is the product surface. |
| Supported surface index | $Cs = \frac{\sum Ss}{Stot}$ | Where $\sum Ss$ is the sum of the surfaces that need supports and *Stot* is the product surface. |
| Hole index | $Ch = \frac{\sum n^{\circ} Ha}{Htot}$ | Where $\sum n^{\circ} Ha$ is the sum of the holes that do not have the axis parallel to the production direction and *Htot* is the number of the total holes. |
| Massive volumes index | $Cn = \frac{\sum Vm}{Vtot}$ | Where $\sum Vm$ is the sum of the massive volumes and *Vtot* is the overall product volume. |
| Height ratio index | $Cl = \frac{Hp}{Lmax}$ | Where *Hp* is the product height and *Lmax* is the maximum length of the part in one of its three principal directions. |
| Small holes index | $Cd = \frac{\sum n^{\circ} Hd}{Htot}$ | Where $\sum n^{\circ} Hd$ is the sum of the holes number with the diameter below a critical value that depends on the specific type of technology and *Htot* is the number of the total holes. |
| Thickness distribution index | $Cd = \frac{\sum Vtn}{Vtot}$ | Where $\sum Stn$ is the sum of the walls volume with thickness below a critical value that depends on the specific type of technology and *Vtot* is the product volume. |
| Surface orientation index | $Ce = \frac{Sp}{Stot}$ | Where *Sp* is the surface parallel to the build direction and *Stot* is the product surface. |
| Layer variation index | $Cl = \frac{\sum_{i=1}^{N} \left\| (Sli - Sl(i-1)) \right\|}{Sltot}$ | Where *Sli* is the surface of the *i*-th layer and *Sl(I* − 1) is surface of the (*i* − 1)-th layer. Sltot is the maximum trace of the product. *N* is the total number of the layers. |

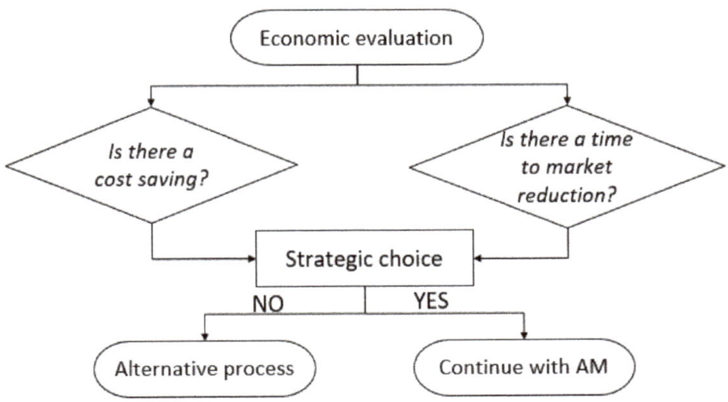


**Figure 6.** Final assessment of cost and time-to-market.

## 4. Case study

In this section, the framework and relative guidelines are exemplified with a case study. An ultralight aircraft part called Propeller Pitch Controller (PPC) has been selected, and it is represented in Figure 7. Given the interest in parts obtainable by metal deposition, which are collected in the industrial field nowadays, a metal component has been selected to demonstrate the proposed design workflow. This product is assembled on the variable pitch aircraft propellers with three blades and it is connected through three connecting rods to them (Figure 7b). It allows the pitch of the blades to change thanks to a translation movement, thus regulating the provided engine thrust.

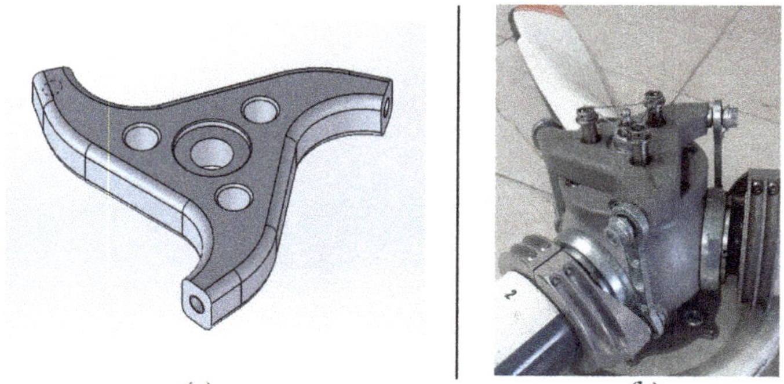

(a) (b)

**Figure 7.** (**a**) 3D CAD model of the Propeller Pitch Controller (PPC); (**b**) Real part assembled with the blades.

Analog design procedures could be applied in case studies employing plastic materials and technologies such as Fused Deposition Modelling or Stereolithography. Some considerations, requirements and data reported in the papers should be modified accordingly to the different application contexts and specificities.

### 4.1. Preliminary Analysis Application

As it is depicted in Figure 7, the PPC bears three main loads transmitted by the connecting rods. Three screws fix the part. The original material is aluminium alloy. In Table 7, all the features and the design requirements are summarized.

**Table 7.** Design requirements for the analysed product.

| Requirement | Value |
|---|---|
| Application | Structural |
| Expected overall size | 130 × 130 × 20 mm |
| Load | 3 × 3000 N |
| Constraints | Rotation blocked by 3 screws M6 |
| Target properties of the material | Current material: Aluminium<br>Current density: 2700 [Kg/m$^3$]<br>Current Stiffness: 50 [GPa]<br>Twork: −30 °C to 130 °C<br>Conductivity E: Not important<br>Conductivity T: Not important<br>Required Hardness: 20 [Rock]<br>Wear resistance: Good resistance<br>Corrosion resistance: Good resistance<br>Current tensile strength: 210 [N/mm$^2$]<br>Fatigue strength: Millions of cycles required<br>Current yield strength: 120 [N/mm$^2$]<br>Maximum expected powder cost: EUR 75/kg |
| Production method | Milling from block |
| Buy-to-fly ratio | Current: 0.347 |
| Batch size | 150 p/years |
| Time-to-Market | Few weeks |
| Production cost | 25 EUR/p |

The three main initial constraints outlined in Section 3.2 are fulfilled by the specific application. Most of the constraints push the choice towards metals, mainly aluminium and steel, which are widely available as powders to be employed in PBF processes. The size of the part is limited and compatible with most of the printers. Indeed, the size allows multiple parts to be processed simultaneously, thus small batch sizes, as required, are definitely feasible.

The approach depicted in Section 3 has been followed for the PPC and the main results are reported here. The process has moved from the computation of the AHP pairwise matrix to assess the relative weights of the material selection indices (Table 8). The table was compiled with the help of the industrial partner and the resulting weights computed according to the AHP method are reported in Table 9.

**Table 8.** AHP pairwise matrix of the indices of materials for the PPC case study. I1 to I7 refers to the seven selected material indices to rank materials, as defined in Section 3.2.

| | I1 | I2 | I3 | I4 | I5 | I6 | I7 |
|---|---|---|---|---|---|---|---|
| **I1** | 1.00 | 0.20 | 0.20 | 0.14 | 0.50 | 0.20 | 0.20 |
| **I2** | 5.00 | 1.00 | 5.00 | 1.00 | 5.00 | 2.00 | 2.00 |
| **I3** | 5.00 | 0.20 | 1.00 | 0.20 | 1.00 | 0.50 | 0.50 |
| **I4** | 7.00 | 1.00 | 5.00 | 1.00 | 5.00 | 2.00 | 2.00 |
| **I5** | 2.00 | 0.20 | 1.00 | 0.20 | 1.00 | 1.00 | 1.00 |
| **I6** | 5.00 | 0.50 | 2.00 | 0.50 | 1.00 | 1.00 | 1.00 |
| **I7** | 5.00 | 0.50 | 2.00 | 0.50 | 1.00 | 1.00 | 1.00 |

**Table 9.** Weights of the material selection indices calculated for the case study.

| | I1 | I2 | I3 | I4 | I5 | I6 | I7 |
|---|---|---|---|---|---|---|---|
| PPC case study | 3.26 | 27.14 | 7.70 | 28.09 | 8.13 | 12.84 | 12.84 |

Table 10 reports a part of the evaluated materials list, including the values of the selected material properties. So, the upper part of the sorted list of materials according to the computed *CI* is reported in Table 11.

**Table 10.** Indices values of the compatible materials (index 3 is omitted since it is not significant for aluminium alloys).

| | I1 = σr/ρ | | I2 = σy/ρ | | I4 = E/ρ | | I5 = E^ (1/2) /ρ | | I6 = σy/(ρ*C) | | I7 = E/(ρ*C) | |
|---|---|---|---|---|---|---|---|---|---|---|---|---|
| **Material** | **MIN** | **MAX** | **MIN** | **MAX** | **MIN** | **MAX** | **MIN** | **MAX** | **MIN** | **MAX** | **MIN** | **MAX** |
| LaserForm AlSi12 | 171,642 | 186,567 | 93,284 | 108,209 | 0 | 0 | 26,119 | 27,985 | 3122 | 3231 | 1333 | 1546 |
| LaserForm AlSi7Mg0.6 | 131,086 | 161,049 | 78,652 | 101,124 | 0 | 0 | 26,217 | 26,966 | 3134 | 3178 | 1269 | 1631 |
| LaserForm AlSi7Mg0.6 (DA) | 149,813 | 172,285 | 108,614 | 123,596 | 0 | 0 | 27,341 | 27,715 | 3200 | 3222 | 1724 | 1962 |
| AlSi10Mg (HT) | 131,086 | 161,049 | 86,142 | 112,360 | 0 | 0 | 22,472 | 24,345 | 2901 | 3020 | 1305 | 1702 |
| CL 30Al (AlSi12) (SR) | 116,105 | 121,723 | 63,670 | 82,397 | 0 | 0 | 28,090 | 28,090 | 3244 | 3244 | 936 | 1212 |
| CL 32Al (AlSi10Mg) (HT) | 126,217 | 132,210 | 73,034 | 88,015 | 0 | 0 | 28,090 | 28,090 | 3244 | 3244 | 1124 | 1354 |
| EOS AlSi10Mg | 164,794 | 179,775 | 86,142 | 104,869 | 0 | 0 | 24,345 | 31,835 | 3020 | 3453 | 1346 | 1639 |
| EOS AlSi10Mg (SR) | 125,468 | 134,831 | 80,524 | 91,760 | 0 | 0 | 18,727 | 29,963 | 2648 | 3350 | 1220 | 1390 |
| AlSi10Mg | 133,708 | 149,064 | 78,277 | 89,513 | 0 | 0 | 26,966 | 31,461 | 3178 | 3433 | 1263 | 1444 |
| Laserform 17-4PH (HT) | 177,333 | 188,000 | 156,000 | 174,667 | 28,000 | 32,000 | 25,333 | 28,000 | 1838 | 1932 | 2600 | 2911 |
| Maraging steel | 137,500 | 137,500 | 116,250 | 116,250 | 42,875 | 76,875 | 17,500 | 17,500 | 1479 | 1479 | 2114 | 2114 |
| CL 91 RW (HT) | 212,500 | 212,500 | 20,000 | 200,000 | 30,000 | 31,250 | 25,000 | 25,000 | 1768 | 1768 | 274 | 2740 |
| 17-4 PH SS | 130,250 | 130,250 | 82,500 | 82,500 | 27,500 | 31,250 | 24,375 | 24,375 | 1746 | 1746 | 1650 | 1650 |
| EOS Maraging steel | 125,000 | 150,000 | 112,500 | 143,750 | 42,875 | 76,875 | 16,875 | 23,125 | 1452 | 1700 | 1607 | 2054 |
| CL 91 RW (HT) | 212,500 | 212,500 | 20,000 | 200,000 | 30,000 | 31,250 | 25,000 | 25,000 | 1768 | 1768 | 274 | 2740 |
| SLM Solution 1.2709 | 121,500 | 131,250 | 95,125 | 114,125 | 91,500 | 91,500 | 17,875 | 23,875 | 1495 | 1728 | 1903 | 2283 |
| ... | ... | ... | ... | ... | ... | ... | ... | ... | ... | ... | ... | ... |

Table 11 provides minimum and maximum values of the *CI* in consideration of the ranges of physical properties of the materials in the market.

The best choices for the PPC application are Laserform 17-4PH (heat treated) and LaserForm AlSi7Mg0.6 (direct aging). Following the formalized path, compatible printers need to be selected according to product dimensions and batch size. As already pointed out many machines are compatible with the application. Considering the main manufacturers some machines are listed in Table 12 to confirm the compatibility with the product geometry

**Table 11.** Sorted list of materials based on Compliance Index *CI* (HT = heat treated; SR = stress relief; DA = direct aging).

| Material | CI MIN % | CI MAX % |
|---|---|---|
| (1) Laserform 17-4PH (HT) | 86.12 | 83.18 |
| (2) LaserForm AlSi7Mg0.6 (DA) | 76.50 | 71.24 |
| (3) CL 91 RW (HT) | 49.07 | 80.00 |
| (4) EOS AlSi10Mg | 66.09 | 73.02 |
| (5) SLM Solution 1.2709 | 66.55 | 72.49 |
| (6) LaserForm AlSi7Mg0.6 | 67.17 | 65.67 |
| (7) EOS Maraging steel | 59.90 | 70.10 |
| (8) AlSi10Mg (SR) | 68.29 | 69.48 |
| (9) LaserForm AlSi12 | 69.21 | 66.78 |
| (10) CL 32Al (AlSi10Mg) (HT) | 67.79 | 63.30 |
| (11) 17-4 PH SS | 68.40 | 61.05 |
| (12) Maraging steel | 65.97 | 60.65 |
| (13) CL 30Al (AlSi12) (SR) | 64.58 | 61.27 |
| (14) AlSi10Mg (HT) | 62.15 | 63.15 |
| (15) EOS AlSi10Mg (SR) | 54.80 | 66.46 |
| (16) LaserForm AlSi7Mg0.6 (SR) | 61.18 | 63.70 |
| (17) ... | ... | ... |

**Table 12.** Extract of the printer database showing the models compatible with the required part size.

| Supplier | Name | Technology | Build Size | Compatible Materials |
|---|---|---|---|---|
| 3D System | DMP Factory 350<br>DMP Felx 350<br>ProX DMP 320 | Powder Bed Fusion (PBF) | 275 × 275 × 380 mm<br>275 × 275 × 380 mm<br>275 × 275 × 420 mm | LaserForm AlSi10Mg<br>LaserForm AlSi10Mg<br>LaserForm AlSi12<br>Laserform 316L (SR)<br>Laserform 17-4PH<br>Maraging steel |
| Concept Laser | X line 2000R<br>M2 Cusing | PBF | 800 × 400 × 500 mm<br>250 × 250 × 280 mm | CL 30Al (AlSi12)<br>CL 32Al (AlSi10Mg)<br>CL 20ES (SS 1.4404)<br>CL 91 RW (HT)<br>17-4 PH SS |
| Renishaw | AM 250<br>AM 400<br>RenAM 500Q | PBF | 250 × 250 × 365 mm<br>250 × 250 × 300 mm<br>245 × 245 × 335 mm | AlSi10Mg (HT)<br>Maraging steel |
| SLM Solution | SLM 280 2.0<br>SLM 500 | PBF | 280 × 280 × 350 mm<br>500 × 280 × 325 mm | SLM Solution AlSi10Mg<br>SLM Solution AlSi9Cu3<br>EOS Maraging steel<br>EOS 316L<br>SLM Solution 1.2709 |

### 4.2. Brainstorming and Potential Analysis Application

The main goal of the PPC redesign process is the mass reduction. The new geometry could reduce the total weight of aircrafts which leads to a reduction in fuel consumption during the flight and a lower inertia in the orientation of the blades. Thanks to a brainstorming session with students and industry experts, the AHP pairwise matrix of the drivers for a generic structural application has been completed (Table 13).

**Table 13.** AHP pairwise matrix to compute relative weights of the drivers for a generic structural application.

| | D1 | D2 | D3 | D4 | D5 | D6 | D7 | D8 | D9 | D10 | D11 | D12 | D13 | D14 |
|---|---|---|---|---|---|---|---|---|---|---|---|---|---|---|
| **D1** | 1.00 | 2.00 | 1.00 | 1.00 | 1.00 | 2.00 | 2.00 | 7.00 | 9.00 | 7.00 | 7.00 | 7.00 | 9.00 | 9.00 |
| **D2** | 0.50 | 1.00 | 0.50 | 0.50 | 0.50 | 1.00 | 0.50 | 1.00 | 5.00 | 2.00 | 2.00 | 2.00 | 1.00 | 7.00 |
| **D3** | 1.00 | 2.00 | 1.00 | 1.00 | 1.00 | 2.00 | 1.00 | 7.00 | 7.00 | 7.00 | 7.00 | 7.00 | 7.00 | 9.00 |
| **D4** | 1.00 | 2.00 | 1.00 | 1.00 | 0.50 | 2.00 | 1.00 | 7.00 | 7.00 | 7.00 | 7.00 | 7.00 | 7.00 | 9.00 |
| **D5** | 1.00 | 2.00 | 1.00 | 2.00 | 1.00 | 2.00 | 1.00 | 7.00 | 7.00 | 7.00 | 7.00 | 7.00 | 7.00 | 9.00 |
| **D6** | 0.50 | 1.00 | 0.50 | 0.50 | 0.50 | 1.00 | 0.50 | 2.00 | 5.00 | 2.00 | 2.00 | 2.00 | 1.00 | 7.00 |
| **D7** | 0.50 | 2.00 | 1.00 | 1.00 | 1.00 | 2.00 | 1.00 | 5.00 | 7.00 | 2.00 | 7.00 | 7.00 | 2.00 | 9.00 |
| **D8** | 0.14 | 1.00 | 0.14 | 0.14 | 0.14 | 0.50 | 0.20 | 1.00 | 2.00 | 1.00 | 1.00 | 1.00 | 1.00 | 5.00 |
| **D9** | 0.11 | 0.20 | 0.14 | 0.14 | 0.14 | 0.20 | 0.14 | 0.50 | 1.00 | 1.00 | 0.50 | 0.50 | 0.50 | 2.00 |
| **D10** | 0.14 | 0.50 | 0.14 | 0.14 | 0.14 | 0.50 | 0.50 | 1.00 | 1.00 | 1.00 | 0.50 | 0.50 | 1.00 | 5.00 |
| **D11** | 0.14 | 0.50 | 0.14 | 0.14 | 0.14 | 0.50 | 0.14 | 1.00 | 2.00 | 2.00 | 1.00 | 1.00 | 1.00 | 5.00 |
| **D12** | 0.14 | 0.50 | 0.14 | 0.14 | 0.14 | 0.50 | 0.14 | 1.00 | 2.00 | 2.00 | 1.00 | 1.00 | 1.00 | 5.00 |
| **D13** | 0.11 | 1.00 | 0.14 | 0.14 | 0.14 | 1.00 | 0.50 | 1.00 | 2.00 | 1.00 | 1.00 | 1.00 | 1.00 | 5.00 |
| **D14** | 0.11 | 0.14 | 0.11 | 0.11 | 0.11 | 0.14 | 0.11 | 0.20 | 0.50 | 0.20 | 0.20 | 0.20 | 0.20 | 1.00 |

Since the PPC case study falls in this category of products, the resulting indices, as reported in Table 14, have been used to compute an overall *PI* value.

**Table 14.** Potential index calculation for the PPC case.

| Drivers | Weight | Given Score |
|---|---|---|
| (D1) Does the use of AM lead to potential cost savings considering the entire life cycle? | 0.1573 | 9 |
| (D2) Can assembly, logistic or procurement cost be reduced? | 0.0593 | 7 |
| (D3) Can the product benefit from a reduced time-to-market? | 0.1431 | 7 |
| (D4) Is the production batch size limited? | 0.1375 | 9 |
| (D5) Is the post processing phase limited? | 0.1520 | 6 |
| (D6) Can material waste be reduced? | 0.0610 | 8 |
| (D7) Would the product benefit from a mass reduction? | 0.1166 | 10 |
| (D8) It is beneficial to increase the complexity of the part? | 0.0294 | 7 |
| (D9) It is feasible to consolidate parts? | 0.0165 | 1 |
| (D10) Is customization required? | 0.0268 | 4 |
| (D11) Would the product benefit from areas with differentiated material properties? | 0.0284 | 5 |
| (D12) Would the product benefit from varied density along the volume? | 0.0284 | 5 |
| (D13) Is a volumetric lattice or a superficial texture required? | 0.0338 | 5 |
| (D14) Is it advantageous to add functionalities in the parts? | 0.0099 | 1 |
| | Potential Index (*PI*) | 74.28 |

The obtained *PI* is 74.28%, which shows good potential in the application of AM. So far, the product has passed the two steps of the early phase: it is possible to invest in a possible design solution and carry out an in-depth analysis where a new geometry is developed.

### 4.3. Topological Optimization Result

Given the structural application, the definition of a new geometry has been based on the TO approach. Altair Inspire 2019 was used as TO software. An optimized design has been developed by selecting the material and imposing the constraints and loads to which the part is subjected. After that, the geometry has been redesigned in Solidworks 2020 to achieve more regular shapes. Standard algorithms to create Non-Uniform Rational B-Spline (NURBS) patches from the tessellated models have not be employed due to the strong irregularity of the surfaces. Instead, a new geometry has been modelled by tracing the OT result. In Figure 8, the new geometry is presented.

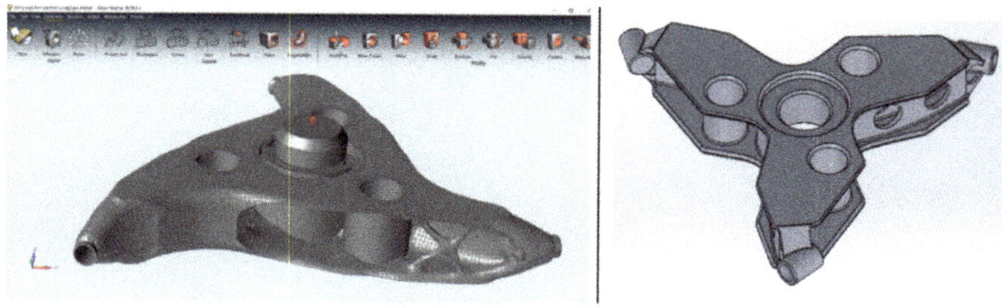

(a) (b)

**Figure 8.** (**a**) One of the geometries obtained by TO in Altair Inspire 2019. (**b**) Redesigned PPC geometry modelled in Solidworks 2020.

The new geometry has been verified by an additional FEM simulation, using the Solidworks 2020 dedicated module. Two solutions have been devised, using aluminium and steel as alternative materials. By using aluminium, the weight can be dramatically reduced, from 250 g of the original design up to 100 g. If steel is employed, the weight can be reduced to 215 g, thus reaching 14% mass reduction. The starting volume was around 90,000 mm$^3$, while the final design volume is around 28,200 mm$^3$ for a steel component and around 40,000 mm$^3$ for the aluminium version. So, the designer has the chance to choose whether to use steel, which is better in terms of mechanical performance and cost, or aluminium, which guarantees a lower weight but more material consumption is required.

The following step of the in-depth analysis has focused on the optimization of the design to evaluate punctual geometric features by means of the proposed indices. The geometric shape has been properly refined to fulfil the DFAM design guidelines and maximise the expected quality of the part. Finally, some printers compatible with the selected materials have been identified. In particular, the ProX 320 is selected as it guarantees the lowest production cost.

For the sake of a complete assessment, alternative manufacturing solutions have been evaluated. The complexity of the new shape is hardly obtainable with other processes, for instance, casting, due to the deep recess and inner features. Indeed, the small production batch hardly justifies the investment for a complicated mould.

The manufacturing costs of the new solution are still higher than the original one, as confirmed by process simulations carried out in collaboration with a printing service supplier. In fact, almost 3 h are required for the printing process, even taking advantage of the possibility of producing multiple parts at the same time. Indeed, additional machining is required to reach acceptable levels of roughness and tolerances in the functional coupling faces. Such costs could be balanced by an improved lightness of the component and better dynamic properties, which are conferred to the aircraft. However, this trade-off needs to be validated accurately, considering a broader design context extended to the whole airplane performance.

## 5. Conclusions

Few works in the literature address the evaluation of AM adoption in early product design. Even fewer are the works that implement multi criteria decision processes to systematically validate the results obtained. Furthermore, a gap that has been highlighted is the absence of a clear distinction between considerations and information that can be extrapolated or calculated in an early phase from those that require a more in-depth study.

The main objective of this paper is to provide a procedural approach that supports the designer to understand the potential offered by additive technologies in a given design context. This paper focuses on the early phase, consisting of the Preliminary Analysis and the Potential Analysis, where multicriteria decision approaches have been leveraged. Since the detailed redesign phase turns out to be the most onerous task, it is limited to only promising situations resulting from a potentiality assessment.

The application of these stages has been exemplified by analysing a propeller pitch controller test case. The proposed flow has proved to be a useful approach as the designer is able to evaluate whether a product can be produced through AM, leveraging the limited information available at an early stage. A new optimized product geometry has then been presented to be realized by PBF. Furthermore, the new part has benefits over the initial design, thus confirming the assumptions of the previous phases. In conclusion, it can be affirmed that the case study is suitable to AM production as foreseen by the application of the preliminary evaluations.

In future work, the approach, as conceived in an algorithmic way, is suitable to be embedded in a software tool to operatively support and speed up the designer activity, especially in the context of configuring customized products [42]. Following the phases described in this work, such a tool would aim at automating the process, such as indices computation and geometry evaluation, also verifying the geometric and production limits of AM and traditional technologies. Finally, a cost assessment phase based on geometric product features needs to be added to complete the analysis for an effective application in companies.

**Author Contributions:** Conceptualization and methodology, R.R.; validation and data curation, J.L.; writing—original draft preparation, J.L., J.S.; writing—review and editing, M.P. (Margherita Peruzzini); supervision and funding acquisition, M.P. (Marcello Pellicciari). All authors have read and agreed to the published version of the manuscript.

**Funding:** This research received no external funding.

**Institutional Review Board Statement:** Not applicable.

**Informed Consent Statement:** Not applicable.

**Data Availability Statement:** All data is contained within the paper.

**Conflicts of Interest:** The authors declare no conflict of interests.

## References

1. Thompson, M.K.; Moroni, G.; Vaneker, T.; Fadel, G.; Campbell, R.I.; Gibson, I.; Bernard, A.; Schulz, J.; Graf, P.; Ahuja, B.; et al. Design for Additive Manufacturing: Trends, opportunities, considerations, and constraints. *CIRP Ann.* **2016**, *65*, 737–760. [CrossRef]
2. Lettori, J.; Raffaeli, R.; Peruzzini, M.; Schmidt, J.; Pellicciari, M. Additive manufacturing adoption in product design: An overview from literature and industry. *Procedia Manuf.* **2020**, *51*, 655–662. [CrossRef]
3. Boothroyd, G.; Dewhurst, P.; Knight, W.A. *Product Design for Manufacture and Assembly*, 3rd ed.; CRC Press: Boca Raton, FL, USA, 2010.
4. Gardan, N.; Schneider, A. Topological optimization of internal patterns and support in additive manufacturing. *J. Manuf. Syst.* **2015**, *37*, 417–425. [CrossRef]
5. Raffaeli, R.; Mengoni, M.; Germani, M. A software system for "Design for X" impact evaluations in redesign processes. *J. Mech. Eng.* **2010**, *56*, 707–717.
6. Campbell, R.; Bourell, D.L.; Gibson, I. Additive manufacturing: Rapid prototyping comes of age. *Rapid Prototyp. J.* **2012**, *18*, 255–258. [CrossRef]
7. Butt, J. Exploring the interrelationship between additive manufacturing and Industry 4.0. *Designs* **2020**, *4*, 13. [CrossRef]
8. Gibson, L.; Rosen, W.D.; Stucker, B. *Additive Manufacturing Technologies*; Springer: New York, NY, USA, 2014; Volume 17.
9. Global Additive Manufacturing Market and Technology Forecast to 2028, Business Wire, Dublin. 2020. Available online: https://www.businesswire.com/news/home/20200914005395/en/Global-Additive-Manufacturing-Market-Generated-12-Billion-Revenue-in-2020-and-is-Forecast-to-Reach-78-Billion-by-2028---ResearchAndMarkets.com (accessed on 17 January 2021).

10. ASTM F2792-12a. *Standard Terminology for Additive Manufacturing Technologies*; ASTM: West Conshohocken, PA, USA, 2015.
11. Duty, C. Pros and Cons of Additive Manufacturing. Composites Manufacturing. 2014. Available online: http://compositesmanufacturingmagazine.com/2014/10/pros-cons-additive-manufacturing/ (accessed on 17 January 2021).
12. Lopes, L.; Da Silva, A.F.; Carneiro, O. Multi-material 3D printing: The relevance of materials affinity on the boundary interface performance. *Addit. Manuf.* **2018**, *23*, 45–52. [CrossRef]
13. Proof, H.; Staffen, A. Challenge of Additive Manufacturing—Why Companies Don't Use Additive Manufacturing in Serial Production. 2019. Available online: https://www2.deloitte.com/content/dam/Deloitte/de/Documents/operations/Deloitte_Challenges_of_Additive_Manufacturing.pdf (accessed on 17 January 2021).
14. Ngo, T.D.; Kashani, A.; Imbalzano, G.; Nguyen, K.; Hui, D. Additive manufacturing (3D printing): A review of materials, methods, applications and challenges. *Compos. Part B Eng.* **2018**, *143*, 172–196. [CrossRef]
15. Kim, D.B.; Witherell, P.; Lipman, R.; Feng, S.C. Streamlining the additive manufacturing digital spectrum: A systems approach. *Addit. Manuf.* **2015**, *5*, 20–30. [CrossRef]
16. Pei, E.; Ressin, M.; Campbell, R.I.; Eynard, B.; Xiao, J. Investigating the impact of additive manufacturing data exchange standards for re-distributed manufacturing. *Prog. Addit. Manuf.* **2019**, *4*, 331–344. [CrossRef]
17. Tang, Y.; Zhao, Y.F. A survey of the design methods for additive manufacturing to improve functional performance. *Rapid Prototyp. J.* **2016**, *22*, 569–590. [CrossRef]
18. ISO/ASTM DIS 20195. *Guide for Design for Additive Manufacturing*; ISO/ASTM: West Conshohocken, PA, USA, 2018.
19. VDI-Guideline NN. *304: Additive Fabrication-Rapid Technologies (Rapid Prototyping)-Fundamentals, Terms and Definitions, Quality Parameter, Supply Agreements*; VDI: Düsseldorf, Germany, 2010.
20. ISO. *DIS 17296—Part 1,2,3,4*; ISO: Geneva, Switzerland, 2015.
21. King, W.; Anderson, A.T.; Ferencz, R.M.; Hodge, N.E.; Kamath, C.; Khairallah, S.A. Overview of modelling and simulation of metal powder bed fusion process at Lawrence Livermore National Laboratory. *Mater. Sci. Technol.* **2015**, *31*, 957–968. [CrossRef]
22. Stewart, S.; Giambalvo, J.; Vance, J.; Faludi, J.; Hoffenson, S. A Product Development Approach Advisor for Navigating Common Design Methods, Processes, and Environments. *Designs* **2020**, *4*, 4. [CrossRef]
23. Kumke, M.; Watschke, H.; Vietor, T. A new methodological framework for design for additive manufacturing. *Virtual Phys. Prototyp.* **2016**, *11*, 3–19. [CrossRef]
24. Lindemann, C.; Reiher, T.; Jahnke, U.; Koch, R. Towards a sustainable and economic selection of part candidates for additive manufacturing. *Rapid Prototyp. J.* **2015**, *21*, 216–227. [CrossRef]
25. Klahn, C.; Leutenecker, B.; Meboldt, M. Design for Additive Manufacturing—Supporting the Substitution of Components in Series Products. *Procedia CIRP* **2014**, *21*, 138–143. [CrossRef]
26. Faludi, J.; Cline-Thomas, N.; Agrawala, S. 3D printing and its environmental implications. In *The Next Production Revolution Implications for Governments and Businesses*; OECD Publishing: Paris, France, 2017.
27. Favi, C.; Raffaeli, R.; Germani, M.; Gregori, F.; Manieri, S.; Vita, A. A life cycle model to assess costs and environmental impacts of different maritime vessel typologies. In Proceedings of the ASME Design Engineering Technical Conference, Cleveland, OH USA, 6–9 August 2017.
28. Greco, S.; Figueira, J.; Ehrgott, M. *Multiple Criteria Decision Analysis*; Springer: New York, NY, USA, 2016.
29. Greco, S.; Matarazzo, B.; Slowinski, R. Rough sets theory for multicriteria decision analysis. *Eur. J. Oper. Res.* **2001**, *129*, 1–47 [CrossRef]
30. Stylidis, K.; Wickman, C.; Söderberg, R. Perceived quality of products: A framework and attributes ranking method. *J. Eng. Des* **2020**, *31*, 37–67. [CrossRef]
31. Mançanares, C.G.; Zancul, E.; Da Silva, J.C.; Miguel, P.A.C. Additive manufacturing process selection based on parts' selection criteria. *Int. J. Adv. Manuf. Technol.* **2015**, *80*, 1007–1014. [CrossRef]
32. Zaman, U.K.U.; Rivette, M.; Siadat, A.; Mousavi, S.M. Integrated product-process design: Material and manufacturing process selection for additive manufacturing using multi-criteria decision making. *Robot. Comput. Manuf.* **2018**, *51*, 169–180. [CrossRef]
33. Saaty, T.L. *Fundamentals of Decision Making and Priority Theory with the Analytic Hierarchy Process*; RWS Publications: Pittsburgh, PA, USA, 2000; Volume 6.
34. Jahan, A.; Edwards, K. A state-of-the-art survey on the influence of normalization techniques in ranking: Improving the materials selection process in engineering design. *Mater. Des.* **2015**, *65*, 335–342. [CrossRef]
35. Pahl, G.; Beitz, W.; Feldhusen, J.; Grote, K. *Engineering Design: A systematic Approach*, 3rd ed.; Springer Science+ Business Media Deutschland GmbH: Berlin/Heidelberg, Germany, 2007.
36. Available online: http://senvol.com/ (accessed on 17 January 2021).
37. Ashby, M.F.; Cebon, D. Materials selection in mechanical design. *J. Phys. IV Fr.* **1993**, *3*, C7-1–C7-9. [CrossRef]
38. Conner, B.; Manogharan, G.P.; Martof, A.N.; Rodomsky, L.M.; Rodomsky, C.M.; Jordan, D.C.; Limperos, J.W. Making sense of 3-D printing: Creating a map of additive manufacturing products and services. *Addit. Manuf.* **2014**, *1*, 64–76. [CrossRef]
39. Raffaeli, R.; Mengoni, M.; Germani, M.; Mandorli, F. An approach to support the implementation of product configuration tools. In Proceedings of the ASME International Design Engineering Technical Conferences and Computers and Information in Engineering Conference DETC 2009, San Diego, CA, USA, 30 August–2 September 2009; pp. 559–570.

40. Joshi, D.; Ravi, B. Quantifying the Shape Complexity of Cast Parts. *Comput. Des. Appl.* **2010**, *7*, 685–700. [CrossRef]
41. Diegel, O.; Nordin, A.; Motte, D. *A Practical Guide to Design for Additive Manufacturing*; Springer: Berlin/Heidelberg, Germany, 2019.
42. Raffaeli, R.; Cicconi, P.; Mengoni, M.; Germani, M. Modular product configuration: An automatic tool for eliciting design knowledge from parametric cad models. In Proceedings of the ASME Design Engineering Technical Conference 2010, Montreal, QC, Canada, 15–18 August 2010; pp. 207–218.

applied sciences

MDPI

Article

# Enhancing Design for Additive Manufacturing Workflow: Optimization, Design and Simulation Tools

Nicolas Alberto Sbrugnera Sotomayor, Fabrizia Caiazzo and Vittorio Alfieri *

Department of Industrial Engineering, University of Salerno, Via Giovanni Paolo II 132, 84084 Fisciano, Italy; n.sbrugnerasotoma@studenti.unisa.it (N.A.S.S.); f.caiazzo@unisa.it (F.C.)
* Correspondence: valfieri@unisa.it; Tel.: +39-089968164

**Abstract:** In the last few decades, complex light-weight designs have been successfully produced via additive manufacturing (AM), launching a new era in the thinking–design process. In addition, current software platforms provide design tools combined with multi-scale simulations to exploit all the technology benefits. However, the literature highlights that several stages must be considered in the design for additive manufacturing (DfAM) process, and therefore, performing holistic guided-design frameworks become crucial to efficiently manage the process. In this frame, this paper aims at providing the main optimization, design, and simulation tools to minimize the number of design evaluations generated through the different workflow assessments. Furthermore, DfAM phases are described focusing on the implementation of design optimization strategies as topology optimization, lattice infill optimization, and generative design in earlier phases to maximize AM capabilities. In conclusion, the current challenges for the implementation of the workflow are hence described.

**Keywords:** additive manufacturing; design for additive manufacturing; topology optimization; generative design; lattice infill optimization

**Citation:** Sbrugnera Sotomayor, N.A.; Caiazzo, F.; Alfieri, V. Enhancing Design for Additive Manufacturing Workflow: Optimization, Design and Simulation Tools. *Appl. Sci.* **2021**, *11*, 6628. https://doi.org/10.3390/app11146628

Academic Editor: Marco Mandolini

Received: 24 June 2021
Accepted: 16 July 2021
Published: 19 July 2021



## 1. Introduction

According to ISO/ASTM standards, additive manufacturing (AM) is defined as the process of joining materials, generally by layer-by-layer approach, to produce parts starting from digital representations [1]. Continuous development of metal AM technologies, such as directed energy deposition (DED) or laser powder bed fusion (LPBF), yield new industrial design perspectives due to several technology benefits.

Indeed, the main AM advantages include: the possibility to manufacture complex internal part-functionalities like cooling channels in turbine blades [2], lattice and gyroids structures in medical implants [3,4]; low buy-to-fly ratio [5]; consolidate part assemblies into fewer components [6]; achieve mechanical properties as tensile and yield strength comparable to bulk materials [7]; and high part-customization. Consequently, AM becomes an economically suitable technology where low-production volumes or customizable production on-demand are required [8].

However, there are still drawbacks related to the complex thermo-physical phenomena involved: repeated cycles of rapid cooling of the melted pool generate anisotropy depending on building direction [9]; control of printing parameters is required to prevent physical problems of balling, key-hole formation and lack of fusion [10]; residual porosity affects end-parts mechanical properties [11]; energy consumption is high to melt or sinter the metal powder [12]; and low surface quality occurs due to layer-by-layer deposition [13], thus reducing fatigue life [9]. Moreover, volumetric restrictions on part size [14], time-consuming jobs and lack of standard certifications for certain industries as aerospace, are still considered limitation factors [15].

In the traditional design for manufacturing (DFM) process, it is required to find a design solution that minimizes manufacturing, assembly and logistic costs [16]. To take into consideration these objectives with the unique AM technology capabilities (e.g., part consolidation, design freedom and highly part-customization), the design for additive manufacturing (DfAM) allows rethinking the whole design process from the digital database to the final printed part. In addition, to take into consideration the geometric printing technology limitations, DfAM guidelines are provided, based on empirical analysis depending on the material, printing parameters and manufacturing technology. Namely, it's possible to recognize [17]: critical angle of self-supported faces, minimum diameters for unsupported holes, optimal position on the building plate, maximum allowable aspect ratio of thin columns, minimum printable wall thickness and minimum feature size. To reduce manufacturing costs related to energy consumption, printing time and post-processing, parts should be designed with minimum support generation. Indeed, supports are usually required for overhanging sections, to prevent collapse during building. However, for metal AM, supports are crucial and cannot be removed at all due to the high-stress gradient generated during printing. In this case, supports act as heat transfer structures preventing excessive distortions and residual stresses.

Initially, the guided-design process performs post-processing checks, while recent advancements on software platforms allow predicting thermal effects and mechanical performance in earlier design phases. However, the physical and manufacturing limitations mentioned above mean that the DfAM process workflow involves several stages to maximize printing capabilities and ensure correct manufacturability. In addition, the process is considered a challenging task to perform, since computer-aided design (CAD), computer aided engineering (CAE), and computer-aided manufacturing (CAM) must be evaluated at different abstraction levels requiring full design, simulation and manufacturing knowledge.

Engineering fields, such as automotive, aerospace, robotics, and medical prosthesis often require the design of lightweight structures for improving the energy efficiency of moving systems enhancing the physical performance of analysis. To achieve this goal structural optimization is used as a suitable design framework based on mathematic formulations consisting of coupling physic system responses such as stress, natural frequencies compliance, or displacements with deterministic or stochastic algorithms to find an optimal layout material by an iterative process. This global definition includes [18]: sizing, shape and topology optimization (TO). Sizing optimization aims at determining the optimum cross-sectional area of structural members, while shape optimization focuses on the optimum boundary domain shape [19]. Particular attention has been addressed in the industry and academia on TO strategies due to more design flexibility generating internal voids in the design domain. In general, TO algorithms result in complex geometries, which are not common by traditional thinking, mainly mimicking nature designs [20]. As a result, AM became the perfect manufacturing technology to be coupled with TO capabilities, and hence improve the DfAM workflow.

In this frame, this manuscript aims to describe the main DfAM workflow assessments, focusing on suitable optimization, design, and simulation tools to reduce the number of required design evaluations. Particular attention was addressed on including topology optimization, lattice infill optimization and generative design in earlier design phases to obtain high-performance parts that could be properly produced via AM. In this field, the main theory behind the optimization algorithm used in commercial software is hence described.

## 2. Holistic DfAM Workflow

As suggested in Figure 1, from a macro point of view, traditional DfAM holistic workflow involves four main phases described by local tasks to be performed [21]: product planning, design optimization, manufacturing optimization and product validation. Product planning activities include modelling preparation of base design, definition of objectives and constraints, finite element analysis and feasibility analysis to include a structural optimization strategy. Design optimization involves an iterative process of: design optimization strategy; design interpretation of optimized results and product simulation. Manufacturing optimization actions imply printing modeling with support generation, optimization of support structures, and additive process simulation. The last phase involves part manufacturing, post-processing, performance validation via mechanical testing and quality inspection.

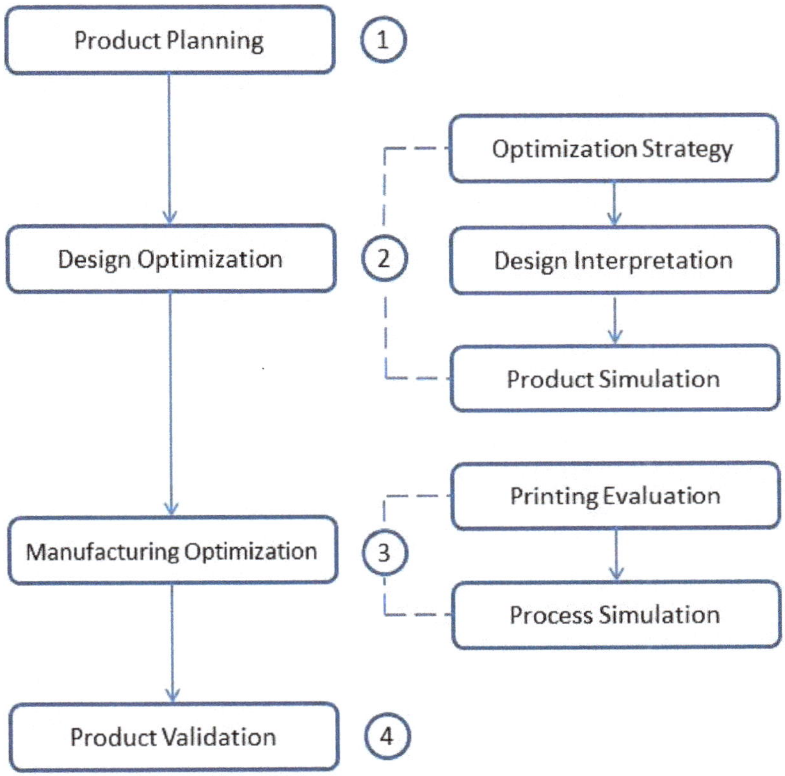


**Figure 1.** Holistic design workflow for AM.

Design and manufacturing optimization represents the digital phases of the DfAM workflow, they are characterized by several design iterations to validate the design proposals. As described in [22], an automatic approach is required with suitable self-parametric design interpretation, coupling AM simulation models with manufacturing settings, cost evaluation and support structures. However, this level of automation was not yet reached. Instead, software platforms as 3DExperience (Dassault Systemes, Velizy-Villacoublay, France) and Ansys (Ansys Inc., Canonsburg, USA) integrates all the required computer aided tools such as geometry preparation, topology optimization, FEM validation, additive preparation and AM process simulation in the same software interface. The main advantage offered by this approach is to facilitate interchange data and manipulation of intermediate results [21].

## 3. Product Planning

Once a candidate part is selected to perform the DfAM process, a CAD file must be created for FEM analysis, to predict the real performance based on the domain distribution of physical responses such as temperature, stresses, displacements and modal frequencies of vibration. Based on this, a qualitative analysis is available for material distribution and feasibility to include a structural optimization strategy. Additionally, the initial safety factor found from the maximum response values and the admissible material resistance enables the viability of proposing any change in material design [23].

In some cases, where a CAD model is not available for starting the process [24], computer-aided reverse engineering (CARE) is performed [25]: the computer model is obtained by surface cloud points through measurements of the object with laser-based scanner or coordinate measure machines. In this operation, a surface representation results in digital formats as standard triangulation language (STL) [26], mesh-part file and graphics exchange specification (IGES). Then, CAD file is obtained with adequate modeling tools offered in software platforms.

This phase is also characterized by the definition of design and non-design domains for applying the structural optimization process. One of the main features of mathematical-based optimization is that solutions are very sensitive to the defined initial domains, with a huge impact on mechanical performance in the simulation validation process [27]. To maximize optimization exploration capabilities to find the optimal material distribution, the definition of a vast design domain is recommended considering space limitation for part-assembly. Additionally, at this stage, to reduce the number of required iterations in the manufacturing-phase optimization, a suitable printing direction is decided, based on the size of the building chamber, with possible additional restrictions via the expedient of frozen regions [28].

## 4. Optimization Strategy

This phase is focused on solving the selected structural optimization process and validating the performance of the proposed design. Given a design domain, boundary constraints and loading conditions, TO describes solid mathematical method implemented through computer-aided engineering (CAE) software. The method allows finding the optimum material layout that maximizes or minimizes an objective function representing a physic response of the system subjected to constraints. In general, real systems are subjected to multiple loading conditions requiring the definition of a multi-objective criteria optimization [29]. One common approach is to solve the multi-objective problem with a single scalar function by associating weighted factors to the different loading cases [30]. This converts the problem into the so-called Pareto-optimality describing a frontier of admissible optimal solutions. In [31] a 199-line Matlab code for tracing the Pareto optimal frontier is proposed: interestingly, no sub-objective can be improved without diminishing another with the main limitation to find a priori adequate values of weight factors [32].

Given this background, TO could be applied to a vast range of physics problems, with the conditional enforcement that the partial differential equation (PDE) of analysis might be reliably discretized and modelled using finite element method (FEM), boundary element method (BEM), finite volume method (FVM) or other domain discretization schemes [33]. Besides classic static structural analysis, the process can be applied to fluid dynamics [34], heat transfer [35], electromagnetism [36], acoustic [37] and Multiphysics combinations between them [38]. In this manner, it is possible to associate a wide number of objectives and constrains formulations including compliance, stress, frequency, displacements, Eigen frequencies, reaction forces, moment of inertia, critical buckling load, mass or volume. Typical optimization tasks for static structural problems are: compliance minimization that is equivalent to stiffness maximization with volume constraint [39]; volume minimization with stress constraint [40]; and minimizing displacement under volume constraint [41].

The global classification of TO methods is based on the fact that the optimization process requires or does not calculate the gradient information of the objective functions. Therefore, two main groups are defined: deterministic (gradient-based) and stochastic (heuristic) methods; the later are inspired by nature and gradient calculus is not required. Gradient-based methods include homogenization, density interpolation schemes as SIMP (Solid Isotropic Material with Penalization) or RAMP (Rational Approximation of Material Properties), and level set methods. Stochastic methods include metaheuristic algorithms like evolutionary approaches as Genetic Algorithm (GA), ESO, BESO and bioinspired algorithms like particle swarm intelligence, human base (tabu-search) and physical-based (Colliding bodies). The reader is kindly referred to [42] for an in-depth review of these TO methods and their numerical formulation. All of the proposed algorithms have in common the definition of a discretized domain and the necessity of implement filtering techniques to ensure smooth convergence and mesh-independency results. Differences between consists in purely mathematical definition with continuous or discrete variables, linear or non-linear programming, heuristic or mathematical derivation, local or global length scale control with implicit-explicit means.

Nowadays, software platforms allow applying several optimization strategies for maximizing performance and obtaining lightweight structures that are suitably produced by AM technologies. In this manuscript, different TO approaches are presented: traditional gradient-based sequence with density interpolation schemes, infill optimization with lattice structures and the innovative generative design approach. All of these are significantly appropriate in the frame of DfAM.

### 4.1. Numerical Instabilities of TO

Main numerical instabilities of TO problems are defined in [43] as checkerboard patterns, mesh dependence and local minima solution. The checkerboard pattern describes a periodic pattern of low and high values for the design variable arranged as a checkerboard. This result is undesirable since the material layout is not appropriate. It was shown in [44] that a finite discretized domain with a patch arrangement of average density of 1/2 provides artificial high stiffness when applied to the TO for minimizing compliance. Additionally, [45] concludes that this numerical instability is caused by FEM where the equilibrium equations are only referred to element nodes. Moreover, the use of lower-order elements promotes checkerboard patterns, while higher-order discretization schemes reduce the global effect. In this case, finite-volume theory accomplishes the task to eliminate checkerboards.

The mesh dependence problem appears when mesh-refinement of the design domain results in different topology instead of better description for boundaries [46]. The solution to both checkerboard-patterns and mesh dependency is the regularization of the domain design by filtering techniques. From the wide spectrum of filtering schemes the most valuable for easy implementation and efficiency are density and sensitivity filters [47]. In the former, the element density is calculated as a weighted average of the neighborhood elements included in the characteristic filter radius; in the latter, standard calculus of element density is used to calculate a weighted average of the neighborhood sensitivities [48].

Nerveless, it is worth noting that regularization schemes fail in mitigating solutions with a grayscale interface. To obtain pure black and white designs, with a minimum interface, post-processing on the design domain is required through projection schemes like the Heaviside threshold [49]. For example, commercial software Comsol Multiphysics (Comsol Inc., Stockholm, Sweden) englobes filtering and projection schemes in the so-called three-field density representation [50]. The chain of transformations starts from the design domain, then a PDE density filter scheme is applied and the last operation includes a Heaviside projection function to improve the contour boundaries.

Solving TO problems with penalized interpolation schemes results in non-convex problems, hence obtaining local minima solutions [51]. To prevent this, continuation schemes are typically employed on material and filter parameters. For the SIMP scheme, the continuation strategies solve an initial convex problem, characterized with uniqueness solution, with low value of the penalization parameter. This solution is used as input for consecutive cycles, where this parameter is increased in steps until arriving at adequate values.

### 4.2. Gradient-Based TO

Available commercial software as Altair Hyperworks (Altair Engineering, Troy, MI, USA) [52], SolidWorks (Dassault Systemes, Velizy-Villacoublay, France), MSC (MSC Software Corporation, Newport Beach, CA, USA) [53], Comsol Multiphysics (Comsol Inc Stockholm, Sweden) [54], Ansys (Ansys Inc., Canonsburg, PA, USA) and Abaqus (Abaqu Inc., Velizy-Villacoublay, France) solves the TO problem by using the gradient-based der sity approach, associated to the density interpolation scheme method SIMP or power-lav approach. The density approach is characterized by a design variable assigned to dis cretized elements representing an artificial density of material [55]. This variable range from 0 to 1 in the case of void elements or solid material, respectively. In general, platform provide as objective function minimizing compliance to find the maximum stiffness layou material distribution, aiming to reduce the computing complexity of gradient-based algo rithms [56]. Alternatively, other optimization algorithms are available through MATLAI or Python codes and integrated into FEM commercial software [57].

The solution of gradient-based TO strategy is found through an iterative process callec nested-formulation [58]. The iterative loop consists of five main steps [59]: FEM analysis sensitivity analysis, filtering techniques, optimality algorithm updating design variable and post-processing.

In general, the stiffness matrix structure and the displacement field are found via FEM Then, the density design variables are assigned into the design domain as constants for each finite element. By using a sensitivity analysis, the process calculates the partial derivatives of the objective function concerning the value of the design variable, at each element. Fo the particular case of compliance objective function, the derivative is always negative [60] thus indicating that increasing element-density yields a decrease of the overall compliance and a stiffer structure. Aiming at reducing the numerical instabilities, the next step is applying density or sensitivity filtering techniques to impose a length-scale restrictior on the design domain, thus limiting the spectrum of possible feasible solutions. The following phase uses an optimality algorithm method like optimality criteria (OC) [61] o the method of moving asymptotes (MMA) [62] that approximate the value of the objective function by using the sensitivity results. In this manner, the optimization task is divided into sub-problems so that solving is more efficient. The last step consists of updating the design variable values. The cycle is repeated until numerical convergence. Additional post-processing via projection filters might be required to improve contour boundaries.

### 4.3. SIMP Approach

One efficient strategy to solve the gradient-based TO formulation consists in a con tinuous representation of the density variable associated with some form of penalty that steers intermediate solutions into discrete values. In the SIMP approach, a penalization parameter is used to penalize intermediate values of the design variable so they became unfavorable in the sense that the stiffness obtained is small compared to the required material volume [46]. Then, the design variable is multiplied onto physical quantities as stiffness, cost or conductivity to evaluate the performance of the material distribution in the FEM analysis.

In density interpolation schemes the physical interpretation of greyscale or intermedi ate values is represented by the homogenization method correlated to porous composite materials [58]. Low values of the penalization parameter cause too much greyscale so the optimum material distribution is improperly defined. On the other hand, high values, lead to a fast convergence solution into local minima, thus reducing the probability to obtain a global optimum distribution. Adequate values are recommended from 3 to 5 based on the verification of composite Hashin–Shtrikman bounds [58].

An evolution of the method has been suggested with the introduction of a minimum stiffness value [47] that prevents singularity problems of the stiffness matrix when comput ing equilibrium equations. In addition, this method presents the advantage of their easy generalization to many filtering techniques. This approach was efficiently introduced in the 88-lines Matlab code for minimizing compliance [63].

### 4.4. Lattice Infill Optimization

Lattice structures are very attractive for industrial lightweight applications [64]. Namely, the hip implant prosthesis is optimized with internal lattice infill to reduce the physical phenomenon of stress shielding and promoting excellent biocompatibility [65]. Indeed, the porous infill pattern gives the advantage of high specific strength, enhanced stiffness, superior capacity of energy absorption and offers the possibility to add internal functionality. Many mathematical models are available to generate porous patterns. Namely, it is possible to recognize periodically distributed porous patterns as the triply periodic minimal surfaces (TPMS) where uniform or functionally graded designs are created [66]; and random porous distribution through stochastic point cloud-based and fractal geometry that mimics the real porous distribution of nature scaffolds [67].

This type of structure is included in the so-called multi-scale design, where the overall performance is evaluated using information from different abstraction levels. In fact, they can be considered as composite material with mechanical properties calculated via homogenization techniques [68]. Homogenization-based structural optimization is even one of the mathematical foundations for density interpolations schemes. This method converts the isotropic material TO analysis into a composite material consisting of infinite small holes, periodically distributed through the material. In this manner, TO is converted into a sizing optimization problem using micromechanical modelling where the design variable is described by several sub-variables to be optimized [69].

To include lattice structures into the TO process, proper steps are required to combine both topology and size optimization [70]. The sequence starts by implementing a TO process to find the density field domain, then a density threshold is applied to preserve areas with high-density regions describing solid boundaries; the remaining domain serves as input to lattice wireframe generation. After the generation of infill lattice contours, a size optimization with adequate meshing and FEM simulation is applied. The result of the process is a graded infill lattice in low-density regions that has better performance with minimum support generation. A similar process is used inside commercial software nTop (nTopology, New York, NY, USA) [71].

### 4.5. Generative Design

Generative design (GD) describes a constrained design exploration process that allows finding multiple convergent solutions by using evolutionary approaches based on nature. This approach describes the artificial intelligence applied to structural optimization and works as an auxiliary tool for traditional TO algorithms, where an exploration strategy is added through a random disturbance to change search design direction [72].

Exploration algorithm works as a black-box where the input information is: materials, manufacturing technology, physical constraints and design restrictions; while output information involves multiple solutions meeting the initial demand. Design solutions are categorized by a ranking performance to select an adequate candidate to manufacture.

The main difference from TO traditional process is that GD does not require the definition of a design domain where the algorithm modifies the material distribution. In fact, domain restrictions associated to the assembly with other parts are defined. Then, the evolutionary algorithm creates a material path between fixed connections and domain limitations, providing more freedom of design.

The main advantages of this design exploration strategy are part consolidation of complex assemblies [73], almost ready parametrized CAD design solutions requiring minimum modifications, multiple solutions obtained simultaneously for different materials and manufacturing technologies. However, as indicated in [74], the strategy implies a high consuming time task not recommended for simple designs where traditional TO is more effective. Available commercial software that includes GD approach includes: MSC Apex Generative Design (MSC Software Corporation, USA) [75], nTop (nTopology, USA) [76] and Autodesk Fusion360 (Autodesk, San Rafael, USA) [77]. In addition, it is possible to differentiate cloud-based GD from real-time GD. The first approach, used by Fusion360, takes advantage of cloud computing to efficiently solve different initial settings as materials, constraints and manufacturing technologies simultaneously; while real time GD describes an exploration analysis limited to fixed initial settings.

### 4.6. TO Constraints

Introducing optimization constraints in TO process improves the efficiency of the DfAM workflow by reducing the number of design iterations to verify the product simulation and printing evaluation. Major structural constrains include: physical limits as maximum stress and displacements and AM geometrical limitations based on fixed printing direction.

In structural applications, maximum stress measure takes relevance to verify material resistance. Solving TO with compliance minimization as objective function takes into account preserving material in regions with elevated strain energy, providing results with elevated stress concentrations inducing several design iterations until the prescribed safety factor. One feasible solution is to include stress constraints in the TO formulation. However, this implies computational problems related to physical phenomena such as stress singularity, local nature and highly non-linear stress behavior [78]. Singularity problem appears in degenerated regions where the design variables tend to zero: for these elements, nonzero values of stress are found promoting the selection of a local minima solution. Stress is a local measure in the design domain, the use of local stress constraints for each element needs a high number of variables, thus increasing the computational effort. Highly non-linear stress behavior is related to stress gradient and design domain, as for reentrant corners where density changes in neighboring regions.

A common approach is to convert the local stress-constrained for each discretization element into a single global stress measure by relaxing methods with p-norm or Kreisselmeier–Steinhauser (KS) functions [79]. This technique lacks physical interpretation, since it provides an approximation of the maximum reached stress by penalization of the local stress measure for all domain elements.

Since stress-constrained TO is well suited to solve the shape optimization problem with parametrized domain boundaries, another approach solves the traditional unconstrained TO problem at first for then applying a design refinement on critical regions by shape optimization technique [80].

To obtain TO designs that are suitable to AM with minimum support generation, manufacturing constraints can be included into the algorithm process by imposing geometrical printing limitations such as minimal angle of unsupported faces related to material and printing technology; and minimum size thickness related to machine printing capabilities [81]. AM constraints limit the algorithm freedom of searching the optimal layout distribution obtaining a solution with reducing mechanical properties concerning the unconstrained one. Therefore, a trade-off between support generation and performance must be set.

## 5. Design Interpretation

The solution of the TO process consists of a density-field representation where solid material and void areas are accepted to the light-weight structure without compromising the structural response. To obtain a boundary representation of the optimized design and reduce the density interface, software platforms provide threshold tools to limit the fraction of density elements that are visualized. This design is available in specific file formats depending on the available software or can be exported as tessellation representation, in general as STL or IGS.

Design interpretation describes the methodology to convert TO results into parameterized CAD model. This procedure is required to verify the physical performance via FEM simulation and regularize the design boundaries. Depending on the software interface, it is possible to find different design approaches such as: traditional parameterized model, implicit modelling or NURBS representation. Among these, NURBS allows the creation of organic shapes with minimum effort thanks to the freedom to manipulate control points. This approach is used by the pioneer aerospace company ArianeGroup to maximize AM constraints [82].

Manual reconstruction is a highly time-consuming operation due to complex shapes generated in the optimization process. However, software platforms as nTop (nTopology, New York, NY, USA), Fusion 360 (Autodesk, San Rafael, CA, USA) and Creo (PTC Creo, Boston, MA, USA) offer an automatic optimization to CAD reconstruction based on B-reps and Boolean operations into a watertight boundary representation [83]. As a result, this technique requires minimum post-processing to preserve the details of the optimization result. In this stage, the designer must consider the DfAM guided rules as well, to reduce the number of attempts to validate the part in the manufacturing optimization sequence.

## 6. Product Simulation

Software simulations allow us to predict the physical performance of the proposed design by analyzing maximum stresses, displacements, absorbed strain energy, natural frequencies of vibration, buckling modes and other physic parameters of interest. This phase consists of checking the performance of the reconstructed design by FEM analysis. To verify the robustness of the proposed mesh, a convergence analysis is required: the size element is reduced until it arrives at convergence of the physical measure. Additionally, a trade-off between element order discretization and computing time must be set [84].

For the special case of lattice structures validation, one common approach uses the homogenization technique where a representative volume element of the repetitive pattern distribution is analyzed to calculate the cell anisotropy properties and then results are extrapolated into a solid isotropic material representing the overall infill domain. In this manner, the required computational time is considerably reduced.

## 7. Printing Evaluation

Printing evaluation involves the analysis of the STL design representation and the support generation depending on the selected building direction. This analysis is performed through a draft angle measure where the minimum unsupported face angle of DfAM guidelines is considered. There are several strategies to optimize the process. Some of these are included in software platforms as Magics (Materialise, Leuven, Belgium) where the optimal part position is found by ranking criteria including printing time, support volume, total mass and center of gravity [85]. Additionally, optimization may involve the support structure design with lattice infill and tree design.

Additionally considering printing set-up parameters as layer-height and infill strategy, it is possible to generate a slice representation of the building process aiming at finding possible failures and collapses.

## 8. Process Simulation

At micro-scale analysis, metal AM involves the complex physical phenomena of rapid cooling [86]. This model represents a multi-scale thermal-transient analysis on which every scan-laser hatching modifies the thermal response of bottom layers with melting-solidification cycles [87]. A constant heat transfer in the building direction to prevent high-stress concentrations is aimed. In this frame, numerical simulation plays an important role by predicting stress concentrations, associated deformations and high plastic strain regions. The last characteristic may lead to crack failures, detachment from the support plate, excessive geometric distortions or high anisotropic microstructure.

The main advantage of the thermal analysis is optimizing the part design, building orientation and support generation to prevent high-stress concentrations. In addition, geometric compensation could be applied on the design to print pre-deformed part to obtain the nominal geometric tolerance after support removal [88]. The main drawback of this approach is that it becomes challenging to model due to the multi-scale behavior and the excessive computational time.

Recently, software platforms such as MSC or Ansys provide supplementary modules to perform AM printing simulation with great accuracy and significant reductio of calculation time [85]. This model is based on the inherent strain approach that wa first developed in academia for welding large components. The main characteristic i that thermo-mechanical simulation is replaced with a quasi-static FEM simulation wher distortions are induced by user defined inherent strains [89]. The method starts with course voxel discretization domain defined by the part with its support structure. Th size of voxel elements is a multiple of real layer thickness, in this way every voxel layer i simulated as a manufactured layer. The material deposition modeling is addressed by a F activation strategy, where new layers are activated with the corresponding inherent strai depending on hatching strategy and building time.

Initial inherent strains can be obtained via simulation [89] and empirical meth ods [90]. The first method is based on the reduce order approach, which describes thermo-mechanical simulation applied to a small-scale volume representation [91]. Th last approach used by MSC Simufact Additive involves the manufacturing of samplec cantilever specimens to measure the maximum deflection after cutting from the buildin plate. Other input parameters include beam width, speed, power and material propertie To calibrate the model, the software uses an iterative procedure to find an adequate inheren strain related to maximum deflections for the sample and the selected printing strateg After convergence, the calibration process is finished and this characteristic strain is fed a input to layer-by-layer deposition FEM static distortion analysis. Crucially, the simulatior must be run under the same printing conditions defined on the calibration process.

## 9. Product Validation

Once the optimized design verifies the physical and manufacturing simulations, i performed a prototype printing including the traditional post-processing sequence o thermal treatment, detachment of the piece from the plate with support removal anc surface finishing with laser polishing or machining via CNC.

Therefore, as indicated in [92], part validation consists in: checking the materia mechanical properties with tensile coupons, dimensional control, non-destructive testing evaluation of density and microstructure. Due to the high anisotropy of AM metal process standard tensile coupons are printed in different building orientations to validate the mechanical properties as ultimate tensile strength, yield strength and elongation for a given building strategy [93]. Evaluation of density and microstructure can be performed via scanning electron microscopy. Dimensional control consists in verifying the allowable tolerances of the part, since thermal distortions and layer-by-layer deposition affects dimensional measures. Nondestructive testing techniques as computed tomography scan and penetrant testing provide additional information about internal porosity and cracks at the surface, respectively. Eventually, depending on the nominal condition of loading the mechanical resistance is assessed. To date, ongoing research is conducted to address a general lack of international standards for AM parts qualification [94,95], although specific references have been published recently [96,97].

## 10. Discussion

The successful management of the DfAM workflow is directly related to the selected design, optimization and simulation tools implemented during the different assessments From the design point of view, it is recommended to conduct the interpretation of optimized results by using software platforms with smooth automated design interpretation that minimizes the design intervention. Instead, if manual reconstruction is performed, the NURBS approach demonstrates full potential to easily reproduce complex organic shapes generated during the optimization phase; also the high design flexibility allows reducing the effort to change design shapes for manufacturing analysis.

Recent DfAM frameworks proposals exploit AM capabilities via size and topology optimization strategies [98–100]. Nevertheless, without including adequate physical and manufacturing optimization constraints the designer might fall in several design iterations. This work empathizes that independent of the selected optimization strategy as topology optimization, lattice infill optimization or generative design, imposing optimization physical and manufacturing constrains becomes therefore fundamental to considerably reduce the number of required evaluations in the product simulation and printing evaluation. However, is necessary to take into consideration that manufacturing constraints imposes geometrical limits on the design exploration. Consequently, finding the material distribution that maximizes mechanical performance with minimum support generation becomes a challenging task to perform and a trade-off between these opposite objectives must be set.

Available simulation tools for predicting thermal distortions and failures, as the thermo-mechanical simulation by the inherent strain method, offers great accuracy with minimum computational time comparing to traditional fluid-flow thermal simulations.

## 11. Conclusions

This paper describes the main DfAM workflow phases focusing on design, optimization and simulation tools to minimize the number of iterative design evaluations. Optimization design strategies were described to maximize AM capabilities, and the main highlights are presented as follows:

1. The guided-design TO strategy improves the workflow efficiency by using optimization constraints for FEM validation and AM printing limitations.
2. Nowadays, software platforms provide automatic CAD reconstructions techniques for TO, requiring minimum post-processing time and modelling expertise. To maximize this technique, TO and FEM validation should be performed via the same software platform, to facilitate data manipulation.
3. In general, TO algorithms works as a black-box inside software platforms. However, the designer must understand the physical interpretation of density fields and check solver convergence to ensure adequate results.
4. The analysis of different TO solutions is recommended to find an adequate trade-off between performance and manufacturing costs.

Unfortunately, the major limitation is the crucial amount of non-automated tasks involving intensive software knowledge in different areas. Therefore, multidisciplinarity is strongly required.

**Author Contributions:** Conceptualization, N.A.S.S.; methodology, F.C. and V.A.; software, N.A.S.S.; validation, N.A.S.S., F.C. and V.A.; formal analysis, N.A.S.S.; investigation, N.A.S.S.; resources, F.C.; data curation, F.C. and V.A.; writing—original draft preparation, N.A.S.S.; writing—review and editing, V.A.; visualization, N.A.S.S.; supervision, F.C.; project administration, F.C.; funding acquisition, F.C. All authors have read and agreed to the published version of the manuscript.

**Funding:** This research was funded by the Italian "Ministero dello Sviluppo Economico", in the framework of "Piano Territoriale Regionale 2019–2021 della Ricerca di Sistema Elettrico Nazionale (RdS)" presented by ENEA.

**Institutional Review Board Statement:** Not applicable.

**Informed Consent Statement:** Not applicable.

**Data Availability Statement:** No new data were created or analyzed in this study. Data sharing is not applicable to this article.

**Conflicts of Interest:** The authors declare no conflict of interest.

## References

1. ISO; ASTM. *ISO/ASTM 52900:2015 (ASTM F2792)—Additive Manufacturing–General Principles—Terminology*; ISO International Organization for Standardization: Geneva, Switzerland; ASTM American Society for Testing and Materials: West Conshohocken, PA, USA, 2015.

2. Magerramova, L.; Vasilyev, B.; Kinzburskiy, V. Novel designs of turbine blades for additive manufacturing. In Proceedings of the ASME Turbo Expo, Seoul, Korea, 13–17 June 2016; ASME: New York, NY, USA. [CrossRef]
3. Caiazzo, F.; Alfieri, V.; Bujazha, B.D. Additive manufacturing of biomorphic scaffolds for bone tissue engineering. *Int. J. Adv. Manuf. Technol.* **2021**, *113*, 2909–2923. [CrossRef]
4. Arabnejad, S.; Johnston, B.; Tanzer, M.; Pasini, D. Fully porous 3D printed titanium femoral stem to reduce stress-shielding following total hip arthroplasty. *J. Orthop. Res.* **2017**, *35*, 1774–1783. [CrossRef]
5. Materialise: An introduction of Buy-To-Fly Ratio Cutting Costs with Metal 3D Printing. Available online: https://www.materialise.com/en/manufacturing/whitepaper-buy-to-fly-ratio-cutting-costs-metal-3d-printing (accessed on 23 June 2021).
6. Yang, S.; Min, W.; Ghibaudo, J.; Zhao, Y.F. Understanding the sustainability potential of part consolidation design supported by additive manufacturing. *J. Clean. Prod.* **2019**, *232*, 722–738. [CrossRef]
7. Rafi, H.K.; Starr, T.L.; Stucker, B.E. A comparison of the tensile, fatigue, and fracture behavior of Ti-6Al-4V and 15-5 PH stainless steel parts made by selective laser melting. *Int. J. Adv. Manuf. Technol.* **2013**, *69*, 1299–1309. [CrossRef]
8. Guo, N.; Leu, M.C. Additive manufacturing: Technology, applications and research needs. *Front. Mech. Eng.* **2013**, *8*, 215–243. [CrossRef]
9. Alafaghani, A.; Qattawi, A.; Jaman, M.S.; Ablat, M.A. Microstructure and mechanical properties of direct metal laser–sintered 15-5PH steel with different solution annealing heat treatments. *Int. J. Adv. Manuf. Technol.* **2019**, *105*, 3499–3520. [CrossRef]
10. Calignano, F.; Galati, M.; Iuliano, L. A Metal Powder Bed Fusion Process in Industry: Qualification Considerations. *Machines* **2019**, *7*, 72. [CrossRef]
11. Sola, A.; Nouri, A. Microstructural porosity in additive manufacturing: The formation and detection of pores in metal parts fabricated by powder bed fusion. *J. Adv. Manuf. Process.* **2019**, *1*, 1–21. [CrossRef]
12. Liu, Z.Y.; Li, C.; Fang, X.Y.; Guo, Y.B. Energy Consumption in Additive Manufacturing of Metal Parts. *Procedia Manuf.* **2018**, *26*, 834–845. [CrossRef]
13. Alfieri, V.; Argenio, P.; Caiazzo, F.; Sergi, V. Reduction of surface roughness by means of laser processing over additive manufacturing metal parts. *Materials* **2017**, *10*, 30. [CrossRef] [PubMed]
14. Gibson, I.; Rosen, D.; Stucker, B.; Khorasani, M. *Additive Manufacturing Technologies*; Springer: Cham, Switzerland, 2021. [CrossRef]
15. Caiazzo, F.; Alfieri, V. Optimization of laser beam welding of steel parts made by additive manufacturing. *Int. J. Adv. Manuf. Technol.* **2021**. [CrossRef]
16. Barroqueiro, B.; Andrade-Campos, A.; Valente, R.A.F.; Neto, V. Metal additive manufacturing cycle in aerospace industry: A comprehensive review. *J. Manuf. Mater. Process.* **2019**, *3*, 52. [CrossRef]
17. Crucible: Design Guidelines for Direct Metal Laser Sintering (DMLS). Available online: https://www.crucibledesign.co.uk/images/uploaded/guides/bs7000-part-2-a-management-guide-download-original.pdf (accessed on 23 June 2021).
18. Bendsøe, M.P. *Optimization of Structural Topology, Shape, and Material*; Springer: New York, NY, USA, 1995.
19. Upadhyay, B.D.; Sonigra, S.S.; Daxini, S.D. Numerical analysis perspective in structural shape optimization: A review post 2000. *Adv. Eng. Softw.* **2021**, *155*, 102992. [CrossRef]
20. Du Plessis, A.; Broeckhoven, C.Y.; Yadroitsev, I.; Hands, I.; Kunju, C.H.; Bhate, R.; Dhruv, B. Beautiful and Functional: A Review of Biomimetic Design in Additive Manufacturing. *Addit. Manuf.* **2019**, *27*, 408–427. [CrossRef]
21. Dalpadulo, E.; Pini, F.; Leali, F. Integrated CAD platform approach for Design for Additive Manufacturing of high performance automotive components. *Int. J. Interact. Des. Manuf.* **2020**, *14*, 899–909. [CrossRef]
22. Wiberg, A.; Persson, J.; Ölvander, J. Design for additive manufacturing—A review of available design methods and software. *Rapid Prototyp. J.* **2019**, *25*, 1080–1094. [CrossRef]
23. Mohiuddin, M.V.; Khan, M.M.A. Re-design of an Aircraft Bracket Using Topology Optimization Technique. *Int. J. Mech. Eng.* **2020**, *7*, 42–53. [CrossRef]
24. Pang, T.Y.; Fard, M. Reverse engineering and topology optimization for weight-reduction of a bell-crank. *Appl. Sci.* **2020**, *10*, 8568. [CrossRef]
25. Page, D.; Koschan, A.; Abidi, M. Methodologies and Techniques for Reverse Engineering—The Potential for Automation with 3-D Laser Scanners. In *Reverse Engineering*; Springer Series in Advanced Manufacturing; Raja, V., Fernandes, K., Eds.; Springer: London, UK, 2008. [CrossRef]
26. Szilvási-Nagy, M.; Mátyási, G. Analysis of STL Files. *Math. Comput. Model.* **2003**, *38*, 945–960. [CrossRef]
27. Berrocal, L.; Fernandez, R.; Gonzalez, S.; Peñiran, A.; Tudela, S.; Villanova, J.; Rubio, L.; Marquez, J.M.; Guerrero, J.; Lasagni, F. Topology optimization and additive manufacturing for aerospace components. *Prog. Addit. Manuf.* **2019**, *4*, 83–95. [CrossRef]
28. Renishaw: Is Topological Optimization Really Optimal? Case Study: Suspension Bell-Crank. Available online: https://resources.renishaw.com/en/details/--101324 (accessed on 23 June 2021).
29. Christensen, P.W.; Klarbring, A. *An Introduction to Structural Optimization*; Springer Science + Business Media: Basingstoke, UK, 2009. [CrossRef]
30. Iqbal, T.; Wang, L.; Li, D.; Dong, E.; Fan, H.; Fu, J.; Hu, C. A general multi-objective topology optimization methodology developed for customized design of pelvic prostheses. *Med. Eng. Phys.* **2019**, *69*, 8–16. [CrossRef]
31. Suresh, K. A 199-line Matlab code for Pareto-Optimal tracing in topology optimization. *Struct. Multidiscip. Optim.* **2010**. [CrossRef]

32. Luo, Z.; Chen, L.P.; Yang, J.; Zhang, Y.Q. Multiple stiffness topology optimizations of continuum structures. *Int. J. Adv. Manuf. Technol.* **2006**, *30*, 203–214. [CrossRef]
33. Sigmund, O. EML webinar overview: Topology Optimization—Status and Perspectives. *Extrem. Mech. Lett.* **2020**, *39*, 100855. [CrossRef]
34. Pietropaoli, M.; Montomoli, F.; Gaymann, A. Structural and Multidisciplinary Optimization Three-dimensional fluid topology optimization for heat transfer. *Struct. Multidiscip. Optim.* **2018**, *59*, 801–812. [CrossRef]
35. Høghøj, L.C.; Nørhave, D.R.; Alexandersen, J.; Sigmund, O.; Andreasen, C.S. Topology optimization of two fluid heat exchangers. *Int. J. Heat Mass Transf.* **2020**, *163*, 120543. [CrossRef]
36. Deng, Y.; Korvink, J.G. Topology optimization for three-dimensional electromagnetic waves using an edge element-based finite-element method. *Proc. R. Soc. A Math. Phys. Eng. Sci.* **2016**, *472*. [CrossRef] [PubMed]
37. Zhao, W.; Zheng, C.; Chen, H. Acoustic topology optimization of porous material distribution based on an adjoint variable FMBEM sensitivity analysis. *Eng. Anal. Bound. Elem.* **2019**, *99*, 60–75. [CrossRef]
38. Corbera Caraballo, S.; Olazagoitia Rodríguez, J.L.; Lozano Ruiz, J.A.; Álvarez Fernández, R. Optimization of a butterfly valve disc using 3D topology and genetic algorithms. *Struct. Multidiscip. Optim.* **2017**, *56*, 941–957. [CrossRef]
39. Deaton, J.D.; Grandhi, R.V. A survey of structural and multidisciplinary continuum topology optimization: Post 2000. *Struct. Multidiscip. Optim.* **2014**, *49*, 1–38. [CrossRef]
40. Da Silva, G.A.; Aage, N.; Beck, A.T.; Sigmund, O. Three-dimensional manufacturing tolerant topology optimization with hundreds of millions of local stress constraints. *Int. J. Numer. Methods Eng.* **2021**, *122*, 548–578. [CrossRef]
41. Shi, G.; Guan, C.; Quan, D.; Wu, D.; Tang, L.; Gao, T. An aerospace bracket designed by thermo-elastic topology optimization and manufactured by additive manufacturing. *Chin. J. Aeronaut.* **2020**, *33*, 1252–1259. [CrossRef]
42. Sigmund, O.; Maute, K. Topology optimization approaches: A comparative review. *Struct. Multidiscip. Optim.* **2013**, *48*, 1031–1055. [CrossRef]
43. Sigmund, O.; Petersson, J. Numerical instabilities in topology optimization: A survey on procedures dealing with checkerboards, mesh-dependencies and local minima. *Struct. Optim.* **1998**, *16*, 68–75. [CrossRef]
44. Díaz, A.; Sigmund, O. Checkerboard patterns in layout optimization. *Struct. Optim.* **1995**, *10*, 40–45. [CrossRef]
45. Araujo, M.V.O.; Lages, E.N.; Cavalcante, M.A.A. Checkerboard free topology optimization for compliance minimization applying the finite-volume theory. In Proceedings of the XL Ibero-Latin-American Congress on Computational Methods in Engineering, ABMEC, Natal, Brazil, 9 November 2020. [CrossRef]
46. Bendsoe, M.P.; Sigmund, O. *Topology Optimization Theory—Methods and Applications*; Springer Science + Business Media: New York, NY, USA, 2003.
47. Sigmund, O. Morphology-based black and white filters for topology optimization. *Struct. Multidiscip. Optim.* **2007**, *33*, 401–424. [CrossRef]
48. Hu, S.B.; Chen, L.P.; Zhang, Y.Q.; Yang, J.; Wang, S.T. A crossing sensitivity filter for structural topology optimization with chamfering, rounding, and checkerboard-free patterns. *Struct. Multidiscip. Optim.* **2009**, *37*, 529–540. [CrossRef]
49. Wang, F.; Lazarov, B.S.; Sigmund, O. On projection methods, convergence and robust formulations in topology optimization. *Struct. Multidiscip. Optim.* **2011**, *43*, 767–784. [CrossRef]
50. Lazarov, B.S.; Wang, F.; Sigmund, O. Length scale and manufacturability in density-based topology optimization. *Arch. Appl. Mech.* **2016**, *86*, 189–218. [CrossRef]
51. Li, L.; Khandelwal, K. Volume preserving projection filters and continuation methods in topology optimization. *Eng. Struct.* **2015**, *85*, 144–161. [CrossRef]
52. Altair University: Practical Aspects of Structural Optimization a Study Guide. Available online: https://altairuniversity.com/free-ebooks/free-ebook-practical-aspects-of-structural-optimization-a-study-guide/ (accessed on 23 June 2021).
53. MSC Software Corporation: Design Sensitivity and Optimization User's Guide. Available online: https://simcompanion.mscsoftware.com/infocenter/index?page=content&id=DOC10014 (accessed on 23 June 2021).
54. Comsol Multiphysics: Optimization Module 5.4. Available online: https://doc.comsol.com/5.6/docserver/#!/com.comsol.help.comsol/helpdesk/helpdesk.html (accessed on 23 June 2021).
55. Bendsøe, M.P. Optimal shape design as a material distribution problem. *Struct. Optim.* **1989**, *1*, 193–202. [CrossRef]
56. Sigmund, O. On the usefulness of non-gradient approaches in topology optimization. *Struct. Multidiscip. Optim.* **2011**, *43*, 589–596. [CrossRef]
57. Lima, C.; Reis, M. A Topology Optimization Solver Applied to 3D Compliant Mechanism. In Proceedings of the 24th ABCM International Congress of Mechanical Engineering, Curitiba, PR, Brazil, 3–8 December 2017. [CrossRef]
58. Bendsøe, M.P.; Sigmund, O. Material interpolation schemes in topology optimization. *Arch. Appl. Mech.* **1999**, *69*, 635–654. [CrossRef]
59. Kaminakis, N.T.; Stavroulakis, G.E. Topology optimization for compliant mechanisms, using evolutionary-hybrid algorithms and application to the design of auxetic materials. *Compos. Part B Eng.* **2012**, *43*, 2655–2668. [CrossRef]
60. Lazarov, O.S. Filters in topology optimization based on Helmholtz-type differential equations. *Int. J. Numer. Methods Eng.* **2011**, *86*, 765–781. [CrossRef]
61. Molter, A.; dos Santos Fernandez, L.; Lauz, J.B. An optimality criteria-based method for the simultaneous optimization of the structural design and placement of piezoelectric actuators. *Struct. Multidiscip. Optim.* **2019**, *59*, 1125–1141. [CrossRef]

62. Svanberg, K. The method of moving asymptotes—A new method for structural optimization. *Int. J. Numer. Methods Eng.* **1987**, *2* 359–373. [CrossRef]
63. Andreassen, E.; Clausen, A.; Schevenels, M.; Lazarov, B.S.; Sigmund, O. Efficient topology optimization in MATLAB using 8 lines of code. *Struct. Multidiscip. Optim.* **2011**, *43*, 1–16. [CrossRef]
64. Maconachie, T.; Leary, M.; Lozanovski, B.; Zhang, X.; Qian, M.; Faruque, O.; Brandt, M. SLM lattice structures: Propertie performance, applications and challenges. *Mater. Des.* **2019**, *183*, 108137. [CrossRef]
65. He, Y.; Burkhalter, D.; Durocher, D.; Gilbert, J.M. Solid-Lattice Hip Prosthesis Design: Applying Topology and Lattice Optimizatio to Reduce Stress Shielding from Hip Implants. In Proceedings of the 2018 Design of Medical Devices Conference, Minneapoli MN, USA, 9–12 April 2018. [CrossRef]
66. Al-Ketan, O.; Lee, D.; Rowshan, R.; Abu Al-Rub, R. Functionally graded and multi-morphology sheet TPMS lattices: Desig manufacturing, and mechanical properties. *J. Mech. Behav. Biomed. Mater.* **2020**, *102*. [CrossRef]
67. Ullah, A.; D'Addona, D.; Seto, Y.; Yonehera, S.; Kubo, A. Utilizing Fractals for Modeling and 3D Printing of Porous Structure *Fractal Fract.* **2021**, *5*, 40. [CrossRef]
68. Wu, J.; Sigmund, O.; Groen, J.P. Topology optimization of multi-scale structures: A review. *Struct. Multidiscip. Optim.* **2021**, *6* 1455–1480. [CrossRef]
69. Groen, J.P.; Sigmund, O. Homogenization-based topology optimization for high-resolution manufacturable microstructures. *In J. Numer. Methods Eng.* **2018**, *113*, 1148–1163. [CrossRef]
70. Dong, G.; Tang, Y.; Li, D.; Zhao, Y.F. Design and optimization of solid lattice hybrid structures fabricated by additive manufactu ing. *Addit. Manuf.* **2020**, *33*, 101116. [CrossRef]
71. Groen, J.P.; Thomsen, C.R.; Sigmund, O. Multi-scale topology optimization for stiffness and de-homogenization using implici geometry modeling. *Struct. Multidiscip. Optim.* **2021**. [CrossRef]
72. Sun, H.; Ma, L. Generative design by using exploration approaches of reinforcement learning in density-based structural topolog optimization. *Designs* **2020**, *4*, 10. [CrossRef]
73. Autodesk: How GM and Autodesk Are Using Generative Design for Vehicles of the Future. Available online: https://adsknews autodesk.com/news/gm-autodesk-using-generative-design-vehicles-future (accessed on 23 June 2021).
74. Vlah, D.; Žavbi, R.; Vukašinović, N. Evaluation of Topology Optimization and Generative Design Tools As Support for Conceptua Design. *Proc. Des. Soc. Des. Conf.* **2020**, *1*, 451–460. [CrossRef]
75. MSC Software Corporation: MSC Apex Generative Design. Available online: https://www.mscsoftware.com/product/msc apex-generative-design (accessed on 23 June 2021).
76. NTopology: nTopology Generative Design. Available online: https://ntopology.com/generative-design-software/ (accessed on 23 June 2021).
77. Autodesk: Autodesk Fusion 360. Available online: https://www.autodesk.com/solutions/generative-design/manufacturing (accessed on 23 June 2021).
78. Le, C.; Norato, J.; Bruns, T.; Ha, C.; Tortorelli, D. Stress-based topology optimization for continua. *Struct. Multidiscip. Optim.* **2009** *41*, 605–620. [CrossRef]
79. Lee, K.; Ahn, K.; Yoo, J. A novel P-norm correction method for lightweight topology optimization under maximum stress constraints. *Comput. Struct.* **2016**, *171*, 18–30. [CrossRef]
80. Lian, H.; Christiansen, A.N.; Tortorelli, D.A.; Sigmund, O.; Aage, N. Combined shape and topology optimization for minimization of maximal von Mises stress. *Struct. Multidiscip. Optim.* **2017**, *55*, 1541–1557. [CrossRef]
81. Mhapsekar, K.; McConaha, M.; Anand, S. Additive Manufacturing Constraints in Topology Optimization for Improved Manufac turability. *J. Manuf. Sci. Eng. Trans.* **2018**, *140*, 1–16. [CrossRef]
82. Schelhorn, L.; Gosch, M.; Debeugny, L.; Schröter, P.; Schwarz, W.; Soller, S. Optimal Design and Process Simulation for Additive Manufacturing. In Proceedings of the 8th European Conference for Aeronautics and Space Sciences, Madrid, Spain, 1–4 July 2019 [CrossRef]
83. Marinov, M.; Amagliani, M.; Barback, T.; Flower, J.; Barley, S.; Furuta, S.; Charrot, P.; Henley, I.; Santhanam, N.; Finnigan, G.T. et al. Generative Design Conversion to Editable and Watertight Boundary Representation. *CAD Comput. Aided Des.* **2019**, *115* 194–205. [CrossRef]
84. Schneider, T.; Hua, Y.; Gao, X.; Dumas, J.; Zorin, D.; Panozzo, D. A Large-Scale Comparison of Tetrahedral and Hexahedral Elements for Finite Element Analysis. *arXiv* **2019**, arXiv:1903.09332.
85. Pagac, M.; Hajnys, J.; Halama, R.; Aldabash, T.; Mesicek, J.; Jancar, L.; Jansa, J. Prediction of model distortion by fem in 3d printing via the selective laser melting of stainless steel aisi 316l. *Appl. Sci.* **2021**, *11*, 1656. [CrossRef]
86. Cheng, B.; Loeber, L.; Willeck, H.; Hartel, U.; Tuffile, C. Computational Investigation of Melt Pool Process Dynamics and Pore Formation in Laser Powder Bed Fusion. *J. Mater. Eng. Perform.* **2019**, *28*, 6565–6578. [CrossRef]
87. Carraturo, M.; Jomo, J.; Kollmannsberger, S.; Reali, A.; Auricchio, F.; Rank, E. Modeling and experimental validation of an immersed thermo-mechanical part-scale analysis for laser powder bed fusion processes. *Addit. Manuf.* **2020**, *36*, 101498. [CrossRef]
88. Afazov, S.; Denmark, W.A.D.; Lazaro Toralles, B.; Holloway, A.; Yaghi, A. Distortion prediction and compensation in selective laser melting. *Addit. Manuf.* **2017**, *17*, 15–22. [CrossRef]

89. Chen, Q.; Liang, X.; Hayduke, D.; Liu, J.; Cheng, L.; Oskin, J.; Whitmore, R.; To, A.C. An inherent strain based multiscale modeling framework for simulating part-scale residual deformation for direct metal laser sintering. *Addit. Manuf.* **2019**, *28*, 406–418. [CrossRef]
90. Setien, I.; Chiumenti, M.; Veen, S.D.; San Sebastian, M.; Garciandía, F.; Echeverría, A. Empirical methodology to determine inherent strains in additive manufacturing. *Comput. Math. Appl.* **2019**, *78*, 2282–2295. [CrossRef]
91. Liang, X.; Chen, Q.; Cheng, L.; Hayduke, D.; To, A.C. Modified inherent strain method for efficient prediction of residual deformation in direct metal laser sintered components. *Comput. Mech.* **2019**, *64*, 1719–1733. [CrossRef]
92. Orme, M.; Madera, I.; Gschweitl, M.; Ferrari, M. Topology optimization for additive manufacturing as an enabler for light weight flight hardware. *Designs* **2018**, *2*, 51. [CrossRef]
93. Caiazzo, F.; Alfieri, V.; Corrado, G.; Argenio, P. Laser powder-bed fusion of Inconel 718 to manufacture turbine blades. *Int. J. Adv. Manuf. Technol.* **2017**, *93*, 4023–4031. [CrossRef]
94. Seifi, M.; Gorelik, M.; Waller, J.; Hrabe, N.; Shamsaei, N.; Lewandowski, J. Progress Towards Metal Additive Manufacturing Standardization to Support Qualification and Certification. *Miner. Met. Mat. Soc.* **2017**, *69*, 3. [CrossRef]
95. Bourell, D.L.; Rosen, D.W.; Leu, M.C. The Roadmap for Additive Manufacturing and Its Impact. *3D Print. Addit. Manuf.* **2014**, *1*, 6–9. [CrossRef]
96. ISO; ASTM. *ISO/ASTM 17296:2014—Additive Manufacturing–General Principles—Part 3: Main Characteristics and Corresponding Test Methods;* ISO International Organization for Standardization: Geneva, Switzerland; ASTM American Society for Testing and Materials: West Conshohocken, PA, USA, 2014.
97. ISO; ASTM. *ISO/ASTM 52904:2019—Additive Manufacturing–Process Characteristics and Performance—Practice for Metal Powder Bed Fusion Process to Meet Critical Applications;* ISO International Organization for Standardization: Geneva, Switzerland; ASTM American Society for Testing and Materials: West Conshohocken, PA, USA, 2019.
98. Rosso, S.; Uriati, F.; Grigolato, L.; Meneghello, R.; Concheri, G.; Savio, G. An optimization workflow in design for additive manufacturing. *Appl. Sci.* **2021**, *11*, 2572. [CrossRef]
99. McEwen, I.; Cooper, D.E.; Warnett, J.; Kourra, N.; Williams, M.A.; Gibbons, G.J. Design & manufacture of a high-performance bicycle crank by Additive Manufacturing. *Appl. Sci.* **2018**, *8*, 1360. [CrossRef]
100. Nieto, D.M.; Sánchez, D.M. Design for additive manufacturing: Tool review and a case study. *Appl. Sci.* **2021**, *11*, 1571. [CrossRef]

applied sciences

MDPI

Article

# An Optimization Workflow in Design for Additive Manufacturing

Stefano Rosso [1,*], Federico Uriati [2], Luca Grigolato [3,4], Roberto Meneghello [1], Gianmaria Concheri [4] and Gianpaolo Savio [4]

1 Department of Management and Engineering, University of Padova, Stradella S. Nicola, 3, 36100 Vicenza, Italy; roberto.meneghello@unipd.it
2 Department of Engineering and Architecture, University of Parma, Parco Area delle Scienza, 181/A, 43124 Parma, Italy; federico.uriati@unipr.it
3 Department of Industrial Engineering, University of Padova, Via Gradenigo, 6/a, 35131 Padova, Italy; luca.grigolato@phd.unipd.it
4 Department of Civil, Environmental, and Architectural Engineering, University of Padova, Via Venezia, 1, 35131 Padova, Italy; gianmaria.concheri@unipd.it (G.C.); gianpaolo.savio@unipd.it (G.S.)
* Correspondence: stefano.rosso.3@phd.unipd.it

**Abstract:** Additive Manufacturing (AM) brought a revolution in parts design and production. It enables the possibility to obtain objects with complex geometries and to exploit structural optimization algorithms. Nevertheless, AM is far from being a mature technology and advances are still needed from different perspectives. Among these, the literature highlights the need of improving the frameworks that describe the design process and taking full advantage of the possibilities offered by AM. This work aims to propose a workflow for AM guiding the designer during the embodiment design phase, from the engineering requirements to the production of the final part. The main aspects are the optimization of the dimensions and the topology of the parts, to take into consideration functional and manufacturing requirements, and to validate the geometric model by computer-aided engineering software. Moreover, a case study dealing with the redesign of a piston rod is presented, in which the proposed workflow is adopted. Results show the effectiveness of the workflow when applied to cases in which structural optimization could bring an advantage in the design of a part and the pros and cons of the choices made during the design phases were highlighted.

**Keywords:** DfAM; design for additive manufacturing; size optimization; topology optimization; design workflow; computational geometry; geometric modeling

**Citation:** Rosso, S.; Uriati, F.; Grigolato, L.; Meneghello, R.; Concheri, G.; Savio, G. An Optimization Workflow in Design for Additive Manufacturing. *Appl. Sci.* **2021**, *11*, 2572. https://doi.org/10.3390/app11062572

Academic Editor: Marco Mandolini

Received: 22 February 2021
Accepted: 11 March 2021
Published: 13 March 2021



## 1. Introduction

From the works of the early pioneers, additive manufacturing (AM) technologies were characterized by great growth in the last 35 years [1]. According to ISO/ASTM standards "AM is the general term for those technologies that, based on a geometrical representation, create physical objects by successive addition of material" [2]. Depending on the method of layer manufacturing, it is possible to organize the AM technologies in the following categories: vat photopolymerization, material jetting, binder jetting, powder bed fusion, material extrusion, directed energy deposition, and sheet lamination [2].

This technology brings new opportunities especially in design freedom, allowing very complex shapes, integrating cinematics and multi-material parts, reducing the number of components through part consolidation, and increasing mass customization. On the other hand, to fully exploit the AM technologies' potential, many needs in different sub-fields were highlighted [1,3–8], as summarized in Figure 1. For example, a highly skilled workforce is required, file formats for exchanging the data related to the AM workflow need enhancements [8,9], and design methods and tools for complex structures, multi-material parts, and functionally graded materials need to be improved [10,11]. The concerns over the structural integrity of these complex parts require static and dynamic mechanical characterization [12,13]; also, experimental tests help to mechanically characterize the materials and the obtained information is used in numerical simulations to predict the different mechanical behavior between the products obtained through additive manufacturing and the ones obtained by traditional techniques of material subtraction [14,15]. More, dedicated qualification standards for AM are needed to guarantee an adequate quality of the printed parts [16,17] and their representation in 2D drawings [18].

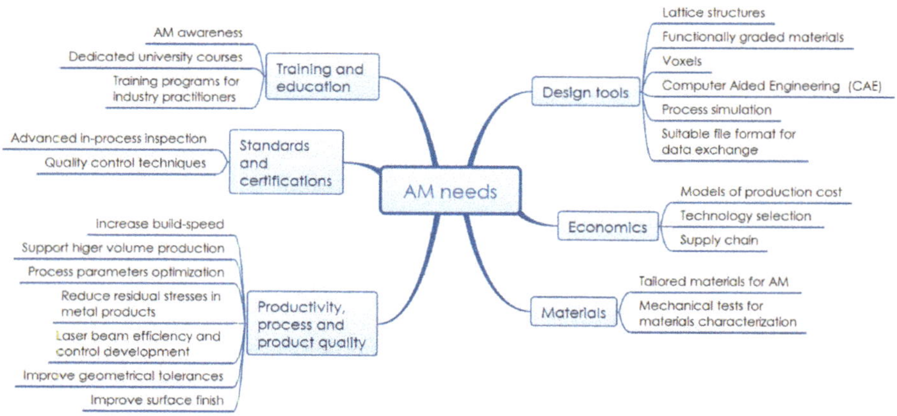


**Figure 1.** Additive Manufacturing needs.

In a recent work, *The Economist* claims that the value of AM products will no longer be in the physical item, but in its design [19]. These reflections lead to a reconsideration of the design approaches for AM. According to Ullman [20], the design process can be divided into six major phases: product discovery, project planning, product definition, conceptual design, product development, and product support. During the conceptual design phase, several concepts are generated and evaluated; however, the knowledge of the concepts is limited and the goal in this phase is to select the best alternatives with the least expenditure of time and other resources. In the product development phase instead, once the product is generated, it is fundamental to compare its performance to the engineering specifications. This is done both with virtual simulations and physical prototypes, often resulting in a time-consuming iterative process due to part modifications and redesigns. A different approach that exploits the higher computational capabilities available nowadays is the computational design synthesis [21], where the tasks needed to obtain a solution are divided into four main steps: "representation" deals with the creation of a mental model of the object, "generation" deals with the object creation, "evaluation" verifies if the constraints and design goals are met, and "guidance" gives a feedback for the design improvements [22]. The last three phases are iteratively repeated until a final design is obtained. In design synthesis, optimization is performed in the representation and generation phases, where the design has not a specific topology yet. Usually, stochastic methods are applied to obtain different designs that satisfy the requirements [23].

Furthermore, when designing parts that are going to be produced by AM technologies, several thoughts must be considered to "maximize product performance through the synthesis of shapes, sizes, hierarchical structures, and material compositions, subject to the capabilities of AM technologies" [24]. All these considerations can be grouped within the Design for Additive Manufacturing (DfAM) concept. Gibson et al. [24] distinguished between opportunistic and restrictive DfAM; the former allows to take advantages of the unique capabilities of AM, such as cellular solids, part consolidation, and multi-material, whereas the latter focuses on the restrictions and limitations of the AM technologies, such as the minimum feature size and the need of support structures. Rosen [25] used a Process–Structure–Property–Behavior framework to describe and model a design and proposed a DfAM system organized in several modules dealing with the modeling, manufacturing simulation, and design behavior analysis phases. Ponche et al. [26] presented a DfAM methodology that takes into account the design requirements and the manufacturing specificities, adopting a redesign strategy based on the functional surface approach; they concentrated on metallic components produced by Additive Laser Manufacturing. Vayre et al. [27] applied a four-step designing methodology consisting of initial shape generation, set of geometrical parameter definition, shape optimization through the tuning of the parameters, and final validation of a metallic part produced by direct metal deposition and electron beam melting AM technologies. Briard et al. [28] presented a four-step methodology to maximize the potential of generative design coupled with DfAM; the first phase deals with the translation of the problem to a suitable input for generative design, whereas the following three phases deal with an unconstrained iterative optimization, an iterative optimization driven by the AM guidelines, and a final iterative optimization refining the part including AM-enabled structures, such as lattices. Duro-Royo et al. [29] presented a computational workflow for the design and the fabrication of multi-material and multi-scale structured objects; they focused on water-based heterogeneous materials based on polysaccharide hydrogels in 1% to 12% concentrations in *w/v* of 1% acetic acid aqueous solutions and these gels were also mixed with cellulose microfiber to obtain volumetric composites. They created a model that considers the input data, like materials and geometry, and calculates all the instructions for the object fabrication via a pneumatic extruder mounted on a six-axes robotic arm. Boddeti et al. [30] presented a digital design and manufacturing workflow able to design both the macroscopic topology and the microstructure of an object; the workflow is divided into three steps: a design automation process that optimizes the material distribution and its microstructure, a material compilation process that creates a material layout and generates the code for fabrication, and a digital fabrication step with multi-material photopolymer material jetting technology. Zhang et al. [31] proposed an evaluation framework to assess the design from the perspective of process planning for AM; two sets of indicators were used to check whether the part is suitable to be produced by AM manufacturing and to verify the design's utilization of the characteristics of an AM process. Similarly, Lettori et al. [32] proposed an approach to assess the compatibility and suitability of a product for the AM production through a set of reference questions and a compliance index; then, they validated the method with case studies found in the literature. Motyl and Filippi [33] reviewed the scientific literature to explore the relationship between AM processes and product design, concentrating on the conceptual design phase and the theory of inventive problem solving (TRIZ) [34]; some of the analyzed works use the TRIZ for the definition of the DfAM guidelines.

Nonetheless, the literature highlights the lack of exhaustive frameworks that describe the design process and take full advantage of the possibilities offered by AM. Seepersad [35] stated that advances are still needed to couple computer-aided design (CAD) software and computer-aided engineering (CAE) tools to incorporate the DfAM knowledge into the design process. Kumke et al. [36] highlighted some limitations on the existing DfAM frameworks too: they do not cover the entire design process steps, they focus on the utilization of a single AM potential, and they are often too specific for a single case study.

In this contribution, a heuristic design workflow for AM is proposed aimed at exploiting the new possibilities offered by AM technologies and the high computational resources available nowadays. The workflow focuses on the product design phase, also referred to as embodiment design [20,37], where the design is developed up to the production. It specifically concentrates on cases in which mechanical performances are required, together with a reduction of the weight of the parts. Different geometric modeling opportunities and structural optimization techniques are presented: commercial software is used to perform the topology optimization and the redesign of the optimized results. As an alternative, a method developed by the research group designs conformal lattice structures with size optimization performed on the beams and allows to automatically obtain a smooth mesh model. The proposed workflow is then validated on a test case, adopting different design methods based on lattice structures, PolyNurbs, and parametric CAD to reach innovative solutions.

## 2. Design Workflow

The proposed workflow helps the designer during the product development process of AM components, guiding him throughout decisions that allow to fully exploit AM potential. AM-related engineering requirements and technological constraints are considered in the first phases of the workflow and simulation tools are used to optimize and validate the geometric model. In particular, the workflow can be adopted during the embodiment design phase of parts in which the mechanical performance needs to be maximized and the weight needs to be as low as possible. Structural optimization approaches such as size and topology optimization perfectly suit this scenario.

Figure 2 shows the proposed design workflow for AM. First, the design space is identified. The design space is a volume where the material distribution is going to be optimized; it can be obtained from an existing model or it can be specifically designed considering the maximum allowable size of the component. Then, two paths can be followed: the first one performs a topology optimization on the entire design space, whereas the second one performs a size optimization on a lattice structure. Regardless of the selected approach, a finite element (FE) model is created taking into consideration not only the "usual" boundary conditions such as the material, the loads, and the constraints but also the constraints and conditions strictly related to the design for AM. For instance, the technological constraints could include a limit for the inclination of the structure to avoid overhang angles (if required by the manufacturing technology) and the upper and lower limits for the most critical features, i.e., hole size, strut dimension, wall thickness, etc. Furthermore, since AM allows the production of complex geometrical shapes, it is easier to create parts resulting from multi-objective optimization; the optimization goals like targeted mass and natural frequencies, or desired heat exchanging properties can be considered as engineering requirements. Including this information in the first part of the workflow enables to obtain a design with the desired functionalities that is likely to be produced without the need of stepping back to the product development phase.

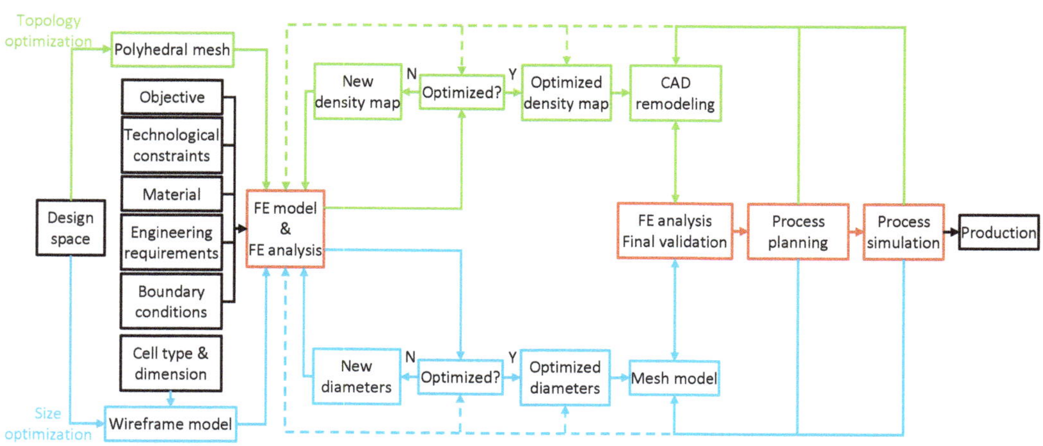


**Figure 2.** The proposed design workflow for additive manufacturing. Black blocks: steps related to the boundary conditions; green blocks: steps related to the topology optimization; blue blocks: steps related to the size optimization; red blocks: steps related to computer-aided engineering (CAE) simulation software.

If the topology optimization approach is chosen, the design space is discretized and a polyhedral mesh is obtained; then, the topology optimization is performed. During this process, the material is arranged inside the design space to find the best distribution of material under a set of boundary conditions and respecting the structural and dimensional performance requirements. According to the literature, several topology optimization algorithms can be used [38]. Homogenization methods use the mathematical theory of homogenization [39,40] to study a complex domain previously divided into microstructures, i.e., the finite elements, as a continuum domain made up of a virtual material called effective material [41]. Density methods consider the density as the only design variable for each finite element, and the variable can assume a value between 0 and 1; since the optimal solutions would consist of elements with values mostly between 0 and 1, the results would be far from a solid (1)–void (0) situation. The most popular numerical method for suppressing intermediate densities is the Solid Isotropic Microstructure with Penalization (SIMP) method, proposed by Bendsoe [42], which penalizes elements with intermediate densities exploiting a power law. To avoid intermediate densities in the optimized solution, discrete methods, also called "hard-kill" methods, can be used; in the Evolutionary Structural Optimization (ESO) method proposed by Xie and Steven [43], a criterion parameter is calculated for each element and at each iteration the elements with the lowest criterion parameter value are eliminated. Furthermore, Bi-directional Evolutionary Structural Optimization (BESO) method [44] is an extension of ESO, in which new elements can be added next to those elements with a high criterion parameter at each iteration. Level-set Methods (LSMs) are another category of topology optimization methods where the iso-contours of a level-set function implicitly define the interface between the material phases [45,46]; this approach allows to obtain a sharp transition between void and solid regions. In the proposed workflow, a SIMP density method is adopted because this method requires relatively few iterations, is suitable for a combination of a wide range of design constraints, multiple load conditions, multi-physics problems, and extremely large (often 3D) systems, is extensively used in industrial software [38], and can be easily implemented with a simple code [47,48]. As a result of the SIMP method, a density map is obtained, which is contoured to a specific level of density (threshold), obtaining a mesh surface. Often, the resulting mesh cannot be directly used for the production phase due to the lack of connection zones or the presence of coarse regions due to the process discretization. The remodeling of the topology optimization mesh is a research topic of interest, indeed. Zegard and Paulino [49] presented a tool that generates suitable outputs for AM by using filters and the continuation approach on the penalization parameter; Jiu et al. [50] proposed a CAD-oriented topology optimization method able to perform the optimization directly on the CAD model instead of on the mesh. Most of the time the procedure is operated manually, and the part is modeled in a CAD environment using the mesh as guideline during the modeling. Alternatively, software tools for automatic remodeling are available; these mainly adopt quad-remesh and subdivision surface approaches.

If the size optimization approach is followed, given a cell type and the unit cell dimension, a wireframe model is obtained filling the design space with a conformal lattice structure. In a conformal structure, the unit cell can deform to adapt to the boundary of the part or the lines of the stress field; this feature eliminates weakness at boundaries and provides stiffness and resistance to the entire model [51]. The size optimization is then performed on the wireframe [52,53]. The diameter of the beams iteratively varies until all the beams reach the target utilization, defined as the ratio between the maximum Von Mises stress measured on the beam and the admissible stress; as previously highlighted, the size of the beam is controlled to ensure that the beam diameter is thick enough to be manufactured and that it does not exceed the upper bound to avoid interferences with the surrounding beams. The optimized wireframe is then modeled with a boundary representation mesh-based approach as proposed by Savio et al. [54]. The obtained coarse mesh is then smoothed adopting the Catmull–Clark subdivision surface algorithm [55]: each quad face is subdivided into four smaller quad faces at every iteration. The algorithm produces a surface with continuity in curvature (C2 surface), except at extraordinary vertices where they are C1. This allows to reduce stress concentration, especially at nodal points, enhancing the mechanical properties and the fatigue life of the lattice [56–58].

Alternatively, it is possible to combine topology and size optimization. The literature shows how the density map of the topology optimization can be used to assign an optimized dimension to the diameter of a beam-like lattice structure [59,60] and to the thickness of a shell-like lattice structure [61].

Once the optimized part is modeled, it is necessary to perform an additional FE analysis to mechanically validate the final model. This step is mandatory in topology optimization because during the modeling phase weak zones could arise, especially if manual remodeling is adopted. After that, process-related considerations are done. Process planning deals with all the necessary operations needed once the AM production technology and machine have been selected. The orientation of the part inside the manufacturing machine, the generation of the supports (if needed), and the generation of the print path strategy affect the quality of the printed part, i.e., the surface texture and the mechanical properties [26], the material and energy consumption, and the production time [62–64]. Then, process simulation helps to predict residual stress and geometric distortion of the printed parts, avoiding time-consuming and expensive experimental campaigns based on trial and error. The evaluation of residual stresses and thermal distortions allows compensating the geometrical CAD model obtaining parts with the desired dimensional specifications and mechanical properties, reducing the probability of defects that lead to crack initiation, propagation, and failure both during the printing and the utilization of the product, especially when dealing with metallic components [65]. Process simulation methods that concentrate not only on metals AM techniques [66–69] but also on material extrusion [70,71] and powder bed fusion [72,73] of polymers can be found. If the FE analysis fails to validate the component or if the manufacturing process simulation highlights dimensional deformation and residual stresses higher than requirements, it is necessary to remodel the part or to step back to the optimization phase, changing the boundary conditions of the model.

When all these steps are successfully completed, the part is ready to be produced. The component is optimized for the intended use and it is likely to not encounter manufacturing issues during the printing phase. An important consideration is whether to use the topology optimization or the size optimization approach. One option could be to apply both and compare the solutions, choosing the one that best suits the application, but it is computationally demanding. Some a priori thoughts can help to decide as well. If the part is metallic, the struts of the lattice could act as internal supports, heat dissipators, and prevent thermal warping. Furthermore, being less bulky, the lattice structure could present lower residual stress. The mechanical validation of a lattice structure requires high computational resources and it is time consuming due to the high number of 3D elements needed to mesh the structure; at the same time, while the FE validation analysis is mandatory for the topology optimized part because the manual remodeling phase can introduce weak areas, it is less necessary for the lattice structure model because the diameters of the beams were previously optimized through a FE analysis and in the presented mesh modeling approach the optimized value is adopted at the middle of the beam, whereas the diameter tends to increase towards the nodal points.

## 3. Case Study

The proposed workflow was applied to the remodeling of a piston rod. Reducing the weight of a piston rod while maintaining the mechanical performance is an important goal, especially in competitions in the automotive fields, but also in the industry where weight reduction leads to less inertia and to a reduction of energy consumption.

The part is currently produced with a pressure die-casting process and is intended to be produced with the Selective Laser Melting (SLM) powder bed fusion AM technology. The material is an aluminum AlSi10Mg with properties as in Table 1.

**Table 1.** Aluminum AlSi10Mg properties.

| | |
|---|---|
| Density | 2700 kg/m$^3$ |
| Young modulus | 68 GPa |
| Yield strength | 190 MPa |
| Ultimate tensile strength | 335 MPa |
| Poisson Ratio | 0.30 |

Overall dimensions, loads, and constraints applied to the piston rod are shown in Figure 3a,b, respectively. The loads and constraints are summarized in Table 2.

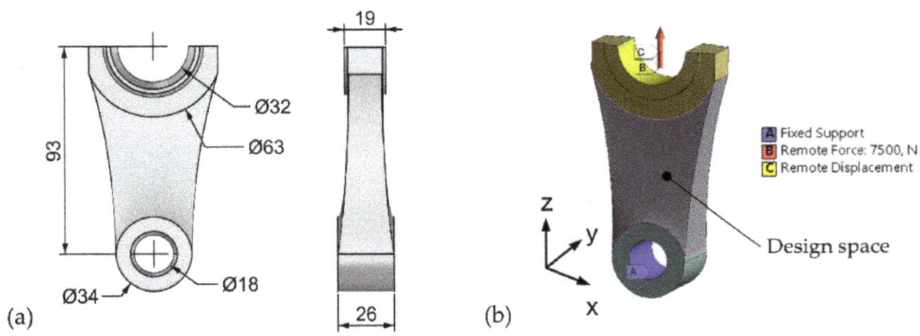


**Figure 3.** Piston rod: (**a**) overall dimension, (**b**) boundary conditions and design space.

**Table 2.** Load and constraints applied to the piston rod.

| | |
|---|---|
| Load | 7.5 kN axial traction along $z$-axis.<br>Applies to big rod's end face. |
| Constraints | All the displacements and rotations locked.<br>Applies to inner face of the small rod's end.<br>Displacements along $x$- and $y$-directions locked.<br>Applies to big rod's end face. |

The topology optimization of the design space was performed in SolidWorks 2019 (Dassault Systèmes) using the SolidWorks Simulation module; the software adopts the SIMP method for solving the optimization problem. The "best stiffness to weight ratio" goals was set, using the constraint of a final mass equal to 25% of the original part. The symmetry of the final part with respect to the YZ and ZX plane was imposed. Figure 4a shows the mesh resulting from the topology optimization. This mesh was then used as a starting point for the manual remodeling phase, performed in Inspire Studio CAD 3D software (Altair), as in Figure 4b.

The size optimization section of the workflow was performed in Rhinoceros 6 (Robert McNeel & Associates) inside Grasshopper environment. The design space was filled with a conformal wireframe based on the simple cubic unit cell; the number of instances is 10, 4, and 15 along the $x$-, $y$-, and $z$-axis, respectively, and the minimum element size equals 3 mm. The FE beams model was set-up using Karamba3D plugin [74]. The loads and the constraints were directly applied at the nodes of the beams placed at the interface between the design space and the big and small rod's ends; the 7.5 kN load was equally distributed on each node of the upper part of the wireframe, so as the rotation and translation constraints on each node of the lower and upper part of the wireframe. The target utilization ratio was set to $(90 \pm 1)\%$ with respect to the yield strength. The upper and lower bound for the diameter of the beams were defined as 1.5 mm and 0.5 mm, respectively. Figure 5 shows the conformal wireframe and the utilization ratio of the optimized structure. Some beams do not reach the required utilization ratio; indeed, they present a utilization ratio lower than the target because the optimized diameter was smaller than the minimum allowable size (i.e., 0.5 mm), so, being assigned the 0.5 mm diameter, they are under-utilized.

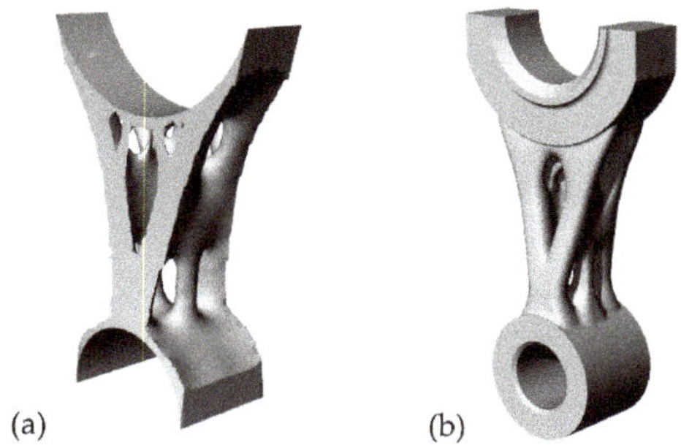

**Figure 4.** Topology optimization: (**a**) SolidWorks Simulation result (mesh), (**b**) Inspire Studio manual remodeling result (NURBS).

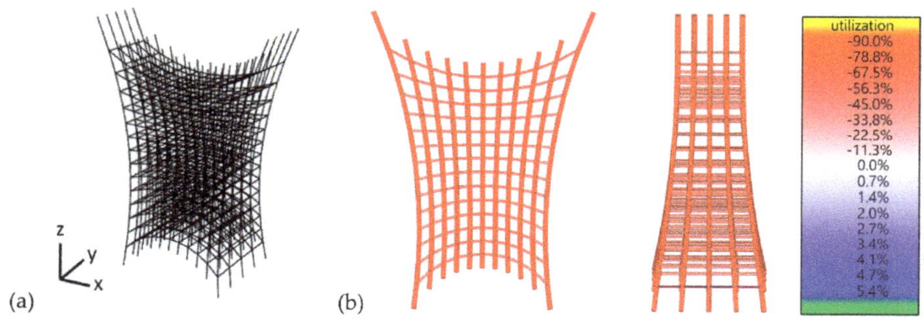


**Figure 5.** Size optimization: (**a**) wireframe model, (**b**) utilization ratio of the optimized structure (compression is positive).

The lattice structure was then modeled adopting the mesh approach and the Catmull–Clark subdivision surface algorithm. As can be seen in Figure 6, smooth surfaces are obtained, especially at nodal points.

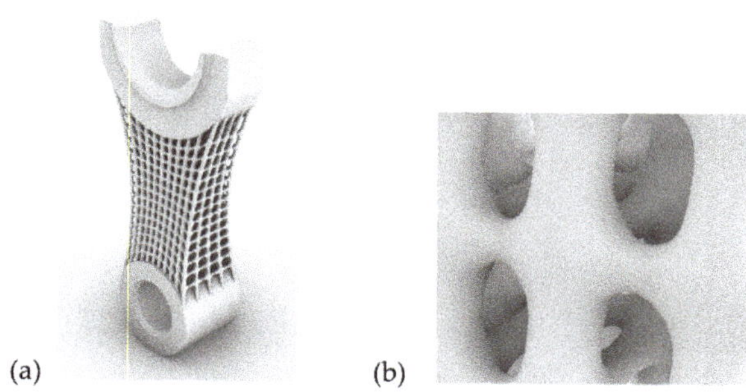

**Figure 6.** Modeled lattice structure: (**a**) lattice connected to the piston rod's ends, (**b**) detail on the smooth surfaces at nodal points.

Then, the two models obtained from the topology and the size optimization were validated through FE analyses in Ansys Mechanical 2019 R1. The parts were meshed with tetrahedron elements using an element minimum size of 0.1 mm and an element maximum size of 1 mm. The meshing method is patch-independent and includes automatic refinement in curvature and proximity. The symmetry of the topology optimized part with respect to the ZX and YZ plane was exploited to simulate only one quarter of the model. The results are represented in Figure 7. In the size optimized model (Figure 7b), the higher value of stress observed on the legend is related to the absence of fillet between the lattice structure and the body and does not depend on the modeling method of the lattice.

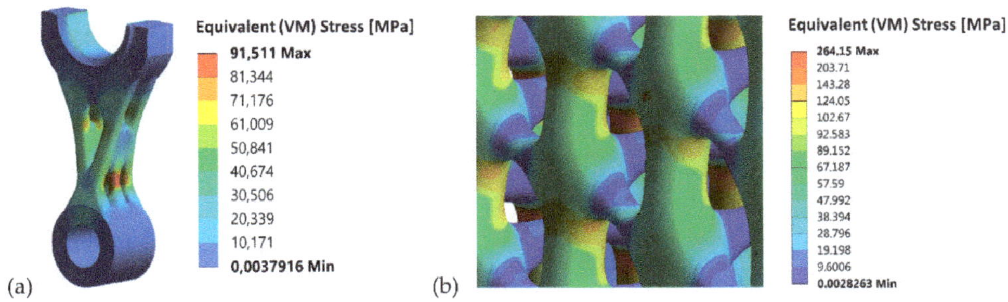


**Figure 7.** FE analyses validation, equivalent Von Mises stress: (**a**) topology optimization, (**b**) size optimization.

Netfabb Premium 2020.3 (Autodesk) was used for preliminary process planning, choosing for the best part orientation. Supports too were created inside the software. The Renishaw AM 400 SLM machine was selected, together with the default material configuration for Aluminum AlSi10Mg-0403 printed with a 25 μm layer thickness, as suggested by the powders manufacturer [75]. First, the topology optimized model was oriented. Among the proposed orientations, three are reported in Figure 8. The final decision is driven by several considerations. The orientation in Figure 8a has the lowest height, resulting in faster printing time, but presents the highest supported area, requiring more time for supports removal and post-process such as sandblasting to avoid the lower quality of the surface finish in the supported areas; the configuration in Figure 8b has the less supported area, but has the highest support volume and height, resulting in a long build time and high material waste; the configuration in Figure 8c has the lower support volume and a relatively low supported area and build time. Moreover, since the configuration in Figure 8a lays on the platform, the circular functional surfaces on the rod's ends will have the best dimensional and geometrical accuracy. Further milling operations could be needed to comply with the requested tolerances.

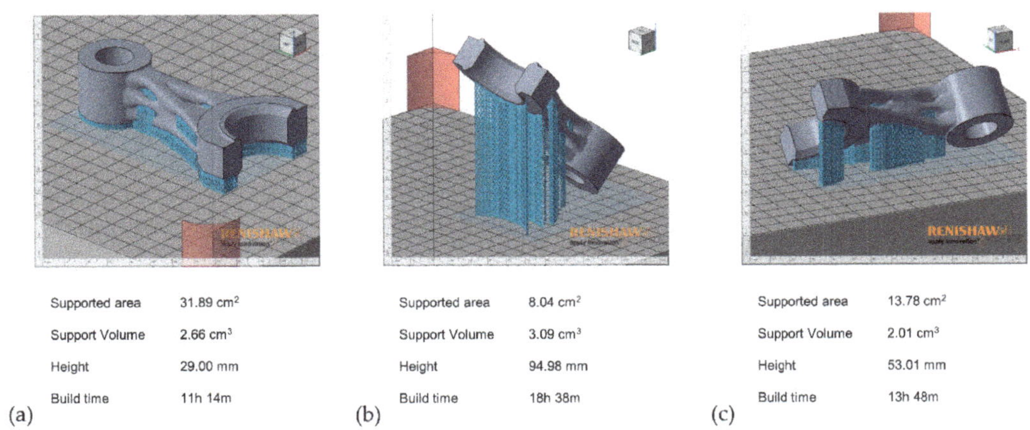


**Figure 8.** Part orientation and support generation for the topologically optimized piston rod. (**a**–**c**) show three possible orientations.

The same procedure was followed for the orientation of the piston rod with the lattice structure. This time, due to the extreme difficulty of removing the supports between the beams of the lattice, the only orientation that did not present supports in the central part of the model was chosen, as shown in Figure 9.

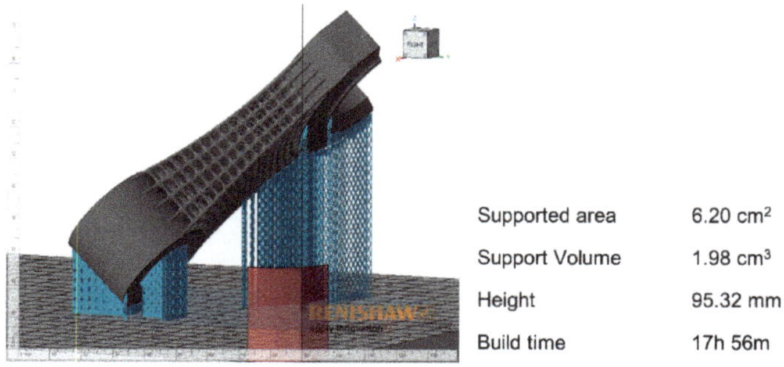


**Figure 9.** Part orientation and support generation for the size optimized piston rod.

Finally, the next step would be to perform the process simulation of the oriented part to assess the printability of the models, the geometrical distortions, and the residual stresses. According to the results, the designer can consider geometrically compensating the part, changing the part orientation, or going back to the modeling phase and remodel the part.

The application of the workflow for redesigning the piston rod enabled to obtain two optimized parts and a considerable reduction of the mass, as shown in Table 3. The two methods present some drawbacks too. The topology optimized model requires manual remodeling, and it is more prone to failure during the final FE analysis verification, due to human errors during the remodeling phase. The proposed method for the size optimization is in its prototypal stage and it still has a limited choice of unit cells and does not correctly manage the connection between the lattice structure and the adjacent objects. Numerical analyses on the lattice model are computationally demanding but less necessary since the wireframe was previously optimized and the mesh modeling method does not alter the diameter of the beams.

**Table 3.** Mass reduction of the optimized models. The mass only considers the design space volume and not the rod's ends, which were not optimized.

| Model/Approach | Mass [g] | % of Mass Reduction |
|---|---|---|
| Starting design space | 104.7 | |
| Topology optimization | 29.99 | −71% |
| Size optimization | 20.98 | −80% |

## 4. Conclusions

In this work, a design workflow for Additive Manufacturing was proposed, trying to overcome the limits highlighted in the literature, where it is stated that the available frameworks do not exploit all the advantages offered by AM and do not cover the entire design process. The presented workflow considers the embodiment design phase, from the definition of a design volume to the production of the part, integrating both CAD tools for the geometric modeling of the part and CAE tools for the optimization and simulation phases; more, it considers the possibility to use the size optimization to obtain lattice structures with optimized beams, and the topology optimization to obtain more organic shapes. The workflow was then applied to the remodeling and optimization of a piston rod in which both commercial and custom tools were adopted, showing its ease and universality of implementation.

As future works, the process simulation will be performed on the oriented parts. Then, a hybrid method that combines size and topology optimization is going to be developed to further expand the workflow possibilities; the 0–1 density parameter will drive the dimension of the beams, shell, or solid elements. The connection between the lattice structures and the adjacent objects will be addressed to obtain smoother links and enhance the mechanical properties of the parts. Moreover, aspects related to hybrid manufacturing technologies will be addressed.

**Author Contributions:** Conceptualization, S.R., F.U., L.G., G.S., R.M., and G.C.; Formal analysis, S.R., F.U., and L.G.; Funding acquisition, G.S., R.M., and G.C.; Investigation, S.R.; Methodology, S.R., and G.S.; Supervision, G.S., R.M., and G.C.; Writing—original draft, S.R.; Writing—review and editing, S.R., F.U., L.G., G.S., R.M., and G.C. All authors have read and agreed to the published version of the manuscript.

**Funding:** This research was funded by a Ph.D. grant of Fondazione Cassa di Risparmio di Padova e Rovigo (CARIPARO), by the grant "FSE 2105-37-11-2018: Hybrid manufacturing strategies through selective laser sintering for the customization of series components" by Regione Veneto, and by the grant "BIRD 190850—Design, geometric modeling and analysis of components with functionally graded variation for additive manufacturing technologies" by the Department of Civil, Environmental, and Architectural Engineering, University of Padova, Italy.

**Institutional Review Board Statement:** Not applicable.

**Informed Consent Statement:** Not applicable.

**Data Availability Statement:** The raw/processed data required to reproduce these findings can be partially shared upon request.

**Conflicts of Interest:** The authors declare no conflict of interest.

## References

1. Pei, E. Editorial PIAM October 2019. *Prog. Addit. Manuf.* **2019**, *4*, 355–356. [CrossRef]
2. *ISO/ASTM, ISO/ASTM 52900:2015 (ASTM F2792)—Additive Manufacturing–General Principles—Terminology*; ISO International Organization for Standardization, ASTM American Society for Testing and Materials: Geneva, Switzerland; West Conshohocken, PA, USA, 2015.
3. AM Platform, Additive Manufacturing Strategic Research Agenda. 2014. Available online: https://www.rm-platform.com/linkdoc/AM%20SRA%20-%20February%202014.pdf (accessed on 12 March 2021).
4. SASAM. Additive Manufacturing: SASAM Standardisation Roadmap. 2015. Available online: https://www.rm-platform.com/downloads2/summary/50-strategic-research-agenda/608-sasam-standardisation-roadmap-open-june-2015 (accessed on 12 March 2021).
5. FoFAM. Additive Manufacturing Roadmap: Gaps and Actions on Market Driven Value Chains. 2016. Available online: https://portal.effra.eu/result/show/1080 (accessed on 12 March 2021).
6. America Makes, ANSI, Standardization Roadmap for Additive Manufacturing—Version 1.0. 2017. Available online: https://share.ansi.org/Shared%20Documents/Standards%20Activities/AMSC/AMSC_Roadmap_February_2017.pdf (accessed on 12 March 2021).
7. America Makes, ANSI, Standardization Roadmap for Additive Manufacturing—Version 2.0. 2018. Available online: https://www.americamakes.us/america-makes-ansi-standardization-roadmap-additive-manufacturing-presented-formnext-conference/ (accessed on 12 March 2021).
8. Savio, G.; Meneghello, R.; Rosso, S.; Concheri, G. 3D model representation and data exchange for additive manufacturing. In *Lecture Notes in Mechanical Engineering*; Springer: Cham, Switzerland, 2019; pp. 412–421. [CrossRef]
9. Bacciaglia, A.; Ceruti, A.; Liverani, A. A systematic review of voxelization method in additive manufacturing. *Mech. Ind.* **2019**, *20*, 630. [CrossRef]
10. Grigolato, L.; Rosso, S.; Meneghello, R.; Concheri, G.; Savio, G. Heterogeneous objects representation for Additive Manufacturing: A review. *Instant J. Mech. Eng.* **2019**, 14–23. [CrossRef]
11. Thompson, M.K.; Moroni, G.; Vaneker, T.; Fadel, G.; Campbell, R.I.; Gibson, I.; Bernard, A.; Schulz, J.; Graf, P.; Ahuja, B.; et al. Design for Additive Manufacturing: Trends, opportunities, considerations, and constraints. *CIRP Ann.-Manuf. Technol.* **2016**, *65*, 737–760. [CrossRef]
12. Benedetti, M.; du Plessis, A.; Ritchie, R.; Dallago, M.; Razavi, S.; Berto, F. Architected cellular materials: A review on their mechanical properties towards fatigue-tolerant design and fabrication. *Mater. Sci. Eng. R Rep.* **2021**, *144*, 100606. [CrossRef]

13. Raghavendra, S.; Molinari, A.; Dallago, M.; Zappini, G.; Zanini, F.; Carmignato, S.; Benedetti, M. Uniaxial static mechanical properties of regular, irregular and random additively manufactured cellular materials: Nominal vs. real geometry. *Forces Mech.* **2021**, *2*, 100007. [CrossRef]
14. Calignano, F.; Lorusso, M.; Roppolo, I.; Minetola, P. Investigation of the Mechanical Properties of a Carbon Fibre-Reinforced Nylon Filament for 3D Printing. *Machines* **2020**, *8*, 52. [CrossRef]
15. Allevi, G.; Capponi, L.; Castellini, P.; Chiariotti, P.; Docchio, F.; Freni, F.; Marsili, R.; Martarelli, M.; Montanini, R.; Pasinetti, S.; et al. Investigating Additive Manufactured Lattice Structures: A Multi-Instrument Approach. *IEEE Trans. Instrum. Meas.* **2020**, *69*, 2459–2467. [CrossRef]
16. Bourell, D.L.; Rosen, D.W.; Leu, M.C. The roadmap for additive manufacturing and its impact. *3D Print. Addit. Manuf.* **2014**, *1*, 6–9. [CrossRef]
17. Seifi, M.; Gorelik, M.; Waller, J.; Hrabe, N.; Shamsaei, N.; Daniewicz, S.; Lewandowski, J.J. Progress Towards Metal Additive Manufacturing Standardization to Support Qualification and Certification. *JOM* **2017**, *69*, 439–455. [CrossRef]
18. Bacciaglia, A.; Ceruti, A.; Liverani, A. Proposal of a standard for 2D representation of bio-inspired lightweight lattice structures in drawings. *Proc. Inst. Mech. Eng. Part C J. Mech. Eng. Sci.* **2020**, 095440622095159. [CrossRef]
19. The Economist Intelligence Unit. *Adding It Up: The Economic Impact of Additive Manufacturing*; Agency for Science, Technology and Research (A* STAR): Singapore, 2018; 24p.
20. Ullman, D.G. *The Mechanical Design Process*, 4th ed.; McGraw-Hill: New York, NY, USA, 2010. [CrossRef]
21. Chakrabarti, A. *Engineering Design Synthesis: Understanding, Approaches, and Tools*; Springer: London, UK, 2002.
22. Cagan, J.; Campbell, M.I.; Finger, S.; Tomiyama, T. A Framework for Computational Design Synthesis: Model and Applications. *J. Comput. Inf. Sci. Eng.* **2005**, *5*, 171. [CrossRef]
23. Shea, K.; Cagan, J. Innovative dome design: Applying geodesic patterns with shape annealing. *Artif. Intell. Eng. Des. Anal. Manuf.* **1997**, *11*, 379. [CrossRef]
24. Gibson, I.; Rosen, D.; Stucker, B.; Khorasani, M. *Additive Manufacturing Technologies*; Springer International Publishing: Cham, Switzerland, 2021. [CrossRef]
25. Rosen, D.W. Computer-Aided Design for Additive Manufacturing of Cellular Structures. *Comput. Aided Des. Appl.* **2007**, *4*, 585–594. [CrossRef]
26. Ponche, R.; Kerbrat, O.; Mognol, P.; Hascoet, J.Y. A novel methodology of design for Additive Manufacturing applied to Additive Laser Manufacturing process. *Robot. Comput. Integr. Manuf.* **2014**, *30*, 389–398. [CrossRef]
27. Vayre, B.; Vignat, F.; Villeneuve, F. Designing for Additive Manufacturing. *Procedia CIRP* **2012**, 632–637. [CrossRef]
28. Briard, T.; Segonds, F.; Zamariola, N. G-DfAM: A methodological proposal of generative design for additive manufacturing in the automotive industry. *Int. J. Interact. Des. Manuf.* **2020**, *14*, 875–886. [CrossRef]
29. Duro-Royo, J.; Mogas-Soldevila, L.; Oxman, N. Flow-based fabrication: An integrated computational workflow for design and digital additive manufacturing of multifunctional heterogeneously structured objects. *Comput. Des.* **2015**, *69*, 143–154. [CrossRef]
30. Boddeti, N.; Ding, Z.; Kaijima, S.; Maute, K.; Dunn, M.L. Simultaneous Digital Design and Additive Manufacture of Structures and Materials. *Sci. Rep.* **2018**, *8*, 15560. [CrossRef]
31. Zhang, Y.; Bernard, A.; Gupta, R.K.; Harik, R. Evaluating the design for additive manufacturing: A process planning perspective. *Procedia CIRP* **2014**, *21*, 144–150. [CrossRef]
32. Lettori, J.; Raffaeli, R.; Peruzzini, M.; Schmidt, J.; Pellicciari, M. Additive manufacturing adoption in product design: An overview from literature and industry. *Procedia Manuf.* **2020**, *51*, 655–662. [CrossRef]
33. Motyl, B.; Filippi, S. Investigating the Relationships between Additive Manufacturing and TRIZ: Trends and Perspectives. In *Lecture Notes in Mechanical Engineering*; Springer: Cham, Switzerland, 2019; pp. 903–911. [CrossRef]
34. Altshuller, G. And Suddenly the Inventor Appeared: TRIZ, the Theory of Inventive Problem Solving, Worcester. 1996. Available online: http://books.google.com/books?hl=en&lr=&id=s7Qk_6WELWUC&oi=fnd&pg=PA1&dq=And+Suddenly+the+Inventor+Appeared&ots=2Fc3TJGicr&sig=b5C0LEyulNyiMQVfz-cOx0FPdrU (accessed on 19 November 2020).
35. Seepersad, C.C. Challenges and Opportunities in Design for Additive Manufacturing. *3D Print. Addit. Manuf.* **2014**, *1*, 10–13. [CrossRef]
36. Kumke, M.; Watschke, H.; Vietor, T. A new methodological framework for design for additive manufacturing. *Virtual Phys. Prototyp.* **2016**, *11*, 3–19. [CrossRef]
37. Pahl, G.; Beitz, W.; Feldhusen, J.; Grote, K.-H. *Engineering Design: A Systematic Approach*; Springer: London, UK, 2007. [CrossRef]
38. Rozvany, G.I.N. A critical review of established methods of structural topology optimization. *Struct. Multidiscip. Optim.* **2009**, *37*, 217–237. [CrossRef]
39. Bendsøe, M.P.; Kikuchi, N. Generating optimal topologies in structural design using a homogenization method. *Comput. Methods Appl. Mech. Eng.* **1988**, *71*, 197–224. [CrossRef]
40. Hassani, B.; Hinton, E. *Homogenization and Structural Topology Optimization*; Springer: London, UK, 1999. [CrossRef]
41. Savio, G.; Curtarello, A.; Rosso, S.; Meneghello, R.; Concheri, G. Homogenization driven design of lightweight structures for additive manufacturing. *Int. J. Interact. Des. Manuf.* **2019**. [CrossRef]
42. Bendsøe, M.P. Optimal shape design as a material distribution problem. *Struct. Optim.* **1989**, *1*, 193–202. [CrossRef]
43. Xie, Y.; Steven, G. A simple evolutionary procedure for structural optimization. *Comput. Struct.* **1993**, *49*, 885–896. [CrossRef]

34. Querin, O.; Steven, G.; Xie, Y. Evolutionary structural optimisation (ESO) using a bidirectional algorithm. *Eng. Comput.* **1998**, *15*, 1031–1048. [CrossRef]
35. Haber, R.; Bendsøe, M. Problem formulation, solution procedures and geometric modeling: Key issues in variable-topology optimization. In *Proceedings of the 7th AIAA/USAF/NASA/ISSMO Symposium on Multidisciplinary Analysis and Optimization, St. Louis, MO, USA, 2–4 September 1998*; American Institute of Aeronautics and Astronautics Inc., AIAA: Reston, VA, USA; pp. 1864–1873. [CrossRef]
36. van Dijk, N.P.; Maute, K.; Langelaar, M.; van Keulen, F. Level-set methods for structural topology optimization: A review. *Struct. Multidiscip. Optim.* **2013**, *48*, 437–472. [CrossRef]
37. Sigmund, O. A 99 line topology optimization code written in matlab. *Struct. Multidiscip. Optim.* **2001**, *21*, 120–127. [CrossRef]
38. Andreassen, E.; Clausen, A.; Schevenels, M.; Lazarov, B.S.; Sigmund, O. Efficient topology optimization in MATLAB using 88 lines of code. *Struct. Multidiscip. Optim.* **2011**, *43*, 1–16. [CrossRef]
39. Zegard, T.; Paulino, G.H. Bridging topology optimization and additive manufacturing. *Struct. Multidiscip. Optim.* **2016**, *53*, 175–192. [CrossRef]
40. Jiu, L.; Zhang, W.; Meng, L.; Zhou, Y.; Chen, L. A CAD-oriented structural topology optimization method. *Comput. Struct.* **2020**, *239*, 106324. [CrossRef]
41. Wang, H.V. A Unit Cell Approach for Lightweight Structure and Compliant Mechanism. Ph.D. Dissertation, Georgia Institute of Technology, Atlanta, GA, USA, 2005.
42. Savio, G.; Gaggi, F.; Meneghello, R.; Concheri, G. Design method and taxonomy of optimized regular cellular structures for additive manufacturing technologies. *Proc. Int. Conf. Eng. Des. ICED* **2015**, *4*, 235–244. Available online: https://www.designsociety.org/publication/37788/design_method_and_taxonomy_of_optimized_regular_cellular_structures_for_additive_manufacturing_technologies (accessed on 12 March 2021).
43. Savio, G.; Meneghello, R.; Concheri, G. Optimization of lattice structures for additive manufacturing technologies. In *Advances on Mechanics, Design Engineering and Manufacturing. Lecture Notes in Mechanical Engineering*; Springer: Cham, Switzerland, 2016; pp. 213–222. [CrossRef]
44. Savio, G.; Meneghello, R.; Concheri, G. Geometric modeling of lattice structures for additive manufacturing. *Rapid Prototyp. J.* **2018**, *24*, 351–360. [CrossRef]
45. Catmull, E.; Clark, J. Recursively generated B-spline surfaces on arbitrary topological meshes. *Comput. Des.* **1978**, *10*, 350–355. [CrossRef]
46. Rosso, S.; Meneghello, R.; Biasetto, L.; Grigolato, L.; Concheri, G.; Savio, G. In-depth comparison of polyamide 12 parts manufactured by Multi Jet Fusion and Selective Laser Sintering. *Addit. Manuf.* **2020**, *36*, 101713. [CrossRef]
47. Rosso, S.; Meneghello, R.; Concheri, G.; Savio, G. Scale and Shape Effects on the Fatigue Behaviour of Additively Manufactured SS316L Structures: A Preliminary Study. In *Lecture Notes in Mechanical Engineering*; Springer: Cham, Switzerland, 2019; pp. 879–890. [CrossRef]
48. Savio, G.; Rosso, S.; Curtarello, A.; Meneghello, R.; Concheri, G. Implications of modeling approaches on the fatigue behavior of cellular solids. *Addit. Manuf.* **2019**, *25*, 50–58. [CrossRef]
49. Wu, Z.; Xia, L.; Wang, S.; Shi, T. Topology optimization of hierarchical lattice structures with substructuring. *Comput. Methods Appl. Mech. Eng.* **2019**, *345*, 602–617. [CrossRef]
50. Alzahrani, M.; Choi, S.K.; Rosen, D.W. Design of truss-like cellular structures using relative density mapping method. *Mater. Des.* **2015**, *85*, 349–360. [CrossRef]
51. Savio, G.; Meneghello, R.; Concheri, G. Design of variable thickness triply periodic surfaces for additive manufacturing. *Prog. Addit. Manuf.* **2019**, *4*, 281–290. [CrossRef]
52. Jiang, J.; Xu, X.; Stringer, J. Optimization of process planning for reducing material waste in extrusion based additive manufacturing. *Robot. Comput. Manuf.* **2019**, *59*, 317–325. [CrossRef]
53. Ahsan, A.N.; Habib, A.; Khoda, B. Resource based process planning for additive manufacturing. *CAD Comput. Aided Des.* **2015**, *69*, 112–125. [CrossRef]
54. Cicconi, P.; Mandorli, M.; Favi, C.; Campi, F.; Germani, M. Metal Additive Manufacturing for the Rapid Prototyping of Shaped Parts: A Case Study. *Comput. Des. Appl.* **2021**, *18*, 1061–1079. [CrossRef]
55. Song, X.; Feih, S.; Zhai, W.; Sun, C.-N.; Li, F.; Maiti, R.; Wei, J.; Yang, Y.; Oancea, V.; Brandt, L.; et al. Advances in additive manufacturing process simulation: Residual stresses and distortion predictions in complex metallic components. *Mater. Des.* **2020**, *193*, 108779. [CrossRef]
56. Schoinochoritis, B.; Chantzis, D.; Salonitis, K. Simulation of metallic powder bed additive manufacturing processes with the finite element method: A critical review. *Proc. Inst. Mech. Eng. Part B J. Eng. Manuf.* **2016**, *231*, 96–117. [CrossRef]
57. Gong, X.; Cheng, B.; Price, S.; Chou, K. Powder-bed electron-beam-melting additive manufacturing: Powder characterization, process simulation and metrology. In Proceedings of the 2013 ASME Early Career Technical Conference (ECTC), Birmingham, AL, USA, 2–3 November 2013; pp. 55–66. Available online: https://www.researchgate.net/publication/275960612 (accessed on 23 December 2020).
58. Martukanitz, R.; Michaleris, P.; Palmer, T.; DebRoy, T.; Liu, Z.-K.; Otis, R.; Heo, T.W.; Chen, L.-Q. Toward an integrated computational system for describing the additive manufacturing process for metallic materials. *Addit. Manuf.* **2014**, *1*, 52–63. [CrossRef]

69. Jayanath, S.; Achuthan, A. A computationally efficient hybrid model for simulating the additive manufacturing process of metal *Int. J. Mech. Sci.* **2019**, *160*, 255–269. [CrossRef]
70. Watanabe, N.; Shofner, M.L.; Treat, N.; Rosen, D.W. A model for residual stress and part warpage prediction in material extrusio with application to polypropylene. In *Solid Freeform Fabrication 2016, Proceedings of the 27th Annual International Solid Freefor Fabrication Symposium—An Additive Manufacturing Conference, Austin, TX, USA, 8–10 August 2016*; TMS: Pittsburgh, PA, US 2016; pp. 2437–2455.
71. Brenken, B.; Barocio, E.; Favaloro, A.; Kunc, V.; Pipes, R.B. Development and validation of extrusion deposition additiv manufacturing process simulations. *Addit. Manuf.* **2019**, *25*, 218–226. [CrossRef]
72. Fradl, D.; Panditaratne, J.; Bi, J.; Fu, R.; Oancea, V. Finite Element Simulation of the Multi Jet Fusion (MJFTM) Process using Abaqu In Proceedings of the SIMULIA Science in the Age of Experience, Chicago, IL, USA, 15–18 May 2017; pp. 440–469. Availabl online: https://pdfs.semanticscholar.org/fb66/17aab5ff74f47709ebfecb694d32f23ceff2.pdf (accessed on 23 January 2020).
73. Kolossov, S.; Boillat, E.; Glardon, R.; Fischer, P.; Locher, M. 3D FE simulation for temperature evolution in the selective lase sintering process. *Int. J. Mach. Tools Manuf.* **2004**, *44*, 117–123. [CrossRef]
74. Preisinger, C. Linking Structure and Parametric Geometry. *Arch. Des.* **2013**, *83*, 110–113. [CrossRef]
75. Renishaw, Data Sheets—Additive Manufacturing—AlSi10Mg-0403 (400 W). 2021. Available online: https://www.renishaw.com en/data-sheets-additive-manufacturing--17862 (accessed on 19 February 2021).

*applied sciences*

MDPI

*Article*

# Development Workflow for Manifolds and Fluid Components Based on Laser Powder Bed Fusion

**Nicolas Rolinck, Matthias Schmitt *, Matthias Schneck, Georg Schlick and Johannes Schilp**

Department of Processing Technologies, Fraunhofer Institute for Casting, Composite and Processing Technology IGCV, 86159 Augsburg, Germany; Nicolas.Rolinck@igcv.fraunhofer.de (N.R.); matthias.schneck@igcv.fraunhofer.de (M.S.); georg.schlick@igcv.fraunhofer.de (G.S.); johannes.schilp@igcv.fraunhofer.de (J.S.)

* Correspondence: matthias.schmitt@igcv.fraunhofer.de

**Featured Application: This article proposes a universal development methodology for hydraulic manifolds and fluid components manufactured by laser powder bed fusion. The proposed process leads systematically to a part with increased performance characteristics, economic efficiency, and reliable manufacturability.**

**Abstract:** Through its unique characteristics, additive manufacturing yields great potential for designing fluid components with increased performance characteristics. These potentials in advanced design, functional structure, and manufacturing are not easily realized. Therefore, the present study proposes a holistic development methodology for fluid components with a specific focus on hydraulic manifolds. The methodology aims to lead the designer from the specification of the task, through a step-by-step embodied design, to a technical and economic evaluation of the optimized, first-time manufactured part. A case study applies the proposed methodology to a part of a rail-vehicle braking application. Through its application, a significant reduction in weight, size, as well as significant contributions to the company's AM strategy can be assigned to the part. At the same time, increased direct manufacturing costs are identified. Based on the increased performance characteristics of the resulting design and the holistic foundation of the subsequent economic decisions, a satisfying efficiency can be allocated to the proposed methodology.

**Keywords:** manifold; hydraulic component; additive manufacturing; design methodology; development workflow; laser powder bed fusion

**Citation:** Rolinck, N.; Schmitt, M.; Schneck, M.; Schlick, G.; Schilp, J. Development Workflow for Manifolds and Fluid Components Based on Laser Powder Bed Fusion. *Appl. Sci.* **2021**, *11*, 7335. https://doi.org/10.3390/app11167335

Academic Editors: Marco Mandolini, Patrick Pradel and Paolo Cicconi

Received: 30 June 2021
Accepted: 5 August 2021
Published: 9 August 2021



## 1. Introduction

Additive manufacturing (AM) describes all manufacturing processes in which the part is fabricated by adding volume elements layer by layer to produce the desired geometry. The added elements are directly derived from the 3D data [1]. AM technologies have been advancing steadily through the years. Especially the sector of metal AM matured from niche applications to mainstream manufacturing and is being accepted as a valuable production process for demanding engineering tasks. For metal applications, laser powder bed fusion (LPBF) is the most widespread manufacturing principle [2]. When designing parts specifically for AM, these technologies yield unique characteristics in advanced geometric freedom, part consolidation, and product customization in comparison to other manufacturing processes. Through the given geometric freedom, new potentials in the performance and economic efficiency of the product are enabled, which can be realized through advanced design. If such a design can consolidate multiple parts of an assembly, additional improvements in size, weight, and assembly effort are achievable [3]. In the LPBF manufacturing process, differences in the individual geometry of parts manufactured in one build job have minor influence on the economics of the build job compared to conventional manufacturing methods. This leads to an economic feasibility of a lot size of one and makes the technology therefore practical for highly customized products [4]. However, these potentials are not easy to identify nor to realize. It requires the designer to have a fundamental understanding of the technology and to re-think the system-level design of the product [5]. Existing, universal product development methodologies are applicable for AM parts, but can hardly exploit the given potentials efficiently due to the design complexity and multitude of possible solutions [6]. Nonetheless, as a development of a completely new design process is neither efficient nor possible, development guidelines for AM products should be based on established methodologies and further recognize the specific potentials and limitations of the technology. As the VDI 2221 [7,8] guideline is widely accepted as a core standard for the development of technical products and was previously suggested for an AM specific methodology [9]. Kumke bases the general structure of his comprehensive method for the development of AM parts onto it [5].

As for any other manufacturing method, design guidelines have to be followed for a qualitative and economic fabrication of the designed part. The established methods of design for manufacturing (DfM) aim for design solutions with minimum manufacturing costs for a given manufacturing technology. Even though such guidelines for favorable manufacturing are still an elementary part of a design for additive manufacturing (DfAM), these methods shall further support the exploration of AM-enabled potentials and possibilities. Laverne et al. propose to differentiate opportunistic DfAM, restrictive DfAM, and dual DfAM [10]. Opportunistic DfAM methods shall enable the designer to explore innovative solutions enabled by the possible geometric complexity of AM products. The solution space of such methods is not narrowed by the limitations of the AM technology as in the definition of conventional DfM. These limitations are covered in the class of restrictive DfAM methods. In these, design rules are defined, that guide the designer to create a part with sufficient manufacturing quality. The majority of such design rules are developed in university research through test series [11–14]. These restrictions can be quantified as for minimum radii or wall thicknesses, as well as general guidelines for a sufficient manufacturing result [14]. They are more and more transferred into official design guidelines [15] and textbooks [4]. Finally, methods utilizing a combination of both, restrictive and opportunistic DfAM, are considered to be the most efficient for a holistic product design manufactured by AM technologies. With these, innovative concepts are designed for a qualitative manufacturing outcome [5]. Therefore, for each classification, different tools as depicted in Figure 1 are suggested.

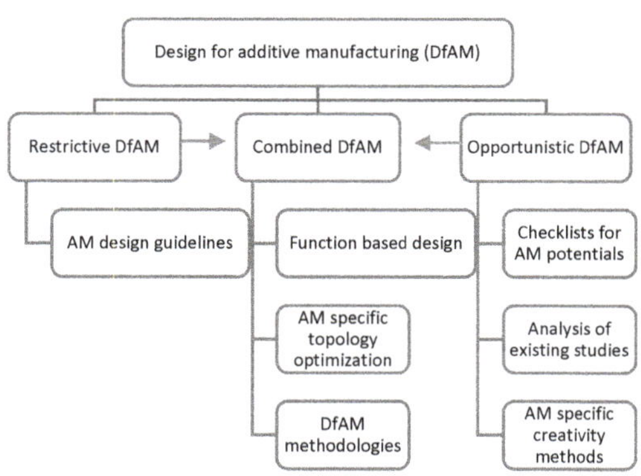


**Figure 1.** Differentiation of DfAM methods and specific tools derived from [5].

Manifold bodies in pressure-loaded applications are traditionally manufactured from a solid material block, creating the hydraulic schematic by intersecting, gun-drilled holes. For manufacturing efficiency, these channels are usually manufactured with a 90° orientation towards the outside surface of the semi-finished block. All unused openings are closed by sealing stoppers in a subsequent assembly process. As a result, complex hydraulic schematics require a larger manifold due to the inflexible channel design, a greater number of drilled holes, and therefore a greater number of sealing points. Furthermore, the perpendicular, sharp corners in the fluid flow decrease the flow efficiency of the hydraulic medium. Through the AM-enabled geometric complexity, manifold bodies hold great potential for increased product performance through advanced design [4].

A comprehensive approach to the development of hydraulic manifolds fabricated by LPBF systems is the work of Kausch [16]. After extensive research on various topics concerning the development of AM manifolds, the generated knowledge is transferred onto two different manifolds, which are manufactured by LPBF. The evaluation of the part proves a great potential for weight reduction of high-pressure manifolds and suggests similar production costs to the conventional counterpart with further process and productivity improvements of the technology. The research of Diegel et al. [17] follows an existing AM manifold design and focuses on the further optimization of the manufacturability. This includes primarily the reduction in support structure to minimize manufacturing and post-processing time. Schmelzle et al. [18] present a specific flowchart for the re-design of hydraulic manifolds. The study places special focus on the channel cross-section geometry as well as a detailed description of manufacturing, inspection, and testing of the resulting manifold. Zhang et al. [19] approach a manifold re-design with the motivation of an optimized fluid flow and adapt each phase of the development process accordingly. As in the previous studies, a significant optimization in weight, size, and fluid flow efficiency is achieved. Cooper et al. [20] examine the pressure, flow, porosity, surface and micro-hardness characteristics of a titanium pipe manufactured by LPBF. The research offers detailed findings on the pressure resistance of thin-walled pipes, their internal surface roughness, their as-built geometric accuracy as well as the characterization of the porosity of the material. Table 1 depicts an overview of manifold-specific information in reviewed studies.

**Table 1.** Overview of conducted research on the re-design of manifolds for Additive Manufacturing (AM).

| | Schmelzle et al. | Kausch | Diegel et al. | Zhu et al. | Cooper et al. |
|---|---|---|---|---|---|
| Design methodology covered | yes | yes | yes | no | no |
| Functional optimization | yes | yes | yes | yes | yes |
| Manifold dimensions | Approximately 127 mm × 127 mm × 70 mm | no | 176.2 mm × 125.5 mm × 120 mm | 246 mm × 140 mm × 66.5 mm | no |
| Maximum pressure | 127.9 MPa | 31.4 MPa | 98.5 MPa | 21 MPa | 24 MPa |
| AM material | 17-4 steel | TiAl6Va | AlSi10Mg, 316L steel | 316L steel | Ti6Al4V |
| Number of installation spaces | 3 | 6 | 6 | 10 | 0 |
| Number of additional connectors | 3 | 7 | 14 | 4 | 2 |
| Process parameters covered | yes | no | no | yes | yes |
| Fatigue characteristics | no | yes | yes | no | no |
| DfAM approaches | Restrictive, opportunistic, combined | Restrictive | Restrictive | Restrictive | - |

The review of conducted research on hydraulic manifolds shows that the re-design process systematic is an important topic for most studies. The models and flowcharts presented for a re-design are often rudimentary for the task of developing an additively manufactured manifold with increased performance while ensuring qualitative and economic manufacturability. Most of the reviewed studies only utilize restrictive DfAM methods for their design. The very detailed approach of Kumke [5] on the complete design process for AM, applying restrictive and opportunistic DfAM, has not been applied for a hydraulic manifold application. Furthermore, the number of hydraulic components and the resulting complexity of the reviewed manifolds remain in a range of 3 to 10 installation spaces.

Based on the state of the art in the development of AM manifolds, a research defici on the systematic development of manifolds with advanced hydraulic complexity anc additional functional requirements is identified. Product improvements through the appl cation of opportunistic DfAM are often left unused. Consequently, the present article aim to propose a universal development and design methodology that leads systematicall to a manifold with improved product performance while being optimally designed fo qualitative and economic manufacturing by LPBF. After the methodology is presented, it applicability and efficiency are assessed with a case study on a safety-critical manifold o the rail industry.

For the methodology development, the following requirements are defined:

- The universal applicability of the process must be given. The design methodolog shall serve hydraulic applications independently of the specific product.
- The process shall facilitate opportunistic DfAM potentials by introducing specific metl ods and tools when necessary.
- Restrictive DfAM methods and tools are to be introduced to the process when necessar
- The process shall lead to an AM geometry that embodies the optimum quality anc profitability of the build job.
- The process shall systematically lead to a manifold with increased performance cha acteristics such as improved weight, size, or flow efficiency.
- The process shall allow an iterative shift between the different phases to utilize a experience gain of the user for upstream phases.

## 2. Methodology Development

As previously identified, Kumke [5] offers a comprehensive process model for th universal development of AM parts. To guide the designer to a design that respect technological as well as economic aspects, this process constitutes the general structur of the proposed methodology. This structure is, however, supplemented with manifol specific design phases and tasks. Additionally, suitable tools for the specific tasks ar proposed. In that way, the methodology facilitates a complete exploration of the potential enabled through advanced design while acknowledging the economic efficiency anc reliable manufacturability of the resulting part.

As depicted in Figure 2, the proposed development methodology starts with th planning and conceptual phase. These phases, including their individual steps, are trans ferred analogously from the VDI 2221 [7]. Based on the high significance of a detailec description of requirements, an independent list of requirements for the part is to be createc in the first task. This list should describe all boundary conditions concerning the par available AM technologies, material availability, and loads. This is a mandatory proces step, which is executed for re-design tasks as well as new development tasks. Even thoug the requirements shall be formulated detached from the manufacturing technology use AM-specific potentials and resulting requirements, for example, lightweight design, ca already be addressed [5]. In the second task, the functional structure of the part is to b arranged hierarchically into main and sub-functions (cf. Figure 2) [21].

The exploration of design freedom in new- as well as re-designs benefits significantl from function-based thinking, being detached from conventional manufacturing restric tions. In doing so, all surfaces which serve a specific engineering function shall be identifie and their individual boundary conditions, for example, freedom of position, is specifie This facilitates a subsequent, flux of force-oriented design. Based on this analysis, differen conceptual solutions for the identified, functional structure are to be developed in the thir task. To assist the designer in exploring unconventional solutions, tools like "AM-potentia checklists", "the analysis of existing studies" or "cross-departmental brainstorming" can b utilized during this task. It is important to conduct such an extensive conceptual phase fo re-designs as well. Existing solutions may include conventional thinking barriers, wherea the presented steps and tools create a large solution space for innovative concepts [5].

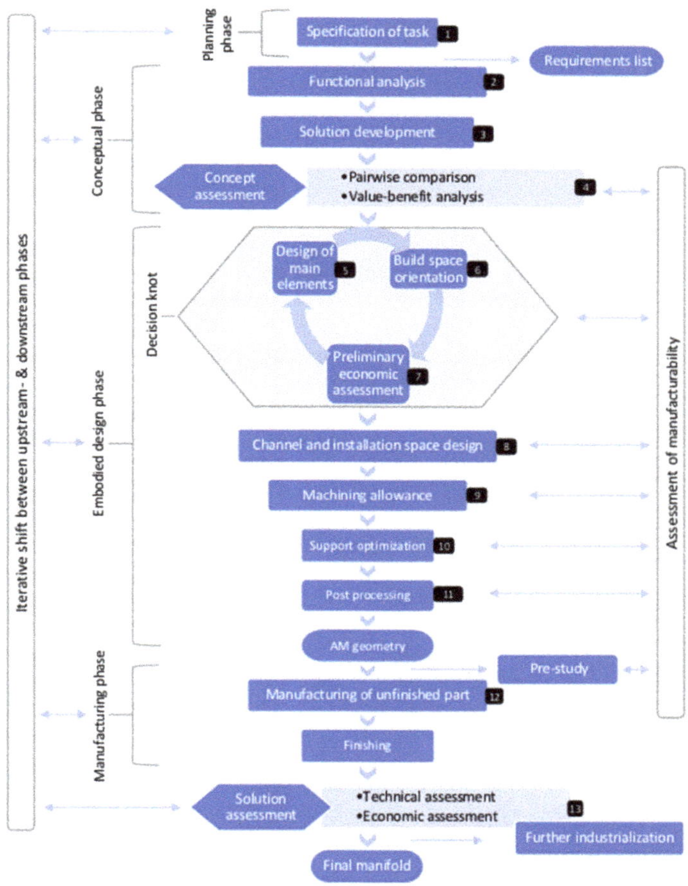


**Figure 2.** Development methodology for Additive Manufacturing (AM) manifolds. Squared boxes: process-tasks; round boxes: process-outputs; diamond-shaped boxes: assessment-tasks.

In the following task four, the concept assessment, the acquired concepts are evaluated by "pairwise comparison" and "value benefit analysis", which are specified/recommended in VDI guideline [7]. The assessment criteria for the value benefit analysis need to be adapted to consider the new AM aspects [5]. The following criteria are proposed for the assessment of manifold concepts:

- Potential for size reduction
- Potential for weight reduction
- Potential for improved hydraulic efficiency
- Technical feasibility/ development risk
- Degree of innovation
- AM material efficiency
- Profitability of the build job

Based on the quantitative assessment, one concept is selected for the embodied design phase. In this phase, potentials for improved product performance can be realized through advanced design. The chosen concept is introduced to the decision knot, in which tasks five to seven are executed iteratively. First, the parts main elements are designed and positioned according to the specified boundary conditions [5]. For manifold bodies, these are primarily installation spaces, hydraulic connectors, fixation points for the motor, and pump. If the position of an installation space for a hydraulic component was specified as variable in the analysis of functional surfaces, its position in the AM design shall regard the following three aspects. Its position within the hydraulic logic, the assembly and accessibility with the required tools, and finally an efficient nesting of the valves, sensors, and motor. There are several software solutions [22] available, which create two-dimensional nesting solutions for a set of defined geometries. For a re-design, it is important to question existing component positions, as conventional manufacturing constraints may have lost relevance through AM.

Once these elements are positioned, the orientation of the part within the build space of the LPBF system is to be chosen. This decision has a fundamental influence on the design of subsequent geometries, the manufacturing quality of the part, the necessary amount of support material, and finally, the economic efficiency of the build job. After the technical feasibility of the selected orientation is proven by a manufacturing simulation, a preliminary economic assessment is conducted in task seven. The outcome of this decision sequence is often a compromise of minimum manufacturing time, best manufacturing quality, mechanical properties, necessary support material, and manufacturing costs [17]. It shall therefore be repeated iteratively until the best compromise for the given boundary conditions is found. Additionally, if no such compromise can be found, the decision knot at this early stage of the development process prevents the usage of further resources on unpromising concepts and questions the suitability of AM for the specific project [5].

As the main elements are usually standardized installation spaces and threads, the DfAM potential lies mainly in the design of the hydraulic interconnection of the installation spaces. The channel diameter is to be selected according to the present fluid flow of the application [23]. Circular channels with a diameter greater than 7 mm require internal support structure [16,18,24]. As the complex channel design makes these structures inaccessible, they must be avoided. Multiple studies provide restrictive design guidelines for non-circular channel cross-sections [16,18,25]. Additionally, an appropriate wall thickness must be selected, which resists high, cyclic pressure load over the complete lifetime as specified in the requirements list [16,26–28]. The design complexity enables channels that can be designed conformal to existing walls and installation spaces, sharing a single wall, and saving thereby AM-material and manufacturing process time. Through this design flexibility, the size of a manifold is not limited by the channel layout anymore.

As most functional surfaces require a high surface quality and geometric precision, they must be post-processed with subtractive processes [18]. In task nine, the necessary machining offset, registration features, and clamping positions are added to the geometry to enable precise machining. Again, these features shall be designed according to restrictive design guidelines presented in the literature [25,29]. Further, referencing strategies for subtractive operations on defined registration points shall be established at this point.

All geometric elements designed in tasks five to nine serve a specific function. Even though these elements are designed under consideration of restrictive DfAM guidelines and the chosen orientation, the geometric complexity requires compromises in the usage of support for certain elements. However, it is a major intention of DfAM to minimize the usage of supports, as they require extensive manual labor for their removal [25]. Consequently, most of the necessary support structures shall be replaced by design elements, for example, thin walls, which are not removed and serve as additional stiffening elements. Support-critical areas are identified by build-preparation software [30] and then replaced by geometric elements in the design software iteratively.

As the final part of the embodied design phase, task eleven takes into account the post-processing of the part. For manifold bodies, the powder removal from the complex channels is of high importance. Therefore, the entrance and exit of every channel must be designed with an opening in the unfinished manifold. Further, considerations on the deburring of the machined installation spaces and the final cleaning of all channels are taken into account. These processes must remove all chips, coolant-, oil-, or powder residuals from the manifold in- and outside.

The proposed sequence of eleven tasks leads to the unfinished AM geometry, which is manufactured on the available LPBF system. Due to the definition of all boundary conditions at the beginning of the development, the resulting design shall be suitable for the given AM system and its limitations. To verify the decisions taken in the previous steps without the financial and time-consuming risk of a complete, fully utilized build job, this study proposes to manufacture a section of the complete manifold, which contains critical features, under similar fabrication circumstances. In a second manufacturing step of the pre-study, the efficiency of the projected machining strategy is assessed, and necessary alterations to upstream tasks can be implemented. If the pre-study does not indicate issues in the additive manufacturing and subtractive finishing, the complete, first build is to be manufactured and finished according to the specifications in the manufacturing phase.

To enable comprehensive product development, it is the final task of the proposed methodology, to assess the developed manifold by technological as well as economic aspects. For the technological assessment, firstly, all resulting product characteristics such as weight, size, fluid flow, or mechanical resistance are to be identified and matched with the requirements list. Secondly, all noticeable problems identified during the build job, or by visual inspection of the part and their cause are to be described. Finally, as the internal shape of the manifold is of high importance for its functionality but may be inaccessible by endoscopes due to its complex shape, industrial computer tomographic (CT) scanning is suggested to inspect the manifold-inside without damaging the part. Furthermore, the global shape accuracy of the part can be examined with the results of the CT scan [31].

The economic evaluation of the part is conducted in multiple stages. As there is no single formula that can represent manufacturing costs, lifecycle costs, customer value, and corporate value of the part, the study applies a four-stage evaluation based on the findings of Lakomiec [32]. The first stage specifies the geometric complexity of the part, and therefore its suitability for AM. As the previous tasks guided the designer towards a geometry specifically designed for AM, there should be little doubt about its suitability. The second stage calculates the manufacturing costs of the AM manifold. As most manifold bodies require post-processing of functional surfaces, a classification of the costs for the unfinished part by AM and the subtractive finishing should be taken into account. Further, if part consolidation was achieved, differences in the assembly costs are to be introduced into the calculation. The third stage of the evaluation is the most important for AM parts. This stage regards the added value and increased product performance through advanced design. For manifold bodies being designed for reduced weight and improved fluid flow, this consists primarily of a reduction in lifecycle costs, which are quantifiable in many cases. Added customer value in contrast, such as a reduced installation size of the assembly, is only monetarily rateable if related to an existing customer project. With an increased level of detail, aspects such as impacts on the process and supply-chain or effects on sustainability are to be considered [33]. Thereby it is important to distinguish if the added value can be assigned to the single product, or whether it has a strategic value for the complete enterprise. An exemplary structure of the presented stages is depicted in Table 2.

After all aspects are summarized, strategies on further proceedings and industrialization are derived according to the individual project or the general strategy of the organization. For a numeric decision finding, the individual aspects can be weighted and summarized in a target figure [32].

The execution of the proposed development methodology leads to a manifold body with enhanced performance characteristics that ensures economic and reliable manufacturing by the available LPBF system. Based on the complete prototype manufacturing, and a holistic economic evaluation, a profound basis for actions towards further industrialization of the developed part is created. Through the iterative information flow from any point of the process, conclusions based on the solution assessment can be introduced to upstream tasks.

**Table 2.** Four-stage evaluation scheme for AM parts derived from [32].

| 1. Check geometric complexity | |
|---|---|
| Construction complexity<br>Consolidated parts | |
| **2. Calculation production costs** | |
| Costs unfinished part<br>Costs finishing | Comparison conventional part<br>Assessment process steps<br>Reduction factor to conventional process steps |
| **3. Evaluation value add** | |
| Attributable to part | Lifecycle costs<br>Customer value<br>Supply-chain impact |
| Overall factors | Market and technology portfolio<br>Business case calculation |
| **4. Summary consideration and weighting** | |

## 3. Case Study

The proposed development methodology was applied to a safety-critical brake application of Knorr-Bremse rail vehicle systems (KB), which is a founding member of the innovation network "Mobility goes Additive". Of all application fields of AM, new customer products with advanced design require the highest effort for technological qualification while entailing the highest customer value. Through the application of the proposed methodology for a re-design of the manifold of an electro-hydraulic unit, a next step is taken towards the production of end-use parts by AM.

### *3.1. Planning Phase (Step 1 of Development Methodology)*

The manifold to be re-designed, as shown in Figure 3, is the centerpiece of the control unit for a hydraulic braking system. The AM re-design shall be suitable to be manufactured by the KB-internal AM infrastructure. Therefore, information and boundary conditions as shown in Table 3 are identified. For urban rail applications, the installation size of the assembly in the bogie of the vehicle is of high importance. Therefore, it is the central goal to reduce the manifold size. The second priority of the study lies in the weight reduction of the part. Even though the re-design may lead to an improved fluid flow, such optimization potential is not further investigated in this study, due to the low volume flow of the application.

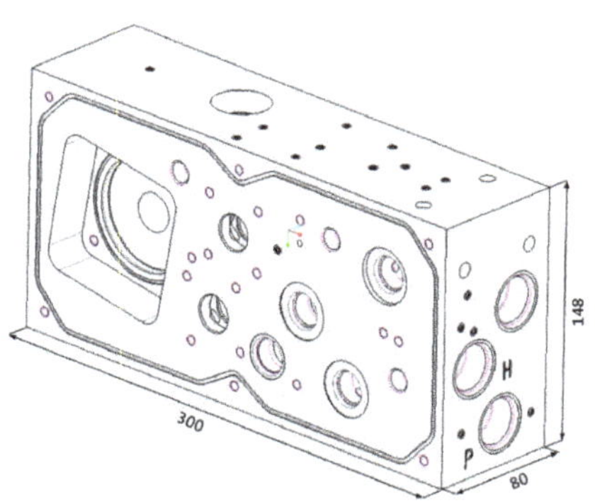


**Figure 3.** Isometric view of the conventional manifold (Dimensions in mm).

**Table 3.** Specifications and boundary conditions for the manifold.

| Type | Value | Specification |
|---|---|---|
| Size | 300 mm × 148 mm × 80 mm | |
| Weight | 8.55 kg | Including 37 sealing stoppers |
| Pressure load | 0–16 MPa | |
| Lifetime | >$10^7$ load cycles | |
| Available AM build space | 250 mm × 250 mm × 325 mm | |
| Available AM material | $AlSi_{10}Mg$ [34] | |

### 3.2. Conceptional Phase (Steps 2–4 of Development Methodology)

At the beginning of the conceptual phase, the functional structure of the manifold is visualized. In addition to the primary hydraulic functionality, the manifold fulfills several further functions like the fixation of the assembly onto the vehicle, creating a sealing surface for the oil tank, and providing attachment points for the tank and electric cover. Once the functional structure is described, multiple conceptual solutions are introduced as to be seen in Figure 4. These solutions include only the main elements to create multiple concepts in a time-efficient manner. The topology optimized (TO) concept shown in (a) places all main elements within a TO frame structure, which may create a sufficient stiffness of the part, while large surfaces can be made of a thin wall. The L-shape concept aims to position all service components onto one side and takes advantage of a perpendicular orientation of valve elements. Lastly, concept three realizes a radial sealing for the oil tank which, in contrast to the existing flange sealing implicates improved sealing characteristics.

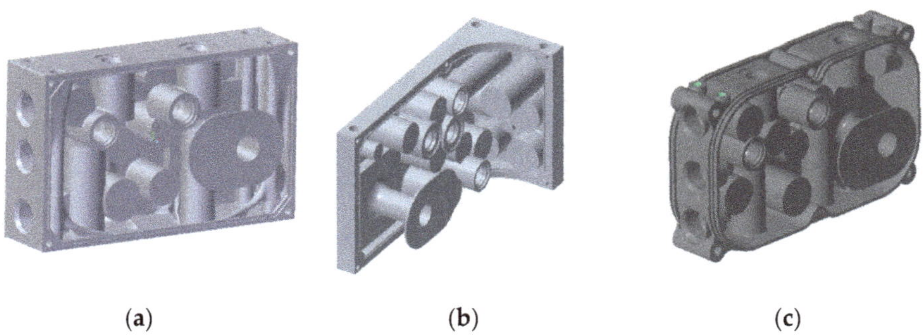

**Figure 4.** (**a**) Topology optimized concept; (**b**) L-shape array concept; (**c**) Radial sealing concept.

Based on the introduced concepts, a value-benefit analysis regarding the evaluation criteria proposed in Section 2 is conducted (cf. Table 4). Weight reduction is most probable with concept a), whereas concept b) has the potential for short hydraulic channels. The build job profitability is rated by comparing them with each other. An analysis compared to the serial part was not conducted since the goal of the development study was to gain knowledge and experience with the AM workflow. An early evaluation is feasible with existing AM costs models, e.g., from Schneck et al. [35]. As a result, concept a) is recognized as being the most suitable concept for the boundary conditions of the project.

### 3.3. Embodied Design Phase (Steps 5–11 of Development Methodology)

With a concept for the further development selected, the embodied design starts with the positioning of the main elements. For the efficient nesting of all components, the Software Deepnest [22] is used to propose nesting solutions for the cross-section geometries of the motor and hydraulic components, which are to be positioned on the manifold. Figure 5 depicts one nesting solution and the resulting assembly, designed with PTC Creo.

**Table 4.** Evaluation of the proposed concepts.

| Criteria and Weighting | | Concept (a) | Concept (b) | Concept (c) |
|---|---|---|---|---|
| Potential for manifold size reduction | 0.1 | 4 | 2 | 2 |
| Potential for weight reduction | 0.15 | 4 | 2 | 2 |
| Potential for improved hydraulic efficiency | 0.1 | 2 | 3 | 1 |
| Technical feasibility/low development risk | 0.3 | 2 | 1 | 3 |
| Degree of innovation | 0.1 | 4 | 3 | 1 |
| Material efficiency | 0.05 | 2 | 4 | 2 |
| Profitability of the build job | 0.2 | 3 | 4 | 3 |
| | | 2.90 | 2.40 | 2.30 |

Rating: 4—very high; 3—high; 2—medium; 1—low; 0—very low.

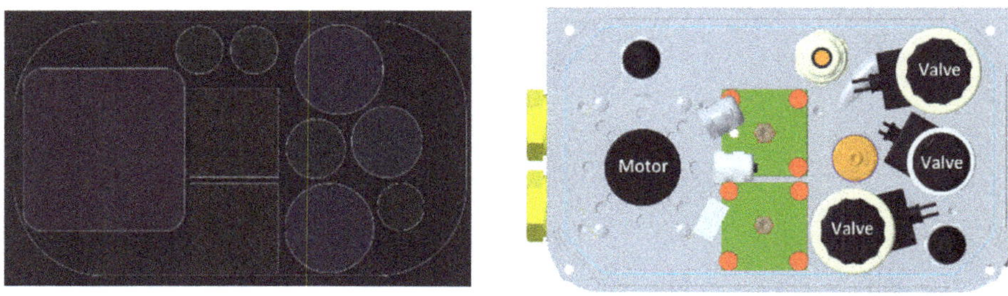


(**a**) (**b**)

**Figure 5.** (**a**) Proposed nesting solution; (**b**) resulting arrangement of the assembly.

As the orientation of the main elements has strictly defined boundary conditions, the build-space orientation is chosen with regard to these elements. For the manufacturing with the available LPBF system, the economics, the support utilization, and the quality of key features are assessed for three orientations as depicted in Table 5. A flat, 0° orientation of the concept, as depicted in Table 4 can only manufacture one manifold per build job. A 90° orientation may enable the manufacturing of three manifolds but leads to excessive usage of support structure and an unfavorable orientation of key features. Finally, a tilted orientation can facilitate two manifolds per build job, while requiring the least amount of support structure of the three investigated orientations. Therefore, and under consideration of the resulting build job costs, a 49° orientation of the manifold within the build space is selected. The actual orientation was chosen in regard to support structure, achievable geometry resolution, surface quality, and build height through an in-depth analysis (among others laser scans, experience from previous build jobs) of angles from 45 to 55°.

**Table 5.** Comparison of build space orientations.

| | 0° Orientation | 49° Orientation | 90° Orientation |
|---|---|---|---|
| | 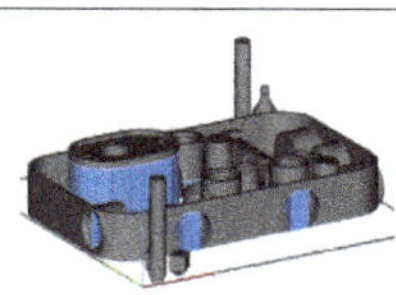 | 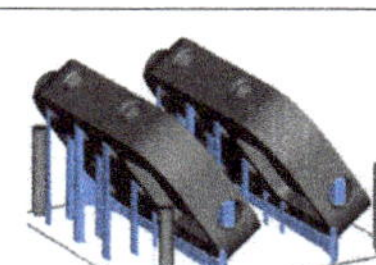 | 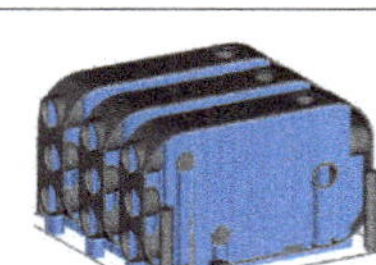 |
| $V_{Part}/V_{Support}$ Ratio | 10.71 | 28.33 | 2.6 |
| Build height | 81.9 mm | 154.17 mm | 152.19 mm |
| Build time | 22 h 28 min | 49 h 19 min | 87 h 55 min |
| Build time per part | 22 h 28 min | 24 h 40 min | 29 h 18 min |

Following, the embodied design steps eight to eleven are executed as proposed in Figure 2. As depicted in Figure 6a, hydraulic channels are designed based on spline curves, with a special focus on conformal positioning onto existing features, a homogenous channel cross-section, and large radii for flow efficiency. Through the freedom in channel design, the integration of the hydraulic layout into the defined boundary conditions implicates little difficulties. The unfinished geometry of installation spaces is designed as uniform, cylindrical geometry, with a machining offset of at least 1 mm.

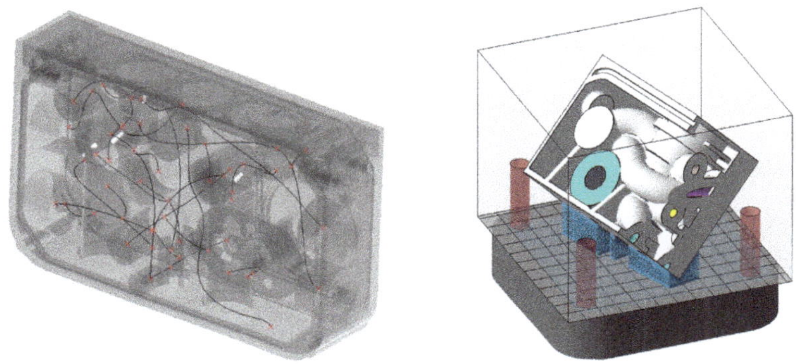

(**a**) (**b**)

**Figure 6.** (**a**) Manifold main elements with spline curves between starting points (red marks) for channel design; (**b**) pre-study part: Cut-out from the hydraulic manifold.

Even though the minimization of the support structure is an elementary part of restrictive DfAM, the complex hydraulic layout requires compromises toward the usage of support. Critical elements, which may require support are identified with the AM preparations software Materialise Magics [30]. To reduce post-processing time, supportive structures are designed to be a structural part of the manifold. Therefore, several walls of equal thickness are designed in locations, which require support. The need for support is determined iteratively with the preparation software until all temporary structures with difficult accessibility are removed. All channels and installation spaces are designed as open to facilitate a complete powder removal.

For post-processing, the unfinished manifold is media-blasted, and all functional surfaces are machined in two clamping positions with polycrystalline diamond (PCD) tools. As all machined surfaces are easily accessible, no advanced deburring process, such as paste deburring, is performed. This is an additional improvement to the conventional design, as the internal intersections of the gun-drilled holes require deburring actions. Finally, the part is washed and rinsed thoroughly, to remove all machining chips as well as powder- and coolant-residues from the out- and inside.

Before the time- and cost-intensive manufacturing of the fully utilized build job is initiated, a pre-study is conducted by manufacturing a representative section of the designed manifold. The chosen section, as to be seen in Figure 6b, contains three installation spaces on three different sides, as well as multiple, constricted channels. The part is manufactured on an SLM Solutions 125 HL system, utilizing the equal $AlSi_{10}Mg$ aluminum as for the complete part. All post-processing is conducted analogously to the complete manifold.

### *3.4. Results and Solution Assessment*

#### 3.4.1. Technical Assessment (Step 13 of Development Methodology)

Based on the successful fabrication of the pre-study in the first and of the complete two-manifold build job in the second step, the following findings are recognized. The designed elements replacing temporary support structures work in combination with the 49° orientation of the part. No defects due to critically overhanging features can be identified through visual inspection and CT scan. Only seven support structures are to be removed manually. The required sealing groove cannot be manufactured by the LPBF system and is to be machined. No issues regarding warping and insufficient connection onto the build plate are detected. Under consideration of the AM-material volume, the previously assumed manufacturing reliability is approved by a first-time right manufacturing of a two-manifold build job as to be seen in Figure 7a.

(a) (b)

**Figure 7.** (**a**) The unfinished 2-manifold build job with tensile- and density-test-specimen; (**b**) finished manifold.

Through its thin-wall design, the manifold offers additional oil-reservoir volume on its inside. Though not further specified in this study, improved fluid flow characteristics are expected, as no sharp, 90° corners are introduced to the flow path. The finished part meets the equal dimensional requirements as the conventionally manufactured design. Table 6 offers further details on the part.

**Table 6.** Comparison of product characteristics.

| Characteristic | AM Design | Conventional Design |
|---|---|---|
| Manifold size | 245 mm × 148 mm × 55 mm | 300 mm × 148 mm × 80 mm |
| Weight | 2.06 kg | 8.55 kg |
| Sealing stopper | 0 | 37 |
| Additional oil reservoir | 930.7 cm$^3$ | - |
| Material volume | 771.54 cm$^3$ | 3166.67 cm$^3$ |
| Necessary support structure in [%] of total build job material | 0.39% | - |

3.4.2. Economic Assessment (Step 13 of Development Methodology)

The economic evaluation is conducted based on four stages. Firstly, as the part is specifically designed for fabrication by LPBF, its suitability for the process is clearly given. Further, the design consolidates 37 sealing stoppers. Secondly, the AM costs are significantly higher in comparison to the conventional manifold body. Therefore, a mere substitution with the AM part is not economic. The AM design is to be classified as an additional product for the acquisition of projects with special requirements (e.g., size or lightweight). Quotes of an external supplier show that external AM production costs are more than five times higher than current serial costs. However, increasing industrialization and adoption of AM are expected to decrease the AM production costs in the near future. Thirdly, the added value and its impact on the lifecycle of the part and the complete organization are assessed. The design-based reduction in the size of the assembly may serve customer projects with highly limited space requirements. Further, the lightweight design reduces the part's energy consumption during operation in an often-accelerated rail vehicle and therefore decreases its operational costs and $CO_2$ emissions. Finally, the experience gained in the design, manufacturing, and industrialization of AM products is universally applicable through the company and a valuable contribution to the AM strategy of Knorr-Bremse. On this basis, the roadmap depicted in Table 7 is presented.

**Table 7.** Roadmap for economic evaluation.

| 1. Check geometric complexity | |
|---|---|
| Construction complexity<br>Consolidated parts | Only to be manufactured through AM<br>37 sealing stoppers |
| 2. Calculation production costs | |
| Costs unfinished part | Significantly Higher |
| Costs finishing | Reduced finishing costs<br>Total manufacturing costs significantly higher |
| 3. Evaluation value add | |
| Attributable to part | 76 % weight reduction, reduced operational costs<br>55 mm reduced installation space<br>Unique selling proposition |
| Overall factors | Universal gain of AM experience<br>Contribution to Knorr-Bremse Eco design |

As the market does not demand a direct substitution of the existing manifold, the added value for customers and the company may be prioritized over the higher manufacturing costs when considering further steps to a serial application of an AM manifold. Overall, the design and manufacturing project is perceived as a success by Knorr-Bremse even if economic break-even is not reached yet. The gain in experience in design, software, printing, and post-processing is seen as highly valuable and can be transferred to other parts. With decreasing AM costs in the future, the economic assessment is expected to change.

## 4. Discussion

The universal applicability of the proposed study is a central requirement of the development methodology. However, in the present study, the methodology was only applied to one use case. To fully confirm its universal validity, it must be applied to applications of various industries. The achieved reduction in weight and size of the given application as well as the clear, economic foundation for future decisions indicate a positive efficiency of the proposed development methodology. Through the step-by-step process, many design-enabled potentials are realized, while the application of restrictive DfAM guidelines leads to reliable manufacturability. The resulting part embodies an optimum utilization of the given LPBF system in terms of manufacturing quality and economic efficiency. Its added value for the customer and the manufacturer is considered in decisions about further steps. As the presented case study is closed with the successful fabrication of the first prototype part, the desired iterative information flow is only utilized partly. During the execution of the embodied design phase, small iterations based on experience gain during the development and pre-study were possible. However, most of the information derived from the technical and economic assessment can only be utilized for a second, future version of the manifold.

## 5. Summary and Outlook

In this study, a holistic development methodology for hydraulic manifolds and fluid components manufactured by LPBF is proposed. The methodology aims to lead systematically to a manifold design, which embodies improved product characteristics through advanced design while ensuring reliable and economic manufacturability by a given LPBF system. After a general introduction of the development methodology, the process is applied to re-design a hydraulic manifold of the rail industry. After the specification of the re-design task, an extensive conceptual phase explores opportunistic design potentials. A numeric assessment of possible concepts leads to a specific concept for the embodied design. In this phase, the individual elements of the manifold are designed step-by-step under consideration of restrictive DfAM guidelines and tools. Through this, the weight of the manifold is reduced by 76% and its width by 18%. Furthermore, the manifold inside provides additional oil-reservoir volume, with which, either a greater variety of actuators can be served, or the assembly size can be reduced.

The acquired roadmap, which assesses multiple economic aspects, provides a simple, yet holistic foundation for strategic decisions concerning the further industrialization of the developed part.

Based on the findings of the solution assessment, areas of improvement can be derived. Further research and development regarding data management, software-supported channel design, as well as fatigue resistance of the designed geometries needs to be conducted. Additionally, the improved flow characteristics through short channels without 90° corners have to be quantified in a suitable test bench.

**Author Contributions:** Conceptualization, N.R., M.S. (Matthias Schmitt) and M.S. (Matthias Schneck); methodology, N.R. and M.S. (Matthias Schneck); software, N.R.; validation, N.R. and M.S. (Matthias Schneck); writing—original draft preparation, N.R. and M.S. (Matthias Schmitt); writing—review and editing, M.S. (Matthias Schneck), G.S., J.S.; supervision, J.S. All authors have read and agreed to the published version of the manuscript.

**Funding:** This research was partly funded by Knorr-Bremse rail vehicle systems among the provision of the use-case and its manufacturing by LPBF.

**Data Availability Statement:** The raw/processed data required to reproduce these findings cannot be shared at this time as the data also forms part of an ongoing study.

**Acknowledgments:** The authors gratefully acknowledge the help of Knorr-Bremse for providing the AM infrastructure and specifically Torsten Stickel for his contribution of product and development knowledge to the project.

**Conflicts of Interest:** The authors declare no conflict of interest. The funders had no role in the design of the study; in the collection, analyses, or interpretation of data; in the writing of the manuscript, or in the decision to publish the results.

## References

1. *DIN EN ISO/ASTM 52900: Additive Fertigung—Grundlagen—Terminologie*; Beuth Verlag GmbH: Berlin, Germany, 2018.
2. Wohlers, T.T.; Campbell, I.; Diegel, O.; Huff, R.; Kowen, J. *Wohlers Report 2021: 3D Printing and Additive Manufacturing Global State of the Industry*; Wohlers Associates: Fort Collins, CO, USA, 2021.
3. Thompson, M.K.; Moroni, G.; Vaneker, T.; Fadel, G.; Campbell, R.I.; Gibson, I.; Bernard, A.; Schulz, J.; Graf, P.; Ahuja, B.; et al. Design for Additive Manufacturing: Trends, opportunities, considerations, and constraints. *CIRP Ann.* **2016**, *65*, 737. [CrossRef]
4. Gibson, I.; Rosen, D.W.; Stucker, B. *Additive Manufacturing Technologies: 3D Printing, Rapid Prototyping, and Direct Digital Manufacturing*; Springer: Berlin/Heidelberg, Germany, 2015.
5. Kumke, M. *Methodisches Konstruieren von Additiv Gefertigten Bauteilen*; Springer: Berlin/Heidelberg, Germany, 2018.
6. Kamps, T.; Gralow, M.; Seidel, C.; Reinhart, G. Systematische Bionische Bauteilgestaltung zur Ausschöpfung des AM-Designpotenzials/Leveraging AM-induced Design Potential through Systematic Biomimetic Part Design. In *Rapid.Tech—International Trade Show & Conference for Additive Manufacturing, Proceedings of the 13th Rapid.Tech Conference Erfurt, Erfurt, Germany, 14–16 June 2016*; Hanser: München, Germany, 2016; p. 171.
7. VDI 2221 Blatt 1. Design of Technical Products and Systems—Model of Product Design. 2019. Available online: https://www.vdi.de/richtlinien/details/vdi-2221-blatt-1-entwicklung-technischer-produkte-und-systeme-modell-der-produktentwicklung#:~{}:text=Die%20Richtlinie%20behandelt%20Grundlagen%20der,in%20der%20industriellen%20Praxis%20darstellen (accessed on 18 January 2021).
8. VDI 2221 Blatt 2. Design of Technical Products and Systems—Configuration of Individual Product Design Processes. 2019. Available online: https://www.vdi.de/richtlinien/details/vdi-2221-blatt-2-entwicklung-technischer-produkte-und-systeme-gestaltung-individueller-produktentwicklungsprozesse (accessed on 18 January 2021).
9. Beyer, C. Strategic Implications of Current Trends in Additive Manufacturing. *J. Manuf. Sci. Eng.* **2014**, *136*. [CrossRef]
10. Laverne, F.; Segonds, F.; Anwer, N.; Le Coq, M. Assembly Based Methods to Support Product Innovation in Design for Additive Manufacturing: An Exploratory Case Study. *J. Mech. Des.* **2015**, *137*. [CrossRef]
11. Kranz, J.; Herzog, D.; Emmelmann, C. Design guidelines for laser additive manufacturing of lightweight structures in $TiAl_6V_4$. *J. Laser Appl.* **2015**, *27*. [CrossRef]

12. Adam, G.A.O. *Systematische Erarbeitung von Konstruktionsregeln für die Additiven Fertigungsverfahren Lasersintern, Laserschmelzen und Fused Deposition Modeling*, 1st ed.; Shaker: Aachen, Germany, 2015.
13. Adam, G.A.O.; Zimmer, D. On design for additive manufacturing: Evaluating geometrical limitations. *Rapid Prototyp. J.* **2015**, *21*, 662. [CrossRef]
14. Thomas, D. *The Development of Design Rules for Selective Laser Melting: The Development of Design Rules for Selective Laser Melting*; University of Wales: Wales, UK, 2009.
15. VDI 3405. Additive Manufacturing Processes, Rapid Manufacturing—Basics, Definitions, Processes. 2014. Available online: https://www.vdi.de/richtlinien/details/vdi-3405-additive-fertigungsverfahren-grundlagen-begriffe-verfahrensbeschreibungen (accessed on 5 February 2021).
16. Kausch, M. *Entwicklung Hochbelasteter Leichtbaustrukturen aus Lasergenerierten Metallischen Komponenten mit Faserverbundverstärkung*, 1st ed.; Verl. Wiss. Scripten: Auerbach, Germany, 2013.
17. Diegel, O.; Schutte, J.; Ferreira, A.; Chan, Y.L. Design for additive manufacturing process for a lightweight hydraulic manifold. *Addit. Manuf.* **2020**, *36*, 101446. [CrossRef]
18. Schmelzle, J.; Kline, E.V.; Dickman, C.J.; Reutzel, E.W.; Jones, G.; Simpson, T.W. (Re)Designing for Part Consolidation: Understanding the Challenges of Metal Additive Manufacturing. *J. Mech. Des.* **2015**, *137*. [CrossRef]
19. Zhang, C.; Wang, S.; Li, J.; Zhu, Y.; Peng, T.; Yang, H. Additive manufacturing of products with functional fluid channels: A review. *Addit. Manuf.* **2020**, *36*. [CrossRef]
20. Cooper, D.E.; Stanford, M.; Kibble, K.A.; Gibbons, G.J. Additive Manufacturing for product improvement at Red Bull Technology. *Mater. Des.* **2012**, *41*. [CrossRef]
21. Ehrlenspiel, K.; Meerkamm, H. *Integrierte Produktentwicklung: Denkabläufe, Methodeneinsatz, Zusammenarbeit*, 5th ed.; Hanser: München, Germany, 2013.
22. Deepnest. Opensource Nesting Software. Available online: https://deepnest.io/ (accessed on 1 June 2021).
23. Wittel, H.; Jannasch, D.; Voßiek, J.; Spura, C. *Roloff/Matek Maschinenelemente: Normung, Berechnung, Gestaltung*, 24th ed.; Springer Vieweg: Wiesbaden, Germany; Springer Fachmedien Wiesbaden: Wiesbaden, Germany, 2019.
24. Motte, D.; Diegel, O.; Nordin, A. *A Practical Guide to Design for Additive Manufacturing*; Springer: Singapore, 2020.
25. Snyder, J.C.; Stimpson, C.K.; Thole, K.A.; Mongillo, D. Build Direction Effects on Additively Manufactured Channels. *J. Turbomach.* **2016**, *138*, 051006. [CrossRef]
26. Schrandt, C.; Schulz, A.; Beckert, M.; Koppa, P. Berstdruckbestimmung an additiv gefertigten Bauteilen. In *Additive Fertigung von Bauteilen und Strukturen*; Springer Fachmedien Wiesbaden: Wiesbaden, Germany, 2017; p. 75.
27. Brandao, A.D.; Gumpinger, J.; Gschweitl, M.; Seyfert, C.; Hofbauer, P.; Ghidini, T. Fatigue Properties of Additively Manufactured AlSi10Mg—Surface Treatment Effect. *Procedia Struct. Integr.* **2017**, *7*, 58. [CrossRef]
28. Blinn, B.; Klein, M.; Gläßner, C.; Smaga, M.; Aurich, J.C.; Beck, T. An Investigation of the Microstructure and Fatigue Behavior of Additively Manufactured AISI 316L Stainless Steel with Regard to the Influence of Heat Treatment. *Metals* **2018**, *8*, 220. [CrossRef]
29. Zimmermann, M.; Müller, D.; Kirsch, B.; Greco, S.; Aurich, J.C. Analysis of the machinability when milling AlSi10Mg additively manufactured via laser-based powder bed fusion. *Int. J. Adv. Manuf. Technol.* **2021**, *112*, 989. [CrossRef]
30. Materialise Magics. Additive Manufacturing Data and Preperation Software. Available online: https://www.materialise.com/en/software/magics (accessed on 1 June 2021).
31. Zhu, Y.; Wang, S.; Zhang, C.; Yang, H. *Am-Driven Design of Hydraulic Manifolds: Enhancing Fluid Flow and Reducing Weight; Volume 1—Symposium*; Technische Universität Dresden: Dresden, Germany, 2020; p. 155.
32. Lakomiec, M. *Economic and Technical Assessment of Additive Manufactured Jet Engine Components for Serial Production*; Shaker Verlag: Düren, Germany, 2018.
33. Peng, T.; Wang, Y.; Zhu, Y.; Yang, Y.; Yang, Y.; Tang, R. Life cycle assessment of selective-laser-melting-produced hydraulic valve body with integrated design and manufacturing optimization: A cradle-to-gate study. *Addit. Manuf.* **2020**, *36*, 101530. [CrossRef]
34. EOS GmbH. EOS Aluminium AlSi10Mg: Materialdatenblatt. Available online: https://lightway-3d.de/wp-content/uploads/2019/11/LIGHTWAY_EOS_Aluminium_AlSi10Mg_de_Datenblatt.pdf (accessed on 19 March 2021).
35. Schneck, M.; Schmitt, M.; Schlick, G. Supply Chain and Cost Evaluation for Laser Powder Bed Fusion. In Proceedings of the 2020 IEEE 10th International Conference Nanomaterials: Applications & Properties (NAP), IEEE, Sumy, Ukraine, 9–13 November 2020.

applied sciences

MDPI

Article

# Integration of Topology Optimisation and Design Variants Selection for Additive Manufacturing-Based Systematic Product Redesign

**Enrico Dalpadulo, Francesco Gherardini *, Fabio Pini and Francesco Leali**

Department of Engineering "Enzo Ferrari", University of Modena and Reggio Emilia, Via Vivarelli 10, 41125 Modena, Italy; enrico.dalpadulo@unimore.it (E.D.); fabio.pini@unimore.it (F.P.); francesco.leali@unimore.it (F.L.)

* Correspondence: francesco.gherardini@unimore.it

Received: 14 October 2020; Accepted: 3 November 2020; Published: 5 November 2020

**Featured Application: This paper presents a redesign approach integrating design for additive manufacturing and topology optimisation, which supports the designer in product and process modification. Thanks to the integration of a systematic concept selection-based approach for product optimisation, the method provides an objective assessment of the developed configurations from a functional and production point of view. As a main goal, the method leads to cost and material reduction, without affecting product functionality. All these tools are implemented into a commercial computer aided technologies (CAx) platform, leading to a higher level of integration.**

**Abstract:** The development of additive manufacturing allows the transformation of technological processes and the redesign of products. Among the most used methods to support additive manufacturing, the design can be optimised through the integration of topology optimisation techniques, allowing for creating complex shapes. However, there are critical issues (i.e., definition of product and process parameters, selection of redesign variants, optimised designs interpretation, file exchange and data management, etc.) in identifying the most appropriate process and set-ups, as well as in selecting the best variant on a functional and morphological level. Therefore, to fully exploit the technological potentials and overcome the drawbacks, this paper proposes a systematic redesign approach based on additive manufacturing technologies that integrate topology optimisation and a tool for selecting design variants based on the optimisation of both product and process features. The method leads to the objective selection of the best redesigned configuration in accordance with the key performance indicators (KPIs) (i.e., functional and production requirements). As a case study, the redesign of a medical assistive device is proposed, previously developed in fused filament fabrication and now optimised for being 3D printed with selective laser melting.

**Keywords:** design for additive manufacturing; topology optimisation; design variants selection; redesign; design method; selective laser melting; assistive device; design optimisation

---

## 1. Introduction

Additive manufacturing (AM) technologies enable the construction of complex parts, such as high-performance parts or highly customised and specific parts [1,2]. Production of parts providing features of almost every function and shape becomes feasible [3]. Since part complexity and geometrical features have low impact on product cost and production time, construction of batches of parts different from the other becomes feasible. Moreover, the direct link between a computer aided design (CAD) model and a built part can be currently exploited in order to produce in a fast way both prototypes and end-use parts [4].

Therefore, several technologies have been replaced by AM, even if this replacement requires dedicated redesign of product and process. In the past, other technological changes affected the production of parts and components: as an example, in the 2000s, the metal replacement (or metal-to-plastic replacement) started to be one of the main industrial trends [5,6], involving all the industrial fields, and it is still an ongoing trend. Similarly, but most recently, we can state that the same role is played by the AM technologies. These trends share the main common goal of making parts lighter and more efficient to be manufactured. Moreover, AM can further benefit from its integration with topology optimisation (TO), leading to complex morphologies and free form models, with an increase in product customisation.

Many industrial sectors have been positively influenced by AM, e.g., aerospace and automotive, as well as the medical field, which can mainly benefit from parts produced specifically for individual patient's needs. Biomedical devices account for many applications [7] as well as challenges concerning prosthetics and implants development, surgical and diagnostic aids and even tissue and organ engineering [8,9], occupational therapy and rehabilitation [10,11]. A further example is the development of customised assistive devices (ADs), with positive integration of AM and co-design approaches [12]. All these changes open new frontiers as well as new drawbacks. With particular regards to a redesign process based on AM, the following main issues may occur [13]:

- Selection of redesign variants: The outputs of the TO consist of a great number of variants, which need to be managed and selected.
- Optimised designs interpretation: A wide number of topologically optimised design alternatives usually occur due to the redesign loops during the workflow iterations, requiring the designer to evaluate and interpret them into CAD models.
- Definition of product and process parameters: The designer is used to evaluate and optimise one parameter at a time or a single parameter category, considering separately functional and production requirements.
- File exchange and data management during the workflow iterations: The entire workflow, from the first model to final job creation, involves different software tools that need different file formats (as inputs), requiring multiple conversions of file formats due to iterations during the design and the industrialisation tasks. An STL file is used as interchange file between different software: as a support for CAD modelling parts starting from the results of TO, for finite elements analysis (FEA) simulation, for industrialisation tasks in the AM preparation software, for AM process simulation.

These practises, common in the industry, highlight the need for approaches, methods and tools to support the product optimisation and, at the same time, the process optimisation, by identifying the best-developed solution. Therefore, to exploit the benefits offered by the AM implementation, the concurrent development of technology and its specific knowledge is urged. Thus, design for additive manufacturing (DfAM) of proper methods, systematic tools, objective guidelines and approaches, as well as their implementation and application, is fundamental [14–16].

In this work, we share our experience with the development of a product redesign approach when replacing/changing the manufacturing technology and, then, its guiding design principles. As a case study, we focus on the development of an assistive device (AD), originally co-designed with patients, which failed some functional goals due to the original material and production process.

The paper is organised as follows: Firstly, the systematic method for the redesign of product is introduced in Section 2; Section 3 reports the case study and the tools involved; in Section 4, the results of the redesign method implementation are discussed; finally, the conclusions are summarised in Section 5.

## 2. Method

A product optimisation approach can be considered as a key phase of a general DfAM workflow [17], as synthesised in Figure 1. The workflow represents an iterative process based on product and process optimisation subjected to design and industrialisation constraints, aiming to exploit the potentials of the AM technology.

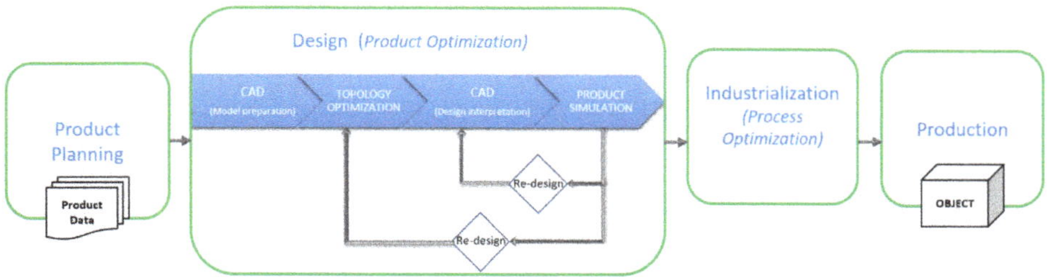


**Figure 1.** General design for additive manufacturing workflow.

The workflow consists of four main phases: The product planning phase leads to the complete definition of the product data, considering design targets and constraints, and including testing results and users' suggestions. Concerning the design phase, the product optimisation and validation tasks aim to achieve the established targets. In the industrialisation phase, process study and 3D printing preparation are mandatory, while additional optimisation can be performed. The production phase covers 3D printing and post-processing operations necessary to provide the user with the optimised product version.

Moreover, product optimisation can be obtained by implementing TO techniques [14,18] and is generally achieved through a series of redesigns. Both these operations are included among the most time-consuming activities and represent a complex research area due to automation needs [19,20]. Therefore, this phase can benefit from a systematic approach to better exploit the potentials offered by AM, while clarifying the decisional steps and speeding up the workflow. Figure 2 proposes a workflow that provides a design optimisation through a systematic concept selection-based approach [21]. The aim is to perform a design solution exploration by using TO and subsequently defining results selection criteria. Results represent conceptual solutions generated from TO. Finally, by integrating the decisional process into a structured system called trade-off study, just one final redesign can occur in order to perform product optimisation.

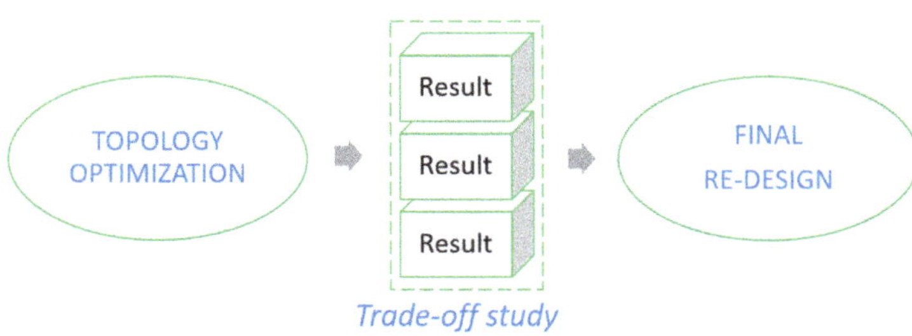


**Figure 2.** Systematic concept selection-based approach for product optimisation.

To perform the concept selection step, data collected from analyses and simulation are processed into spreadsheets to create key performance indicators (KPIs) matrices. The evaluated elements can be related not only to the product features but also to the associated processes. By assigning scores and weights to KPIs and depending on the design specifications, the systematic concept selection-based approach can be performed and integrated into the general DfAM workflow. Furthermore, the approach can be extended from DfAM to product redesign and design optimisation, therefore considering in general different product variants, manufacturing technologies or further studies.

## 3. Case Study—Can Opener Assistive Device

In this section, we present the integration of the proposed method into a co-design approach for the development of assistive devices (ADs) and the selected case study is described and analysed, defining design objectives and constraints and detailing the optimisation approach.

### 3.1. Method Integration

The method adopted to exploit AM potentials for development of customised ADs is based on a systematic approach [12]. Product customisation is made possible thanks to the patient's involvement through a co-design approach. The workflow is composed by sundry different tasks to be performed in sequence. These include mapping of the patient's needs, co-design phase, parametric modelling of the AD, validation process, finalisation of models and documentation (Figure 3). Two iterative redesign loops can occur to optimise the AD design. The former is based on results of the patient's preliminary validation of the first prototype. The latter can be performed after the 6-month follow-up and long-term use verification. Redesign loops provide implementation of design modifications required on morphological and/or functional features. A possible execution can be based on direct patient's inclusion for AD improvement [22], especially if functional targets are achieved, whereas ergonomic or aesthetic ones are not satisfied. Otherwise, at each of these stages, a redesign for AM that provides product optimisation (including process/material replacement) can be required to achieve the expected functional targets. For example, after prototyping and testing, some critical issues and suggestions of improvements can emerge and, at this point, product optimisation can occur.

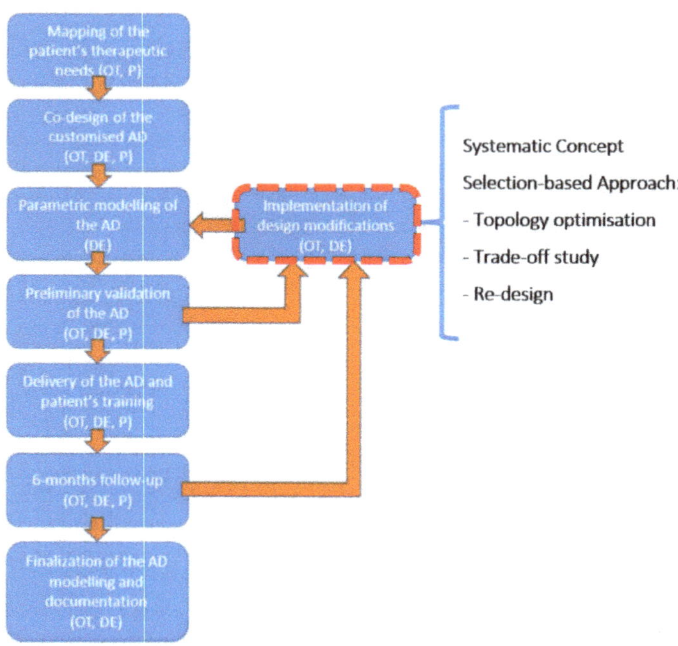


**Figure 3.** Workflow for the co-design of assistive devices, involving patients (P), occupational therapists (OT) and design engineers (DE). The dashed-line box highlights the iterative phases in which topology optimisation (TO), trade-off study and redesign may occur.

### 3.2. Original Part Analysis: Possible Issues

The original can opener may be functionally decomposed into two parts. The first interfacing with the can and its tab, and the second with the patient's hand (i.e., the handle) [12]. The first part is aimed at opening the can tab, with a lifter able to be inserted under the tab and lift it, and a geometry acting as a lever fulcrum. The second part of the can opener acts as a handle for the patient, and it can be resized in its section sizes and length in accordance with the patient's anatomy.

The parts had been made by fused filament fabrication (FFF) of poly-lactic acid (PLA), with the aim of containing production costs. This material is generally used for rapid prototyping applications due to its manufacturability (easiness of 3D printing), aesthetical quality, and low cost. Mechanical properties are quite good in terms of allowable stress, but 3D-printed PLA has poor impact resistance and elongation at break, and worse heat resistance. Nevertheless, the physical test of the AD and mostly its daily use over time brought some cases to part failure. The main issue regards the front hook detachment from the item head, but also the low wear resistance of PLA (FFF) affected the product durability.

To overcome these limitations, the device is required to be improved in its design working on structural strength. An FEA simulation of the can opener shows that the part failure is related to a high stress concentration on the front hook. In particular, Figure 4 shows that a force of 50 N (about 5 kg) applied at the handle causes a stress on the region that is over the threshold of the tensile strength (about 59 MPa) of the material.

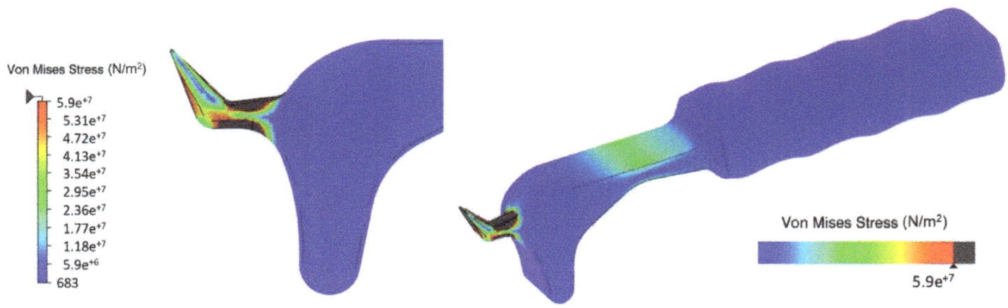


**Figure 4.** Structural analysis performed on the original assistive device (AD).

### 3.3. Design Objectives and Constraints

Firstly, to define Product Data for design optimisation, objectives and constraints are analysed:

1. The part general shape should be kept to preserve the AD functionality. In particular, the can opener had been functionally optimised in lifter and fulcrum size and shape, so these elements should not be modified.
2. To improve structural behaviour, a material change is required at least for the stressed elements, which have to be made of metal. Mean values of mechanical properties of some of the strongest materials processed by FFF have been analysed and compared to a stainless steel obtained by powder bed fusion (PBF), as shown in Table 1. Special filaments like Sabic Ultem® PEI [23] or PEEK [24] can reach optimal strength, with the issue of very high costs.
3. Since the AD has to be provided free of charge, costs have to be minimised. Minimum metal material usage is fundamental, according to eventual technological constraints to reduce production cost. Moreover, the material saving leads to lower costs for required energy and machine time.
4. To create the redesign of a part to be produced by AM, aiming at improving part strength while reducing its mass, freeform TO has been chosen as the key step for product optimisation.
5. The redesign is based on splitting the AD in two parts, which are the already the optimised handle, made of PLA, and a structural head featuring a beam insert, made of 316L Stainless Steel

(316L SS) (Figure 5). The parts have to be additively manufactured, respectively, by FFF and PBF technologies, and then they can be coupled and glued.

**Table 1.** Material comparison for additive manufacturing process: fused filament fabrication (FFF) and powder bed fusion (PBF).

| Material | PLA (FFF) | PC (FFF) | Special (FFF) | 316L (PBF) |
|---|---|---|---|---|
| Density (kg/m$^3$) | 1.24 | 1.2 | 1.32 | 7.91 |
| Young's modulus (GPa) | 2.15 | 2.25 | 3.5 | 180 |
| Ultimate tensile strength (MPa) | 59 | 68 | 100 | 530 |
| Specific Cost (EUR/kg) | 25 | 50 | 500 | 150 |

**Figure 5.** Layout of the product to be optimised.

### 3.4. Optimisation Approach and Tools

A general DfAM approach based on TO can be implemented using an integrated platform to perform design optimisation, since it could provide benefits speeding up the DfAM workflow due to data management and exchange minimisation [13]. The platform allows integration of TO, product simulation, printing preparation and process simulation tools into the CAD environment [17]. Moreover, it provides a specific feature to generate data that can be manipulated in order to perform the aforementioned trade-off study step. We selected the Dassault Systèmes 3DExperience platform.

Once objectives and constraints have been defined, a concept development phase based on optimisation computing, structural validation and process-related remarks is required. Subsequently, the trade-off study based on results of TO can be used to select the final concept model. Finally, detail modelling is required to obtain the optimised component and then physical prototyping can start.

## 4. Results

### 4.1. Model Setup

To simulate part behaviour during the can opening, the structural modelling consists of a load positioned at the centre of the handle with vertical direction (load condition), a hinge allowing rotation on the fulcrum and a fixed displacement restraint in the vertical direction on the top of the lifter (restraints), as shown in Figure 6.

The Design Space for computation originates from the original can opener model: it should be as large as possible to avoid over-constraining the optimisation. Moreover, three functional elements are introduced as Non-Design Space volumes: the beam insert introduced to couple with the plastic handle and, according to the functional design constraints, the optimised lifter and fulcrum size and shape.

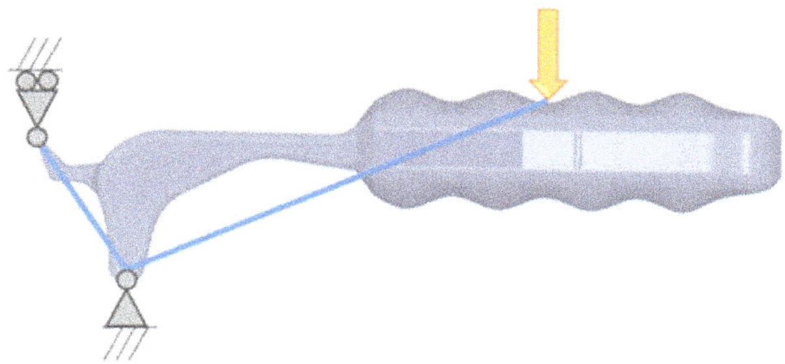

**Figure 6.** Structural model for shape computation and simulation.

### 4.2. Topology Optimisation

A computational algorithm is set up to maximise part stiffness for a fixed target mass. Firstly, sunder iterations between optimisation and structural validation lead to the establishment of target mass value, as depicted in Figure 7 (top). First computations are based on a rough mesh in order to reduce computational cost whereas subsequent models are refined by creating a finer mesh in order to obtain more accurate results.

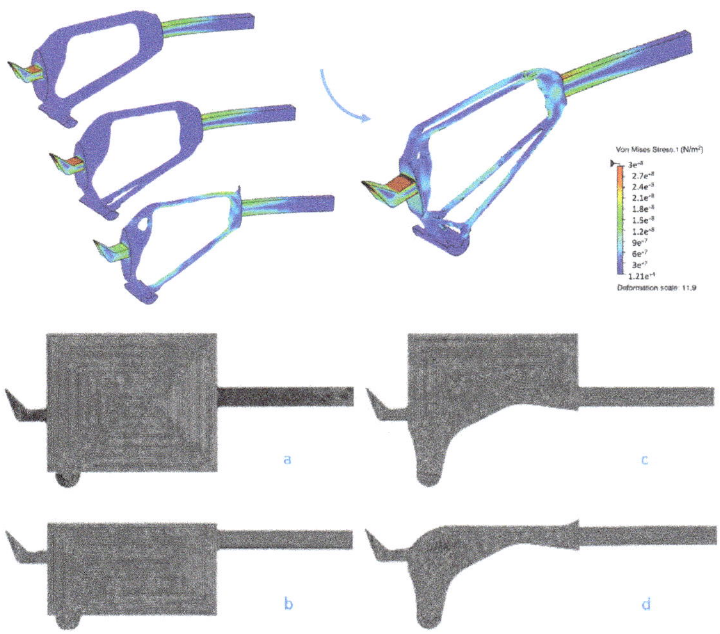


**Figure 7.** (Top): Topology optimisation refinement of the concept (**a**) and (bottom): a comparison of the enlarged (**a**), the top-limited (**b**), the bottom-limited (**c**) and the fully limited (**d**) design spaces.

TO is run on four design spaces, ranging between the design space shape constrained by the original part profile and an enlarged design space in order to study the ideal shape from a structural point of view. Figure 7 (bottom) shows firstly the enlarged design space (a), which corresponds to a volume that includes the original can opener model but is limited on three sides by the functional elements, which cannot be modified. Then, the top-limited design space (b) considers only the volume below the upper face of the original model while, conversely, the bottom-limited one (c) cannot exceed the can opener's lower profile. Finally, the fully limited design space (d) cannot exceed the original part profile. The design space should bring to the result the best structural performance. It is fundamental to preserve the lower profile (i.e., the fulcrum), since it is given by the part envelope during the can opening. Conversely, for the upper shape, further constraints are not necessary, and, in this way, optimisation can be less constrained, so both results from design space (c) and (d) are feasible. TO setup is based on a target mass around 25 g, and a process-related constraint of minimum thickness for computed elements of 2.5 mm.

### *4.3. Concept Simulation*

The results from TO have to be validated from a structural point of view. The use of an integrated platform allows one to obtain a finite element simulation directly on the optimisation output. A tool for automatic generation of a solid model starting from the computed shape is used for this purpose. Since a manual modelling for geometry interpretation is not required before having a structural validation, the exploration process of possible solutions is sped up. Figure 8 depicts results from the finite element analyses of concept models, respectively, from the enlarged (a), the top-limited (b), the bottom-limited (c) and the fully limited (d) design spaces. Maximum permissible stress is 260 MPa, calculated with a safety factor = 2 from the 316L SS tensile strength.

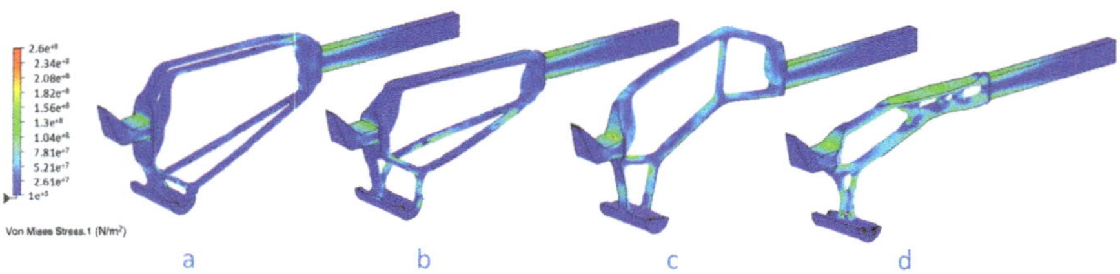


**Figure 8.** Concepts generated from TO results starting from the enlarged (**a**), the top-limited (**b**), the bottom-limited (**c**) and the fully limited (**d**) design spaces.

Concept models are validated since the stress of each one is below the maximum permissible. As expected, the unconstrained model (a) shows the best performance, while the fully limited one (d) shows the worst stiffness. Figure 9 depicts part deformations and shows that concept (a) has a maximum displacement less than half of concept (d). Figure 9 depicts in particular a high level of strain energy on its upper surface—a sign that the shape constraint in that region badly affects the part behaviour.

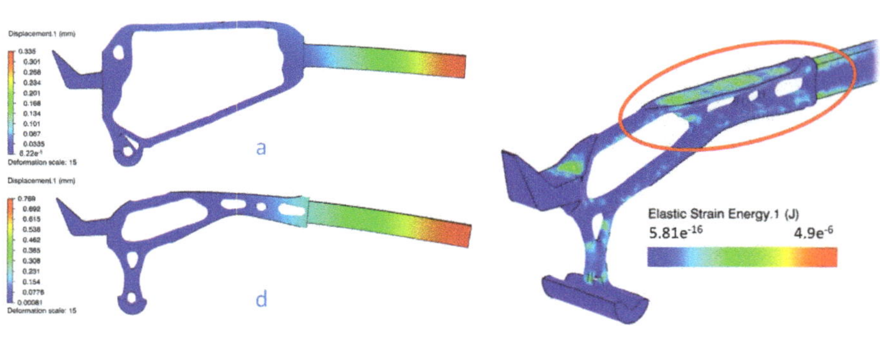


**Figure 9.** Deformation of concepts (**a**) and (**d**) and elastic strain energy of concept (**d**).

### *4.4. Concept Industrialisation*

An industrialisation study on concept shapes is necessary to evaluate the amount of supports required to produce the parts, since their introduction leads to higher material usage (as well as energy and machine time) and increases AD costs. To do that, as shown in Figure 10, printing preparation of models has been performed, including proper part orientation and generation of wire type perforated supports. The model's orientation is selected for both component's strength maximisation (considering material anisotropy) and support material minimisation (considering process costs).

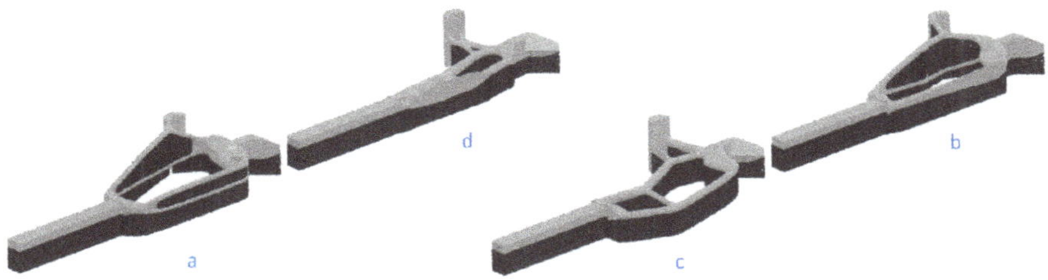


**Figure 10.** Industrialisation study of concepts (**a**–**d**).

The volume of each part, including its supports, has been calculated to quantitatively evaluate variation in usage of material, energy and machine time, under the hypothesis of the linear link between build time and material to be processed.

## 4.5. Trade-Off Study

Concept models have to be evaluated on the basis of part mass, its structural features, but also on a technological point of view. Many actual design efforts aim at combining both structural and economical aspects [25]. In this way, a first analysis includes the design performance (structural aspect) and a second one, the industrialisation (economical aspect). Figure 11 summarises the performed trade-off study based on both the product and process elements, considering the four concepts (a in its third refinement version (V3), b, c, and d).

| Trade-off study | | | | |
|---|---|---|---|---|
| Concept | a (V3) | b | c | d |
| 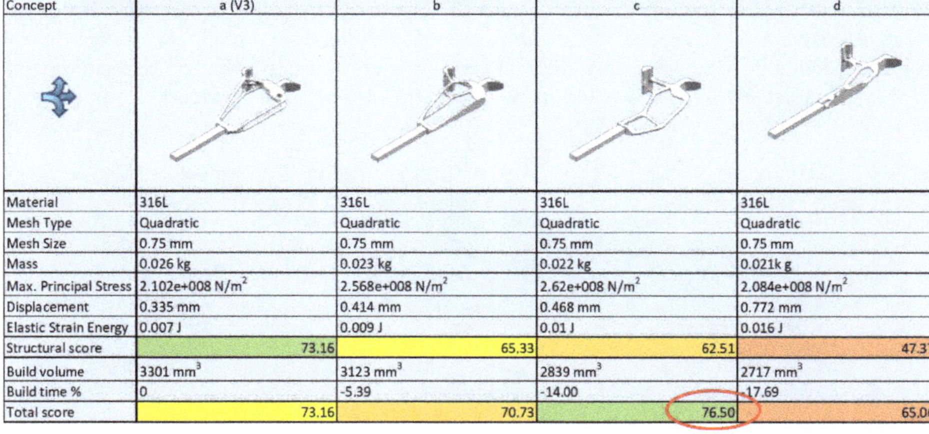 | | | | |
| Material | 316L | 316L | 316L | 316L |
| Mesh Type | Quadratic | Quadratic | Quadratic | Quadratic |
| Mesh Size | 0.75 mm | 0.75 mm | 0.75 mm | 0.75 mm |
| Mass | 0.026 kg | 0.023 kg | 0.022 kg | 0.021k g |
| Max. Principal Stress | 2.102e+008 N/m$^2$ | 2.568e+008 N/m$^2$ | 2.62e+008 N/m$^2$ | 2.084e+008 N/m$^2$ |
| Displacement | 0.335 mm | 0.414 mm | 0.468 mm | 0.772 mm |
| Elastic Strain Energy | 0.007 J | 0.009 J | 0.01 J | 0.016 J |
| Structural score | 73.16 | 65.33 | 62.51 | 47.37 |
| Build volume | 3301 mm$^3$ | 3123 mm$^3$ | 2839 mm$^3$ | 2717 mm$^3$ |
| Build time % | 0 | -5.39 | -14.00 | -17.69 |
| Total score | 73.16 | 70.73 | 76.50 | 65.06 |

**Figure 11.** Key performance indicators (KPIs) matrix involved in the trade-off study step: the four concepts (**a**–**d**) are compared in the columns.

The first analysis is based on KPIs such as mass, part compliance and elastic strain energy minimisations; in addition, a structural score has been calculated. The result, as expected, is that concept (a) shows the best topology and the best performance: while not the lightest in mass, it has the best stress state, stiffness and the lowest strain energy. Subsequently, the volume of each part, including its supports, has been considered to quantitatively evaluate changes in material, energy and machine time usage. Then, the calculated score has been added to the first to obtain a total score for concept selection, taking into account even production costs. Results from the first analysis change, for example concept (a), which showed the best structural score, is the one with the maximum material usage. In particular, concept (c), despite having quite lower features with regard to stiffness, stress state and strain energy, shows lower mass (just higher than concept (d)) and strongly requires less support material, so the estimated build time is reduced by 14%. Considering both analyses, the best total score is assigned to (c) and thus it has been selected for the final design.

### 4.6. Final Design

Once the final concept selection is performed, redesign operations are required. Usually, the application of the general DfAM workflow using specific stand-alone tools requires many redesigns. Conversely, through a CAD platform it can be possible to carry out the whole design solutions exploration without making geometry interpretations and thus limiting the redesign step to the selected concept solution. This can lead to high effectiveness of the method and reduction in product development time. Figure 12 depicts the final design for the topology optimised structural head and the model of the AD with the handle assembled.

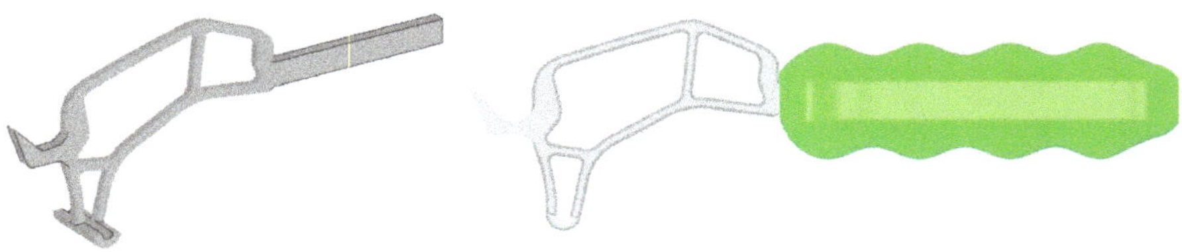

**Figure 12.** Final models of structural head and assistive device.

### 4.7. Production and Testing

Concerning the production phase, the process setup of the structural metal head to be built by PBF defined in the industrialisation study is achieved. Conversely, the plastic handle is built by a rapid prototyping standard FFF process. As previously mentioned, the head is made of 316L SS while the handle is made of PLA. Subsequently, the post-processing phase is performed on both components and they are glued. Figure 13 shows the printed and assembled optimised product.

**Figure 13.** Three-dimensional printed optimised assistive device.

Finally, a physical test for final validation of the optimised product is required. Subsistence of part functionality is verified, and structural strength and wear resistance specifications are respected. Figure 14 shows the AD usage operations.

**Figure 14.** Optimised assistive device testing.

## 5. Conclusions

This paper presents a systematic workflow for the redesign of a part or component and its additive manufacturing production, after being topologically optimised. Starting from a traditional manufacturing technology, the workflow leads to technology change and material replacement.

Today, AM is used in all the fields requiring single-batch production and maximum customisation, thanks to its inner production potentials. Moreover, the use of TO leads to lightweight structures, even if it generates more than one design variants with similar performance, so design selection criteria are urged, such as the one presented here.

The main pros of the workflow are:

- It can be integrated into other existing methods for product and process optimisation.
- It explores all the solutions developable thanks to AM but, at the same time, it offers their structured evaluation process.
- KPIs can be integrated allowing one to respond to specific design needs.
- The method requires dedicated tools and is favoured by an integrated platform because it addresses the entire workflow, but it can also be implemented with different and stand-alone tools.
- The use of an integrated platform leads to shortened times, reduced validation/optimisation times because it is managed systematically.
- The method is independent from the geometry to be redesigned, so it can fit with different fields of applications; in this paper, it is applied to the redesign of a medical AD but can easily be extended to other sectors.

Regarding the specific case study, the results show:

- Increased performance and achievement of the functional goals.
- On the one hand, the new technology replaces the previous process but, on the other hand, it integrates with it by splitting the component into two parts (the method acts only on the most critical one).
- The redesign preserves and indeed improves the primary functionality of the AD (structural aspect) and responds to further needs, e.g., durability over time.

As a main con, the construction costs are high and not convenient for large-scale production, even if the workflow allows for cost minimisation thanks to TO and the cost itself is partly compensated by the part durability, since the original part requires frequent replacements.

**Author Contributions:** Conceptualisation, E.D., F.G., F.P. and F.L.; methodology, E.D., F.P. and F.G.; validation, E.D., F.G. and F.P.; writing—original draft preparation, E.D. and F.G.; writing—review and editing, E.D., F.G. and F.P.; supervision, F.L. All authors have read and agreed to the published version of the manuscript.

**Funding:** This research and the article processing charge ere funded by University of Modena and Reggio Emilia—UNIMORE (Italy), Fondo Ateneo Ricerca Progetti FAR Dipartimentale 2019-PROT. 192218 del 20/09/19-CdA 23/07/19-Bando interno.

**Acknowledgments:** The authors gratefully acknowledge Elena Bassoli and Riccardo Groppo for their valuable support, and Francesco Prochilo for the photographs of Figures 13 and 14.

**Conflicts of Interest:** The authors declare no conflict of interest.

## References

1. Gao, W.; Zhang, Y.; Ramanujan, D.; Ramani, K.; Chen, Y.; Williams, C.B.; Wang, C.C.L.; Shin, Y.C.; Zhang, S.; Zavattieri, P.D. The status, challenges, and future of additive manufacturing in engineering. *CAD Comput. Aided Des.* **2015**, *69*, 65–89. [CrossRef]
2. Thompson, M.K.; Moroni, G.; Vaneker, T.; Fadel, G.; Campbell, R.I.; Gibson, I.; Bernard, A.; Schulz, J.; Graf, P.; Ahuja, B.; et al. Design for Additive Manufacturing: Trends, opportunities, considerations, and constraints. *CIRP Ann. Manuf. Technol.* **2016**, *65*, 737–760. [CrossRef]
3. du Plessis, A.; Broeckhoven, C.; Yadroitsava, I.; Yadroitsev, I.; Hands, C.H.; Kunju, R.; Bhate, D. Beautiful and functional: A review of biomimetic design in additive manufacturing. *Addit. Manuf.* **2019**, *27*, 408–427. [CrossRef]
4. Chen, D.; Heyer, S.; Ibbotson, S.; Salonitis, K.; Steingrímsson, J.G.; Thiede, S. Direct digital manufacturing: Definition, evolution, and sustainability implications. *J. Clean. Prod.* **2015**, *107*, 615–625. [CrossRef]
5. Desai, K. Factors influencing success of metal to plastic conversion programs for under-hood applications. *SAE Tech. Pap.* **1996**, 960149. [CrossRef]
6. Marur, S. *Plastics Application Technology for Lightweight Automobiles*; SAE International: Warrendale, PA, USA, 2013; ISBN 978-0-7680-7640-0.
7. Ahangar, P.; Cooke, M.E.; Weber, M.H.; Rosenzweig, D.H. Current biomedical applications of 3D printing and additive manufacturing. *Appl. Sci.* **2019**, *9*, 1713. [CrossRef]
8. Gibson, I.; Rosen, D.; Stucker, B. *Additive Manufacturing Technologies: 3D Printing, Rapid Prototyping, and Direct Digital Manufacturing*; Springer: New York, NY, USA, 2015. [CrossRef]
9. Tack, P.; Victor, J.; Gemmel, P.; Annemans, L. 3D-printing techniques in a medical setting: A systematic literature review. *Biomed. Eng.* **2016**, *15*, 115. [CrossRef] [PubMed]
10. Lunsford, C.; Grindle, G.; Salatin, B.; Dicianno, B.E. Innovations with 3-dimensional printing in physical medicine and rehabilitation: A review of the literature. *PM&R* **2016**, *8*, 1201–1212. [CrossRef]
11. Ganesan, B.; Al-Jumaily, A.; Luximon, A. 3D printing technology applications in occupational therapy. *Phys. Med. Rehabil. Int.* **2016**, *3*, 1085–1089.
12. Gherardini, F.; Mascia, M.T.; Bettelli, V.; Leali, F. A co-design method for the additive manufacturing of customised assistive devices for hand pathologies. *J. Integr. Des. Process Sci.* **2019**, *22*, 21–37. [CrossRef]
13. Dalpadulo, E.; Pini, F.; Leali, F. Assessment of design for additive manufacturing based on CAD platforms. *Lect. Notes Mech. Eng.* **2020**, 970–981. [CrossRef]
14. Plocher, J.; Panesar, A. Review on design and structural optimisation in additive manufacturing: Towards next-generation lightweight structures. *Mater. Des.* **2019**, *183*, 108164. [CrossRef]
15. Laverne, F.; Segonds, F.; Anwer, N.; Le Coq, M. DfAM in the design process: A proposal of classification to foster early design stages. In Proceedings of the Conference 2014 Croatie, Sibenik, Croatie, 3–4 July 2014; pp. 1–12.
16. Kumke, M.; Watschke, H.; Vietor, T. A new methodological framework for design for additive manufacturing. *Virtual Phys. Prototyp.* **2016**, *11*, 3–19. [CrossRef]
17. Dalpadulo, E.; Pini, F.; Leali, F. Integrated CAD platform approach for design for additive manufacturing of high performance components. *Int. J. Interact. Des. Manuf.* **2020**, *14*, 899–909. [CrossRef]
18. Liu, J.; Gaynor, A.T.; Chen, S.; Kang, Z.; Suresh, K.; Takezawa, A.; Li, L.; Kato, J.; Tang, J.; Wang, C.C.L.; et al. Current and future trends in topology optimization for additive manufacturing. *Struct. Multidiscip. Optim.* **2018**, *57*, 2457–2483. [CrossRef]
19. Wiberg, A.; Persson, J.; Ölvander, J. Design for additive manufacturing—A review of available design methods and software. *Rapid Prototyp. J.* **2019**. [CrossRef]
20. Lindemann, C.; Reiher, T.; Jahnke, U.; Koch, R. Towards a sustainable and economic selection of part candidates for additive manufacturing. *Rapid Prototyp. J.* **2015**, *21*, 216–222. [CrossRef]

21. Dalpadulo, E.; Pini, F.; Leali, F. Systematic integration of topology optimization techniques in design for additive manufacturing methodologies applied to automotive settings. In Proceedings of the International Mechanical Engineering Congress and Exposition (IMECE)—2020, Virtual Conference, Portland, OR, USA, 16–19 November 2020.
22. Gherardini, F.; Petruccioli, A.; Dalpadulo, E.; Bettelli, V.; Mascia, M.T.; Leali, F. A methodological approach for the design of inclusive assistive devices by integrating co-design and additive manufacturing technologies. In *Advances in Intelligent Systems and Computing, Proceedings of the IHSI 2020, Modena, Italy, 19–21 February 2020*; Springer: New York, NY, USA, 2020; Volume 1131, pp. 816–822. [CrossRef]
23. Padovano, E.; Galfione, M.; Concialdi, P.; Lucco, G.; Badini, C. Mechanical and thermal behavior of ultem®9085 fabricated by fused-deposition modeling. *Appl. Sci.* **2020**, *10*, 3170. [CrossRef]
24. Rinaldi, M.; Ghidini, T.; Cecchini, F.; Brandao, A.; Nanni, F. Additive layer manufacturing of poly (ether ether ketone) via FDM. *Compos. Part B Eng.* **2018**, *145*, 162–172. [CrossRef]
25. Cicconi, P.; Castorani, V.; Germani, M.; Mandolini, M.; Vita, A. A multi-objective sequential method for manufacturing cost and structural optimization of modular steel towers. *Eng. Comput.* **2020**, *36*, 475–497. [CrossRef]

**Publisher's Note:** MDPI stays neutral with regard to jurisdictional claims in published maps and institutional affiliations.

applied sciences

MDPI

Review

# Topology Optimisation in Structural Steel Design for Additive Manufacturing

Tiago P. Ribeiro [1,2,*], Luís F. A. Bernardo [2] and Jorge M. A. Andrade [2]

[1] Tal Projecto Lda., 1350-252 Lisbon, Portugal
[2] Centre of Materials and Building Technologies (C-MADE), Department of Civil Engineering and Architecture, University of Beira Interior, 6201-001 Covilhã, Portugal; lfb@ubi.pt (L.F.A.B.); jandrade@ubi.pt (J.M.A.A.)
* Correspondence: tiago.pinto.ribeiro@ubi.pt

**Abstract:** Topology Optimisation is a broad concept deemed to encapsulate different processes for computationally determining structural materials optimal layouts. Among such techniques, Discrete Optimisation has a consistent record in Civil and Structural Engineering. In contrast, the Optimisation of Continua recently emerged as a critical asset for fostering the employment of Additive Manufacturing, as one can observe in several other industrial fields. With the purpose of filling the need for a systematic review both on the Topology Optimisation recent applications in structural steel design and on its emerging advances that can be brought from other industrial fields, this article critically analyses scientific publications from the year 2015 to 2020. Over six hundred documents, including Research, Review and Conference articles, added to Research Projects and Patents, attained from different sources were found significant after eligibility verifications and therefore, herein depicted. The discussion focused on Topology Optimisation recent approaches, methods, and fields of application and deepened the analysis of structural steel design and design for Additive Manufacturing. Significant findings can be found in summarising the state-of-the-art in profuse tables, identifying the recent developments and research trends, as well as discussing the path for disseminating Topology Optimisation in steel construction.

**Keywords:** topology optimisation; topology; steel design; steel structures; optimisation; design methods; design for additive manufacturing; connections; civil engineering

**Citation:** Ribeiro, T.P.; Bernardo, L.F.A.; Andrade, J.M.A. Topology Optimisation in Structural Steel Design for Additive Manufacturing. *Appl. Sci.* **2021**, *11*, 2112. https://doi.org/10.3390/app11052112

Academic Editor: Marco Mandolini

Received: 31 January 2021
Accepted: 23 February 2021
Published: 27 February 2021



## 1. Introduction

### *1.1. The Origins of Topology Optimisation*

Notwithstanding historical perspectives, as discussed in [1,2], which root Structural Optimisation and Topology Optimisation (TO) at the very beginning of the classical theory of elasticity, it is usually well accepted that TO had its *de facto* surgency under the name of Optimal Layout Theory and denoted the ability to optimise not only the structural elements shape and size but also its layout. That dates back to successful attempts made by Prager [3,4] and Rozvany [5–8] from the early 1970s to late 1980s, to generalise the 1904s Michell theory for weight optimisation of thin bars (Figure 1a) [9,10], which is based on the Maxwell theorem for frames [11].

Almost simultaneously, Pedersen pioneered the optimal layout design for trusses [12]. In addition, Olhoff [13] managed to optimise Kirchoff equations solutions for finding the plates optimal thickness, leading to ribbed solutions of good value for the aerospace industry. Further developments were endeavoured by Rozvany and Olhoff, among others [14–16].

After these early developments and proven accomplishments in aerospace structures design, TO experienced rapid growth at the beginning of the 1990s as it became easily distinguishable from Shape or Size Optimisation concepts. The latter already well-disseminated the design practice required near-optimal initial topologies and would yield no better than intuitive final layouts, unlike TO [17,18].

Such revolutionary abilities justify the TO massification in several engineering disciplines throughout the product design and manufacture. The automotive industry soon followed the aerospace and applications widened to medical devices and personalised medicine, defence, electronics, several kinds of consumer goods and mechanical engineering endeavours, new materials design, and even arts and architecture. Civil and structural engineering may be latecomers after an auspicious start with truss optimisation investigations [12,19], but also show an accelerating trend in the TO application. Currently, the fourth industrial revolution and its reliance on Additive Manufacturing (AM) processes on the opposition to standard subtractive processes, face significant challenges, as more advanced, scalable, and user-friendly design methods are required to unleash its incredible potential. TO is undoubtedly an adequate answer for such an enterprise. Its systematic use may overcome the knowledge barriers still moving many practitioners away from AM, as better described by Pradel et al. [20].

### 1.2. Topology Optimisation Modern Age

Formerly described developments were only made possible by the advances on the homogenisation method, by Bendsøe and Kikuchi in 1988 [21,22] and Suzuki and Kikuchi [23]. In simple terms, the homogenisation method is deemed to solve a material distribution problem, as previously formulated by Kohn and Strang [24–27], considering either only two states: The presence or the absence of the structural material. A more profound explanation and interesting example can be found in Bendsøe and Rozvany books [28,29].

The homogenisation method, along with Svanberg's Method of Moving Asymptotes (MMA) [30], became the basis of most of the following applications, developments, and TO commercial software. However, in 1993, Xie and Steven [31] proposed the so-called Evolutionary Structural Optimisation (ESO) procedure as a possibility for simplifying TO computations by mimicking natural evolutionary processes in Finite Element Analyses (FEA). This led to significant criticism from many other researchers after some shortcomings were identified [32–35].

As the discussion proceeded on the ESO method, justifications were debated, and enhancements have been proposed [36–40]. However, the results of evolutionary methods continue to drive some intense discussions in Academia [41].

Fundamental studies on methods and approaches continued with the work by Duysinx and Bendsøe [42] on the continua TO under stress constraints, as well as in related numerical issues [32,43]. After these fundamental works, TO had a significant growth as a discipline, and the distinction between its two significant sub-fields, the Discrete Optimisation and the Optimisation of Continua became less evident. While the first of those two concepts is deemed to optimise a finite number of known elements and is preferred in several practical applications [44], the second unrestrictedly optimises topology within a solid.

Meanwhile, applications-focused research flourished, with optimisation early works being published in materials design [45], compliant mechanisms [46,47], electronics [48], connections positioning and design [49], buckling phenomena and truss design [50,51] [52] and alternative approaches for the ribbed plate problem [53].

The new millennium brought significant progress in TO algorithms but also in modelling and freely or cheaply accessible software, such as Sigmund's MATLAB code [54,55] or the Karamba plug-in for the Grasshopper environment [56].

Discrete Optimisation had some essential developments, both with algorithms development [57] and practical applications [58,59]. The former depicts an interesting option for Discrete Optimisation in the automotive industry and a good "how-to" example. However, significant numerical problems can arise when node positioning is also considered an optimisable parameter [60].

Within the Optimisation of Continua, many valuable works could be highlighted. Yet, one cannot omit important developments in filters [61,62], projection methods [63,64] computational methods [65–67], as well as in the controversy evolutionary approach for continua [68]. Furthermore, practical methods, such as Coelho et al.'s useful model with global and local levels of intervention [69], as well as applications in form-finding [70], stiffened plates [71], and TO under load position uncertainty [72], proved the suitability of TO methods for industrial execution. In fact, all the former offer solutions and examples for managing essential aspects of the TO application to structural steel design.

Concomitantly, automotive research centres developed several in-house tools and approaches to introduce TO in industrial Computer-Aided Engineering (CAE). Some examples can be found by Nishigaki et al. [73], Shin et al. [74], Fredricson et al. [44], Aeri and Morrish [75], and Yao et al. [76].

For the TO applications to leverage AM, several notorious works were published from 2010 on. Brackett et al. [77] is an excellent starting point, as it addresses the TO mature approaches, including Solid Isotropic Microstructure (or Material) with Penalisation (SIMP) and Evolutionary Structural Optimisation (ESO), suitable for practical applications. However, much has evolved in the next 10 years, including the consolidation of other approaches, algorithms, and workflows. Therefore, Leary et al. [78] and Nguyen and Vignat [79] worked on design methods offering a good understanding of the theme.

Likewise, insights on AM technologies, as provided in [80,81] are much recommended for fully understanding the role of TO in the fourth industrial revolution.

### 1.3. Topology Optimisation in Civil and Structural Engineering

Within civil and structural engineering, significant TO applications can be found in several sub-fields. While Discrete Optimisation of truss-like structures is, probably, the most straightforward application—and, indeed, has some early and interesting applications such as the Structural Topology and Shape Annealing (STSA) approach to transmission towers by Shea and Smith [82], the Mixed-Integer Non-Linear Programming (MINLP) approach to industrial steel buildings structural cost optimisation complying with Eurocode 3 (EC3) by Kranvanja and Zula [83], the method for trusses optimisation by Torii et al. [84] or the employment of genetic algorithms by He and Wang [85]—there are some works on the TO of structural systems under seismic loads. Such could be regarded as a surprise, since the seismic design is undoubtedly more complex and demanding compared with static loading common cases but may be explained by the resourcefulness of TO addressing complex issues which, otherwise, would hardly be efficiently solved with analytical means. Among the aforementioned works, one can highlight the steel frames optimisation by Memari and Madhkhan [86], more recent studies of space structures under seismic loads with evolutionary approaches [87,88], a comparison of different soft computing algorithms by Liu and Li, taking infrastructures lifelines as the study object [89], as well as Sarkisian et al.'s well-known mastery for innovative solutions, materialised in the use of TO for meeting significant seismic, aesthetical, budgetary, and regulatory demands for a specific building [90].

The conceptual design of tall buildings gathered practitioners and researchers attention (as depicted in Figure 1b), leading to a more practical design process optimisation [91] or algorithms focused research [92]. Nevertheless, the systematic employment of TO for the conceptual design of tall buildings is particularly evident in the Skidmore, Owings, and Merill experience. As reported by Baker with other SOM engineers and academics [93–97], such solutions were developed for impactful projects and competitions (Figure 1c).

Steel sections and the connections design are two other sub-fields where TO has had an impact. Regarding the former, the Tsavdaridis group (Figure 2a) and Lagaros et al. research on steel beams with web openings [98], Yao et al.'s creative employment of evolutionary algorithms for pre-tensioned cable structures design [99], and Leng's book chapter on cold-formed steel members [100] must be referred. Regarding the latter, it is important to stress out the contributions by Oinonen et al. [101] and Elsabbagh [102] on the bolted steel connections geometric optimisation and stiffeners optimisation, respectively.

The bridge design, on the other hand, has had lesser attention from the TO point of view. Nevertheless, Zhang et al.'s work on bridge design accounting for construction constraints with ESO algorithms [103] and Xie et al.'s Bi-Directional Evolutionary Structural Optimisation (BESO) algorithms application to the bridge conceptual design (Figure 2b) [104] must be mentioned.

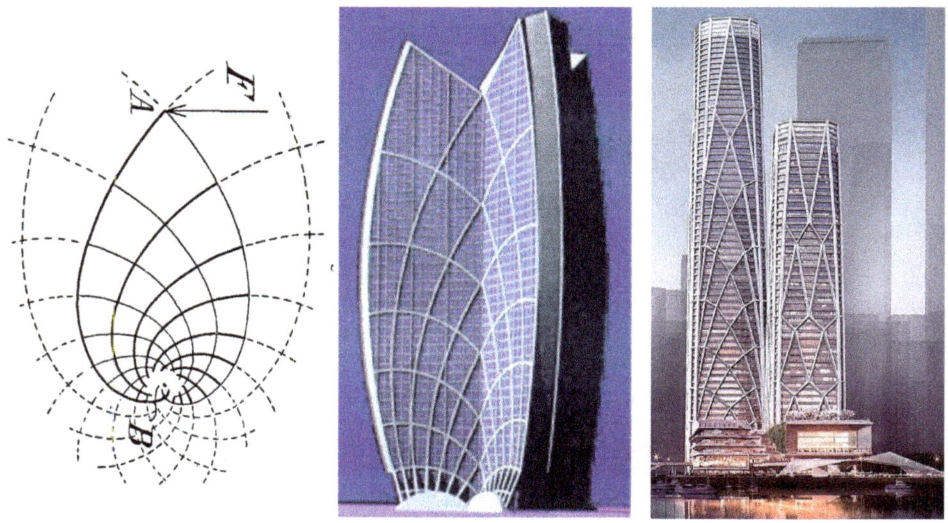


(**a**) (**b**) (**c**)

**Figure 1.** (**a**) Original Michell's minimum frame [9], (**b**) structural design by Zalewski and Zabłocki [105], and (**c**) CITIC financial centre in Shenzhen by SOM [105].

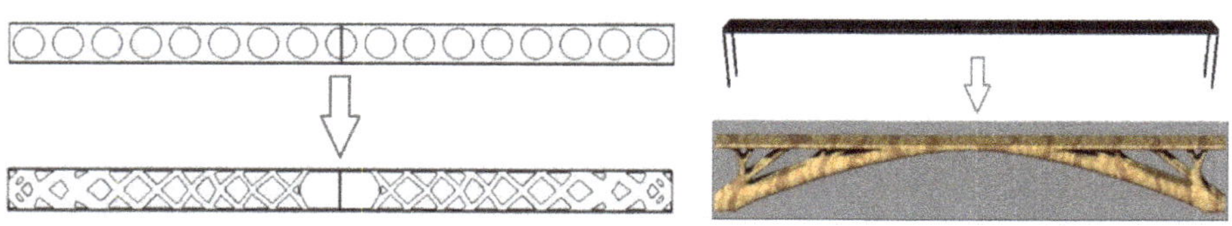

(**a**) (**b**)

**Figure 2.** (**a**) From cellular to topology optimised beam [106] (reproduced under the Creative Commons Attribution 4.0 International License, http://creativecommons.org/licenses/by/4.0/ accessed on 25 February 2021) and (**b**) bridge topology optimisation from problem statement to optimal solution [104].

Beyond steel design but still within structural engineering, concrete TO has had some landmarks, including Lee et al.'s work on frame nodes [107], Briseghella et al.'s fresh look into the classical concrete shell supported bridge design [108], Gaynor et al.'s approach to TO for enhancing strut-and-tie models [109], and the use of TO by Chaves and Cunha [110] in the Carbon Fibre Reinforced Polymers (CFRP) reinforcement design for concrete slabs. Other civil engineering sub-fields for TO include the building's thermal behaviour design [111].

### *1.4. Seminal Works and Systematic Reviews*

Most of the former developments are well documented, explained, and exemplified in extensive and, mostly, user-friendly books. That is the case of Haftka and Gurdal's Elements of Structural Optimisation [112] with several editions, the profusely cited Bendsøe and Sigmund's "Topology Optimisation-Theory, Methods, and Applications" [113] and, more recently, with Arora, Rozvany, and Lewinski books [114,115] and Lewinski et al.'s book [105]. More specific works addressing computational methods [116] or buckling [117] in TO have also recently been made available.

Eschenauer and Olhoff [1] presented an extensive systematic review, encompassing the TO evolution from early structural optimisation to the fundamentals of TO, formulated in Michell structures [9], until the last century's late developments. There, microstructure and macrostructure approaches have been described within a product design context, paving the way for the concepts and approaches developed in the new millennium. The latter include numerical methods such as SIMP, enclosed into the density approaches, and ESO, whose retrospective analysis can be found in another useful Review Article by Rozvany [35]. Moreover, such article is equally crucial for understanding the controversy around ESO (and BESO, its bi-directional development), with several authors referring to it as Sequential Element Rejections and Admissions (SERA) for not employing evolutionary processes nor yielding necessarily an optimal solution.

Before that, Hassani and Hinton [118,119] had summarised the basic concepts and criteria for the emerging TO.

Yet, the history of TO Review Articles is mostly made of sectorial perspectives. Among the first, one can find Fredricson's [120] revision to TO in the automotive industry, focusing on the applications rather than on approaches or methods. Contrariwise, methods-driven reviews have been more frequent, and include assessments about ESO discrete approach developments [121] on Level-Set Methods (LSM or LS) [18]. Broader critical investigations of the former, added to the remaining density approaches, such as Rational Approximation of Material Properties (RAMP), Topological Derivatives (or the Bubble-Method), and Phase-Field Approach, as well as Lagrangian approaches [122,123] are also available.

Optimisation methods in AM have been a theme for a profusion of Review Articles along the last decade. While some assess manufacturing technologies and how optimisation methods can be accounted for in the process [124–128], others analysed the suitability of TO approaches for given AM endeavours, such as construction in Buchanan and Gardner [129], aerospace structures in Plocher and Panesar [130] or the automotive industry in Sehmi et al. [131]. Moreover, optimisation approaches impact the AM customised healthcare, environmental impact, supply chain efficiency, life-cycle analysis, hazards and energy consumption, which is part of Huang et al.'s review [132].

The same decade also brought landmark review articles on particular topics, including TO applications in vibration problems [133], fluid problems [134], and materials design [135]. However, recent reviews on TO-assisted structural engineering are the most meaningful for this article's scope. Those include Kingman et al.'s [136] review on TO employment for perforated steel beams design and tall buildings conceptualisation, a review on buckling by Ferrari and Sigmund [137], Elhegazy's [138] perspective on how TO is a critical part of Value Engineering (VE) and how it benefits the design and life of multi-story buildings, as well as a review by Li and Tsavdaridis on topology optimised and additively manufactured joints for steel structures [139].

Recent reviews on the use of metaheuristic algorithms in civil engineering, by Yang et al. [140] and Bekdaş et al. [141], do not neglect the genetic algorithms of TOs for bridges, roofs, frames, and trusses design.

### *1.5. Potential for Topology Optimisation and Additive Manufacturing in Construction*

The asymmetry in size between the TO related literature in civil and structural engineering and many other fields made clear by the former brief historical overview, suggests that TO applications in construction are far from meeting its potential. Such a fact can only come as a surprise, given the sector size, with a yearly output of USD 10,800,000,000,000 as of 2017 [142], a tradition in analytical structural optimisation and an urgent need for weight reduction along with stiffness and resistance enhancement, as tools for leveraging the continuous race towards higher buildings, greater spans in bridges and roofs, more economic efficiency, and more ambitious sustainability goals.

Steel construction, deploying 450,000,000 to 815,000,000 tonnes [2,143,144] of structural steel per year worldwide pre-COVID-19 era, and with a market size over USD 100,000,000,000 in 2019 [145], is involved in a global effort to meet decarbonisation and energy efficiency goals, including the Paris Agreement pledge, European Green Deal objectives, and further commitments, such as the UN-backed carbon neutrality by 2050, to which most EU countries abide [146,147]. The path for achieving this is narrow and relies mostly upon employing less steel in constructions, while manufacturing a higher-end product, to more stringent sustainability demands. As a result, AM and, therefore, TO will be indispensable.

Apart from some slight oscillations due to technical details, it is commonly accepted that steelmaking consumes over 560 kg of Coal Equivalent [148] and produces an average of 1.85 carbon dioxide tonnes per tonne of structural steel [147]. Thus, reported weight reductions of 18% to 75% in steel connections [139,149–151], by employing TO, are expected to have a critical impact in steel construction goals.

Within steel design, connections detailing is one activity where the potential to optimise Topology is more significant. Connections usually account between 12% and 25% of most of the steel structure's total weight and its conceptual design has a tremendous impact on its weight and efficiency, making it especially prone to TO.

### *1.6. Metrics for Research Output in Topology Optimisation*

Performing a data analysis with the Scopus search tool (www.scopus.com) on 28 November 2020, it has been possible to observe that Topology Optimisation has been referred in scientific literature since the early 1980s, but not until the new millennium has the related scientific output been consistently increasing (Figure 3a). This fact can be related to the well-known recent computing power increase and massification, on which TO has a strong dependence. Furthermore, the ascendant trend has increased over the last 5 years, up to almost 1200 documents per year in 2020, during which more scientific and engineering disciplines have reportedly started employing TO more systematically. An interesting etymological approach lies in noticing an early word choice for *Topological Optimisation* over *Topology Optimisation* which, however, was not able to be employed in more than 10% of the analysed documents, over the recent years (Figure 4).

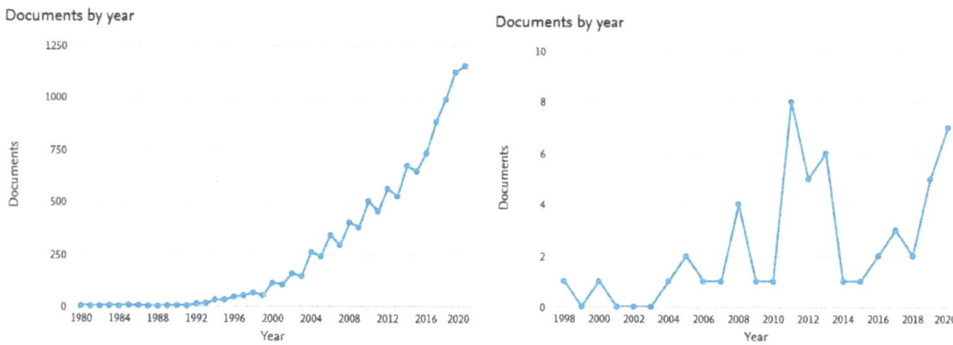


(a) (b)

**Figure 3.** Scopus search data analysis for (**a**) *Topology Optimisation* and (**b**) *Topology Optimisation* AND *Civil Engineering*, in Title, Keywords or Abstract in Research, Review, and Conference articles, on 28 November 2020.

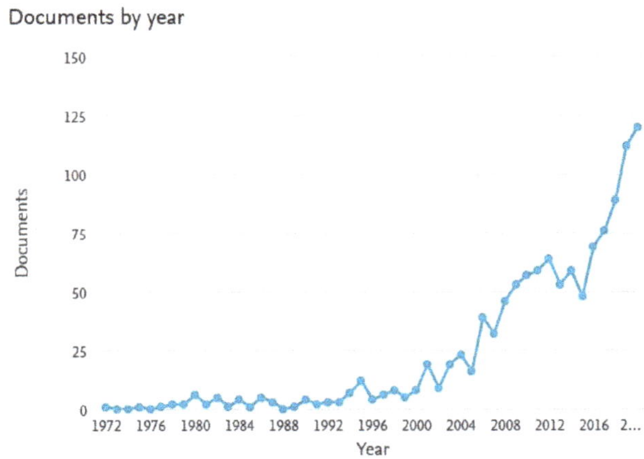


**Figure 4.** Scopus search data analysis for *Topological Optimisation* in Title, Keywords or Abstract in Research, Review, and Conference articles, on 28 November 2020.

A different perspective is attained analysing the coexistence of TO and *Civil Engineering* (Figure 3b), *Structures* (Figure 5a), and *Steel and Connections or Joints* (Figure 5b) in articles title, keywords or abstract. Among these three, TO and *Structures* is the most common, even if under 2.5% of the total TO articles, and with a steady increasing trend. However, many Mechanical and Industrial Engineering documents use the keyword *Structures*. Furthermore, TO and *Civil Engineering* have been practically not coincident until 2010 and, from there, yield a rather inconstant volume of documents not exceeding 10 per year. A similar conclusion can be drawn for the use of TO and *Steel Connections* (or *Joints*), except for the fact that its modern employment seems to have started in 2005 and that it has not peaked above six documents per year. This suggests that the current TO massification as an advanced engineering design method has not yet been brought to Civil and Structural Engineering comprehensively.

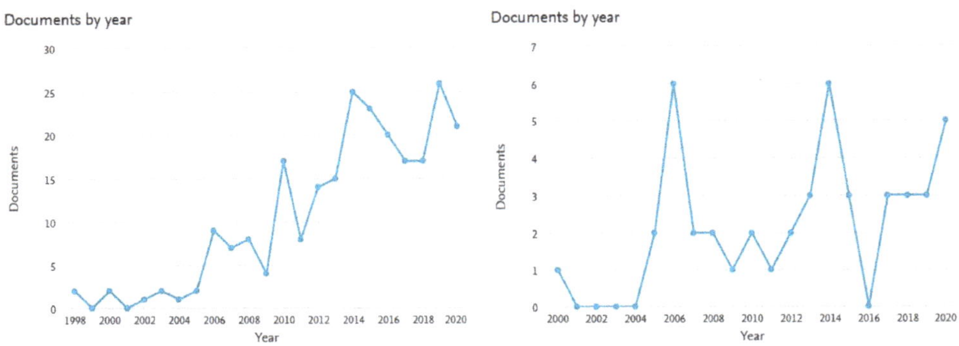


(a) (b)

**Figure 5.** Scopus search data analysis for (**a**) *Topology Optimisation* AND *Structures* and (**b**) *Topology Optimisation* AND *Steel* AND *Connections* OR *Joints*, in Title, Keywords or Abstract in Research, Review, and Conference articles, on 28 November 2020.

Figure 6a as well as Figure 7a,b offer a brief insight on the field leading players. As a result, we observe Sigmund, Nishiwaki, and Xie leading the list of most prolific authors. At the same time, the Dalian University of Technology, Danmarks Tekniske Universitet (DTU) and State Key Laboratory of Structural Analysis for Industrial Equipment are the most productive research centres. The National Natural Science Foundation of China is the primary funding agent for the research on this topic. It is well ahead of the US National Science Foundation and not matched by the remaining entities.

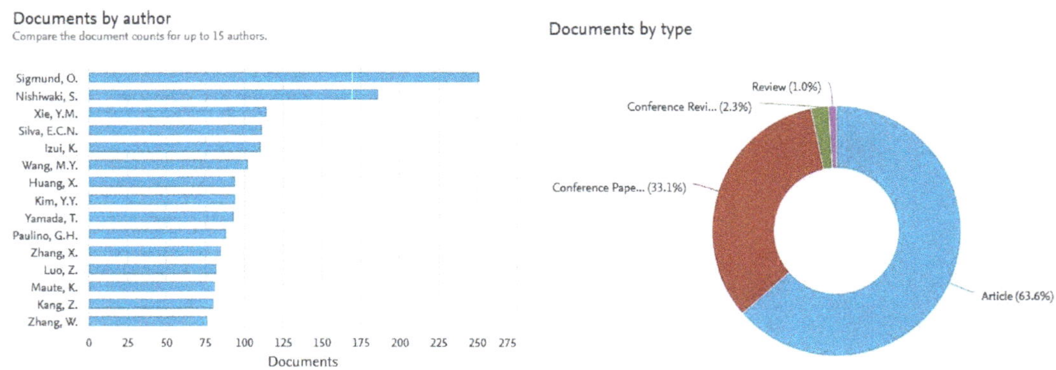


(**a**) (**b**)

**Figure 6.** Scopus search data analysis for *Topology Optimisation* in Title, Keywords or Abstract in Research, Review, and Conference articles without date or discipline constraints, on 28 November 2020, organised by (**a**) author and (**b**) document type.

As depicted in Figure 6b, Research Articles account for almost two-thirds of the analysed documents, with Conference Articles making one-third and leaving Conference Reviews and Journal Review Articles with 2.3% and 1.0% of the literature volume, respectively. This suggests a need for systematic reviews on the topic so that the literature is consolidated as a whole and sectorial reviews help each research discipline adapt and incorporate recent advances in TO, which is being developed in other disciplines.

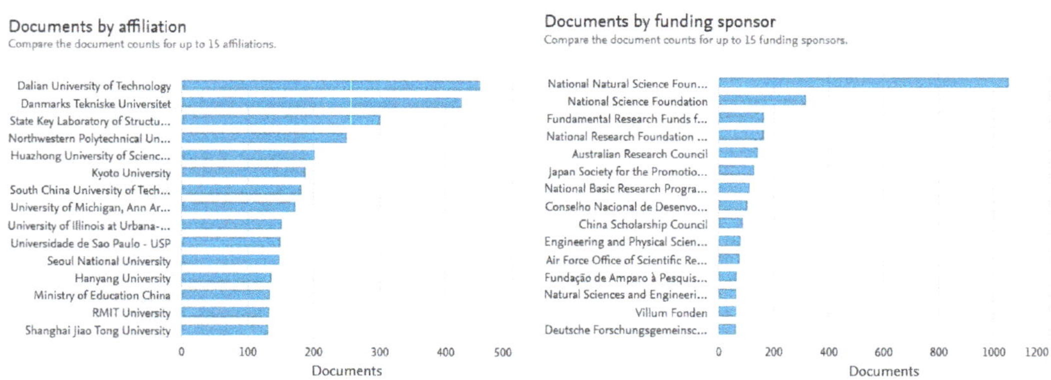


(**a**) (**b**)

**Figure 7.** Scopus search data analysis for *Topology Optimisation* in Title, Keywords or Abstract in Research, Review, and Conference articles without date or discipline constraints, on 28 November 2020, organised by (**a**) research institution and (**b**) funding entity.

### *1.7. Scope of This Document*

This work has been developed to systematise recent advances in the Topology Optimisation for structural steel design. It is focused on the 2015–2020 period, apart from this Introductory Section. It is organised with a dual approach, deemed to depict the relatively modest TO applications in the field in recent years, as well as to systematise significant developments in adjoining fields, which may be inspirational for structural steel design. As a result, it is not expected to collide with any recent Systematic Reviews on TO in other disciplines, while filling the void for a Revision focused on TO recent developments for structural steel design. Therefore, it is aimed to provide a valuable resource and encourage engineers and researchers to embrace TO in a more systematic and sustained manner for structural steel design.

The document structure includes a Methods Section after this Introduction, which will be followed by six sections depicting the literature investigation results, organised in TO fields, approaches, methods, criteria and software, TO in structural steel design, recent advances in other fields with potential for application in structural steel design, TO for AM, and future trends. Afterwards, the most important observations are discussed, and a brief on the attained conclusions is provided.

## 2. Methods

The literature research on Topology Optimisation was conducted between late November and early December 2020, encompassing Identification, Screening, Sorting, Eligibility Assessment, Information Extraction, Qualitative Synthesis, and Discussion Stages, as better depicted and systematised in Figure 8. While attending to this scientific field's case-specificity, compliance with well-established guidelines for systematic reviews, such as in [152] was pursued. Inspiration was found in other recent Review Articles, such as in [153–156].

Scientific literature has been searched with significant broadness. All published and peer-review items were considered, including Journal Research Articles, Conference Proceedings, Review Articles, Peer-reviewed Book Chapters, and Approved Master and Doctoral Theses. On the other hand, Technical Books, which are not necessarily peer-reviewed, were considered only for framing the research topic in the Introductory Section. Many notable books' unavoidability supports such a decision for defining Topology Optimisation, including Bendsøe and Sigmund's "*Topology Optimisation: Theory, Methods, and Applications*" [113] with 7301 citations according to Google Scholar as of 28 November 2020.

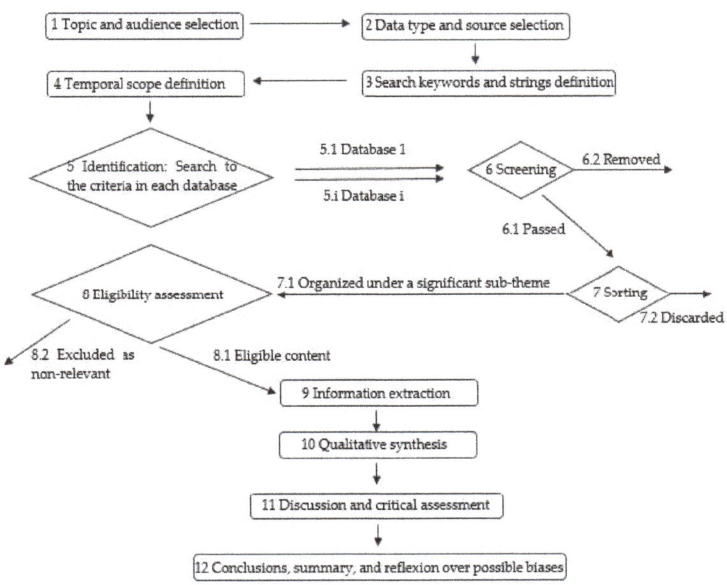


**Figure 8.** Search and systematic review methodology.

For a more in-depth insight into the current research, industry developments, and future trends, EU-funded research projects as well as European, US, and worldwide patents were also investigated.

The temporal scope for literature search has been set for the most recent 5 years (2015-2020), yet considering the Topology Optimisation novelty as a discipline (illustrated in Figure 3a), previous and very significant research items were investigated and considered for conceiving an explanatory Introduction.

An encompassing fabric of data sources was put together for this endeavour. Thus Journal Articles and Conference Proceedings were redundantly searched in Mendeley Desktop, Scopus Online, Springer Online, Taylor and Francis Online, and Willey online search engines. Moreover, most active research groups repositories were also investigated, including the DTU TopOpt group, Loughborough University, and the University of Leeds. Several universities Theses repositories were searched. However, literature findings through reading articles was not negligible and added to the former. Patents were found in Google Patents, Espacenet, and USPTO search engines. Research Projects were found in the European Commission projects database, and books were searched using the ProQuest online tool. Table 1 quantifies the search dimensions.

A group of three basic keywords and 12 keyword strings were systematically used in all databases. The former included *"Topology optimization"*, *"Design for Additive Manufacturing"* and *"Topological optimization"*, while the latter are *"Topology optimization"* AND *"Additive Manufacturing"*, *"Topology optimization"* AND *"Steel Design"*, *"Topology optimization"* AND *"Steel Detailing"*, *"Topology optimization"* AND *"Steel Structures"*, *"Topology optimization"* AND *"Structural Engineering"*, *"Topology optimization"* AND *"Connections"* *"Topology optimization"* AND *"Construction"*, *"Topology optimization"* AND *"Joints"*, *"Topology optimization"* AND *"Review"*, *"Topology optimization"* AND *"Civil Engineering"*, *"Topology optimization"* AND "Multiple *Loading"*, *"Topology optimization"* AND *"Robustness"*. These keywords and strings have been selected to match the articles' Title, Keywords or Abstract where such option is explicitly available, as it is the case of the Scopus search tool.

**Table 1.** Search dimensions.

| Stage | Included | Excluded |
|---|---|---|
| 5 Identification | 5.1 Mendeley ($n$ = 68, of which 27 were eligible)<br>5.1 Scopus ($n$ = 292, of which 253 were eligible)<br>5.2 Springer ($n$ = 55, of which 41 were eligible)<br>5.3 Taylor and Francis ($n$ = 21, of which 15 were eligible)<br>5.4 Wiley ($n$ = 24, of which 16 were eligible)<br>5.5 Research Groups Repositories ($n$ = 51, of which 44 were eligible)<br>5.6 References found in articles ($n$ = 181, of which 174 were eligible)<br>5.7 Approved Theses Repositories ($n$ = 7, of which 4 were eligible)<br>5.8 Google Patents ($n$ = 22, of which 5 were eligible)<br>5.9 Espacenet ($n$ = 12, of which 11 were eligible)<br>5.10 USPTO ($n$ = 19, of which 17 were eligible)<br>5.11 EU-funded Research Projects ($n$ = 5, of which 5 were eligible)<br>5.12 ProQuest ($n$ = 10, of which 10 were eligible) | |
| 6 Screening | $n$ = 729 | $n$ = 38 |
| 7 Sorting | $n$ = 707 | $n$ = 22 |
| 8 Eligibility | $n$ = 622 | $n$ = 85 |

The screening criteria for removal included duplicate items, the mismatch between title and content, corrupted files, and the impossibility of accessing the document. Discarded items under the Sorting stage criteria comprised documents whose content do not adhere to any of the sub-themes previously defined as meaningful for this article scope, and better described in Section 3 to Section 8. The eligibility assessment was focused on the document content, excluding items without a particular relevance, without critical information, with any perceived methodological shortcoming or possible strong bias due to funding.

An effort has been endeavoured for performing inclusive research, avoiding the exclusion of less proficiently written articles and language bias. However, some articles written in Chinese and Japanese, for which machine translation was not successful, could not be included.

## 3. Topology Optimisation—One Concept, Various Fields

### 3.1. The Mathematical Concept of Topology

Topology, described as General Topology and Algebraic Topology under the Mathematics Subject Classification, studies objects' properties which are subjected to continuous deformations [157–159]. Hence, a topological space (or domain), commonly referred to as a topology, maintains its properties, including dimension, compactness, and connectedness if undergone such deformations. Invariance within topological domains accrues in reversibility under continuous deformations or homomorphism, making this abstract concept crucial for the Topology Optimisation's founding principles, as formulated by Maxwell and Michell.

### 3.2. Structural Optimisation and Topological Optimisation in the Context of Structural Steel Design

Structural Optimisation and Topological Optimisation concepts have been used with increasingly unrestrained freedom, even in Academia. Thus, it may be helpful to mention that the most rooted nomenclature employs Structural Optimisation as an umbrella for Topology Optimisation, on the one hand, and Shape and Size Optimisation, on the other hand [44,160]. In the context of Structural Steel Design, while seldomly, the use of Structural Topology Optimisation (STO) has been reported [139], referring to TO in this specific field.

Conceptually, Shape and/or Size Optimisation is easy to define, considering its scope, limited to variables as cross-sectional properties, member types and geometry, and rigidly constrained by predefined configurations. In simple terms, Topology cannot be changed and, therefore, this concept is deemed to improve an existing design, in which it very much depends on closeness to optimal.

Topology Optimisation, also named Layout Optimisation by some pioneers, bears the capacity for topological modification. In other words, it is unrestricted in its ability to create voids or add material in the design domain and act upon the structure's connectivity. That said, it is relatively straightforward that TO contains and exceeds the Shape and Size Optimisation and therefore, fulfils all the possible Structural Optimisation scopes.

It should be highlighted that TO is not constrained to materially homogenous volumes. It may be employed as heterogeneous, including composite or microstructurally designed materials, but also in grid-like "ground structures", made of one-dimensional elements.

In practical terms, differences are significant not so much between concepts, but mostly between applications and engineering fields. Concerning structural steel design, the Size, Shape, and Topology Optimisation can be graded into an intervention freedom continuous scale. The practical constraints and one-dimensional members design may limit optimisation to size and shape, even if the possibility for adding or suppressing members in a predefined configuration is available. The optimisation of continua, either in small volumes as joints or macro elements, as a building or bridge geometrical envelope, find a most suitable tool in TO.

### 3.3. Discrete Optimisation and Optimisation of Continua in the Context of Structural Steel Design

Topology Optimisation applications can be divided into Discrete Optimisation and Optimisation of Continua, based on the Topology considered for the optimisation problem (Figure 9).

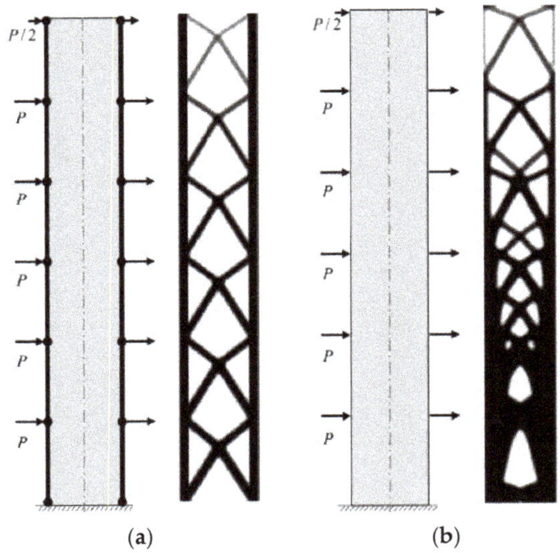


**Figure 9.** Example of (**a**) problem statement and discrete optimisation [94] and (**b**) problem statement (**left**) and optimisation of continua (**right**) [94].

The optimisation of discrete structures has a history of its own. It has been present in structural engineering from its early days, even if with a different scope, and this reason justifies why some researchers find the roots of TO in the classical theory of elasticity. As suggested by its name, Discrete Optimisation is directed to discrete structures. Its objectives lie in finding the optimal number, location, shape, size, and connectivity for structural elements and nodes coordinates. Therefore, one can understand that most early Discrete Optimisation applications have been limited to Shape and Size Optimisation, rather than TO, due to significant limitations on the optimisable variables' domain. For the very same reason, Discrete Optimisation has been used under the names of Truss Topology Optimisation (TTO) or Topology Optimisation of Skeletal Structures (TOSS) and its early applications to finding optimal layouts led to the name Layout Optimisation, regarded by many as a former name for TO.

However, the Discrete Optimisation scope exceeds one-dimensional members. In fact, since the advances by Prager and Rozvany, which used grid-like structures, TO has grown as a merger of techniques and approaches, deemed to find optimal solutions for specific cases. Continuum structures can be modelled as ground structures—complex systems of one-dimensional elements—and therefore, be subjected to Discrete Optimisation.

A major issue of Discrete Optimisation can be attributed to considering nodes coordinates in or out of the optimisation domain. If coordinates are not possible to optimise, solutions can easily result in a plethora of thin, one-dimensional structural element. Conversely, if the nodes positions vary, node superposition will likely occur, leading to significant computational problems.

The Optimisation of Continua is usually applied to solids, shells, or design envelopes (which, in fact, are solids) and is deemed to optimise the design space external boundaries shape, as well as the internal boundaries shape. The latter refers to the newly created boundaries between the material and void.

## 4. Approaches, Methods, Criteria, and Software in Topology Optimisation

### 4.1. Approaches and Optimisation Methods

#### 4.1.1. On the Nomenclature Complexity

An unexpected layer of complexity in TO lies in the terminology. Approaches, methods, methodologies, models, techniques, and algorithms, etc. are used in conflicting ways by an expanding and very diverse research and practice community.

In this article, a choice has been made to refer to the comprehensive strategies for solving a well-defined TO problem as "methods". Such an option adheres to some of the most respected leaders and book-writers in disciplines, such as Bendsøe, Rozvany, and Sigmund, to name a few, with some occasional exceptions. In this manner, the well-known SIMP, RAMP or ESO are considered methods, while others are grouped by similarity. That is the case of OMP and NOM, both nested under the "Homogenisation" methods family. Such an option is, evidently, arguable.

Other options resided in considering six approaches, in which the defining criterion was the ability to group methods by its functional similarity. Thus, Density-Based, Level-Set, Topological Derivatives, Phase-Field, Heuristic and Hybrid approaches were accounted. As a result, other approach classifications, such as Material/Geometrical or Lagrangian/Eulerian, could not be simultaneously considered.

The problem resulting from the possibility of employing different Optimality Criteria (OC) for one given method and possibly using the same Optimality Criterion for different methods could only be solved by creating a diverse group for OC.

However, other problems arise from using different names for the same method or for different authors' applications of the same method. For example, one can consider the ESO method, also referred to as SERA by researchers who were not involved in its original proposition, regarding their views on the method shortcomings. Likewise, the Bubble Method is referred under different names, and recent developments in Level-Set Methods and Topological Derivatives have not yielded universally accepted names for its methods, due to several increments from many researchers and applications to many different problems.

Contrariwise, several well-known methods have remarkable resemblances which could be better regarded as different applications of a single method. That can be found in the extended ESO family, in applying Genetic Algorithms (GA) and Swarm Methods, and in the different options for solving the Hamilton-Jacobi Equation in Level-Set Methods by only a handful of examples. An option was taken to leave all the methods that have some expression in the surveyed literature, regardless of its similitude, and group methods that differ only in using well-known algorithms, such as GA or Swarms.

#### 4.1.2. Framing New Developments into the Body of Knowledge

It was never easy to organise TO in approaches, methods, or sub-fields, as one can observe from the notorious Review Articles' classification discrepancies, such as in [1,35,119,122] or [131]. However, recent research adds complexity in this issue due to so many outputs on particular, yet transversal, issues.

At the current status of accelerated progress and interrelatedness in TO methods and algorithms, it is quite challenging to frame newly published research into categories, so that its value and applicability for one certain problem is evident. For such an end, one can find utility in Table 2, where general approaches and methods were organised to the formerly depicted criteria, and the recent relevant research was inserted. If researchers and practitioners, especially the newly arrived in this field, find it helpful to navigate through the literature and understand the potentially useful contributions, this table will accomplish its purpose.

**Table 2.** Topology optimisation (TO) approaches, methods, and recent developments.

| Approach | References | Non-Exhaustive List of Methods | | References | Recent Developments |
|---|---|---|---|---|---|
| Density-based (also Material Distribution) | [35,122,161,162] | Homogenisation | Optimal Microstructure with Penalisation (OMP) | [163,164] | [165–200] |
| | | | Near-optimal Microstructure (NOM) | [21,201] | |
| | | Rational Approximation of Material Properties (RAMP) | | [202] | |
| | | Solid Isotropic Microstructure (or Material) with Penalisation (SIMP) | | [22,32,35,42] | |
| | | SINH (due to employing the hyperbolic sine function) | | [203] | |
| | | Sum of the Reciprocal Variables (SRV) | | [204] | |
| | | Reliability-Based Topology Optimisation (RBTO) | | [205–212] | [213,214] |
| Level-set (LS) methods | [18,215–222] | Conventional LS for solving the Hamilton-Jacobi Equation | | [164,217,223] | [170,171,224–233] |
| | | Radial-Basis Functions (RBF) for solving the Hamilton-Jacobi Equation | | [234–239] | |
| | | Spectral LS | | [240] | |
| | | Non-Linear Programming | | [234] | |
| Topological Derivatives | [241–243] | Bubble Method | | [1,244,245] | [246,247] |
| | | Topological Sensitivity | | [243,248,249] | |
| Phase-field approach | [250,251] | Cahn-Hilliard Method | | [252–254] | [171,255,256] |
| | | Allen-Cahn Method | | [257] | |
| | | Relaxed Phase-Field Methods | | [250,252,258] | |
| Heuristic (also Non-gradient or Evolutionary) [a] | [35,123,141,259–263] | Evolutionary Structural Optimisation (ESO), also Sequential Elements Rejection and Admission (SERA) | | [31,35,68,264] | [165,265–287] |
| | | (Hard-kill) Bidirectional ESO (BESO) | | [35,121,288–294] | |
| | | Additive ESO (AESO) | | [123,131,295] | |
| | | Soft-kill BESO | | [121,296,297] | |
| | | Swarms, including Particle Swarm Optimisation (PSO), Fish Swarm Optimisation (FSOA), Ant Colony Optimisation (ACO), Stochastic Diffusion Search (SDS), Artificial Swarm Intelligence (ASI), Multi-Swarm Optimisation, Artificial Bee Colony Algorithm (ABC) | | [131,140,298,299] | |
| | | Genetic Algorithms (GA), including Genetic ESO (GESO), Lindenmayer (also map-L) Method | | [300–311] | |
| Hybrid approaches | [122] | Combination of several features and techniques | | | [312,313] |

[a] Another common name for this approach is the "Hard-kill" method. However, it does not account for its current diversity, which includes the "Soft-kill" option.

Recent research is generally less focused on fundamental and theoretical issues and more attentive to computational issues, application details, and case-specificity. There is also a trend to employ and mix concepts from different methods. For these reasons, framing recent research into approaches and methods is increasingly difficult and potentially erroneous.

One other interesting issue concerns SIMP methods. While such a theme dominated the research output for a long time, new publications devoted to it are relatively decreasing (mostly compared to the escalating numbers of TO related papers). Moreover, many research and industrial endeavours still employ SIMP, even if depicting and discussing it ceased to be considered pivotal.

Among the newer approaches, Level-Set Methods appear to be in accelerated development. This can be explained by the migration of density-based, and especially SIMP, researchers and practitioners to an approach with so many standard features.

On the other hand, Phase-Field Approaches are yet to gain momentum and Topological Derivatives, even if with a significant history, have not much recent research output. The latter, however, continues to provide a background for many developments and comparisons in mainstream methods.

With the Heuristics group, it is quite interesting to observe that the last two decades of profuse output in ESO and BESO methods seem to decrease, while Genetic Algorithms are sharply on the rise.

The working-set approach by Verbart and Stolpe [314] has not been included in the previous table due to its versatility, allowing an easy adaptation to several methods and optimality criteria.

Recent developments, such as the Deformable Simplicial Complex (DSC) Method [315] and the Virtual Scalar Field Method (VSFM), which allows considering a connectivity constraint as a thermal effect [316], show interesting features which substantiate mentioning it in this review. However, it is not yet the time to insert it in the Table 2 classification, as further developments will tell whether specific categories are justified.

Analogously, other methods, such as the Moving Morphable Components (MMC) have been proposed in the past [317,318], as alternatives to the more established aforementioned ones, but its applications beyond these authors' works remain not very profuse. However, promising contributes are regularly obtained by employing it, such as the Virtual Component Skeleton (VCS) method by Wang et al. [319], for controlling topologically optimised boundary smoothness, which deserves to be referred.

For the Discrete Optimisation of trusses, the article by Zhang et al. [320] is fundamental for understanding the Ground Structure Method (GSM), as well as its Voronoi and quadrilateral methods of discretisation.

Recent conclusions on the Equivalent Static Loads Method (ESLM) [321] offered clarity to previously reported findings and highlighted the caution needed for analysing the potential of the methods.

For an interesting discussion on preconditioning, its impact upon computing efficiency, and an example on Preconditioned Conjugate Gradients (PCG), the reader is referred to the work by Kaveh et al. [322].

#### 4.1.3. Heuristics as a Source of Controversy

The use of optimisation methods whose solution is not necessarily optimal, referred to as Heuristics, has been in the centre of discussion on the TO theory, since the first so-called Evolutionary methods were proposed. Without entering a discussion already well depicted in [35,121–123], criticism is mostly due to the reported incapacity of evolutionary methods for attaining convergent optimal solutions, to failing to achieve acceptable solutions in some cases, and to the difficulty in generalising the method for real structures constraints.

It is quite interesting that some other researchers highlight the employment of Heuristics in filters and other techniques deemed to avoid local optima, which are used well beyond the evolutionary methods.

As expected, many of those claims have been rebutted, discussed, but also admitted and led to many of the current developments.

As a result, the current discussion on TO methods in Academia is still centred on the SIMP/BESO antagonism, as well as in the new developments in Level-Set, Topological Derivatives, and Phase-Field Methods, while practical applications are mostly employing SIMP methods.

Considering that Meta-Heuristics (high-level procedures for combining or selecting heuristic methods for one given problem solving or adequate approximation) and Artificial Intelligence (AI) based methods are rapidly spreading within civil and structural engineering [323,324], due attention will be given to the Heuristics approaches in structural steel design in this article. However, the current imbalance must be highlighted.

### 4.2. Optimality Criteria

Considering the impracticability in proving that an attained solution is mathematically correct when thousands of variables are involved, the Optimality Criteria (OC) had to be set. For such an end, several precursory intuitive criteria were used, such as the Fully Stressed Design (FSD) and the Simultaneous Failure Mode Design (SFMD) until the so-called rigorous criteria were adopted. The latter epithet is usually given to any criterion complying with Kunh-Tucker optimality conditions.

Table 3 systematises the most common OC, which is applied to the prevalent Density-Based Approach methods.

**Table 3.** Frequently used optimality criteria methods.

| OC Methods | | References |
|---|---|---|
| Discretised, Continuum-type Optimality Criteria technique (DCOC) | | [119,325] |
| Continuum-based Optimality Criteria (COC) | | [119,326] |
| Iterative COC | | [119,327] |
| General optimisation codes | Design Optimisation Tools (DOT) | [123] |
| | Sparse Nonlinear Optimiser (SNOPT) | [328] |
| | Interior Point Optimiser (IPOPT) | [329] |
| | Convex Linearisation Method (CONLIN) | [42,330,331] |
| | Method of Moving Asymptotes (MMA) | [30,332] |
| | Globally Convergent Method of Moving Asymptotes (GCMMA) | [333] |

### 4.3. Practical Methodologies

Unsurprisingly, the literature devoted to TO is profuse in depicting theoretical approaches and methods, as well as in validating it with well-known or trivial cases, but much scarcer in providing comprehensive and holistic methodologies (which contain practical aspects) for implementing a TO strategy into the Engineering design. This is related not only to researchers' tendency to publish their work and industrial practitioners and developers not doing so, but also to the fact that TO is still in a stage of developing and stabilising methods before a universal application by non-experts. Moreover, Topology Optimisation is still much more complex than the regular structural analysis and design in most engineering fields, requiring a significant time, studying and computational resources that limit the number of large-scale projects currently being developed.

One further issue lies in TO objectives. While academic developments must seek assurance of optimisation to the theoretical extrema, industrial applications are usually comfortable with optimisation to a certain pre-defined threshold and value reliability, predictability, reproducibility and, frequently, computing efficiency above all. However, avoiding local extrema is a common goal.

Nevertheless, some examples were found in a recently published literature, which can be referred to as interesting examples for conceiving case-specific methodologies for the systematic application of TO in structural steel design. Table 4 summarises those findings.

**Table 4.** Methodologies or strategies for practical TO implementation in engineering design.

| Methodology/Strategy | Reference |
|---|---|
| Method for the TO of frame structures with flexible joints | [44] |
| Optimisation for Manufacture (OFM) methodology for introducing manufacturing time and cost into the TO problem | [334] |
| Axiomatic Design Method for AM | [335] |
| TO-directed manufacturing methods | [221] |
| Using surrogate models for conceptual design | [336] |
| TO method to mitigate AM-induced anisotropy | [337] |
| Methodology for introducing AM time and cost into the TO problem | [183] |
| Lumped Parameter Model (LPM) for multiphysics problems | [338] |
| Fail-Safe Methodology | [339,340] |
| Sectional Optimisation Method (SOM) | [341] |
| An AM-focused TO strategy using the SIMP method (for automotive parts) | [342] |
| Integrated design optimisation by Skidmore, Owings, and Merrill LLP (SOM) | [343] |
| Methodology for the TO of cellular structures for AM | [344] |
| Projection-based Ground Structure TO Method (P-GSM) as an advance in Ground Structure Methods (GSM) for addressing the issue of complex geometries, as well as small, disconnected, buckling-prone, and non-manufacturable elements | [194] |
| The TO method accounts for AM geometrical, mechanical, and machining constraints | [345] |

### 4.4. Computer Programmes

The availability of computational resources is paramount for ensuring TO applications beyond the research community, which has the ability to produce their software. That is undoubtedly the case of Structural Steel Design researchers and professionals, to whom software development may not be the primer priority.

Fortunately, both the commercial software and code provided by researchers and developers are available. However, while the former frequently comes as a "black-box" and can even be challenging to be aware of the employed approaches and methods, the latter may lack user-friendliness, require pre- and post-processing, lack graphical interfaces, and not provide an adequate tool for applications more complex than trivial examples.

Considering the relevance of computer programmes for developing TO strategies, a review of the currently available and reportedly more used commercial software and computer programmes is summarised in Table 5.

**Table 5.** TO commercial software and computer programmes.

| Programme | Nature | Runs on | Approaches/ Methods | Reference | Advantages | Disadvantages |
|---|---|---|---|---|---|---|
| Altair HyperWorks platform, including OptiStruct solver and Inspire interface | Commercial | Independent | SIMP and LS methods | [35,130,131,345–353] | User-friendliness. Broadness of use and industrial testing. | Cost. Difficulty in controlling the process. Impossibility in modifying the code. |
| NASA's NASTRAN code | Freeware | Written in Fortran, can be adapted to the user's resources | Density-based approaches | (a) | Freeware. Explicit control over the results and code. The oldest and more tested code for industrial applications. | Laborious input and output. Lack of user-friendliness. |
| MSC NASTRAN | Commercial | Independent | SIMP, Density-based approaches | [35], (a) | Uses the oldest and more tested code for industrial applications. | Cost. Difficulty in controlling the process. Impossibility in modifying the code. |

**Table 5.** *Cont.*

| Programme | Nature | Runs on | Approaches/ Methods | Reference | Advantages | Disadvantages |
|---|---|---|---|---|---|---|
| Simcenter 3D (former NX NASTRAN) | Commercial | Independent | Density-based approaches | [354], (a) | User-friendliness. Uses the oldest and more tested code for industrial applications. | Cost. Difficulty in controlling the process. Impossibility in modifying the code. |
| Autodesk Inventor NASTRAN (former NEi NASTRAN) | Commercial | Independent | SIMP | (a) | Uses the oldest and more tested code for industrial applications. | Cost. Difficulty in controlling the process. Impossibility in modifying the code. |
| Autodesk Within (former Within Enhance) | Commercial | Independent | Uses both the Within Enhance and NEi NASTRAN solvers | [355], (a) | User-friendliness. | Cost. Difficulty in controlling the process. Impossibility in modifying the code. |
| ANSYS Mechanical | Commercial | Independent | SIMP | [35,130,356] | User-friendliness. Pre-defined options. Integration in a very reliable FEA package. | Cost. Lesser propensity for user-defined options. Optimisation algorithms lesser disclosure. Impossibility in modifying the code. |
| TOSCA Structure | Commercial | Dassault Systèmes' Abaqus; Dassault Systèmes' SOLIDWORK; ANSYS; MSC Nastran | Formerly ESO, SIMP + MMA in recent editions | [35,345,349,357–360] | User-friendliness. Integration in very reliable FEA packages. | Cost. Difficulty in controlling the process. |
| GENESIS | Commercial | Either independent or for ANSYS | SIMP, RBTO | [35,130], (a) | User-friendliness. Profusely used in industrial applications. | Cost. Difficulty in controlling the process. Impossibility in modifying the code. |
| Intes PERMAS | Commercial | Independent | SIMP, RAMP | [130], (a) | User-friendliness. Attentive support. Profusely used in industrial applications. | Cost. Difficulty in controlling the process. Impossibility in modifying the code. |
| Samtech Boss Quattro | Commercial | Independent | Density-based approaches and Genetic Algorithms | [130,361] | User-friendliness. Profusely used in industrial applications. | Cost. Difficulty in controlling the process. Impossibility in modifying the code. |
| COMSOL | Commercial | Independent | Density-based approaches and LS methods | [130,362,363] | User-friendliness. Profusely used in industrial applications. | Cost. Difficulty in controlling the process. |
| Karamba3D | Commercial | Rhinoceros or Grasshopper | Genetic Algorithms | [56,364] | Cost. User-friendliness. Already used in some Architecture and Structural Engineering applications. | Difficulty in controlling the process. Impossibility in modifying the code. |

**Table 5.** *Cont.*

| Programme | Nature | Runs on | Approaches/ Methods | Reference | Advantages | Disadvantages |
|---|---|---|---|---|---|---|
| DTU TopOpt app | Freeware | Grasshopper | SIMP | (a) | Freeware. User-friendliness. Developed and tested by the scientific community. | Impossibility in modifying the code. |
| DTU TopOpt programme | Freeware | Web browser | SIMP | [132,365] | Freeware. User-friendliness. Developed and tested by the scientific community. | Impossibility in modifying the code. |
| DTU TopOpt Portable and Extendable Toolkit for Scientific Computing (PETSc) | Freeware | Portable code, which can be implemented in Windows, Linux, etc. | Customisable | [366] | Freeware. Very useful when employing significant computational resources. Developed and tested by the scientific community. | Much more difficult to intervene over the code compared to other DTU's freeware codes. Case-specific. |
| DTU TopOpt mobile app | Freeware | Android, iPhone | SIMP | [123,367] | Freeware. User-friendliness. Developed and tested by the scientific community. | Impossibility in modifying the code. |
| DTU TopOpt Shape mobile app | Freeware | iPhone | Hybrid | [312] | Freeware. User-friendliness. Developed and tested by the scientific community. | Impossibility in modifying the code. |
| DTU Sigmund SIMP code for MATLAB | Freeware | MATLAB | SIMP | [54], Appendix of [122,359] | Freeware. Developed and tested by the scientific community. | Code modification is needed for most real cases. Less user-friendliness. More prone to user errors. Difficult to apply in complex geometries. |
| DTU Sigmund SIMP code for MATLAB new (2020) generation | Freeware | MATLAB | SIMP | [186] | Freeware. Developed and tested by the scientific community. | Code modification is needed for most real cases. Less user-friendliness. More prone to user errors. |
| DTU Andreassen SIMP code for MATLAB | Freeware | MATLAB | SIMP | [55], Appendix of [122,359] | Freeware. Developed and tested by the scientific community. | Code modification is needed for most real cases. Less user-friendliness. More prone to user errors. Difficult to apply in complex geometries. |
| DTU Andreassen LS code for MATLAB | Freeware | MATLAB | LS methods | [232], (a) | Freeware. Developed and tested by the scientific community. | Code modification is needed for most real cases. Less user-friendliness. More prone to user errors. |

**Table 5.** *Cont.*

| Programme | Nature | Runs on | Approaches/ Methods | Reference | Advantages | Disadvantages |
|---|---|---|---|---|---|---|
| Python alternatives to DTU (Sigmund, Andreassen et al.) MATLAB codes | Freeware | Python | SIMP | (a) | Freeware. Developed and tested by the scientific community. | Code modification is needed for most real cases. Less user-friendliness. More prone to user errors. |
| Zuo and Xie's BESO code for Python | Freeware | Python | BESO | [359,368] | Freeware. Developed and tested by the scientific community. | Code modification is needed for most real cases. Less user-friendliness. More prone to user errors. Difficult to apply in complex geometries. Some researchers still contest the approach. |
| Liu and Tovar's SIMP code for MATLAB | Freeware | MATLAB | SIMP | [359,369] | Freeware. Developed and tested by the scientific community. | Code modification is needed for most real cases. Less user-friendliness. More prone to user errors. Difficult to apply in complex geometries. |
| Huang and Xie's BESO code for MATLAB | Freeware | MATLAB | Soft-kill BESO | [121] | Freeware. Developed and tested by the scientific community. | Code modification is needed for most real cases. Less user-friendliness. More prone to user errors. Some researchers still contest the approach. Difficult to apply in complex geometries. |
| Suresh's Pareto-optimal tracing code for MATLAB | Freeware | MATLAB | Topological Sensitivity | [370] | Freeware. Developed and tested by the scientific community. | Code modification is needed for most real cases. Less user-friendliness. More prone to user errors. Difficult to apply in complex geometries. |
| Challis' LS code for MATLAB | Freeware | MATLAB | LS Methods | [371] | Freeware. Developed and tested by the scientific community. | Code modification is needed for most real cases. Less user-friendliness. More prone to user errors. Difficult to apply in complex geometries. |
| TOBS (Topology Optimisation of Binary Structures) code for MATLAB | Freeware | MATLAB | Gradient-based method with binary variables | [372] | Freeware. Developed and tested by the scientific community. | Code modification is needed for most real cases. Less user-friendliness. More prone to user errors. The approach undergone recent developments and it is not yet clear if it will gather acceptance. |

**Table 5.** *Cont.*

| Programme | Nature | Runs on | Approaches/ Methods | Reference | Advantages | Disadvantages |
|---|---|---|---|---|---|---|
| Wei et al.'s LS code for Matlab | Freeware | MATLAB | LS methods | [373] | Freeware. Developed and tested by the scientific community. | Code modification is needed for most real cases. Less user-friendliness. More prone to user errors. Difficult to apply in complex geometries. |
| Wang et al.'s TOPLSM for MATLAB | Freeware | MATLAB | LS methods | [369,373] | Freeware. Developed and tested by the scientific community. | Code modification is needed for most real cases. Less user-friendliness. More prone to user errors. Difficult to apply in complex geometries. |
| Schmidt and Schulz's SIMP code for MATLAB | Freeware | MATLAB | SIMP | [373,374] | Freeware. Developed and tested by the scientific community. | Lengthy code, which hinders its edition. Code modification is needed for most real cases. Less user-friendliness. More prone to user errors. Difficult to apply in complex geometries. |
| Polytop for MATLAB | Freeware | MATLAB | SIMP | [373,375] | Freeware. Developed and tested by the scientific community. | Code modification is needed for most real cases. Less user-friendliness. More prone to user errors. Difficult to apply in complex geometries. |
| Zhou et al.'s BESO code for MATLAB | Freeware | MATLAB | BESO | [373,376] | Freeware. Developed and tested by the scientific community. | Code modification is needed for most real cases. Less user-friendliness. More prone to user errors. Difficult to apply in complex geometries. Some researchers still contest the approach. |
| Otomori et al.'s LS code for MATLAB | Freeware | MATLAB | LS methods | [373,377] | Freeware. Developed and tested by the scientific community. | Code modification is needed for most real cases. Less user-friendliness. More prone to user errors. Difficult to apply in complex geometries. |
| Xia and Breitkopf's Homogenisation code for MATLAB | Freeware | MATLAB | Homogenisation | [373,378] | Freeware. Developed and tested by the scientific community. | Code modification is needed for most real cases. Less user-friendliness. More prone to user errors. Difficult to apply in complex geometries. |

**Table 5.** *Cont.*

| Programme | Nature | Runs on | Approaches/ Methods | Reference | Advantages | Disadvantages |
|---|---|---|---|---|---|---|
| Zhang et al.'s MMC code for MATLAB | Freeware | MATLAB | MMC | [373,379] | Freeware. Developed and tested by the scientific community. | Code modification is needed for most real cases. Less user-friendliness. More prone to user errors. Difficult to apply in complex geometries. |
| OpenMDAO | Freeware | Independent (Platform for modular code) | SIMP and LS methods | [380,381] | Freeware. Very useful for developing and adapting code with modularity. | It is not a TO programme and code needs to be developed. Suitable for research and learning, not so much for industrial applications. |
| Allaire's LS code for Scilab | Freeware | Scilab | LS Methods | [373] | Freeware. Developed and tested by the scientific community. | Code modification is needed for most real cases. Less user-friendliness. More prone to user errors. Difficult to apply in complex geometries. Many researchers are not used to Scilab environment. |
| FreeFem + + | Freeware | Can be adapted to the user's resources | LS Methods | [123,382] | Freeware. Developed and tested by the scientific community. | Code modification is needed for most real cases. Less user-friendliness. More prone to user errors. |
| Chisari's TOSCA (Tool for Optimisation in Structural and Civil engineering Analyses) | Freeware | Independent | Genetic Algorithms | [383] | Freeware. User-friendliness. | Difficulty in controlling the process. Recent and not profusely tested by the community. |
| Lagaros' C# code for SAP2000 | Freeware | SAP2000 open application programming interface (OAPI) | SIMP | [192,359] | Freeware. Very useful for civil and structural engineering practitioners. Developed and tested by the scientific community. | Limited user-friendliness for the target group. Once implemented in SAP2000, process control and code modification are limited. |
| He's script for adaptive layout optimisation of trusses | Freeware | Python | Discrete Optimisation; Heuristics | [384] | Freeware. User-friendliness. Allows member adding. Developed and tested by the scientific community. | Constrained to the optimisation of ground structures. |

**Table 5.** *Cont.*

| Programme | Nature | Runs on | Approaches/ Methods | Reference | Advantages | Disadvantages |
|---|---|---|---|---|---|---|
| GRAND (Ground structure analysis and design) | Freeware | MATLAB | Discrete Optimisation | [385,386] | Freeware. User-friendliness. Developed and tested by the scientific community. | Constrained to the optimisation of ground structures. Does not allow Member Adding. |
| Sokól's truss optimisation code for Mathematica | Freeware | Mathematica | Discrete Optimisation | [387] | Freeware. Developed and tested by the scientific community. | Constrained to the optimisation of ground structures. Does not allow Member Adding. Many researchers are not used to Mathematica programming. |

[(a)] Based either on publicly available information or on personal communications with the companies' contacts (which can be disclosed).

A remark shall be made concerning the *Advantages* and *Disadvantages* columns. Not only does the information rely on the consulted literature and commercial software technical detailing, but it also compares very different programmes. Commercial software is fundamentally different from free codes. Therefore, the advantages and disadvantages are essentially focused on each programme's nature and much less on its quality. Regarding the latter, we remain neutral and only reported (mostly) successful applications depicted in the published literature.

## 5. Topology Optimisation in Structural Design of Steel Elements and Joints

### 5.1. Steel Elements Design

Structural design has been pushed into Topology Optimisation for several reasons. Not only does the technology availability in much user-friendlier tools [388] play a critical role in facilitating the centuries-old task of optimising structural design, but also external pressures drove structural engineers into TO. Such pressures have been found both upstream, with an architectural demand for shapes that can only be optimised with extreme computational resources [389,390] and downstream, with the need for design processes able to foster Additive Manufacturing [391].

The former reasoning also explains why structural steel design has shown a particular prospect for successful TO applications [392]. The range of the applications includes Shape and Size Optimisation for steel members, as well as Topology Optimisation for the whole structural envelope. Furthermore, and unlike most of the other engineering disciplines, structural steel engineering still finds a preferred tool in Discrete Optimisation, over the Optimisation of Continua, for several problems with the macro-structures conceptual design.

Among the members design, perforated beams and shear walls are particularly prone to TO. The former has been studied in depth by Tsavdaridis' group at the University of Leeds [346], and the latter had a recent development in Bagherinejad [360], using commercial software. In both cases, a preliminary design with circular holes has been shown to evolve to a lattice-like geometry, with significant material subtraction in well-known less stressed areas. In a larger scale, floor diaphragm members have also been studied by Fischer et al. [393], to understand the optimisation possibilities under in-plane loading.

Still, within the sectional optimisation domain, the Free Material Optimisation (FMO) method extension [394] provided a critical tool for plates, shells, and member's parts design, and an impressive Academia-Industry joint effort allowed developing the Sectional Optimisation Method (SOM), which enabled the design of optimised aluminium members, accounting for fabrication constraints, standards regulations, and local instability phenomena [341].

Concerning structural systems and trusses optimisation, recent improvements include multi-objective optimisation techniques [395], Differential Evolution Algorithms [267], quantile regression for fostering the use of (discontinuous) I-beam cross-sections as design variables [396], and developments with the Interior Point method for non-linear and non-convex truss optimisation problems [397]. However, one of the most impressive developments can be found by Larsen et al. [398], a near-optimal truss design based on the homogenisation-based continuum TO. Practical applications of truss optimisation to bracing the systems design have some interesting contemporary examples in [399–403].

Therefore, high-rise buildings are an ideal ground for employing truss optimisation methods [106]. Specific issues of tall buildings have been addressed by recent TO studies, including the development of a genetic algorithm-based method for optimising outrigger systems [404], using swarm optimisers [279], considering linearised buckling in the TO process [405], developing a method for optimising bracing systems under adaptive multimodal load patterns [406,407], and conceiving structural systems for tall buildings based on the Optimisation of Continua using Reissner–Mindlin (or Mindlin–Reissner) shell elements [407].

Loading is, undeniably, a major challenge for TO in buildings structural systems, especially when multiple actions, combinations, and modes are considered and more so when the structure-load interaction is strong and influences the latter intensity. That is the case of seismic loading, in which specific, and usually extensive, code prescriptions apply. Subjected to such constraints, research in Topology Optimisation of seismically loaded structures is still exiguous. Nevertheless, the works of Kaveh's group on shear walls [408] and on different ductility steel Moment Resisting Frames (MRF) [409] can be highlighted as well as Qiao et al.'s [410] braces optimisation to the non-linear dynamic analyses of earthquakes time-histories.

A similar complexity can be found in fixed offshore structures for oil and gas or wind energy production. Even though such structural systems usually have a simple and intelligible conceptual design, wave and extreme loading require rather complex engineering [411]. For those reasons, recent developments in the conceptual design optimisation of jackets [412], including geotechnical aspects [413,414] and fatigue design [415] are very significant. Likewise, the method by Cicconi et al. [416] for multicriteria optimisation of modular steel towers is expected to have an impact on the industry.

Long-span structures have also been an object of study in TO, even if the research output is scarce in quantity. Within this subject, one can mention the Topology Optimisation of domes through a Colliding Bodies Optimisation (CBO) method by Kaveh [417] and the general space-frame steel roof optimisation method with Genetic Algorithms by Kociecki and Adeli [265]. Concerning bridge engineering, the spotlight is in the DTU TO group steel girder optimisation for super-long suspended spans, employing computational morphogenesis [418] (Figure 10).

Other significant advances, which may have an impact on structural steel design optimisation, include Christensen's work on TO under extensive and non-linear deformations [165] and Kristiansen et al.'s [419] studies on contact pressure and friction.

As steel design and detailing is deeply affected by fabrication and erection procedures, which usually constrain engineering options in a more extensive manner compared to construction with other common materials, a further investigation into recent research papers devoted to considering such aspects into TO is due. For such an end, and regardless of a deeper review on Additive Manufacturing for TO in Section 7, one shall refer to recent reviews [420,421], and keynote [422] on employing welding robotics in the so-called Wire and Arc Additive Manufacturing (WAAM) of topology optimised steel members and connections. However, the most common methods of creating steel in AM use powder as a feedstock and laser or electron beams as binding mechanisms [423]. Even if high-quality surfaces and accurate geometries are achievable with such means [424], steel properties are TO variables with broader uncertainty, due to its complex and repeated heating and cooling cycles [425]. Control, inspection, and testing will have an increasingly paramount role in enabling steel fabrication processes reliant on Topology Optimisation [129] and will add to the well-known hindrances of cost, aversion to change, and lack of technical knowledge [426].

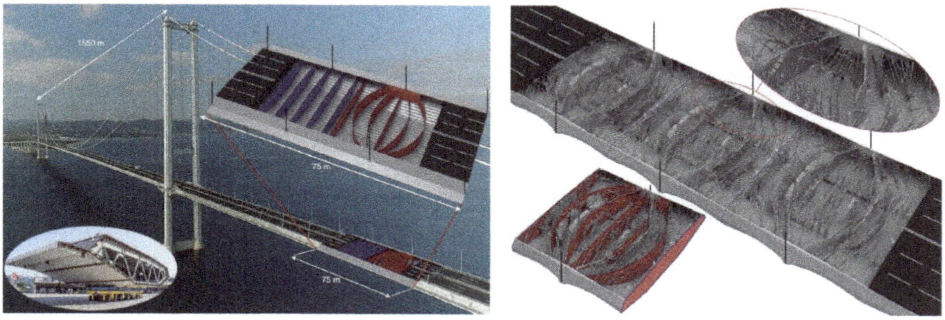

(a) (b)

**Figure 10.** Super-long suspended spans optimisation with computational morphogenesis. (**a**) Case-study initial (blue) and interpreted (red) design [418] and (**b**) computational morphogenesis result [418]; (reproduced under the Creative Commons Attribution 4.0 International License, http://creativecommons.org/licenses/by/4.0/ accessed on 25 February 2021).

## 5.2. Joints Detailing

In the current state-of-the-art steel connections (or joints, since both names are interchangeably used in structural engineering), TO is built of many mechanical engineering originated research endeavours and some developments made in the context of structural engineering research projects. Within the latter domain, innovative connections have been prototyped, breaking barriers of conventional manufacture constraints and delivering outstanding weight reduction [139]. Moreover, joints compactness, as regularly achieved by such designs, is critical for structural steel detailing, which is usually heavily constrained by space limitations.

Some recent examples of topologically optimised and additively manufactured joints can be found in Figure 11. Herein, new concepts and interesting solutions for space structures can be found [422], yet the significant predominance of applications for lattice, reticulated, and generally tensegrity structures is evident. Such shortcoming is mostly due to the limited quantity of load combinations which can reasonably be considered for the optimisation process. Should multi-axial bending and shear add to the tension and compression stresses, the TO procedure would escalate several levels of complexity and yield less intuitive results, in which validation would become a critical issue [427]. In fact, experimental testing of these joint details under a multitude of load cases is an inevitable step towards the broader employment of TO in steel joints detailing [139,427].

A further step towards the application of TO in joints design can be envisaged in Wang et al.'s comprehensive method for optimizing and fabricating joints in three-like structures [428]. As such, the nature-inspired structural concept is particularly prone to TO, in which a solution for optimising and manufacturing its joints is paramount. Other interesting developments in the TO of joints among tubular elements can be found in the work by Kanyilmaz and Berto [429,430]. Concerning the optimisation of spherical nodes in space frames, the work by Hassani et al. [364] is particularly relevant for providing a holistic guideline, including fabrication concerns, practical issues with Grasshopper modelling, as well as results post-processing and interpretation. Likewise, the work by Alberdi et al. [431] on the connections topology optimisation in Moment Resisting Frames (MRF) is deemed to assist structural designers in the task of managing the recursive task of designing frames and its joints.

Nevertheless, several and significant factors still hinder joints TO. Some of the most apparent problems lie in the materials properties uncertainty and fabrication cost. With a strong relation to AM techniques [139], those can only be solved by sensible technology developments, financial investments in the industrial capacity, and workforce know-how, as well as a stronger standards framework.

Problems with the optimisation process have an outstanding issue in the incapacity of some algorithms for accounting non-linearity [422]. This is critical in several connection types, including the bolted ones, where plasticity plays a major role. However, significant advances have been made in the modelling of bolted connections, holes, bolt-hole contact, and friction in the context of TO [432–435].

Furthermore, a very significant research published in the context of mechanical engineering systems is general enough to have a profound impact also on steel joints detailing. That is undoubtedly the case of the technique for creating idealized bolts with the topological derivative approach proposed by Rakotondrainibe et al. [247] in a Renault associated research. Other examples can be found in the synthesis method for mechanisms proposed by Kang et al. [436] with a possible application to the pinned and sliding joints in structures.

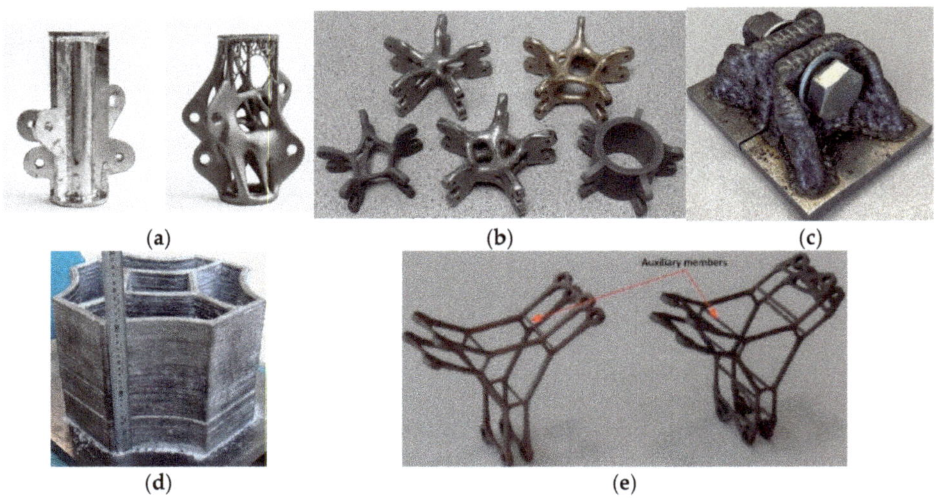


**Figure 11.** Examples of topologically optimised and additively manufactured metallic connections for civil engineering structures. Left to right and top to down: (**a**) Non-TO and TO joint specimens by ARUP [355]; (**b**) several joints designed in the University of Leeds [139]; (**c**) wire and arc additive manufacturing (WAAM) printed bolted joint in the Technical University of Darmstadt [422] (reproduced under the Creative Commons Attribution-NonCommercial-NoDerivs License); (**d**) WAAM printed node in the Technical University of Darmstadt [422] (reproduced under the Creative Commons Attribution-NonCommercial-NoDerivs License); (**e**) joint specimens topologically optimized with the bi-directional evolutionary structural optimisation (BESO) method for axial loads and bending in the RMIT University [427].

### 5.3. Buckling and Local Instability Phenomena

Member buckling phenomena has been addressed in Discrete Optimisation for a long time. However, within the Optimisation of Continua, multiple global and local, real and fictional, buckling modes hampered progress after precursory works undertaken in Instituto Superior Técnico [437]. Some notorious problems have been found in the appearance of buckling associated with low-density regions, high local stresses, which add to the repetition of modes, convergence problems, and the need for an extensive computational capacity for dealing with so many modes [137,388,438].

Recent advances already allow accounting for buckling in TO endeavours in a computationally feasible manner [438–442], mitigating some of the aforementioned problems

Though, those methods rely on linearized buckling, which is generally regarded as an inadequate simplification for many engineering problems [137,443,444] and therefore, face significant opposition from many researchers. Adding to such a discussion, a recent research paper using non-linear pre-buckling analyses in the context of microstructural design [445] found that under certain loading conditions, linear and non-linear buckling analyses yield similar results. Considering the knowledge gathered by structural engineers in linear and non-linear buckling over time, it is expected that such a conclusion may be proven valid for several other cases.

On the other hand, advances have been made in the last few years on the non-linear buckling of topologically optimized continua [446,447]. Such remarkable achievements employ non-incremental analyses and recursive design.

Despite the mentioned problems, under which buckling phenomena can only be regarded as a mostly unsolved problem in the TO of continua, some practical applications found simpler or more sophisticated strategies for modelling instabilities in optimisation methods. That is the case of Tsavdaridis' work on local instability in optimised aluminium cross-sections [341].

Regarding Discrete Optimisation, methods for considering buckling also evolved. Among a recent contribution, one can highlight the work by Weldeyesus and Tugilimana on trusses [397,448], Xu et al. suggested a practical approach for TO in tensegrity structures [449,450], as well as the research by Zhao et al. [451] on methods for mitigating member instability in reticulated structures.

### 5.4. Structural Design Codes Compliance

Bridging research and practice in structural engineering faces some hindrances beyond the simple transfer of knowledge. Unlike many other fields, where product design is strongly bounded with Research and Development, since testing, compliance, and certification will follow pilot production, the design of building and bridge structures must comply with an extensive set of rules, codes of practice, and standards beforehand. Those documents are typically reviewed in a pluriannual basis and not necessarily include the most recent research, since a broad and heterogeneous community of practitioners is not expected to radically change the design methods frequently.

As a result, many calculation approaches based on non-constant members or employing advanced sectional analyses, even if practical and validated, may take long until explicitly defined in structural standards [452]. This is certainly the case for widely adopting TO in civil and structural engineering and, also, a reason underlying the scarce number of recent publications in TO to specific standards prescriptions.

Within the recent literature pertaining to optimisation programmes, exceptions to the former can be found in Tsavdaridis' optimisation of aluminium cross-sections [341], as well as in works of the Discrete Optimisation of trussed structures, including truss design to Eurocode 3 (EN1993-1-1 or, simply, EC3) with Differential Evolution Algorithms [267] and bracing systems optimisation with GA to the AISC-LRFD American standard [402,453].

### 5.5. Multiple Loading and Robustness

Significant obstacles still limit the broader employment of TO in structural design. One of the most important is the susceptibility of optimisation results to loading patterns. Albeit, recent advances onto robust solutions, which show endurance to loading scenarios multiplicity [454] and extreme degrees of TO in structural design usually result in members that efficiently withstand a finite number of loading patterns used in the design, but may be inefficient for other loading patterns, even resulting in less intuitive and less visually appealing topologies. In fact, real structural design optimisation problems are deemed to fulfil the requirements of multiple objective functions [455–458].

Under these circumstances, the TO use for structural engineering is impaired by three reasons: First, loading patterns can be very profuse, up to hundreds or thousands in complex structures. This is not necessarily a research problem, but a practical one since common computational methods are still not suited for delivering TO results for many loading patterns in a reasonable time.

The second problem is related to the absence of accidental and unconventional loading patterns in regular structural design. Without it, the redundancy of structural members may become negligible and an unexpected collapse under non-conventional loading may occur.

The last, and possibly most severe, set of problems is related to the current methods for TO under multiple alternating loads. Notwithstanding some recent advances, current methods still face the inconvenience of non-unique solutions [459] and local extrema, as well as the difficult to account for loads with very different scales of intensity [458].

Multiple-loading and multi-objective optimisations have been handled with a plethora of methods. Both deterministic or physically accurate, and uncertainty-based methods, including notorious fuzzy approaches, have been used [458]. Arguably, the Kreisselmeier–Steinhauser function (KS) stood out in the past decades, mostly in aeronautical engineering studies.

Among recent advances, one can highlight the efforts to account for uncertainties in the loading intensity and position with the method by Wang and Gao [460], and the compliance-function-shape-oriented approach by Csébfalvi [461] and precursory work [462], the generalized material interpolation scheme by Chan et al. [463], as well as the RBTO approach by Nishino and Kato [464].

Previously, Li et al. proposed one more option for multi-objective TO, employing the Normalized Exponential Weighted Criterion (NEWC) and the Fuzzy Multiple-Attribute Group Decision-Making (FMAGDM) theory [458]. Likewise, Yi and Sui introduced the use of the Transplanting Independent Continuous and Mapping Ideas into Materials with the Penalisation (TIMP) method for the TO of plate structures under multiple loading [465]. Its interesting feature lies in adding one more penalty function to a SIMP-like method.

One very different advance can be found in the approach by Tang et al. [466] to the wind loading complexity through integrating Computational Fluid Dynamics (CFD) models into a BESO method for TO.

Much of this research is related to contemporary multiple loading. The problem of non-simultaneous multiple loading implies significant computational efforts to extend the analysis scope, integrate TO results, and deal with very significant practical issues.

Works on dealing with alternating loads can be found early in the TO literature, including the 1970s Prager work [467], but hardly can be considered solutions for the complexity of the currently analysed problem. Recent works with alternating loads include Alkalla et al.'s Revolutionary Superposition Layout (RSL) method [468], as well as Lógó and Pintér contributions, [459,467]. Furthermore, Tsavdaridis et al. managed to use a method to examine and overly stress paths and compose comprehensive layouts, optimized for several sets of loads [341], [469].

As the way ahead is likely to be facilitated by the admirable rate of global computational power increase, only a few TO research centres [169] already possess the means for dealing with multiple and alternating loads and load combinations in more-than-trivial problems.

However, structural engineering has deployed solutions for dealing with uncertain loading and enhancing reliability. One solution lies in the current "Capacity Design" philosophy, favoured in Structural Eurocodes for seismic actions [470,471]. Employing it in TO could provide a minimum threshold for providing the versatility to the structural design of elements.

## 6. Recent Advances in Related Fields with Applicability in Structural Steel Design

Owing to the broadness of Topology and Topology Optimisation concepts, applications are primarily cross-disciplinary within scientific, engineering, and even graphical fields. Thus, neighbouring fields have been thoroughly investigated for recent developments with a perceived or expected potential for usefulness in structural steel design.

Nevertheless, one shall refer to the fact that only a fraction of what lies within the TO umbrella is of value for the theme under scrutiny. Suitability depends mostly on the driven objectives towards optimisation. Therefore, the following synthesis is not deemed to assess TO in other fields, but to identify advances in TO which can be adopted in TO endeavours for structural steel design.

### 6.1. A Broader Look into the Construction Industry

Newly developed tools for cross-disciplinarity, namely involving architectural design into the structural design efforts for TO [343], will have an impact on structural design. One other interesting investigation analyses incremental loads and structural layouts in TO [472]. Even if it is directed towards incremental concrete bridges, its methods can be useful for dealing with steel structures staged construction and, in a broader sense, assist in TO with multiple loading and evolutive layouts. Furthermore, advances in soil-structure interaction for geotechnical structures optimisation [473] may have an impact on the moving loads' issue.

### 6.2. Concrete Structural Design

The recent "Digital Concrete" conference [474] contains strong proofs of concrete design research and practice engagement in TO. While it is undeniable that the current progress is more strongly bounded to concrete Additive Manufacturing, under the name of 3D Concrete Printing (3DCP) [475], Topology Optimisation in concrete structures already advanced far beyond the unreinforced concrete, where AM is already flourishing, paving the way for interesting developments in reinforcement design models.

More than 120 years after the work by Ritter [476] and Mörsch, as well as over 30 years past Schlaich et al.'s notorious contribution [477], the Strut-and-Tie Models (STM) still govern the concrete design with remarkable resemblances to what previous generations of structural engineers mastered. That may well be subject to change as new developments in TO are deemed to reshape our understanding of reinforced concrete design, with contributions by Zhou et al.'s [478], Yang et al.'s [479], Jewett and Carstensen's [480], and Xia et al.'s [481]. However, the promise of fundamental innovation beyond the aforesaid applications remains restricted to only a few distinct approaches. One of those is Pastore et al.'s [482] risk-factor to replace the Von Mises stress criteria for optimizing heavily constraint structural elements.

Other applications of optimisation in structural concrete design include targeting seismic performance objectives [483–485], the optimisation of prestressed concrete members [486], and defining critical concrete structures general topology, from the Optimisation of Continua, as performed by Wu and Wu for bridge pylons [487]. These works will, most certainly, have a contribution also for steel and concrete composite structures.

### 6.3. Aerospace and Defence Industries

Aerospace engineering, with its utter need for weight reduction and frequently generous funding, has been a source of inspiration for TO breakthrough innovations. Over the years, sensible contributions to critical issues in TO, such as fastener design and dynamic analyses, now used in structural steel design, came from this field [488].

Recent advances in TO from this field include optimisation methodologies for additively manufactured parts with enhanced accuracy to strain and displacement [489], experimental analysis of several topologically optimised micro and macro-structural systems for ribs [490], optimisation to cumulative mechanical and thermal loads [491], and advanced materials design for high-quality AM [492]. All these contributions can have a deep impact on the quality of topologically optimised and additively manufactured alloys, leveraging its rapid application also in the construction industry.

### 6.4. Automotive Industry

Practical and fabrication-oriented methods are regularly deployed by the automotive industry. Hence, this field has been a continuous source of inspiration for framing TO developments in the track of meaningful advances. Novel methods include Mantovani et al.'s guidelines for integrating AM requirements into TO, as well as for successfully processing and managing the TO results [493]. Similarly relevant are the studies by Van de Ven et al. [494] and Mass and Amir [495] for mitigating the impact of overhangs for AM as design constraints. Limiting or conveniently positioning overhangs is paramount not only for subtracting complexity to the numerical models, but also may prevent failure mechanisms in steel alloys, such as fatigue-related. Such developments are applicable to the most additively manufactured parts, not only in the automotive industry.

Detailed procedures for optimising vehicles parts can be found in Topaç et al.'s [496] reduction of 63% of a mechanical component mass in Kumar and Sharma [497], in Mantovani et al.'s [498] reduction of a steering column support mass in almost 50% in Li and Kim's approach to topologically optimise a car part with limited information, manufacturability concerns to leverage extrusion and casting processes, as well as a post-processing method for geometry reinterpretation, while achieving almost 40% of weight reduction [499].

### 6.5. New Materials, Composites, and Polymers Design

Materials design has been taking advantage of TO at a microstructural level and, conversely, promoting its development. As Osanov and Guest [135] formulated it, the fundamental question in architected material design optimisation can be resumed to finding which microstructure will deploy sensible enhancements to macrostructural properties. The answer is complex and involves entrenched unit cell modelling, upscaling, and repetition [184,500,501]. Yet, diverse and frequently remarkable solutions can be attained, from useful elastic properties, including auxetic (NPR) materials (those with a negative Poisson's ratio), as popularized in Sigmund's work [502] and currently drawing much attention from the scientific community [503], to extreme thermal properties, optimised fluid permeability, and materials governed by non-linear mechanics for utmost energy absorption.

Within the aforementioned exciting framework, recently published works include modelling methods for designing hierarchical structures employing non-uniform topologically optimised lattices [504], serving as enhanced energy absorbers in sandwich sheets [505] or facilitating manufacturing [506], even with new methods for mitigating non-smooth surfaces, as the Bézier Skeleton Explicit Density (BSED) Representation Algorithm [507]. The microstructural design for avoiding stress peaks has also been recently addressed [508], as uncertainty-resilient design for inter-diffusion interface issues has been brought forward [509], and advances in buckling of microstructures were published by Bluhm et al. [510].

The fabrication of these architected materials with AM techniques is yet another concern. Not seldomly published methods can be incomplete or vague, as Huang et al.'s review pertinently points out, and offers mitigation by congregating some state-of-the-art answers [511].

Even though new materials and, specifically, new material microstructures may still be far from promoting a change in steel alloys, structural steel design can find many important lessons in the formerly cited research. Both the tools and methods created for microstructural optimisation can be employed in the multi-purpose TO, but also solutions developed for a particular material microscale may have an employment in construction macrostructures composed of smaller members. That is also the case of recently developed methods for efficiently handling buckling in polymers optimisation [512], as well as for multimaterial TO (MMTO) [513,514], with its specific problems, as extensive local extrema, and techniques designed to overcome such obstacles. Furthermore, many important advances in TO methods and approaches are being developed within the materials design field, making it especially relevant, also for TO in civil and structural steel design.

At a coarser scale, construction composite materials can also be enhanced under the assistance of TO. That is the case of the types of cement with an enclosure of rubber waste [515] or Carbon Fibre Reinforced Plastics (CFRP) [516].

### 6.6. Industrial Design, Mechanical Engineering, and Multiphysics Endeavours

One of the most prolific research lines in TO relates to compliant mechanisms design. These flexible structures, which are ubiquitous in most high and low technology consumable products nowadays, have been a perfect ground for TO. Owing to its reliance in parts flexibility rather than hinged joints, these components rigid body design had been complex, frequently erroneous, and very dependent on prototype testing, whereas compliant mechanisms TO bridges most of those shortcomings by efficiently addressing the target flexibility.

Upscaling from small consumables, in electronics for example, larger components such as grippers have been produced using the TO of compliant mechanisms [517]. Furthermore, new methods for addressing the manufacture uncertainty and stress constraints [518] show a new maturity level for compliant mechanisms design, which may lead to its use in more perennial applications and structures.

The mechanical design of diverse components and structures has also faced recent developments in TO related subjects. If, on the one hand, the TO of vibration problems remains largely constrained to small amplitudes [133], on the other hand, techniques are being developed for introducing High-Cycle fatigue as an optimisation criterion [519], [198], and concerns over accidental actions are being addressed within the ship design industry [520].

Multiphysics problems in optimisation are usually deemed to address thermal and mechanics conditions [521]. Recently, Cheng et al. employed the Lattice Structure Topology Optimisation (LSTO) to design a cooling channel system [522] and Perumal et al. found a technique to mitigate thermal accumulations by optimising lattice structures locally where such concentrations are more prone due to the metal AM process with a powder-bed fusion [523].

Another very interesting branch lies in the TO of fluid-based problems. Although recent, it congregates the optimisation of flow contacting surfaces, solids transport, heat transfer or porous media [134,363]. Shortly, the herein developed techniques may be efficiently integrated into CFD analyses [524] and significantly enhance the building structures design and optimisation to wind effects. Furthermore, the fascinating *"poor man's"* approach for natural convection problems [525] can be very useful for optimisation in porous media, including geosciences, geotechnics, and hydrogeology.

### 6.7. Medical Devices and Personalised Medicine

The medical devices industry has shown a remarkable appetite for creating high-value products with newly available technologies for the very demanding health sector. Thus, it was no surprise that AM was mastered by customised medical devices from its early days.

Currently, the capacity for three-dimensional printing solutions with high accuracy, efficiency, and adhering to the patient needs is already well-rooted, and that provides practical lessons for many other fields where TO is being used for enabling AM, including structural steel design. Recent examples include Rapid Prototyping (RP) of plastic casts made with human limbs scan data as an efficient replacement for the plaster casts method [526], with further TO for improving resistance, rigidity, and ventilation, as well as personalized aneurysm implants [527], which are expected to be topologically optimised and additively manufactured with NPR materials, so that the clinical use with significant advantages for the patients is attained.

Understandably, the medical devices industry needs to be at the forefront in developing high-quality surfaces for the additively manufactured parts. A recent contribution can be found in Chen et al.'s [528] technique to minimize the number of fabrication supports accounting for the main printing direction.

## 7. Designing for Additive Manufacturing with Topology Optimisation

Metallic materials conventional design has been mostly based on subtractive techniques for centuries. Cutting and drilling flat sheets of various thicknesses and laminated profiles have been the rule in steel construction, in which joining by bolting and welding became a ubiquitous counterpart. As a result, the design is limited to geometrical bounds, stockage is profuse, waste is significant, the material is not necessarily located where it contributes the most, and the whole process is very labour intensive. A notorious exception can be found in casted elements, which had its broadest use in the iron construction and now is mostly limited to steel nodes. However, the need for moulds makes this technique only viable when many similar products must be produced. Such lack of versatility usually limits this option to large space structures with equal nodes.

Contrarily to the former, recent techniques allow producing metallic alloys by addition. Despite some critical technological differences, these techniques have been nested under the informal name of 3D Printing in the context of the current fourth industrial revolution. Other terms include Additive Manufacturing [130], Additive Layer Manufacturing (ALM) [131], to allude to the layer-by-layer nature of the process [423,425] or Solid Freedom Fabrication (SFF) [126].

While AM has been commercially used for more than 20 years, typically for rapid prototyping without commercialisation purposes [354], only now the industrial capacity for widespread and reliable manufacturing of steel, titanium, and aluminium has been achieved [423], offering a tremendous opportunity for the construction industry [129].

Making AM a reality in construction plants and yards will depend on several economic and technical factors. Among the latter, one can highlight the ability to introduce robotics in construction [421] as with Large-Scale Prefabrication (LSP) [426], having consistent and usable-by-practitioners design tools [130,388,529], as well as achieving commercial maturity, including reliability and scale in AM techniques.

Plentiful AM techniques, diverging solely from its name or commercial branding to its nature fundamentally, hinder non-experts the understanding of the available options for manufacturing. Even if an encompassing knowledge on the matter requires consulting comprehensive Review Articles, such as [129,423,425,530].

Table 6, which adheres to the American Society for Testing and Materials (ASTM) classification of AM techniques, may assist in the task. It shall also be noted that distinctions among categories and techniques are usually made concerning its feedstocks and binding mechanisms [423].

However, not all those categories bear AM processes currently suitable for manufacturing of steel alloys. Some, as depicted, do not allow using metal powder or wire, being limited to polymers, ceramics, and other low-strength materials. Regarding the need for post-processing, DED-PA and DED-GMA metal production require machining, DED-EB usually needs surface grinding, DED-L may entail surface treatments, while PBF-L and PBF-EB are less likely to be post-processed [425]. Therefore, PBF and some DED processes are commonly designated direct-to-metal.

**Table 6.** Additive manufacturing (AM) categories and techniques.

| AM Categories to ASTM's Committee F42, ASTM F2792-12a [531] (Withdrawn) and ISO/ASTM 52900:2015 [532] Definitions | AM Techniques | Remarks | References |
|---|---|---|---|
| Vat Photopolymerisation | Stereolithography (SLA) | A liquid resin is solidified by Ultraviolet (UV) light exposure. | [126,530] |
| Material Jetting | Polyjet | Polymer or wax drops are jetted through a nozzle and cured with UV light | [126,530] |
| Binder Jetting | Indirect Inkjet Printing | A print head jets powder-based materials and fluid binder layers. It can be used for metals, but only high-porosity products are usually produced | [126,530] |
| Material Extrusion | Fuse Deposition Modelling (FDM) | In FDM, material is heated and continuously expelled through a nozzle | [126,530,533] |
| | Contour Crafting | | |
| Powder Bed Fusion (PBF) Used in metals | PBF-L, also known as Laser Beam Melting (LBM), Direct Metal Laser Sintering (DMLS), Selective Laser Melting (SLM), Laser Metal Fusion (LMF), LaserCUSING, or industrial 3D printing | Divided by heat source to liquify the powder: L for laser, EB for electron beam. It is essential to mention that while some authors regard DMLS and SLM similarities as sufficient for considering the terms as synonyms, others prefer to separate it. SLS is used for polyamides and polymers. Generally, PBF techniques allow producing metals with both good accuracy and mechanical properties | [126,129,421,423,425,530,534–537] |
| | PBF-EB, also known as Electron Beam Melting (EBM) | | |
| | Selective Laser Sintering (SLS) | | |
| Sheet Lamination | Ultrasonic Additive Manufacturing (UAM). Used in metals | Low temperature joining of metal sheets by ultrasonic welding | [126,129,530] |
| | Laminated Object Manufacturing (LOM) | Usually with paper and glue | |
| Directed Energy Deposition (DED) Used in metals. In such a case the term Metal Deposition (MD) has been employed | DED-L, also known as Laser Metal Deposition (LMD), Laser Engineered Net Shaping (LENS), Direct Metal Deposition (DMD), Laser Engineered Net Shaping (LENS), laser deposition welding or laser cladding | Divided by heat source: L for laser, EB for electron beam, PA for plasma arc, and GMA for gas metal arc. Powder or wire material sources can be used. WAAM technique is similar to conventional automatic welding procedures, using wire and an electric arc, as Plasma Arc Welding (PAW), Gas Metal Arc Welding (GMAW), and Gas Tungsten Arc Welding (GTAW). As a result, WAAM may be nested under the other DED techniques. | [126,129,420,421,423,425,530] |
| | DED-EB | | |
| | DED-PA | | |
| | DED-GMA | | |
| | Wire and Arc Additive Manufacturing (WAAM), also known as 3D welding, Shape Metal Deposition (SMD), Shape Welding (SW), Shape Melting (SM), Rapid Prototyping (RP) or Solid Freeform Fabrication (SFF) | | |

Other aspects of paramount importance for assessing the suitability for steel manu facturing include the fabrication speed, properties reliability, and the alloy cooling rat The latter will have a profound impact not only in residual stresses but also in the ste composition, where the control of martensite and retained austenite is crucial [423].

The former aspects are not always easy to assess, mostly since the equipment, cont and several other industrial manufacturing details play a significant role. Neverthele choice has been made to limit the Table 6 scope to proven and industrially viable techr gies, even if not all of those are applicable to the metal alloys manufacture. Thus, emer technologies such as Electrochemical Additive Manufacturing (ECAM) [129,538] wa included.

On the opposite field, some AM techniques have long been proposed befo concept of AM was defined. That is the case of WAAM, which dates back to 1925 [4 is now highly productive. Hence, some researchers regard WAAM as a leading op steels manufacture [421], provided that shortcomings with high residual stresses, and geometrical properties are successfully addressed (as can be seen in Figur the other hand, PBF-L (also in Figure 12), PBF-EB, and DED-L have been mentior most relevant technologies for producing metallic alloys in the current context of

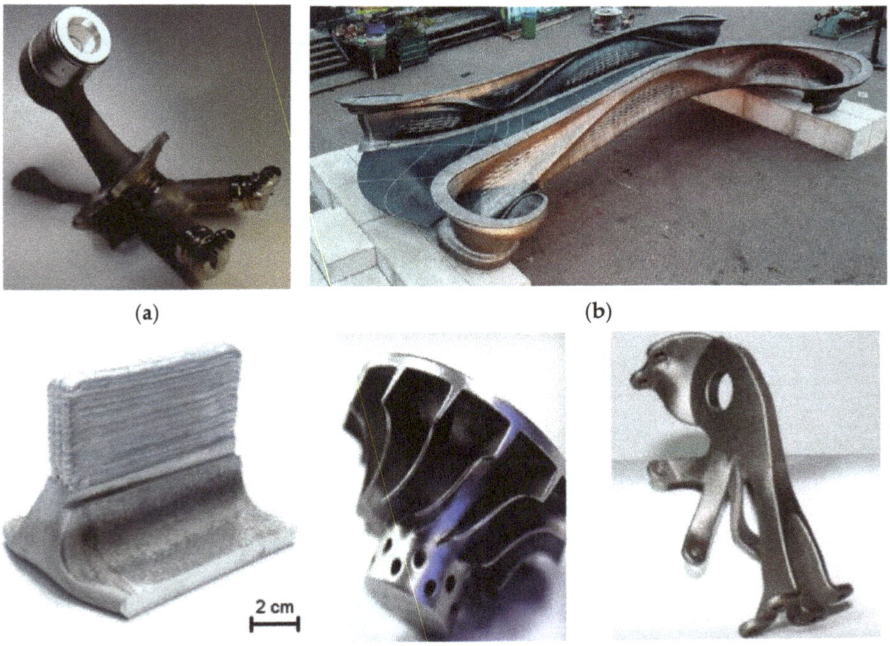


(a) (b) (c) (d) (e)

**Figure 12.** Examples of additively manufactured metallic parts. Left to right and top to down: (**a**) Nozzle manufactured with PBF-L [423]; (**b**) MX3D bridge manufactured with WAAM [129,420]; (**c**) rib manufactured with WAAM [420,539] (reproduced under the Creative Commons Attribution 4.0 International License, http://creativecommons.org/licenses/by/4.0/ accessed on 25 February 2021); (**d**) turbine part manufactured with PBF-L [530]; (**e**) aerospace structure component manufactured with PBF-L [540].

The role of TO as an enabler for AM justifies the unparalleled growth of both [130]. Currently, the Design for Additive Manufacturing (DfAM), which succeeded the Design for Manufacturing (DfM) traditional methods [541,542], employs more or less advanced optimisation tools, spanning practical lattice-based ones, as depicted by Chen et al. [543] and illustrated in Figure 13, to complex approaches [544,545], and is already paving its way into structural steel connections [546,547], albeit mostly for tension-only nodes.

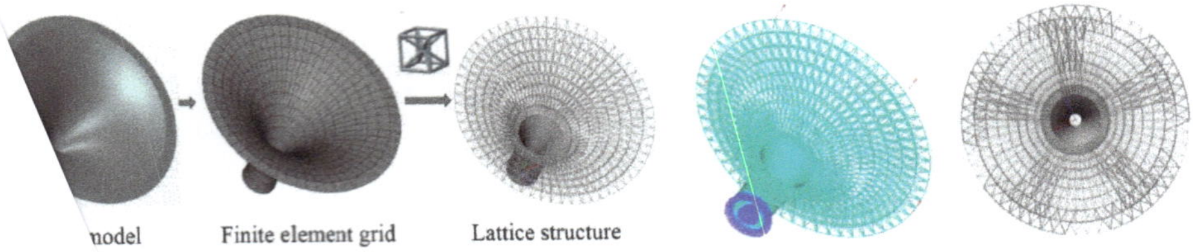


(a) (b) (c)

…based design for additive manufacturing (DfAM). From left to right: (**a**) CAD model; FEM; lattice model; …loading definition; and (**c**) final optimized structure [543] (reproduced under the Creative Commons …ational License, http://creativecommons.org/licenses/by/4.0/ accessed on 25 February 2021).

As highlighted in recent Review Articles [130,424,425,530] as well as Buchanan and Gardner 2019 [421], TO is yet to overcome important challenges to meet AM requirements. A stronger link between the DfAM and production is needed to overcome the current bottleneck, including better accounting for AM metallurgical aspects, such as residual stresses, internal defects, interlayer bonding conditions, and anisotropy [536,548]. In fact, composite materials TO for AM is still afar from the industrial practice.

Other challenges for the current DfAM can be found in the difficulty to avoid overhangs and general support structures [549], which promote waste, require labour, time, and cost to post-process and hinder surfaces' quality. Furthermore, for the AM process convenience, the present DfAM makes extensive employment of open-walled infills which, amidst important contributions for material reduction, have a frequently detrimental effect on the product stiffness [550], and especially on the torsional stiffness.

Another important and often overlooked problem lies in the difficulty in testing additively manufactured parts, especially by Non-Destructive Testing (NDT) techniques [551], which result in a poor understanding of material properties yielded by different AM techniques [552].

On the other hand, recent research brought innovative design techniques to assist AM. Table 7 summarises some selected recent developments, which add to the advances in TO depicted in the previous chapters.

**Table 7.** Recent research on design techniques for AM.

| Development | Reference |
|---|---|
| Multi-Material Topology Optimisation (MMTO) by generalising the single-material Pareto tracing method. | [553] |
| Guidelines for integrating TO and AM to restrain support structures. | [246] |
| A TO approach to minimising support material by optimising an allowable self-supporting angle. | [554] |
| A TO approach to designing self-supporting structures. | [555] |
| Guidelines for TO regarding AM process parameters. | [556] |
| Guidelines for enabling AM newcomers to produce better quality designs. The proposed worksheet has a useful "pass-or-fail" arrangement. | [557] |
| A method for topologically optimised infills in additively manufactured parts. By fostering non-uniform infills, better results have been attained. | [558] |
| TO for AM with EBM (or PBF-EB). | [559] |
| AM-fabricated Materials with Site-Specific Properties (MSP). This design approach allows designing materials with heterogeneous properties, where local material enhancement is possible without requiring an increasing fabrication time and cost for the whole element. | [560] |
| TO approach to minimise overhangs, assisted by its sensitivity analysis. | [561] |
| TO approach to minimise overhangs and generating self-supporting structures. | [562] |
| TO approach to minimise support structures and thin features. | [563] |
| Optimisation of support structures design for AM. | [564] |
| A multi-objective TO approach to minimise supports material consumption and removal time and cost, while minimising parts deformation. For such an end, a repulsion index (RI) is used. | [565] |
| A design method for additively manufacturable free-form periodic metasurfaces. | [566] |
| A multiscale method which includes both TO macro-scale optimisation and AM layers mesoscale. | [567] |
| A new version for the MMC method based on AM data. | [568] |
| A TO model to account for thermal residual stresses and deformations due to the AM processes. | [569] |
| TO approach to minimise overhangs in compliant mechanisms, using the Smallest Univalue Segment Assimilating Nucleus algorithm. | [570] |
| TO approach to minimise overhangs and assure a minimum length scale, both for enhanced printability. Application to a tensegrity connection. | [547] |
| TO for AM with SLM (or PBF-L). | [571] |
| A Non-Probabilistic Reliability-Based Topology Optimisation (NRBTO) method to account for additively manufactured materials' properties uncertainty. | [572] |
| A TO method to account for AM cost and fabrication time. | [183] |
| A TO density-based approach with one field for design parameters and another field for support layout optimisation. | [573] |
| A TO approach for designing and re-designing additively manufacturable parts. | [574] |

**Table 7.** *Cont.*

| Development | Reference |
|---|---|
| A new perspective on overhang control, taking into account AM techniques specific features with a skeleton-based structure decomposition approach. | [575] |
| TO approach to minimise supports in enclosed voids. Employs a Nonlinear Virtual Temperature Method (N-VTM) to find enclosed voids. | [576] |
| A TO approach for re-designing additively manufacturable parts. Integration with a commercial computer-aided design software is discussed, which enforces the study's practical applicability. | [577] |
| A TO approach for parts repair or upgrade with AM after subtractive machining. | [578] |
| A TO approach for merging structural optimisation and WAAM deposition sequence. | [579] |
| TO approach for taking into account the assembly design. | [580] |
| A multi-objective minimisation TO approach which account for AM fabrication time and cost. | [581] |
| Multiresolution Topology Optimisation (MTOP) method for high resolution AM with overhang and minimum length control. | [582] |
| An NRBTO method to account for additively manufactured materials' properties local uncertainty. For such an end, Uncertainty Regions (UR) are defined within the design domain. | [583] |
| TO with kinetic analysis added to the common FEA. | [584] |
| Solid Anisotropic Material with the Penalisation (SAMP) technique better integrating process parameters into TO. | [585] |
| TO of multi-material infills with a systematic multi-phase design method. | [586] |
| Poisson's equation-based scalar field constraint to suppress enclosed voids in TO. | [587] |

The path between a topologically optimised model and a manufacturing input can pose significant obstacles, even if the smoothing of TO model boundaries has already been accomplished [588]. Not only several fabrication parameters must be accounted, but also data transfer can be a complex task. The TO model must be exported to a CAD three-dimensional format, suited to the manufacturing process and sliced into thin layers [423]. Moreover, many available approaches have not been broadly and satisfactorily tested for manufacturability [589].

Regarding this issue, one may refer to the work by Zegard and Paulino [97], yielding TOPslicer, a MATLAB code for transforming matrices or arrays with density values into additively manufacturable digital outputs. Furthermore, ontologies databases may assist designers in enhancing DfAM. Within this domain, Dinar and Rosen [590] contributed to systematise a knowledge base, and to infer the design rules using the Web Ontology Language (OWL)/Resource Description Framework (RDF) Protégé tool must be highlighted.

## 8. Future Trends

In the aftermath of the latest World Congress on Structural and Multidisciplinary Optimisation, held in 2019, surrogate-based optimisation and optimisation under uncertainty were highlighted as two trending issues for the future of TO [591]. In fact, the latter has a significant expression in recent studies by some of the leading researchers in the field, as one can regard in Da Silva et al.'s work with the Augmented Lagrangian method for computing loading uncertainty [214,454].

The prospective dissemination of TO in structural design, based on increasingly user-friendly and reliable software [2], is quite an undisputed prediction. However, the means to enforce such development are subjected to different views. As Sangree et al. reported the experience of inserting TO in engineering education as soon as in freshman levels at Johns Hopkins University both as a design tool and as a mean for understanding the force flows [592], Lagaros [2] alluded to the role of the State to force practitioners to apply new technologies. Concurrently, Gao et al. [530] pointed out to open academic research platforms and intellectual property expiry, but also governmental investments and expertise as catalysts for enforcing the application of emerging standards and fostering DfAM and AM processes integration and interconnectivity, potentially opposed by larger revenue companies.

DfAM or AM-Oriented TO (MOTO) is envisaged to undertake the systematic study of trade-off relationships between TO and AM, unlike what has mostly happened to date [593].

This may also include considering fabrication models which integrate more conventional processes [530], as well as advancing in multi-material products, assisted by progress in simulation algorithms [530]. Nevertheless, an increasingly closer relationship between TO and AM is perceived as vital for the industry's development [594].

On the other hand, significant hurdles are yet to overcome in TO for AM. Meng et al. reported the need for more advanced TO methods for multifunctional products, as well as for progress in multiscale TO [594] and Lim and Wong focused on the need for enhancing TO performance regarding aerodynamics problems [352].

Conclusions drawn from structural steel TO suggest that a higher optimisation grade and further proofs of safe applicability of additively manufactured connections are the future paths for achieving economic viability, compared with labour-intensive traditional manufacturing [355].

Focusing on the AM process, Bañón and Raspall [81] concluded that future directions in three-dimensional printing for architectural purposes include large-scale printing, Artificial Intelligence (AI) embedment into the design process, and computerized assembly [81]. In addition, Liu et al. [595] mentioned the accounting for metal AM residual stresses and the costly post-machining as issues yet to be solved, as Meng et al. [594] stressed out material performance assurance and fabrication speed and resolution as future research lines.

Future research on AM methods is deemed to comprise solutions for lightweight cellular materials design [594] or the validation for AM purposes of several innovative TO approaches [595].

Another perspective on future trends can be attained from analysing current funded projects and recent patents on a subject. Searching within the EU projects database, one can find two projects concerned with TO. An already finished project studied "*Optimization Techniques for MIMO Radar Antenna Systems*" [596] and therefore, is not related to this review scope. On the other hand, the project "*Innovative Re-Design and Validation of Complex Airframe Structural Components Formed by Additive Manufacturing for Weight and Cost Reduction*" has been funded since 2017 until the current year under the framework of "*Design Guide Lines and Simulation Methods for Additive Manufactured Titanium Components*" and deployed significant publications [127,178,597,598]. While the research has been conducted with titanium as the primary material, the innovative approaches are applied in any engineering field where weight reduction is a major objective.

Concerning recent Topology Optimisation patents with the world, Europe, or US coverage, software developers have registered several computational methods. Those include Livermore Software general design methods EP2251805A2 [599] and US8126684 [600], the numerical derivates method US0160078161 [601], and enhanced global design variables method to account for impact events US0170255724 [602]. In addition, Dassault Systèmes' method for designing a mechanical part with TO EP3502931A1 [603], US20190197210A1 [604], EP3647973A1 [605], and JP2020071887A [606], as well as autodesk TO for subtractive manufacturing techniques US 2018/0349531 [607] and Altair's Failsafe TO method US10354024 B2 [608].

Other recently patent-protected design methods include patents US20160140269 [609], US8335668 [610], and WO2020215533A1 with a material-field reduction series expansion [611], EP3285189A1 for flexible hinges [612], EP3292657A1 and US0180139130 for the multi-layer network TO [613,614], US0170161405 employing reduced length boundaries methods [615], US0200134918 for cellular structures [616], and US010613496 for Additive Manufacturing [617].

Unlike what has been depicted for published research, patents show a significantly growing impetus for practical methods, especially in automotive, composite materials manufacturing, and aerospace industries. In fact, while the 2010 to 2015 period has some of the most significant TO industrial patents in Caterpillar's US0100274537 stress-based TO method [618] and US0140156229 fatigue-based TO method [619], the last 5 years brought several and significant advances, such as Toyota's methods for TO with a membership variable WO2019152596A1 [620] and US0090326138 with shape transformation [621]. In addition, the GM US0100035974 system for a plurality of materials [622], several additive manufacturing methods, including microstructure-based US0200180228 [623], Freespace Composites' US0150239178 [624] and US9789652 [625] manufacturing systems, Thales Alenia Space Italia EP3545443A1 adaptive TO for layer AM [626] and Siemens' WO2015106021A1 and US009789651 method for additively manufactured lattice structures [627,628], WO2019178199A1 multi-physics applications [629], WO2020160099A1 machine learning-based TO [630], as well as WO2020159812A1 TO method for additively manufactured thermoelastic structures [631].

## 9. Discussion and Research Perspectives

In opposition to the previous section, where future trends have been collected from distinguished experts' opinions and research articles, the current section aims to analyse the whole revision work and, from a critical assessment, depict a current status and infer future tendencies.

Adding to relatively scarce previous Review Articles, the current article was able to provide an embracing picture of TO, and DfAM developments occurred in the last 5 years. Considering the concentration of cutting-edge research in a few research centres, as well as a significant asymmetry between the leading funding agencies and the remaining ones, the study of patents in TO was regarded as an indispensable step for understanding the role of the industry in pushing TO forward.

Nevertheless, it also has been found that the current surge in TO is partly owed to many newcomers entering the field. Other reasons can be attributed to the close link between TO and AM and the later significant emergence.

Approaches and methods in TO have been found to compose a broad, yet heterogeneous, fabric of resources. Those encompass and blend techniques originated in Discrete Optimisation and Optimisation of Continua and offer an increasing palette of the forthcoming TO massification options. Nevertheless, the latest research trends suggest a post-SIMP era, where such a method's current predominance is now challenged both by very interesting research on Level-Set methods, as well as evolutionary approaches maturity and pervasiveness in many engineering applications, especially fuelled by Genetic Algorithms.

Arguably, this trend may be related to the significant controversy over the ESO and BESO methods, which, despite continuous enhancements, seem to be deprecated to GA methods as the first choice in many new practical applications.

Observing the most prolific research centres recent output, it is possible to speculate that research in TO methods and algorithms is likely to evolve towards computationally demanding solutions, such as multiscale projection and giga-resolution solutions, as well as further developments in surrogate-based optimisation, more complex techniques for addressing stress constraints and optimisation under uncertainty.

Bringing TO into the engineering practice is another issue, as non-experts are particularly dependent on existing programmes, codes, and commercial software. Regarding this matter, the SIMP method is expected to remain preponderant, considering its current dominance in commercial software. However, the herein depicted investigation showed a clear future tendency for the software market leaders to offer hybrid approaches and to patent their own methods.

Another observation from the current revision on programmes and codes is the unveiling of a more profuse set of alternatives, compared to what is frequently referred. Such information may assist both practitioners and early-stage researchers to find convenient alternatives and means for benchmarking solutions.

In structural steel design, TO is now mostly constricted to prototyping and nodes in tensegrity structures. Its large-scale employment depends on several factors, including reliability in additively manufactures alloys properties, the ability to account for multiple alternate loading and multiple local instabilities in TO, better addressing the non-linearity and the existence of comprehensive and code-compliant practical methodologies for practitioners.

Yet, it is interesting to observe that while many researchers find the mass reduction in steel members and connections due to the employment of TO extremely rewarding and unattainable otherwise, others still put the economic viability threshold further than what has been attained so far. Arguably, the type of member and connection seems to play a decisive role in the economic viability of TO.

Notwithstanding, researchers are almost unanimous in pointing significant benefits in using TO for AM in steel structures, namely the waste reduction, sustainability, global weight reduction, which may enhance the performance of big span and earthquake-prone structures, as well as erection speed.

Contributions from other disciplines are envisaged to foster sensible advances in TO for structural steel design. That is the case of more advanced alloys and composites either due to fabrication with leading AM processes provided its cost is reduced, or attained from TO architected microstructures, as well as better procedures and guidelines offered by the industrial practice.

However, the most significant driving force for implementing TO in the design practice is the prospective use of AM. DfAM is reliant on TO, and the massification of AM can only push TO.

Recent advances in DfAM have been mostly centred in taking AM requirements into consideration for TO. Likewise, impressive efforts have been made to attain designs with better surfaces quality, less post-fabrication machining, and avoiding overhangs and general supports.

Notwithstanding some meritorious exceptions, mentioned throughout the text, as one assesses recent research in TO it is evident that Cohn and Dinovitzer's diagnosis of 25 years, mentioning that profuse advances in mathematical algorithms using simple examples, rather than formulating methodologies for real structural engineering problems, led to practitioners' lack of interest in TO [632], still holds its validity.

As any other systematic review, this article is subjected to the risk of unintended bias, incompleteness, etc. To mitigate such risks, efforts were undertaken in performing inclusive research, considering all the possible viewpoints, valuing equally more or less proficiently written articles, and remaining neutral in opinions.

## 10. Conclusions

This article's contribution to the current body-of-knowledge lies in offering a pervasive review of TO methods and applications, developments in the past 5 years, and research trends. It is focused on structural steel design and detailing but encompasses all the adjoining domains, including other fields, the optimisation software, and Additive Manufacturing processes.

Therefore, it is hoped that it may encourage researchers and practitioners, especially newcomers into the field, to endeavour research in a field where it is usually very time-consuming to enter.

Among the herein depicted review work conclusions, one can highlight that SIMP is still the leading method for TO. However, research trends suggest an escalating importance of Level-Set Methods and Genetic Algorithms. On the other hand, commercial software for TO is deemed to continue as SIMP-based, while a trend to offer hybrid and in-house developed methods can be regarded.

Employing TO in steel construction is a clear future trend, either fostered by AM massification or due to its significant benefits in waste reduction, weight reduction, sustainability or as the ultimate optimisation tool.

However, for that to happen, significant advances will be required in the alloys properties quality and reliability, alternate loading, local instabilities, and non-linearity accounting into the design methods, as well as the creation of a holistic and code-compliant practical methodology for practitioners.

Some of the much-needed solutions are expected to be brought from other engineering fields, such as aerospace, automotive, materials or medical devices engineering.

As a further recommendation for the research community, we suggest creating a classification scheme, for example, similar to the Mathematics Subject Classification, in order to better organise and order TO in sub-fields, approaches, and methods.

**Author Contributions:** Conceptualisation, T.P.R.; methodology, T.P.R.; validation, T.P.R.; formal analysis, T.P.R.; investigation, T.P.R.; data curation, T.P.R.; writing—original draft preparation, T.P.R.; writing—review and editing, T.P.R., L.F.A.B., and J.M.A.A.; supervision, L.F.A.B. and J.M.A.A. All authors have read and agreed to the published version of the manuscript.

**Funding:** This research received no external funding.

**Institutional Review Board Statement:** Not applicable (study not involving humans or animals).

**Informed Consent Statement:** Not applicable (study not involving humans or animals).

**Data Availability Statement:** Not applicable (only published literature was considered).

**Acknowledgments:** The authors want to thank Professor Castro Gomes' suggestions and discussion during this article's preparation. Moreover, software developers, representatives, and resellers which provided clarifications about their products are thanked.

**Conflicts of Interest:** The authors declare no conflict of interest.

## Abbreviations

| | |
|---|---|
| 3DCP | 3D Concrete Printing |
| ABC | Artificial Bee Colony Algorithm |
| ACO | Ant Colony Optimisation |
| AESO | Additive Evolutionary Structural Optimisation |
| AI | Artificial Intelligence |
| ALM | Additive Layer Manufacturing |
| AM | Additive Manufacturing |
| ASI | Artificial Swarm Intelligence |
| ASTM | American Society for Testing and Materials |
| BESO | Bidirectional Evolutionary Structural Optimisation |
| BSED | Bézier Skeleton Explicit Density |
| CAE | Computer-Aided Engineering |
| CBO | Colliding Bodies Optimisation |
| CFD | Computational Fluid Dynamics |
| CFRP | Carbon Fibre Reinforced Plastics or Carbon Fibre Reinforced Polymers |
| COC | Continuum-Based Optimality Criteria |
| CONLIN | Convex Linearisation Method |
| DCOC | Discretized Continuum-Type Optimality Criteria Technique |
| DED | Directed Energy Deposition |
| DED-EB | Electron Beam Directed Energy Deposition |
| DED-GMA | Gas Metal Arc Directed Energy Deposition |
| DED-L | Laser Directed Energy Deposition |
| DED-PA | Plasma Arc Directed Energy Deposition |
| DfAM | Design for Additive Manufacturing |
| DfM | Design for Manufacturing |
| DMD | Direct Metal Deposition |
| DMLS | Direct Metal Laser Sintering |
| DOT | Design Optimisation Tools |
| DSC | Deformable Simplicial Complex |
| EBM | Electron Beam Melting |
| EC3 | EN1993-1-1 or Eurocode 3 |
| ECAM | Electrochemical Additive Manufacturing |

| | |
|---|---|
| ESLM | Equivalent Static Loads Method |
| ESO | Evolutionary Structural Optimisation |
| FDM | Fuse Deposition Modelling |
| FEA | Finite Element Analyses |
| FMAGDM | Fuzzy Multiple-Attribute Group Decision-Making |
| FSD | Fully Stressed Design |
| FSOA | Fish Swarm Optimisation |
| GA | Genetic Algorithms |
| GCMMA | Globally Convergent Method of Moving Asymptotes |
| GESO | Genetic Evolutionary Structural Optimisation |
| GMAW | Gas Metal Arc Welding |
| GRAND | Ground Structure Analysis and Design |
| GSM | Ground Structure Method |
| GTAW | Gas Tungsten Arc Welding |
| IPOPT | Interior Point Optimiser |
| KS | Kreisselmeier–Steinhauser Function |
| LBM | Laser Beam Melting |
| LENS | Laser Engineered Net Shaping |
| LMD | Laser Metal Deposition |
| LMF | Laser Metal Fusion |
| LOM | Laminated Object Manufacturing |
| LPM | Lumped Parameter Model |
| LS | Level-Set Methods |
| LSM | Level-Set Methods |
| LSP | Large-Scale Prefabrication |
| LSTO | Lattice Structure Topology Optimisation |
| MD | Metal Deposition |
| MINLP | Mixed-Integer Non-Linear Programming |
| MMA | Method of Moving Asymptotes |
| MMC | Moving Morphable Components |
| MMTO | Multi-Material Topology Optimisation |
| MOTO | Additive Manufacturing-Oriented Topology Optimisation |
| MRF | Moment Resisting Frames |
| MSP | Materials with Site-Specific Properties |
| MTOP | Multiresolution Topology Optimisation |
| NDT | Non-Destructive Testing |
| NEWC | Normalised Exponential Weighted Criterion |
| NOM | Near-Optimal Microstructure |
| NPR | Negative Poisson's Ratio Materials |
| NRBTO | Non-Probabilistic Reliability-Based Topology Optimisation |
| N-VTM | Nonlinear Virtual Temperature Method |
| OAPI | Open Application Programming Interface |
| OC | Optimality Criteria |
| OFM | Optimisation for Manufacture |
| OMP | Optimal Microstructure with Penalisation |
| OWL | Web Ontology Language |
| PAW | Plasma Arc Welding |
| PBF | Powder Bed Fusion |
| PBF-EB | Electron Beam Powder Bed Fusion |
| PBF-L | Laser Powder Bed Fusion |
| PCG | Preconditioned Conjugate Gradients |
| PETSc | Portable and Extendable Toolkit for Scientific Computing |
| P-GSM | Projection-Based Ground Structure Topology Optimisation Method |
| PSO | Swarms including Particle Swarm Optimisation |
| RAMP | Rational Approximation of Material Properties |
| RBF | Radial-Basis Functions |
| RBTO | Reliability-Based Topology Optimisation |
| RDF | Resource Description Framework |

| | |
|---|---|
| RI | Repulsion Index |
| RP | Rapid Prototyping |
| RSL | Revolutionary Superposition Layout |
| SAMP | Solid Anisotropic Material with Penalisation |
| SDS | Stochastic Diffusion Search |
| SERA | Sequential Elements Rejection and Admission |
| SFF | Solid Freeform Fabrication or Solid Freedom Fabrication |
| SFMD | Simultaneous Failure Mode Design |
| SIMP | Solid Isotropic Microstructure (or Material) with Penalisation |
| SLA | Stereolithography |
| SLM | Selective Laser Melting |
| SLS | Selective Laser Sintering |
| SM | Shape Melting |
| SMD | Shape Metal Deposition |
| SNOPT | Sparse Nonlinear Optimiser |
| SOM | Sectional Optimisation Method |
| SOM | Skidmore Owings and Merrill LLP |
| SRV | Sum of the Reciprocal Variables |
| STM | Strut-and-Tie Models |
| STO | Structural Topology Optimisation |
| STSA | Structural Topology and Shape Annealing |
| SW | Shape Welding |
| TIMP | Transplanting Independent Continuous and Mapping Ideas into Material with Pe-nalisation |
| TO | Topology Optimisation |
| TOBS | Topology Optimisation of Binary Structures |
| TOSS | Topology Optimisation of Skeletal Structures |
| TTO | Truss Topology Optimisation |
| UAM | Ultrasonic Additive Manufacturing |
| UR | Uncertainty Regions |
| UV | Ultraviolet Light |
| VCS | Virtual Component Skeleton |
| VE | Value Engineering |
| VSFM | Virtual Scalar Field Method |
| WAAM | Wire and Arc Additive Manufacturing |

## References


1. Eschenauer, H.A.; Olhoff, N. Topology optimization of continuum structures: A review. *Appl. Mech. Rev.* **2001**, *54*, 331–390. [CrossRef]
2. Lagaros, N.D. The environmental and economic impact of structural optimization. *Struct. Multidiscip. Optim.* **2018**, *58*, 1751–1768. [CrossRef]
3. Prager, W. A note on discretized michell structures. *Comput. Methods Appl. Mech. Eng.* **1974**. [CrossRef]
4. Prager, W.; Rozvany, G.I.N. *Optimization of Structural Geometry*; Academic Press Inc.: New York, NY, USA, 1977.
5. Rozvany, G.I.N. Grillages of maximum strength and maximum stiffness. *Int. J. Mech. Sci.* **1972**, *14*, 651–666. [CrossRef]
6. Rozvany, G.I.N. Optimal load transmission by flexure. *Comput. Methods Appl. Mech. Eng.* **1972**, *1*, 253–263. [CrossRef]
7. Rozvany, G.I.N.; Zhou, M.; Rotthaus, M.; Gollub, W.; Spengemann, F. Continuum-type optimality criteria methods for large finite element systems with a displacement constraint. Part I. *Struct. Optim.* **1989**, *1*, 47–72. [CrossRef]
8. Rozvany, G.I.N.; Zhou, M.; Gollub, W. Continuum-type optimality criteria methods for large finite element systems with a displacement constraint. Part II. *Struct. Optim.* **1990**, *2*, 77–104. [CrossRef]
9. Michell, A.G.M. The limits of economy of material in frame-structures. *Lond. Edinb. Dublin Philos. Mag. J. Sci.* **1904**, *8*, 589–597. [CrossRef]
10. Hegemier, G.A.; Prager, W. On Michell trusses. *Int. J. Mech. Sci.* **1969**, *11*, 209–215. [CrossRef]
11. Maxwell, J.C. *The Scientific Papers of James Clerk Maxwell*; Cambridge Library Collection: Cambridge, UK, 1870.
12. Pedersen, P. On the optimal layout of multi-purpose trusses. *Comput. Struct.* **1972**, *2*, 695–712. [CrossRef]
13. Olhoff, N. On Singularities, Local Optima, and Formation of Stiffeners in Optimal Design of Plates. In *Optimization in Structural Design*; 1975.
14. Cheng, K.T.; Olhoff, N. An Investigation Concerning Optimal Design of Solid Elastic Plates. *Int. J. Solids Struct.* **1981**, *17*, 305–323. [CrossRef]

15. Olhoff, N.; Lurie, K.A.; Cherkaev, A.V.; Fedorov, A.V. Sliding regimes and anisotropy in optimal design of vibrating axisymmetric plates. *Int. J. Solids Struct.* **1981**, *17*, 931–948. [CrossRef]
16. Rozvany, G.I.N.; Olhoff, N.; Cheng, K.T.; Taylor, J.E. On the Solid Plate Paradox in Structural Optimization. *J. Struct. Mech.* **1982**, *10*, 1–32. [CrossRef]
17. Haftka, R.T.; Gurdal, Z. Elements of Structural Optimization. *Struct. Optim.* **1992**, *11*, 6221.
18. Van Dijk, N.P.; Maute, K.; Langelaar, M.; Van Keulen, F. Level-set methods for structural topology optimization: A review. *Struct. Multidiscip. Optim.* **2013**, *48*, 437–472. [CrossRef]
19. Kirsch, U. Optimal topologies of truss structures. *Comput. Methods Appl. Mech. Eng.* **1989**, *72*, 15–28. [CrossRef]
20. Pradel, P.; Zhu, Z.; Bibb, R.; Moultrie, J. Investigation of design for additive manufacturing in professional design practice. *J. Eng. Des.* **2018**, *29*, 165–200. [CrossRef]
21. Bendsoe, M.P.; Kikuchi, N. Generating optimal topologies in structural design using a homogenization method. *Comput. Methods Appl. Mech. Eng.* **1988**, *71*, 197–224. [CrossRef]
22. Bendsøe, M.P. Optimal shape design as a material distribution problem. *Struct. Optim.* **1989**, *1*, 193–202. [CrossRef]
23. Suzuki, K.; Kikuchi, N. Shape and topology optimization by a homogenization method. *Am. Soc. Mech. Eng. Appl. Mech. Div. Amd.* **1990**, *115*, 15–30. [CrossRef]
24. Kohn, R.V.; Strang, G. Optimal Design and Relaxation of Variational Problems, I. *Commun. Pure Appl. Math.* **1986**, *39*, 113–137. [CrossRef]
25. Kohn, R.V.; Strang, G. Optimal design and relaxation of variational problems, II. *Commun. Pure Appl. Math.* **1986**, *39*, 139–182. [CrossRef]
26. Kohn, R.V.; Strang, G. Optimal design and relaxation of variational problems, III. *Commun. Pure Appl. Math.* **1986**, *39*, 113–137. [CrossRef]
27. Strang, G.; Kohn, R. Optimal Design in Elasticity and Plasticity. *Int. J. Numer. Methods Eng.* **1986**, *22*. [CrossRef]
28. Bendsoe, M.P. *Optimization of Structural Topology, Shape, and Material*; GmbH Springer: Berlin/Heidelberg, Germmany, 1995; ISBN 9783662031179.
29. Rozvany, G.I.N. *Topology Optimization in Structural Mechanics*; GmbH Springer: Vienna, Austria, 1997; ISBN 9783211829073.
30. Svanberg, K. The method of moving asymptotes—A new method for structural optimization. *Int. J. Numer. Methods Eng.* **1987**, *24*, 359–373. [CrossRef]
31. Xie, Y.M.; Steven, G.P. A simple approach to Structural Optimization. *Compurers Struct.* **1993**, *49*, 885–896. [CrossRef]
32. Sigmund, O.; Petersson, J. Numerical instabilities in topology optimization: A survey on procedures dealing with checkerboards, mesh-dependencies and local minima. *Struct. Optim.* **1998**, *16*, 68–75. [CrossRef]
33. Zhou, M.; Rozvany, G.I.N. On the validity of ESO type methods in topology optimization. *Struct. Multidiscip. Optim.* **2001**, *21*, 80–83. [CrossRef]
34. Rozvany, G.I.N.; Querin, O.M. Theoretical foundations of sequential element rejections and admissions (SERA) methods and their computational implementation in topology optimization. In Proceedings of the 9th AIAA/ISSMO Symposium on Multidisciplinary Analysis and Optimization, Atlanta, GA, USA, 4–6 September 2002; pp. 9–12. [CrossRef]
35. Rozvany, G.I.N. A critical review of established methods of structural topology optimization. *Struct. Multidiscip. Optim.* **2009**, *37*, 217–237. [CrossRef]
36. Yang, X.Y.; Xie, Y.M.; Steven, G.P.; Querin, O.M. Topology Optimization for Frequencies Using an Evolutionary Method. *J. Struct. Eng.* **1999**, 1432–1438. [CrossRef]
37. Liang, Q.Q.; Steven, G.P. A performance-based optimization method for topology design of continuum structures with mean compliance constraints. *Comput. Methods Appl. Mech. Eng.* **2002**, *191*, 1471–1489. [CrossRef]
38. Tanskanen, P. A multiobjective and fixed elements based modification of the evolutionary structural optimization method. *Comput. Methods Appl. Mech. Eng.* **2006**, *196*, 76–90. [CrossRef]
39. Edwards, C.S.; Kim, H.A.; Budd, C.J. An evaluative study on ESO and SIMP for optimising a cantilever tie-beam. *Struct. Multidiscip. Optim.* **2007**, *34*, 403–414. [CrossRef]
40. Huang, X.; Xie, Y.M. *Evolutionary Topology Optimization of Continuum Structures*; John Wiley & Sons Ltd.: Hoboken, NJ, USA, 2010; ISBN 9780470746530.
41. Rozvany, G.I.N. Author's reply to a discussion by Gengdong Cheng and Xiaofeng Liu of the review article "on symmetry and non-uniqueness in exact topology optimization" by George, I.N. Rozvany (2011, Struct Multidisc Optim 43:297-317). *Struct. Multidiscip. Optim.* **2011**, *44*, 719–721. [CrossRef]
42. Duysinx, P.; Bendsøe, M.P. Topology optimization of continuum structures with local stress constraints. *Int. J. Numer. Methods Eng.* **1998**, *43*, 1453–1478. [CrossRef]
43. Bendsøe, M.P.; Sigmund, O. Material interpolation schemes in topology optimization. *Arch. Appl. Mech.* **1999**, *69*, 635–654. [CrossRef]
44. Fredricson, H.; Johansen, T.; Klarbring, A.; Petersson, J. Topology optimization of frame structures with flexible joints. *Struct. Multidiscip. Optim.* **2003**, *25*, 199–214. [CrossRef]
45. Sigmund, O.; Torquato, S. Design of materials with extreme thermal expansion using a three-phase topology optimization method. *J. Mech. Phys. Solids* **1997**, *45*, 1037–1067. [CrossRef]

46. Sigmund, O. On the design of compliant mechanisms using topology optimization. *Mech. Struct. Mach.* **1997**, *25*, 493–524. [CrossRef]
47. Bruns, T.E.; Tortorelli, D.A. Topology optimization of geometrically nonlinear structures and compliant mechanisms. In Proceedings of the 7th AIAA/USAF/NASA/ISSMO Symposium on Multidisciplinary Analysis and Optimization, St. Louis, MO, USA, 2–4 September 1998; Volume 190, pp. 1874–1882. [CrossRef]
48. Jonsmann, J.; Sigmund, O.; Bouwstra, S. Compliant thermal microactuators. *Sens. Actuatorsa Phys.* **1999**, *76*, 463–469. [CrossRef]
49. Jiang, T.; Chirehdast, M. A Systems Approach to Structural Topology Optimization: Designing Optimal Connections. *J. Mech. Des.* **1997**, *119*, 40–47. [CrossRef]
50. Klarbring, A.; Petersson, J.; Rönnqvist, M. Truss topology optimization including unilateral contact. *J. Optim. Theory Appl.* **1995**, *87*, 1–31. [CrossRef]
51. Da Silva Smith, O. Topology optimization of trusses with local stability constraints and multiple loading conditions—A heuristic approach. *Struct. Optim.* **1997**, *13*, 155–166. [CrossRef]
52. Mijar, A.R.; Swan, C.C.; Arora, J.S.; Kosaka, I. Continuum Topology Optimization for Concept Design of Frame Bracing Systems. *J. Struct. Eng.* **1998**, *124*, 541–550. [CrossRef]
53. Kravanja, S.; Kravanja, Z.; Bedenik, B.S. The Minlp Optimization Approach To Structural Synthesis Part III: Synthesis of Roller and Sliding Hydraulic. *Int. J. Numer. Meth. Engng* **1998**, *43*, 329–364. [CrossRef]
54. Sigmund, O. A 99 line topology optimization code written in matlab. *Struct. Multidiscip. Optim.* **2001**, *21*, 120–127. [CrossRef]
55. Andreassen, E.; Clausen, A.; Schevenels, M.; Lazarov, B.S.; Sigmund, O. Efficient topology optimization in MATLAB using 88 lines of code. *Struct. Multidiscip. Optim.* **2011**, *43*, 1–16. [CrossRef]
56. Preisinger, C. Linking structure and parametric geometry. *Arch.. Des.* **2013**, *83*, 110–113. [CrossRef]
57. Jivotovski, G. Gradient based heuristic algorithm and its application to discrete optimization of bar structures. *Struct. Multidiscip. Optim.* **2000**, *19*, 237–248. [CrossRef]
58. Cameron, T.M.; Thirunavukarasu, A.C.; El-Sayed, M.E.M. Optimization of frame structures with flexible joints. *Struct. Multidiscip. Optim.* **2000**, *19*, 204–213. [CrossRef]
59. Fredricson, H. Topology optimization of frame structures-joint penalty and material selection. *Struct. Multidiscip. Optim.* **2005**, *30*, 193–200. [CrossRef]
60. Descamps, B.; Filomeno Coelho, R. A lower-bound formulation for the geometry and topology optimization of truss structures under multiple loading. *Struct. Multidiscip. Optim.* **2013**, *48*, 49–58. [CrossRef]
61. Bourdin, B. Filters in topology optimization. *Int. J. Numer. Methods Eng.* **2001**, *50*, 2143–2158. [CrossRef]
62. Sigmund, O. Morphology-based black and white filters for topology optimization. *Struct. Multidiscip. Optim.* **2007**, *33*, 401–424. [CrossRef]
63. Guest, J.K.; Prévost, J.H.; Belytschko, T. Achieving minimum length scale in topology optimization using nodal design variables and projection functions. *Int. J. Numer. Methods Eng.* **2004**, *61*, 238–254. [CrossRef]
64. Wang, F.; Lazarov, B.S.; Sigmund, O. On projection methods, convergence and robust formulations in topology optimization. *Struct. Multidiscip. Optim.* **2011**, *43*, 767–784. [CrossRef]
65. Lee, D.; Shin, S.; Park, S. Computational morphogenesis based structural design by using material topology optimization. *Mech. Based Des. Struct. Mach.* **2007**, *35*, 39–58. [CrossRef]
66. Martini, K. Harmony Search Method for Multimodal Size, Shape, and Topology Optimization of Structural Frameworks. *J. Struct. Eng.* **2011**, *137*, 1332–1339. [CrossRef]
67. Yi, J.J.; Zeng, T.; Rong, J.H.; Li, Y.M. A topology optimization method based on element independent nodal density. *J. Cent. South. Univ.* **2014**, *21*, 558–566. [CrossRef]
68. Rong, J.H.; Jiang, J.S.; Xie, Y.M. Evolutionary structural topology optimization for continuum structures with structural size and topology variables. *Adv. Struct. Eng.* **2007**, *10*, 681–695. [CrossRef]
69. Coelho, P.G.; Fernandes, P.R.; Guedes, J.M.; Rodrigues, H.C. A hierarchical model for concurrent material and topology optimisation of three-dimensional structures. *Struct. Multidiscip. Optim.* **2008**. [CrossRef]
70. Rahmatalla, S.; Swan, C.C. Form Finding of Sparse Structures with Continuum Topology Optimization. *J. Struct. Eng.* **2003**, *129*, 1707–1716. [CrossRef]
71. Afonso, S.M.B.; Sienz, J.; Belblidia, F. Structural optimization strategies for simple and integrally stiffened plates and shells. *Eng. Comput.* **2005**, *22*, 429–452. [CrossRef]
72. Lógó, J. SIMP type topology optimization procedure considering uncertain load position. *Period. Polytech. Civ. Eng.* **2012**, *56*, 213–219. [CrossRef]
73. Nishigaki, H.; Nishiwaki, S.; Amago, T.; Kojima, Y.; Kikuchi, N. First order analysis-New CAE tools for automotive body designers. *SAE Tech. Pap.* **2001**. [CrossRef]
74. Shin, J.K.; Lee, K.H.; Song, S.I.; Park, G.J. Automotive door design with the ULSAB concept using structural optimization. *Struct. Multidiscip. Optim.* **2002**, *23*, 320–327. [CrossRef]
75. Aeri, P.; Morrish, M. On the optimization of a steering Hanger Beam component. *SAE Tech. Pap.* **2008**. [CrossRef]
76. Yao, Y.; Shi, Y.; Fei, J. Topology optimum design of steel bodywork based on Genetic Algorithms. *Appl. Mech. Mater.* **2011**, *43*, 84–88. [CrossRef]

77. Brackett, D.; Ashcroft, I.; Hague, R. Topology optimization for additive manufacturing. In Proceedings of the 22nd Annual International Solid Freeform Fabrication Symposium—An Additive Manufacturing Conference; SFF 2011. 2011.
78. Leary, M.; Merli, L.; Torti, F.; Mazur, M.; Brandt, M. Optimal topology for additive manufacture: A method for enabling additive manufacture of support-free optimal structures. *Mater. Des.* **2014**. [CrossRef]
79. Nguyen, D.S.; Vignat, F. Topology optimization as an innovative design method for additive manufacturing. In Proceedings of the IEEE International Conference on Industrial Engineering and Engineering Management, Singapore, 10–13 December 2017; pp. 304–308. [CrossRef]
80. Vayre, B.; Vignat, F.; Villeneuve, F. Designing for additive manufacturing. *Procedia CIRP* **2012**, *3*, 632–637. [CrossRef]
81. Bañón, C.; Raspall, F. *3D Printing Architecture Workflows, Applications and Trends*; Springer: Berlin/Heidelberg, Germany, 2021; ISBN 9789811583872.
82. Shea, K.; Smith, I.F.C. Improving Full-Scale Transmission Tower Design through Topology and Shape Optimization. *J. Struct. Eng.* **2006**, *132*, 781–790. [CrossRef]
83. Kravanja, S.; Žula, T. Cost optimization of industrial steel building structures. *Adv. Eng. Softw.* **2010**, *41*, 442–450. [CrossRef]
84. Torii, A.J.; Lopez, R.H.; Biondini, F. An approach to reliability-based shape and topology optimization of truss structures. *Eng. Optim.* **2012**, *44*, 37–53. [CrossRef]
85. He, J.N.; Wang, Z. Study on topology optimization design of steel frame bracing system based on discrete model. *Adv. Mater. Res.* **2012**, *446–449*, 3191–3196. [CrossRef]
86. Memari, A.M.; Madhkhan, M. Optimal design of steel frames subject to gravity and seismic codes' prescribed lateral forces. *Struct. Optim.* **1999**, *18*, 56–66. [CrossRef]
87. Rajasekaran, S.; Chitra, J.S. Ant colony optimisation of spatial steel structures under static and earthquake loading. *Civ. Eng. Env. Syst.* **2009**, *26*, 339–354. [CrossRef]
88. Huang, J.; Wang, Z. Topology optimization of bracing systems for multistory steel frames under earthquake loads. *Adv. Mater. Res.* **2011**, *255–260*, 2388–2393. [CrossRef]
89. Liu, W.; Li, J. Comparison of algorithms for seismic topology optimisation of lifeline networks. *Struct. Infrastruct. Eng.* **2014**, *10*, 1357–1368. [CrossRef]
90. Sarkisian, M.; Beghini, A.; Long, E.; Shook, D.; Krebs, A.; Diaz, A. Innovation in the sustainable design process of Earthquake resistant buildings: From topology optimization to staged construction analysis. *Eng. Prog. Nat. People* **2014**, *102*, 1192–1199. [CrossRef]
91. Yung, C.T.; Nianquan, Z. An innovative approach to structural design of tall buildings. *Hkie Trans. Hong Kong Inst. Eng.* **2003**, *10*, 14–21. [CrossRef]
92. Kicinger, R.; Obayashi, S.; Arciszewski, T. Evolutionary multiobjective optimization of steel structural systems in tall buildings. In *Evolutionary Multi-Criterion Optimization*; Lecture Notes in Computer Science; Springer: Berlin, Heidelberg, 2007; Volume 4403, pp. 604–618. [CrossRef]
93. Stromberg, L.L.; Beghini, A.; Baker, W.F.; Paulino, G.H. Application of layout and topology optimization using pattern gradation for the conceptual design of buildings. *Struct. Multidiscip. Optim.* **2011**, *43*, 165–180. [CrossRef]
94. Stromberg, L.L.; Beghini, A.; Baker, W.F.; Paulino, G.H. Topology optimization for braced frames: Combining continuum and beam/column elements. *Eng. Struct.* **2012**, *37*, 106–124. [CrossRef]
95. Baker, W.F.; Beghini, A.; Mazurek, A. Applications of Structural Optimization in Architectural Design. In Proceedings of the 20th Analysis & Computation Specialty Conference, Chicago, IL, USA, 29–31 March 2012; pp. 207–218.
96. Beghini, A.; Beghini, L.L.; Baker, W.F. Applications of structural optimization in architectural design. In Proceedings of the Structures Congress 2013: Bridging Your Passion with Your Profession, Pittsburgh, PA, USA, 2–4 May 2013; pp. 2499–2507.
97. Zegard, T.; Paulino, G.H. Bridging topology optimization and additive manufacturing. *Struct. Multidiscip. Optim.* **2016**, *53*, 175–192. [CrossRef]
98. Lagaros, N.D.; Psarras, L.D.; Papadrakakis, M.; Panagiotou, G. Optimum design of steel structures with web openings. *Eng. Struct.* **2008**, *30*, 2528–2537. [CrossRef]
99. Yao, L.; Gao, Y.X.; Yang, H.J. Topology optimization design of pre-stressed plane entity steel structure with the constrains of stress and displacement. *Adv. Mater. Res.* **2014**, *945–949*, 1216–1222. [CrossRef]
100. Leng, J. *Optimization Techniques for Structural Design of Cold-Formed Steel Structures*; Elsevier: Amsterdam, The Netherlands, 2016; ISBN 9780081001578.
101. Oinonen, A.; Tanskanen, P.; Björk, T.; Marquis, G. Pattern optimization of eccentrically loaded multi-fastener joints. *Struct. Multidiscip. Optim.* **2010**, *40*, 597–609. [CrossRef]
102. Elsabbagh, A. Size optimization of stiffeners in bending plates. *Mech. Adv. Mater. Struct.* **2013**, *20*, 764–773. [CrossRef]
103. Zhang, W.; Zhang, Y.; Li, G. Evolutionary structural topology optimization for cantilever construction of continuous rigid-frame bridge. *Appl. Mech. Mater.* **2011**, *90–93*, 18–22. [CrossRef]
104. Xie, Y.M.; Zuo, Z.H.; Huang, X.; Black, T.; Felicetti, P. Application of topological optimisation technology to bridge design. *Struct. Eng. Int. J. Int. Assoc. Bridg. Struct. Eng.* **2014**, *24*, 185–191. [CrossRef]
105. Lewiński, T.; Sokół, T.; Graczykowski, C. *Michell Structures*; Springer: Berlin/Heidelberg, Germany, 2019; ISBN 9783319951799.
106. Kingman, J.J.; Tsavdaridis, K.D.; Toropov, V.V. Applications of topology optimization in structural engineering: High-rise buildings and steel components. *Jordan J. Civ. Eng.* **2015**, *9*, 335–357. [CrossRef]

107. Lee, D.K.; Yang, C.J.; Starossek, U. Topology design of optimizing material arrangements of beam-to-column connection frames with maximal stiffness. *Sci. Iran.* **2012**, *19*, 1025–1032. [CrossRef]
108. Briseghella, B.; Fenu, L.; Feng, Y.; Mazzarolo, E.; Zordan, T. Topology optimization of bridges supported by a concrete shell. *Struct. Eng. Int. J. Int. Assoc. Bridg. Struct. Eng.* **2013**, *23*, 285–294. [CrossRef]
109. Gaynor, A.T.; Guest, J.K.; Moen, C.D. Reinforced Concrete Force Visualization and Design Using Bilinear Truss-Continuum Topology Optimization. *J. Struct. Eng.* **2013**, *139*, 607–618. [CrossRef]
110. Chaves, L.P.; Cunha, J. Design of carbon fiber reinforcement of concrete slabs using topology optimization. *Constr. Build. Mater.* **2014**, *73*, 688–698. [CrossRef]
111. Sousa, L.C.; Castro, C.F.; António, C.C.; Sousa, H. Topology optimisation of masonry units from the thermal point of view using a genetic algorithm. *Constr. Build. Mater.* **2011**, *25*, 2254–2262. [CrossRef]
112. Haftka, R.T.; Gurdal, Z. *Elements of Structural Optimization*; Kluwer Academic Publishers: Norwell, MA, USA, 2002.
113. Bendsoe, M.P.; Sigmund, O. *Topology Optimization-Theory, Methods, and Applications*; Springer: Berlin/Heidelberg, Germany, 2012; ISBN 9783642076985.
114. Arora, J. *Introduction to Optimum Design*; Elsevier: Amsterdam, The Netherlands, 2012; ISBN 9780123813756.
115. Rozvany, G.I.N.; Lewiński, T. *Topology Optimization in Structural and Continuum Mechanics*; Springer: Berlin/Heidelberg, Germany, 2014; ISBN 978-3-7091-1642-5.
116. Querin, O.M.; Victoria, M.; Alonso, C.; Ansola, R.; Martí, P. *Topology Design Methods for Structural Optimization*; Elsevier: Amsterdam, The Netherlands, 2017; ISBN 9780081009161.
117. Bian, B. *Topological Optimization of Buckling*; de Gruyter: Veghel, The Netherlands, 2018; ISBN 9783110461169.
118. Hassani, B.; Hinton, E. A review of homogenization and topology optimization I-Homogenization theory for media with periodic structure. *Comput. Struct.* **1998**, *69*, 707–717. [CrossRef]
119. Hassani, B.; Hinton, E. A review of homogenization and topology optimization III-Topology optimization using optimality criteria. *Comput. Struct.* **1998**, *69*, 739. [CrossRef]
120. Fredricson, H. Structural topology optimisation: An application review. *Int. J. Veh. Des.* **2005**, *37*, 67–80. [CrossRef]
121. Huang, X.; Xie, Y.M. A further review of ESO type methods for topology optimization. *Struct. Multidiscip. Optim.* **2010**, *41*, 671–683. [CrossRef]
122. Sigmund, O.; Maute, K. Topology optimization approaches: A comparative review. *Struct. Multidiscip. Optim.* **2013**, *48*, 1031–1055. [CrossRef]
123. Deaton, J.D.; Grandhi, R.V. A survey of structural and multidisciplinary continuum topology optimization: Post 2000. *Struct. Multidiscip. Optim.* **2014**, *49*, 1–38. [CrossRef]
124. Wong, K.V.; Hernandez, A. A Review of Additive Manufacturing. *Isrn Mech. Eng.* **2012**. [CrossRef]
125. Frazier, W.E. Metal additive manufacturing: A review. *J. Mater. Eng. Perform.* **2014**, *23*, 1917–1928. [CrossRef]
126. Shashi, G.M.; Laskar, A.R.; Biswas, H.; Saha, A.K. A Brief Review of Additive Manufacturing with Applications. In Proceedings of the 14th Global Engineering and Technology Conference, Dhaka, Bangladesh, 29–30 December 2017.
127. Wiberg, A.; Persson, J.; Ölvander, J. Design for additive manufacturing–A review of available design methods and software. *Rapid Prototyp. J.* **2019**, *25*, 1080–1094. [CrossRef]
128. Alfaify, A.; Saleh, M.; Abdullah, F.M.; Al-Ahmari, A.M. Design for Additive Manufacturing: A Systematic Review. *Sustain* **2020**, *10*, 3043–3054. [CrossRef]
129. Buchanan, C.; Gardner, L. Metal 3D printing in construction: A review of methods, research, applications, opportunities and challenges. *Eng. Struct.* **2019**, *180*, 332–348. [CrossRef]
130. Plocher, J.; Panesar, A. Review on design and structural optimisation in additive manufacturing: Towards next generation lightweight structures. *Mater. Des.* **2019**, *183*, 108164. [CrossRef]
131. Sehmi, M.; Christensen, J.; Bastien, C.; Kanarachos, S. Review of topology optimisation refinement processes for sheet metal manufacturing in the automotive industry. *Struct. Multidiscip. Optim.* **2018**, *58*, 305–330. [CrossRef]
132. Huang, S.H.; Liu, P.; Mokasdar, A.; Hou, L. Additive manufacturing and its societal impact: A literature review. *Int. J. Adv. Manuf. Technol.* **2013**, *67*, 1191–1203. [CrossRef]
133. Zargham, S.; Ward, T.A.; Ramli, R.; Badruddin, I.A. Topology optimization: A review for structural designs under vibration problems. *Struct. Multidiscip. Optim.* **2016**, *53*, 1157–1177. [CrossRef]
134. Alexandersen, J.; Andreasen, C.S. A review of topology optimisation for fluid-based problems. *Fluids* **2020**, *5*, 29. [CrossRef]
135. Osanov, M.; Guest, J.K. Topology Optimization for Architected Materials Design. *Annu. Rev. Mater. Res.* **2016**, *46*, 211–233. [CrossRef]
136. Kingman, J.J.; Tsavdaridis, K.D.; Toropov, V.V. Applications of topology optimization in structural engineering. In Proceedings of the Civil Engineering for Sustainability and Resilience International Conference, Amman, Jordan, 24–27 April 2014.
137. Ferrari, F.; Sigmund, O. Structural and Multidisciplinary Optimization Revisiting Topology Optimization with Buckling Constraints. 2019. Available online: https://link.springer.com/article/10.1007/s00158-019-02253-3.
138. Elhegazy, H. State-of-the-art review on benefits of applying value engineering for multi-story buildings. *Intell. Build. Int.* **2020**, 1–20. [CrossRef]
139. Li, Z.; Tsavdaridis, K.D. A Review of Optimised Additively Manufactured Steel Connections for Modular Building Systems. *Ind. Addit. Manuf.* **2021**, *1*, 357–373. [CrossRef]

140. Yang, X.S.; Bekdas, G.; Nigdeli, S.M. *Review and Applications of Metaheuristic Algorithms in Civil. Engineering*; Springer: Berlin/Heidelberg, Germany, 2016; Volume 7, ISBN 9783319262451.
141. Bekdaş, G.; Nigdeli, S.M.; Kayabekir, A.E.; Yang, X.S. Optimization in civil engineering and metaheuristic algorithms: A review of state-of-the-art developments. *Comput. Intell. Optim. Inverse Probl. Appl. Eng.* **2018**, 111–137. [CrossRef]
142. GlobalData Global. *Construction Outlook to 2022*; GlobalData Global: London, UK, 2018.
143. World Steel Association. *World Steel in Figures 2017*; World Steel Association: Brussels, Belgium, 2017.
144. EUROFER. *Eurofer Annual Report 2020*; EUROFER: Brussels, Belgium, 2020.
145. Grand View Research. *Structural Steel Market Size, Share & Trends Analysis Report by Application (Non-residential (Industrial, Commercial, Institutional), Residential), by Region, And Segment Forecasts, 2020–2027*, 2020.
146. Fivel, J.B. *Achieving a Decarbonised European Steel Industry in a Circular Economy*; KTH Royal Institute of Technology: Stockholm, Sweden, 2019.
147. Hoffmann, C.; Van Hoey, M.; Zeumer, B. *Decarbonization Challenge for Steel*; 2020.
148. He, K.; Wang, L.; Li, X. Review of the energy consumption and production structure of China's steel industry: Current situation and future development. *Metals* **2020**, *10*, 302. [CrossRef]
149. Galjaard, S.; Hofman, S.; Ren, S. Optimizing Structural Building Elements in Metal by using Additive Manufacturing. In Proceedings of the International Association for Shell and Spatial Structures (IASS), Amsterdam, The Netherlands, 17–20 August 2015.
150. Smith, C.J. Application of layout optimization to the design of additively manufactured metallic components. Ph.D. Thesis, University of Sheffield, Sheffield, UK, 2016; pp. 1297–1313.
151. Smith, C.J.; Gilbert, M.; Todd, I.; Derguti, F. Application of layout optimization to the design of additively manufactured metallic components. *Struct. Multidiscip. Optim.* **2016**, *54*, 1297–1313. [CrossRef]
152. Moher, D.; Liberati, A.; Tetzlaff, J.; Altman, D.G.; Altman, D.; Antes, G.; Atkins, D.; Barbour, V.; Barrowman, N.; Berlin, J.A.; et al. Preferred reporting items for systematic reviews and meta-analyses: The PRISMA statement. *PLoS Med.* **2009**. [CrossRef]
153. Manso, M.; Castro-Gomes, J. Green wall systems: A review of their characteristics. *Renew. Sustain. Energy Rev.* **2015**, *41*, 863–871. [CrossRef]
154. Pradel, P.; Zhu, Z.; Bibb, R.; Moultrie, J. A framework for mapping design for additive manufacturing knowledge for industrial and product design. *J. Eng. Des.* **2018**, *29*, 291–326. [CrossRef]
155. Almeida, J.; Ribeiro, A.B.; Silva, A.S.; Faria, P. Overview of mining residues incorporation in construction materials and barriers for full-scale application. *J. Build. Eng.* **2020**, *29*, 101215. [CrossRef]
156. Ginga, C.P.; Ongpeng, J.M.C.; Daly, M.K.M. Circular economy on construction and demolition waste: A literature review on material recovery and production. *Materials* **2020**, *13*, 2970. [CrossRef] [PubMed]
157. Grayson, R.J. Concepts of General Topology in Constructive Mathematics And In Sheaves. *Ann. Math. Log.* **1981**, *20*, 1–41. [CrossRef]
158. Mendelson, B. *Introduction to Topology*; Dover Publications, Inc.: New York, NY, USA, 2012; ISBN 9780486663524.
159. Simmons, G.F. *Introduction to Topology and Modern Analysis*; McGraw-Hill: New York, NY, USA, 2003.
160. Nishiwaki, S.; Terada, K. Advanced topology optimization. *Int. J. Numer. Methods Eng.* **2018**, *113*, 1145–1147. [CrossRef]
161. Mlejnek, H.P. Some aspects of the genesis of structures. *Struct. Optim.* **1992**, *5*, 64–69. [CrossRef]
162. Lazarov, B.S.; Wang, F.; Sigmund, O. Length scale and manufacturability in density-based topology optimization. *Arch. Appl. Mech.* **2016**, *86*, 189–218. [CrossRef]
163. Allaire, G.; Kohn, R.V. Topology Optimization and Optimal Shape Design Using Homogenization. In *Topology Design of Structures*; Springer: Berlin/Heidelberg, Germany, 1993.
164. Allaire, G.; Jouve, F.; Toader, A.M. A level-set method for shape optimization. *Comptes Rendus Math.* **2002**, *334*, 1125–1130. [CrossRef]
165. Christensen, J. Topology Optimisation of Structures Exposed to Large (Non-Linear) Deformations. Ph.D. Thesis, Coventry University, Coventry, UK, 2015. [CrossRef]
166. Andreassen, E.; Jensen, J.S. A practical multiscale approach for optimization of structural damping. *Struct. Multidiscip. Optim.* **2016**, *53*, 215–224. [CrossRef]
167. Garaigordobil, A.; Ansola, R.; Veguería, E. Study of topology optimization parameters and scaffold structures in additive manufacturing. In Proceedings of the ECCOMAS Congress 2016—7th European Congress on Computational Methods in Applied Sciences and Engineering, Crete Island, Greece, 5–10 June 2016; Volume 2, pp. 3700–3710.
168. Bruggi, M.; Parolini, N.; Regazzoni, F.; Verani, M. Finite element approximation of a time-dependent topology optimization problem. In Proceedings of the ECCOMAS Congress 2016—7th European Congress on Computational Methods in Applied Sciences and Engineering; 2016; Volume 2, pp. 3711–3723.
169. Aage, N.; Andreassen, E.; Lazarov, B.S.; Sigmund, O. Giga-voxel computational morphogenesis for structural design. *Nature* **2017**, *550*, 84–86. [CrossRef]
170. Gregersen, N.; De Lasson, J.R.; Frandsen, L.H.; Hayrynen, T.; Lavrinenko, A.; Mork, J.; Wang, F.; Sigmund, O.; Kim, O.S.; Breinbjerg, O.; et al. Benchmarking five computational methods for analyzing large photonic crystal membrane cavities. In Proceedings of the International Conference on Numerical Simulation of Optoelectronic Devices, Copenhagen, Denmark, 24–28 July 2017; pp. 89–90. [CrossRef]

171. Rojas-Labanda, S.; Sigmund, O.; Stolpe, M. A short numerical study on the optimization methods influence on topology optimization. *Struct. Multidiscip. Optim.* **2017**, *56*, 1603–1612. [CrossRef]
172. Groen, J.P.; Sigmund, O. Homogenization-based topology optimization for high-resolution manufacturable microstructures. *Int. J. Numer. Methods Eng.* **2018**, *113*, 1148–1163. [CrossRef]
173. Jain, N.; Saxena, R. Effect of self-weight on topological optimization of static loading structures. *Alex. Eng. J.* **2018**, *57*, 527–535. [CrossRef]
174. Han, Y.; Lu, W.F. A novel design method for nonuniform lattice structures based on topology optimization. *J. Mech. Des. Trans. Asme* **2018**, *140*, 1–10. [CrossRef]
175. Bruggi, M.; Parolini, N.; Regazzoni, F.; Verani, M. Topology optimization with a time-integral cost functional. *Finite Elem. Anal. Des.* **2018**, *140*, 11–22. [CrossRef]
176. da Silva, G.A.; Beck, A.T.; Sigmund, O. Stress-constrained topology optimization considering uniform manufacturing uncertainties. *Comput. Methods Appl. Mech. Eng.* **2019**, *344*, 512–537. [CrossRef]
177. Vantyghem, G.; De Corte, W.; Steeman, M.; Boel, V. Density-based topology optimization for 3D-printable building structures. *Struct. Multidiscip. Optim.* **2019**, *60*, 2391–2403. [CrossRef]
178. Thore, C.-J.; Grundstrom, H.A.; Torstenfelt, B.; Klarbring, A. Penalty regulation of overhang in topology optimization for additive manufacturing. *Struct. Multidiscip. Optim.* **2019**, *60*, 59–67. [CrossRef]
179. Liu, Y.; Li, Z.; Wei, P.; Chen, S. Generating support structures for additive manufacturing with continuum topology optimization methods. *Rapid Prototyp. J.* **2019**, *25*, 232–246. [CrossRef]
180. Rostami, S.A.L.; Ghoddosian, A. Topology optimization under uncertainty by using the new collocation method. *Period. Polytech. Civ. Eng.* **2019**, *63*, 278–287. [CrossRef]
181. Keshavarzzadeh, V.; James, K.A. Robust multiphase topology optimization accounting for manufacturing uncertainty via stochastic collocation. *Struct. Multidiscip. Optim.* **2019**, *60*, 2461–2476. [CrossRef]
182. Fu, J.; Huang, J.; Liu, J. Topology Optimization with Selective Problem Setups. *Ieee Access* **2019**, *7*, 180846–180855. [CrossRef]
183. Ryan, L.; Kim, I.Y. A multiobjective topology optimization approach for cost and time minimization in additive manufacturing. *Int. J. Numer. Methods Eng.* **2019**, *118*, 371–394. [CrossRef]
184. Allaire, G.; Geoffroy-Donders, P.; Pantz, O. Topology optimization of modulated and oriented periodic microstructures by the homogenization method. *Comput. Math. Appl.* **2019**, *78*, 2197–2229. [CrossRef]
185. Groen, J.P.; Wu, J.; Sigmund, O. Homogenization-based stiffness optimization and projection of 2D coated structures with orthotropic infill. *Comput. Methods Appl. Mech. Eng.* **2019**, *349*, 722–742. [CrossRef]
186. Ferrari, F.; Sigmund, O. A new generation 99 line Matlab code for compliance topology optimization and its extension to 3D. *Struct. Multidiscip. Optim.* **2020**, *62*, 2211–2228. [CrossRef]
187. Lüdeker, J.K.; Sigmund, O.; Kriegesmann, B. Inverse homogenization using isogeometric shape optimization. *Comput. Methods Appl. Mech. Eng.* **2020**, *368*, 113170. [CrossRef]
188. Zhang, S.; Li, H.; Huang, Y. An improved multi-objective topology optimization model based on SIMP method for continuum structures including self-weight. *Struct. Multidiscip. Optim.* **2020**. [CrossRef]
189. da Silva, G.A.; Beck, A.T.; Sigmund, O. Topology optimization of compliant mechanisms considering stress constraints, manufacturing uncertainty and geometric nonlinearity. *Comput. Methods Appl. Mech. Eng.* **2020**, *365*, 112972. [CrossRef]
190. da Silva, G.A.; Aage, N.; Beck, A.T.; Sigmund, O. Three-dimensional manufacturing tolerant topology optimization with hundreds of millions of local stress constraints. *Int. J. Numer. Methods Eng.* **2020**, 1–32. [CrossRef]
191. Da, D.; Xia, L. Design of heterogeneous mesostructures for non-separated scales and analysis of size effects†. *Int. J. Numer. Methods Eng.* **2020**, *122*, 1333–1351. [CrossRef]
192. Sotiropoulos, S.; Kazakis, G.; Lagaros, N.D. High performance topology optimization computing platform. *Procedia Manuf.* **2020**, *44*, 441–448. [CrossRef]
193. Greifenstein, J.; Stingl, M. Topology optimization with worst-case handling of material uncertainties. *Struct. Multidiscip. Optim.* **2020**, *61*, 1377–1397. [CrossRef]
194. Deng, H.; To, A.C. Linear and nonlinear topology optimization design with projection-based ground structure method (P-GSM). *Int. J. Numer. Methods Eng.* **2020**, *121*, 2437–2461. [CrossRef]
195. Olsen, J.; Kim, I.Y. Design for additive manufacturing: 3D simultaneous topology and build orientation optimization. *Struct. Multidiscip. Optim.* **2020**, *62*, 1989–2009. [CrossRef]
196. Suresh, S.; Thore, C.J.; Torstenfelt, B.; Klarbring, A. Topology optimization accounting for surface layer effects. *Struct. Multidiscip. Optim.* **2020**. [CrossRef]
197. Fu, Y.F.; Rolfe, B.; Chiu, L.N.S.; Wang, Y.; Huang, X.; Ghabraie, K. Smooth topological design of 3D continuum structures using elemental volume fractions. *Comput. Struct.* **2020**, *231*, 106213. [CrossRef]
198. Zhao, L.; Xu, B.; Han, Y.; Xue, J.; Rong, J. Structural topological optimization with dynamic fatigue constraints subject to dynamic random loads. *Eng. Struct.* **2020**, *205*, 110089. [CrossRef]
199. Groen, J.P.; Stutz, F.C.; Aage, N.; Bærentzen, J.A.; Sigmund, O. De-homogenization of optimal multi-scale 3D topologies. *Comput. Methods Appl. Mech. Eng.* **2020**, *364*, 112979. [CrossRef]
200. Stutz, F.C.; Groen, J.P.; Sigmund, O.; Bærentzen, J.A. Singularity aware de-homogenization for high-resolution topology optimized structures. *Struct. Multidiscip. Optim.* **2020**, *62*, 2279–2295. [CrossRef]

201. Bendsoe, M.P.; Diaz, A.; Kikuchi, N. Topology and Generalized Layout Optimization of Elastic Structures. In *Topology Design of Structures*; Springer: Berlin/Heidelberg, Germany, 1993; pp. 159–205.
202. Stolpe, M.; Svanberg, K. An alternative interpolation scheme for minimum compliance topology optimization. *Struct. Multidiscip. Optim.* **2001**, *22*, 116–124. [CrossRef]
203. Bruns, T.E. A reevaluation of the SIMP method with filtering and an alternative formulation for solid-void topology optimization. *Struct. Multidiscip. Optim.* **2005**, *30*, 428–436. [CrossRef]
204. Fuchs, M.B.; Jiny, S.; Peleg, N. The SRV constraint for 0/1 topological design. *Struct. Multidiscip. Optim.* **2005**, *30*, 320–326. [CrossRef]
205. Maute, K.; Frangopol, D.M. Reliability-based design of MEMS mechanisms by topology optimization. *Comput. Struct.* **2003**, *81*, 813–824. [CrossRef]
206. Kharmanda, G.; Olhoff, N.; Mohamed, A.; Lemaire, M. Reliability-based topology optimization. *Struct. Multidiscip. Optim.* **2004**, *26*, 295–307. [CrossRef]
207. Jung, H.S.; Cho, S. Reliability-based topology optimization of geometrically nonlinear structures with loading and material uncertainties. *Finite Elem. Anal. Des.* **2004**, *41*, 311–331. [CrossRef]
208. Wang, S.; Moon, H.; Kim, C.; Kang, J.; Choi, K.K. Reliability-based topology optimization (RBTO). In *IUTAM Symposium on Topological Design Optimization of Structures, Machines and Materials: Status and Perspectives*; Springer: Berlin/Heidelberg, Germany, 2006; pp. 493–504.
209. Kang, Z.; Luo, Y. Non-probabilistic reliability-based topology optimization of geometrically nonlinear structures using convex models. *Comput. Methods Appl. Mech. Eng.* **2009**, *198*, 3228–3238. [CrossRef]
210. Silva, M.; Tortorelli, D.A.; Norato, J.A.; Ha, C.; Bae, H.R. Component and system reliability-based topology optimization using a single-loop method. *Struct. Multidiscip. Optim.* **2010**, *41*, 87–106. [CrossRef]
211. Rozvany, G.I.N.; Maute, K. Analytical and numerical solutions for a reliability-based benchmark example. *Struct. Multidiscip. Optim.* **2011**, *43*, 745–753. [CrossRef]
212. Nguyen, T.H.; Song, J.; Paulino, G.H. Single-loop system reliability-based topology optimization considering statistical dependence between limit-states. *Struct. Multidiscip. Optim.* **2011**, *44*, 593–611. [CrossRef]
213. Keshavarzzadeh, V.; Fernandez, F.; Tortorelli, D.A. Topology optimization under uncertainty via non-intrusive polynomial chaos expansion. *Comput. Methods Appl. Mech. Eng.* **2017**, *318*, 120–147. [CrossRef]
214. da Silva, G.A.; Beck, A.T. Reliability-based topology optimization of continuum structures subject to local stress constraints. *Struct. Multidiscip. Optim.* **2018**, *57*, 2339–2355. [CrossRef]
215. Osher, S.; Sethian, J.A. Fronts propagating with curvature-dependent speed: Algorithms based on Hamilton-Jacobi formulations. *J. Comput. Phys.* **1988**, *79*, 12–49. [CrossRef]
216. Wang, M.Y.; Wang, X.; Guo, D. A level set method for structural topology optimization. *Comput. Methods Appl. Mech. Engrg.* **2003**, *192*, 227–246. [CrossRef]
217. Allaire, G.; Jouve, F.; Toader, A.M. *Structural Optimization Using Sensitivity Analysis and a Level-Set Method*; Elsevier: Amsterdam, The Netherlands, 2004; Volume 194, ISBN 3316933301.
218. Chen, S.; Chen, W. A new level-set based approach to shape and topology optimization under geometric uncertainty. *Struct. Multidiscip. Optim.* **2011**, *44*, 1–18. [CrossRef]
219. Guo, X.; Zhang, W.; Zhang, L. Robust structural topology optimization considering boundary uncertainties. *Comput. Methods Appl. Mech. Eng.* **2013**, *253*, 356–368. [CrossRef]
220. Villanueva, C.H.; Maute, K. Density and level set-XFEM schemes for topology optimization of 3-D structures. *Comput. Mech.* **2014**, *54*, 133–150. [CrossRef]
221. Liu, J.; Ma, Y. A survey of manufacturing oriented topology optimization methods. *Adv. Eng. Softw.* **2016**, *100*, 161–175. [CrossRef]
222. Zhou, Y.; Zhang, W.; Zhu, J. Concurrent shape and topology optimization involving design-dependent pressure loads using implicit B-spline curves. *Int. J. Numer. Methods Eng.* **2019**, *118*, 495–518. [CrossRef]
223. Osher, S.J.; Santosa, F. Level Set Methods for Optimization Problems Involving Geometry and Constraints I. Frequencies of a Two-Density Inhomogeneous Drum. *J. Comput. Phys.* **2001**, *171*, 272–288. [CrossRef]
224. Vogiatzis, P.; Chen, S.; Zhou, C. An Open Source Framework for Integrated Additive Manufacturing and Level-Set-Based Topology Optimization. *J. Comput. Inf. Sci. Eng.* **2017**, *17*, 1–10. [CrossRef]
225. Geiss, M.J.; Maute, K. Topology optimization of active structures using a higher-order level-set-XFEM-density approach. In Proceedings of the 2018 Multidisciplinary Analysis and Optimization Conference, Atlanta, GA, USA, 25–29 June 2018; pp. 1–13.
226. Picelli, R.; Townsend, S.; Kim, H.A. Stress and strain control via level set topology optimization. *Struct. Multidiscip. Optim.* **2018**, *58*, 2037–2051. [CrossRef]
227. Geiss, M.J.; Boddeti, N.; Weeger, O.; Maute, K.; Dunn, M.L. Combined Level-Set-XFEM-Density Topology Optimization of Four-Dimensional Printed Structures Undergoing Large Deformation. *J. Mech. Des. Trans. ASME* **2019**, *141*. [CrossRef]
228. Li, L.; Zhu, X. Design of compliant revolute joints based on mechanism stiffness matrix through topology optimization using a parameterization level set method. *Struct. Multidiscip. Optim.* **2019**, *60*, 1475–1489. [CrossRef]
229. Fu, J.; Li, H.; Gao, L.; Xiao, M. Design of shell-infill structures by a multiscale level set topology optimization method. *Comput. Struct.* **2019**, *212*, 162–172. [CrossRef]

230. Andreasen, C.S.; Aage, N. Robust shape and topology optimization using CutFEM. In Proceedings of the 13th World Congress of Structural and Multidisciplinary Optimization; 2019.
231. Li, Y.; Zhu, J.; Wang, F.; Zhang, W.; Sigmund, O. Shape preserving design of geometrically nonlinear structures using topology optimization. *Struct. Multidiscip. Optim.* **2019**, *59*, 1033–1051. [CrossRef]
232. Andreasen, C.S.; Elingaard, M.O.; Aage, N. Level set topology and shape optimization by density methods using cut elements with length scale control. *Struct. Multidiscip. Optim.* **2020**, *62*, 685–707. [CrossRef]
233. Jansen, M.; Pierard, O. A hybrid density/level set formulation for topology optimization of functionally graded lattice structures. *Comput. Struct.* **2020**, *231*, 106205. [CrossRef]
234. De Ruiter, M.J.; Van Keulen, F. Topology optimization using a topology description function. *Struct. Multidiscip. Optim.* **2004**, *26*, 406–416. [CrossRef]
235. Norato, J.; Haber, R.; Tortorelli, D.; Bendsøe, M.P. A geometry projection method for shape optimization. *Int. J. Numer. Methods Eng.* **2004**, *60*, 2289–2312. [CrossRef]
236. Wang, S.; Wang, M.Y. Radial basis functions and level set method for structural topology optimization. *Int. J. Numer. Methods Eng.* **2006**, *65*, 2060–2090. [CrossRef]
237. Luo, Z.; Tong, L.; Wang, M.Y.; Wang, S. Shape and topology optimization of compliant mechanisms using a parameterization level set method. *J. Comput. Phys.* **2007**, *227*, 680–705. [CrossRef]
238. Pingen, G.; Waidmann, M.; Evgrafov, A.; Maute, K. A parametric level-set approach for topology optimization of flow domains. *Struct. Multidiscip. Optim.* **2010**, *41*, 117–131. [CrossRef]
239. Kreissl, S.; Pingen, G.; Maute, K. An explicit level set approach for generalized shape optimization of fluids with the lattice Boltzmann method. *Int. J. Numer. Methods Fluids* **2011**, *65*, 236–253. [CrossRef]
240. Gomes, A.A.; Suleman, A. Application of spectral level set methodology in topology optimization. *Struct. Multidiscip. Optim.* **2006**, *31*, 430–443. [CrossRef]
241. Sokolowski, J.; Zochowski, A. On the topological derivative in shape optimization. *Siam J. Control. Optim.* **1999**, *37*, 1251–1272. [CrossRef]
242. Novotny, A.A.; Feijóo, R.A.; Taroco, E.; Padra, C. Topological sensitivity analysis. *Comput. Methods Appl. Mech. Eng.* **2003**, *192*, 803–829. [CrossRef]
243. Norato, J.A.; Bendsøe, M.P.; Haber, R.B.; Tortorelli, D.A. A topological derivative method for topology optimization. *Struct. Multidiscip. Optim.* **2007**, *33*, 375–386. [CrossRef]
244. Eschenauer, H.A.; Kobelev, V.V.; Schumacher, A. Bubble method for topology and shape optimization of structures. *Struct. Optim.* **1994**, *8*, 42–51. [CrossRef]
245. Kim, D.H.; Lee, S.B.; Kwank, B.M.; Kim, H.G.; Lowther, D.A. Smooth boundary topology optimization for electrostatic problems through the combination of shape and topological design sensitivities. *IEEE Trans. Magn.* **2008**, *44*, 1002–1005. [CrossRef]
246. Mirzendehdel, A.M.; Suresh, K. Support structure constrained topology optimization for additive manufacturing. *CAD Comput. Aided Des.* **2016**, *81*, 1–13. [CrossRef]
247. Rakotondrainibe, L.; Allaire, G.; Orval, P. Topology optimization of connections in mechanical systems. *Struct. Multidiscip. Optim.* **2020**, *61*, 2253–2269. [CrossRef]
248. Céa, J.; Garreau, S.; Guillaume, P.; Masmoudi, M. The shape and topological optimizations connection. *Comput. Methods Appl. Mech. Eng.* **2000**, *188*, 713–726. [CrossRef]
249. Amstutz, S. Connections between topological sensitivity analysis and material interpolation schemes in topology optimization. *Struct. Multidiscip. Optim.* **2011**, *43*, 755–765. [CrossRef]
250. Bourdin, B.; Chambolle, A. Design-dependent loads in topology optimization. *Esaim Control. Optim. Calc. Var.* **2003**, *9*, 247–273. [CrossRef]
251. Blank, L.; Garcke, H.; Sarbu, L.; Srisupattarawanit, T.; Styles, V.; Voigt, A. Phase-field approaches to structural topology optimization. *Int. Ser. Numer. Math.* **2012**, *160*, 245–256. [CrossRef]
252. Wang, M.Y.; Zhou, S. Phase field: A variational method for structural topology optimization. *Comput. Model. Eng. Sci.* **2004**, *6*, 547–566. [CrossRef]
253. Zhou, S.; Wang, M.Y. Multimaterial structural topology optimization with a generalized Cahn-Hilliard model of multiphase transition. *Struct. Multidiscip. Optim.* **2007**, *33*, 89–111. [CrossRef]
254. Wallin, M.; Ristinmaa, M.; Askfelt, H. Optimal topologies derived from a phase-field method. *Struct. Multidiscip. Optim.* **2012**, *45*, 171–183. [CrossRef]
255. Xia, L.; Da, D.; Yvonnet, J. Topology optimization for maximizing the fracture resistance of quasi-brittle composites. *Comput. Methods Appl. Mech. Eng.* **2018**, *332*, 234–254. [CrossRef]
256. Ferro, N.; Micheletti, S.; Perotto, S. POD-assisted strategies for structural topology optimization. *Comput. Math. Appl.* **2019**, *77*, 2804–2820. [CrossRef]
257. Takezawa, A.; Nishiwaki, S.; Kitamura, M. Shape and topology optimization based on the phase field method and sensitivity analysis. *J. Comput. Phys.* **2010**, *229*, 2697–2718. [CrossRef]
258. BURGER, M.; STAINKO, R. Phase-Field Relaxation of Topology Optimization with Local Stress Constraints. *Siam J. Control. Optim.* **2006**, *45*, 1447–1466. [CrossRef]

259. Mattheck, C.; Burkhardt, S. A new method of structural shape optimization based on biological growth. *Int. J. Fatigue* **1990**, *12*, 185–190. [CrossRef]
260. Soh, C.K.; Yang, Y. Genetic Programming-Based Approach for Structural Optimization. *J. Comput. Civ. Eng.* **2000**, *14*, 31–37. [CrossRef]
261. Fraternali, F.; Marino, A.; El Sayed, T.; Della Cioppa, A. On the structural shape optimization through variational methods and evolutionary algorithms. *Mech. Adv. Mater. Struct.* **2011**, *18*, 225–243. [CrossRef]
262. Zuo, Z.H.; Xie, Y.M.; Huang, X. Evolutionary topology optimization of structures with multiple displacement and frequency constraints. *Adv. Struct. Eng.* **2012**, *15*, 359–372. [CrossRef]
263. Garcia-Lopez, N.P.; Sanchez-Silva, M.; Medaglia, A.L.; Chateauneuf, A. An improved robust topology optimization approach using multiobjective evolutionary algorithms. *Comput. Struct.* **2013**, *125*, 1–10. [CrossRef]
264. Liang, Q.Q.; Xie, Y.M.; Steven, G.P. Optimal topology selection of continuum structures with displacement constraints. *Comput. Struct.* **2000**, *77*, 635–644. [CrossRef]
265. Kociecki, M.; Adeli, H. Shape optimization of free-form steel space-frame roof structures with complex geometries using evolutionary computing. *Eng. Appl. Artif. Intell.* **2015**, *38*, 168–182. [CrossRef]
266. Ekici, B.; Chatzikonstantinou, I.; Sariyildiz, S.; Tasgetiren, M.F.; Pan, Q.K. A multi-objective self-adaptive differential evolution algorithm for conceptual high-rise building design. In Proceedings of the 2016 IEEE Congress on Evolutionary Computation, Vancouver, BC, Canada, 24–29 July 2016; pp. 2272–2279. [CrossRef]
267. Fiore, A.; Marano, G.C.; Greco, R.; Mastromarino, E. Structural optimization of hollow-section steel trusses by differential evolution algorithm. *Int. J. Steel Struct.* **2016**, *16*, 411–423. [CrossRef]
268. Feng, R.Q.; Liu, F.; Cheng, X.; Jia, W.; Ma, M.; Liu, Y. Topology optimization method of lattice structures based on a genetic algorithm. *Int. J. Steel Struct.* **2016**, *16*, 743–753. [CrossRef]
269. Serpik, I.N.; Alekseytsev, A.V.; Balabin, P.Y. Mixed approaches to handle limitations and execute mutation in the genetic algorithm for truss size, shape and topology optimization. *Period. Polytech. Civ. Eng.* **2017**, *61*, 471–482. [CrossRef]
270. Babaei, M.; Sheidaii, M.R. Desirability-Based Design of Space Structures Using Genetic Algorithm and Fuzzy Logic. *Int. J. Civ. Eng.* **2017**, *15*, 231–245. [CrossRef]
271. Pholdee, N.; Bureerat, S. A Comparative Study of Eighteen Self-adaptive Metaheuristic Algorithms for Truss Sizing Optimisation. *Ksce J. Civ. Eng.* **2018**, *22*, 2982–2993. [CrossRef]
272. Li, Y.; Lian, S. Improved Fruit Fly Optimization Algorithm Incorporating Tabu Search for Optimizing the Selection of Elements in Trusses. *Ksce J. Civ. Eng.* **2018**, *22*, 4940–4954. [CrossRef]
273. Safonov, A.A. 3D topology optimization of continuous fiber-reinforced structures via natural evolution method. *Compos. Struct.* **2019**, *215*, 289–297. [CrossRef]
274. Talaslioglu, T. Optimal design of steel skeletal structures using the enhanced genetic algorithm methodology. *Front. Struct. Civ. Eng.* **2019**, *13*, 863–889. [CrossRef]
275. Lynch, M.E.; Sarkar, S.; Maute, K. Machine learning to aid tuning of numerical parameters in topology optimization. *J. Mech. Des. Trans. Asme* **2019**, *141*. [CrossRef]
276. Cucinotta, F.; Raffaele, M.; Salmeri, F. A stress-based topology optimization method by a Voronoi tessellation Additive Manufacturing oriented. *Int. J. Adv. Manuf. Technol.* **2019**, *103*, 1965–1975. [CrossRef]
277. Rezayat, H.; Bell, J.R.; Plotkowski, A.J.; Babu, S.S. Multi-solution nature of topology optimization and its application in design for additive manufacturing. *Rapid Prototyp. J.* **2019**, *25*, 1475–1481. [CrossRef]
278. Han, Y.S.; Xu, B.; Zhao, L.; Xie, Y.M. Topology optimization of continuum structures under hybrid additive-subtractive manufacturing constraints. *Struct. Multidiscip. Optim.* **2019**, *57*, 2399–2409. [CrossRef]
279. Talatahari, S.; Azizi, M. Optimal design of real-size building structures using quantum-behaved developed swarm optimizer. *Struct. Des. Tall Spec. Build.* **2020**, *29*, 1–28. [CrossRef]
280. Wang, C.; Yao, S.; Wang, Z.; Hu, J. Deep super-resolution neural network for structural topology optimization. *Eng. Optim.* **2020**. [CrossRef]
281. Kallioras, N.A.; Kazakis, G.; Lagaros, N.D. Accelerated topology optimization by means of deep learning. *Struct. Multidiscip. Optim.* **2020**, *62*, 1185–1212. [CrossRef]
282. Bigham, A.; Gholizadeh, S. Topology optimization of nonlinear single-layer domes by an improved electro-search algorithm and its performance analysis using statistical tests. *Struct. Multidiscip. Optim.* **2020**, *62*, 1821–1848. [CrossRef]
283. Fairclough, H.; Gilbert, M. Layout optimization of simplified trusses using mixed integer linear programming with runtime generation of constraints. *Struct. Multidiscip. Optim.* **2020**, *61*, 1977–1999. [CrossRef]
284. Elhoone, H.; Zhang, T.; Anwar, M.; Desai, S. Cyber-based design for additive manufacturing using artificial neural networks for Industry 4.0. *Int. J. Prod. Res.* **2020**, *58*, 2841–2861. [CrossRef]
285. Després, N.; Cyr, E.; Setoodeh, P.; Mohammadi, M. Deep Learning and Design for Additive Manufacturing: A Framework for Microlattice Architecture. *Jom* **2020**, *72*, 2408–2418. [CrossRef]
286. Glebov, A.O.; Karpov, S.V.; Malygin, E.N. Comparison of topological optimization methods on the example of column press traverse. *Iop Conf. Ser. Mater. Sci. Eng.* **2020**, *709*. [CrossRef]
287. Bi, M.; Tran, P.; Xie, Y.M. Topology optimization of 3D continuum structures under geometric self-supporting constraint. *Addit. Manuf.* **2020**, *36*, 101422. [CrossRef]

288. Yang, X.Y.; Xie, Y.M.; Steven, G.P.; Querin, O.M. Bi-directional evolutionary method for stiffness optimisation. In Proceedings of the 7th AIAA/USAF/NASA/ISSMO Symposium on Multidisciplinary Analysis and Optimization, St. Louis, MO, USA, 2–4 September 1998; Volume 37, pp. 1449–1457. [CrossRef]
289. Querin, O.M.; Steven, G.P.; Xie, Y.M. Evolutionary structural optimisation (ESO) using a bidirectional algorithm. *Eng. Comput.* **1998**, *15*, 1031–1048. [CrossRef]
290. Young, V.; Querin, O.M.; Steven, G.P.; Xie, Y.M. 3D and multiple load case bi-directional evolutionary structural optimization (BESO). *Struct. Optim.* **1999**, *18*, 183–192. [CrossRef]
291. Kim, H.; Querin, O.M.; Steven, G.P.; Xie, Y.M. Determination of an optimal topology with a predefined number of cavities. *8th Symp. Multidiscip. Anal. Optim.* **2000**, *40*. [CrossRef]
292. Querin, O.M.; Young, V.; Steven, G.P.; Xie, Y.M. Computational efficiency and validation of bi-directional evolutionary structural optimization. *Comput. Methods Appl. Mech. Eng.* **2000**, *189*, 559–573. [CrossRef]
293. Yang, X.Y.; Xie, Y.M.; Liu, J.S.; Parks, G.T.; Clarkson, P.J. Perimeter control in the bidirectional evolutionary optimization method. *Struct. Multidiscip. Optim.* **2003**, *24*, 430–440. [CrossRef]
294. Huang, X.; Xie, Y.M. Convergent and mesh-independent solutions for the bi-directional evolutionary structural optimization method. *Finite Elem. Anal. Des.* **2007**, *43*, 1039–1049. [CrossRef]
295. Querin, O.M.; Steven, G.P.; Xie, Y.M. Evolutionary structural optimisation using an additive algorithm. *Finite Elem. Anal. Des.* **2000**, *34*, 291–308. [CrossRef]
296. Zhu, J.H.; Zhang, W.H.; Qiu, K.P. Bi-directional evolutionary topology optimization using element replaceable method. *Comput. Mech.* **2007**, *40*, 97–109. [CrossRef]
297. Huang, X.; Xie, Y.M. Bi-directional evolutionary topology optimization of continuum structures with one or multiple materials. *Comput. Mech.* **2009**, *43*, 393–401. [CrossRef]
298. Christensen, J. Topology Optimisation of Structures Exposed to Large (Non-Linear) Deformations. Ph.D. Thesis, Coventry University, Coventry, UK, 2016.
299. Kanarachos, S.; Griffin, J.; Fitzpatrick, M.E. Efficient truss optimization using the contrast-based fruit fly optimization algorithm. *Comput. Struct.* **2017**, *182*, 137–148. [CrossRef]
300. Hajela, P. Genetic search-An approach to the nonconvex optimization problem. *Aiaa J.* **1990**, *28*, 1205–1210. [CrossRef]
301. Hajela, P.; Lin, C.-Y. Genetic search strategies in multicriterion optimal design. *Struct. Optim.* **1992**, *107*, 99–107. [CrossRef]
302. Hajela, P.; Lee, E.; Lin, C.-Y. Genetic Algorithms in Structural Topology Optimization. *Topol. Des. Struct.* **1993**, 117–133. [CrossRef]
303. Hajela, P.; Lee, E. Genetic algorithms in truss optimization. *Int. J. Solids Struct.* **1995**, *32*, 3341–3357. [CrossRef]
304. Safari, D.; Maheri, M. Genetic Algrithm Search for Optimal Brace Positions in Steel Frames. *Adv. Steel Constr.* **2006**, *2*, 400–415.
305. Wang, S.Y.; Tai, K.; Wang, M.Y. An enhanced genetic algorithm for structural topology optimization. *Int. J. Numer. Methods Eng.* **2006**, *65*, 18–44. [CrossRef]
306. Bureerat, S.; Limtragool, J. Performance enhancement of evolutionary search for structural topology optimisation. *Finite Elem. Anal. Des.* **2006**, *42*, 547–566. [CrossRef]
307. Liu, X.; Yi, W.J.; Li, Q.S.; Shen, P.S. Genetic evolutionary structural optimization. *J. Constr. Steel Res.* **2008**, *64*, 305–311. [CrossRef]
308. Guest, J.K.; Genut, L. Reducing dimensionality in topology optimization using adaptive design variable fields. *Int. J. Numer. Methods Eng.* **2009**. [CrossRef]
309. Zuo, Z.H.; Xie, Y.M.; Huang, X. Combining genetic algorithms with BESO for topology optimization. *Struct. Multidiscip. Optim.* **2009**, *38*, 511–523. [CrossRef]
310. Pedro, H.T.C.; Hude, C.; Kobayashi, M.H. Topology optimization using Map L-Systems. In Proceedings of the 47th AIAA Aerospace Sciences Meeting including The New Horizons Forum and Aerospace Exposition, Orlando, FL, USA, 5–8 January 2009. [CrossRef]
311. Liu, X.; Yi, W.J. Michell-like 2D layouts generated by genetic ESO. *Struct. Multidiscip. Optim.* **2010**, *42*, 111–123. [CrossRef]
312. Nguyen, T.T.; Bærentzen, J.A.; Sigmund, O.; Aage, N. Efficient hybrid topology and shape optimization combining implicit and explicit design representations. *Struct. Multidiscip. Optim.* **2020**. [CrossRef]
313. Fiuk, G.; Mrzygłód, M.W. Topology optimization of structures with stress and additive manufacturing constraints. *J. Appl. Mech.* **2020**, *58*, 459–468. [CrossRef]
314. Verbart, A. Comment on "A working-set approach for sizing optimization of frame-structures subjected to time-dependent constraints". *Struct. Multidiscip. Optim.* **2020**. [CrossRef]
315. Lian, H.; Christiansen, A.N.; Tortorelli, D.A.; Sigmund, O.; Aage, N. Combined shape and topology optimization for minimization of maximal von Mises stress. *Struct. Multidiscip. Optim.* **2017**, *55*, 1541–1557. [CrossRef]
316. Li, Q.; Chen, W.; Liu, S.; Tong, L. Structural topology optimization considering connectivity constraint. *Struct. Multidiscip. Optim.* **2016**, *54*, 971–984. [CrossRef]
317. Guo, X.; Zhang, W.; Zhong, W. Doing topology optimization explicitly and geometrically-a new moving morphable components based framework. *J. Appl. Mech. Trans. Asme* **2014**, *81*, 1–12. [CrossRef]
318. Guo, X.; Zhou, J.; Zhang, W.; Du, Z.; Liu, C.; Liu, Y. Self-supporting structure design in additive manufacturing through explicit topology optimization. *Comput. Methods Appl. Mech. Eng.* **2017**, *323*, 27–63. [CrossRef]

319. Wang, R.; Zhang, X.; Zhu, B. Imposing minimum length scale in moving morphable component (MMC)-based topology optimization using an effective connection status (ECS) control method. *Comput. Methods Appl. Mech. Eng.* **2019**, *351*, 667–693. [CrossRef]
320. Zhang, X.; Maheshwari, S.; Ramos, A.S.; Paulino, G.H. Macroelement and Macropatch Approaches to Structural Topology Optimization Using the Ground Structure Method. *J. Struct. Eng.* **2016**, *142*, 04016090. [CrossRef]
321. Stolpe, M.; Verbart, A.; Rojas-Labanda, S. The equivalent static loads method for structural optimization does not in general generate optimal designs. *Struct. Multidiscip. Optim.* **2018**, *58*, 139–154. [CrossRef]
322. Kaveh, A.; Pishghadam, M.; Jafarvand, A. Topology optimization of repetitive near-regular shell structures using preconditioned conjugate gradients method. *Mech. Based Des. Struct. Mach.* **2020**, 1–22. [CrossRef]
323. Abambres, M.; Rajana, K.; Tsavdaridis, K.D.; Ribeiro, T.P. Neural Network-Based Formula for the Buckling Load Prediction of I-Section Cellular Steel Beams. *Computers* **2018**, *8*, 2. [CrossRef]
324. Sousa, A.L.; Ribeiro, T.P. Using Machine Learning for enhancing the understanding of bullwhip effect in the oil and gas industry. *Mach. Learn. Knowl. Extr.* **2019**, *1*, 57. [CrossRef]
325. Zhou, M.; Rozvany, G.I.N. DCOC: An optimality criteria method for large systems Part I: Theory. *Struct. Optim.* **1992**, *5*, 12–25. [CrossRef]
326. Zhou, M.; Rozvany, G.I.N. The COC algorithm, Part II: Topological, geometrical and generalized shape optimization. *Comput. Methods Appl. Mech. Eng.* **1991**, *89*, 309–336. [CrossRef]
327. Rozvany, G.I.N.; Zhou, M. Layout and Generalized Shape Optimization by Iterative COC Methods. *Optim. Large Struct. Syst.* **1993**, *I*, 103–120. [CrossRef]
328. Gill, P.E.; Murray, W.; Saunders, M.A. SNOPT: An SQP algorithm for large-scale constrained optimization. *Siam Rev.* **2005**, *47*, 99–131. [CrossRef]
329. Wächter, A. Lorenz Biegler On the implementation of an interior-point filter line-search algorithm for large-scale nonlinear programming. *Math. Program.* **2006**, *57*, 25–57. [CrossRef]
330. Fleury, C.; Braibant, V. Structural optimization: A new dual method using mixed variables. *Int. J. Numer. Methods Eng.* **1986**, *23*, 409–428. [CrossRef]
331. Fleury, C. CONLIN: An efficient dual optimizer based on convex approximation concepts. *Struct. Optim.* **1989**, *1*, 81–89. [CrossRef]
332. Xiao, M.; Lu, D.; Breitkopf, P.; Raghavan, B.; Dutta, S.; Zhang, W. On-the-fly model reduction for large-scale structural topology optimization using principal components analysis. *Struct. Multidiscip. Optim.* **2020**, *62*, 209–230. [CrossRef]
333. Svanberg, K. A Class of Globally Convergent Optimization Methods Based on Conservative Convex Separable Approximations. *Siam J. Optim* **2002**, *12*, 555–573. [CrossRef]
334. Liu, J.; Ma, Y.S. 3D level-set topology optimization: A machining feature-based approach. *Struct. Multidiscip. Optim.* **2015**, *52*, 563–582. [CrossRef]
335. Salonitis, K. Design for additive manufacturing based on the axiomatic design method. *Int. J. Adv. Manuf. Technol.* **2016**, *87*, 989–996. [CrossRef]
336. Tseranidis, S.; Brown, N.C.; Mueller, C.T. Data-driven approximation algorithms for rapid performance evaluation and optimization of civil structures. *Autom. Constr.* **2016**, *72*, 279–293. [CrossRef]
337. Mirzendehdel, A.M.; Rankouhi, B.; Suresh, K. Strength-based topology optimization for anisotropic parts. *Addit. Manuf.* **2018**, *19*, 104–113. [CrossRef]
338. Nielsen, D.G.; Søndergaard Jensen, J.; Cutanda Henriquez, V.; Agerkvist, F.T. Finite element model coupled with lumped parameter elements. In Proceedings of the 14th International Conference on Theoretical and Computational Acoustics, ICTCA 2019, Beijing, China, 28 July–1 August 2019; pp. 297–304.
339. Stolpe, M.; Dou, S. Models and numerical methods for optimal design of fail-safe structures. In Proceedings of the IASS Annual Symposium 2019—Structural Membranes 2019, Barcelona, Spain, 7–10 October 2019; pp. 1–2.
340. Stolpe, M. Fail-safe truss topology optimization. *Struct. Multidiscip. Optim.* **2019**, *60*, 1605–1618. [CrossRef]
341. Tsavdaridis, K.D.; Efthymiou, E.; Adugu, A.; Hughes, J.A.; Grekavicius, L. Application of structural topology optimisation in aluminium cross-sectional design. *Thin-Walled Struct.* **2019**. [CrossRef]
342. Mantovani, S.; Campo, G.A.; Ferrari, A. Additive manufacturing and topology optimization: A design strategy for a steering column mounting bracket considering overhang constraints. *Proc. Mech. Eng. Part C J. Mech. Eng. Sci.* **2020**. [CrossRef]
343. Zegard, T.; Hartz, C.; Mazurek, A.; Baker, W.F. Advancing building engineering through structural and topology optimization. *Struct. Multidiscip. Optim.* **2020**, *62*, 915–935. [CrossRef]
344. Wang, Y.; Hu, D.; Wang, H.; Zhang, T.; Yan, H. Practical design optimization of cellular structures for additive manufacturing. *Eng. Optim.* **2020**, *52*, 1887–1902. [CrossRef]
345. Benoist, V.; Arnaud, L.; Baili, M. A new method of design for additive manufacturing including machining constraints. *Int. J. Adv. Manuf. Technol.* **2020**, *111*, 25–36. [CrossRef]
346. Tsavdaridis, K.D.; Kingman, J.J.; Toropov, V.V. Application of structural topology optimisation to perforated steel beams. *Comput. Struct.* **2015**. [CrossRef]
347. Kuczek, T. Application of manufacturing constraints to structural optimization of thin-walled structures. *Eng. Optim.* **2016**, *48*, 351–360. [CrossRef]

348. Gebisa, A.W.; Lemu, H.G. A case study on topology optimized design for additive manufacturing. In *IOP Conference Series: Materials Science and Engineering*; IOP Publishing: Bristol, UK, 2017; Volume 276.
349. Saadlaoui, Y.; Milan, J.L.; Rossi, J.M.; Chabrand, P. Topology optimization and additive manufacturing: Comparison of conception methods using industrial codes. *J. Manuf. Syst.* **2017**, *43*, 178–186. [CrossRef]
350. Siva Rama Krishna, L.; Mahesh, N.; Sateesh, N. Topology optimization using solid isotropic material with penalization technique for additive manufacturing. *Mater. Today Proc.* **2017**, *4*, 1414–1422. [CrossRef]
351. Önnermalm, K.R.; Fredriksson, L.; Vorberg, T.; Bromberger, M. Strategic Development of Lightweight Platforms Made of Steel. *Light. Des.* **2018**, *11*, 48–53. [CrossRef]
352. Lim, S.T.; Wong, T.T. Unleash the potential of additive manufacturing with topology optimization. In Proceedings of the AIP Conference, Kuala Lumpur, Malaysia, 13–14 August 2018; Volume 2035.
353. Ahmad, A.; Raza, M.A.; Campana, F. Simulation Based Topology Optimization Assessment with Manufacturing Constraints. In Proceedings of the 17th International Bhurban Conference on Applied Sciences and Technology, IBCAST 2020, Islamabad, Pakistan, 14–18 January 2020; pp. 174–182. [CrossRef]
354. Sedlacek, F.; Lasova, V. *Optimization of Additive Manufactured Components Using Topology Optimization*; Springer International Publishing: Berlin/Heidelberg, Germany, 2019; Volume 5, ISBN 9783319919898.
355. Galjaard, S.; Hofman, S.; Ren, S. New Opportunities to Optimize Structural Designs in Metal by Using Additive Manufacturing. In *Advances in Architectural Geometry 2014*; Springer: Cham, Switzerland, 2015; pp. 79–93. [CrossRef]
356. Jankovics, D.; Gohari, H.; Tayefeh, M.; Barari, A. Developing Topology Optimization with Additive Manufacturing Constraints in ANSYS®. *IFAC-PapersOnLine* **2018**, *51*, 1359–1364. [CrossRef]
357. Pedersen, C.B.W.; Allinger, P. Industrial implementation and applications of topology optimization and future needs. *Solid Mech. Its Appl.* **2006**, *137*, 229–238. [CrossRef]
358. Garcia-Granada, A.A.; Catafal-Pedragosa, J.; Lemu, H.G. Topology optimization through stiffness/weight ratio analysis for a three-point bending test of additive manufactured parts. In *IOP Conference Series: Materials Science and Engineering*; IOP Publishing: Bristol, UK, 2019; Volume 700.
359. Lagaros, N.D.; Vasileiou, N.; Kazakis, G. A C# code for solving 3D topology optimization problems using SAP2000. *Optim. Eng.* **2019**, *20*. [CrossRef]
360. Bagherinejad, M.H.; Haghollahi, A. Study on Topology Optimization of Perforated Steel Plate Shear Walls in Moment Frame Based on Strain Energy. *Int. J. Steel Struct.* **2020**, *20*, 1420–1438. [CrossRef]
361. Santer, M.; Pellegrino, S. Topological optimization of compliant adaptive wing structure. *Aiaa J.* **2009**, *47*, 523–534. [CrossRef]
362. Haertel, J.; Engelbrecht, K.; Lazarov, B.; Sigmund, O. Topology Optimization of Thermal Heat Sinks. In Proceedings of the COMSOL Conference, Grenoble, France, 14 October 2015; p. 6.
363. Andreasen, C.S. A framework for topology optimization of inertial microfluidic particle manipulators. *Struct. Multidiscip. Optim.* **2020**, *61*, 2481–2499. [CrossRef]
364. Hassani, V.; Khabazi, Z.; Mehrabi, H.A.; Gregg, C.; O'Brien, R.W. Rationalization algorithm for a topologically-optimized multi-branch node for manufacturing by metal printing. *J. Build. Eng.* **2020**, *29*, 101146. [CrossRef]
365. Tcherniak, D.; Sigmund, O. A web-based topology optimization program. *Struct. Multidiscip. Optim.* **2001**, *22*, 179–187. [CrossRef]
366. Aage, N.; Andreassen, E.; Lazarov, B.S. Topology optimization using PETSc: An easy-to-use, fully parallel, open source topology optimization framework. *Struct. Multidiscip. Optim.* **2015**, *51*, 565–572. [CrossRef]
367. Aage, N.; Nobel-Jørgensen, M.; Andreasen, C.S.; Sigmund, O. Interactive topology optimization on hand-held devices. *Struct. Multidiscip. Optim.* **2013**, *47*, 1–6. [CrossRef]
368. Zuo, Z.H.; Xie, Y.M. A simple and compact Python code for complex 3D topology optimization. *Adv. Eng. Softw.* **2015**, *85*, 1–11. [CrossRef]
369. Liu, K.; Tovar, A. An efficient 3D topology optimization code written in Matlab. *Struct. Multidiscip. Optim.* **2014**, *50*, 1175–1196. [CrossRef]
370. Suresh, K. A 199-line Matlab code for Pareto-optimal tracing in topology optimization. *Struct. Multidiscip. Optim.* **2010**, *42*, 665–679. [CrossRef]
371. Challis, V.J. A discrete level-set topology optimization code written in Matlab. *Struct. Multidiscip. Optim.* **2010**, *41*, 453–464. [CrossRef]
372. Picelli, R.; Sivapuram, R.; Xie, Y.M. A 101-line MATLAB code for topology optimization using binary variables and integer programming. *Struct. Multidiscip. Optim.* **2020**. [CrossRef]
373. Wei, P.; Li, Z.; Li, X.; Wang, M.Y. An 88-line MATLAB code for the parameterized level set method based topology optimization using radial basis functions. *Struct. Multidiscip. Optim.* **2018**, *58*, 831–849. [CrossRef]
374. Schmidt, S.; Schulz, V. A 2589 line topology optimization code written for the graphics card. *Comput. Vis. Sci.* **2011**, *14*, 249–256. [CrossRef]
375. Talischi, C.; Paulino, G.H.; Pereira, A.; Menezes, I.F.M. PolyTop: A Matlab implementation of a general topology optimization framework using unstructured polygonal finite element meshes. *Struct. Multidiscip. Optim.* **2012**, *45*, 329–357. [CrossRef]
376. Zhou, S.; Cadman, J.; Chen, Y.; Li, W.; Xie, Y.M.; Huang, X.; Appleyard, R.; Sun, G.; Li, Q. Design and fabrication of biphasic cellular materials with transport properties-A modified bidirectional evolutionary structural optimization procedure and MATLAB program. *Int. J. Heat Mass Transf.* **2012**, *55*, 8149–8162. [CrossRef]

377. Otomori, M.; Yamada, T.; Izui, K.; Nishiwaki, S. Matlab code for a level set-based topology optimization method using a reaction diffusion equation. *Struct. Multidiscip. Optim.* **2015**, *51*, 1159–1172. [CrossRef]
378. Xia, L.; Breitkopf, P. Design of materials using topology optimization and energy-based homogenization approach in Matlab. *Struct. Multidiscip. Optim.* **2015**, *52*, 1229–1241. [CrossRef]
379. Zhang, W.; Yuan, J.; Zhang, J.; Guo, X. A new topology optimization approach based on Moving Morphable Components (MMC) and the ersatz material model. *Struct. Multidiscip. Optim.* **2016**, *53*, 1243–1260. [CrossRef]
380. Chung, H.; Hwang, J.T.; Gray, J.S.; Alicia Kim, H. Implementation of topology optimization using openMDAO. *AIAA 2018-0653. 2018 AIAA/ASCE/AHS/ASC Structures, Structural Dynamics, and Materials Conference*, 2018. [CrossRef]
381. Chung, H.; Hwang, J.T.; Gray, J.S.; Kim, H.A. Topology optimization in OpenMDAO. *Struct. Multidiscip. Optim.* **2019**, *59*, 1385–1400. [CrossRef]
382. Allaire, G.; Pantz, O. Structural optimization with FreeFem++. *Struct. Multidiscip. Optim.* **2006**, *32*, 173–181. [CrossRef]
383. Chisari, C.; Amadio, C. TOSCA: A Tool for Optimisation in Structural and Civil engineering Analyses. *Int. J. Adv. Struct. Eng.* **2018**, *10*, 401–419. [CrossRef]
384. He, L.; Gilbert, M.; Song, X. A Python script for adaptive layout optimization of trusses. *Struct. Multidiscip. Optim.* **2019**, *60*, 835–847. [CrossRef]
385. Zegard, T.; Paulino, G.H. GRAND—Ground structure based topology optimization for arbitrary 2D domains using MATLAB. *Struct. Multidiscip. Optim.* **2014**, *50*, 861–882. [CrossRef]
386. Zegard, T.; Paulino, G.H. GRAND3—Ground structure based topology optimization for arbitrary 3D domains using MATLAB. *Struct. Multidiscip. Optim.* **2015**, *52*, 1161–1184. [CrossRef]
387. Sokół, T. A 99 line code for discretized Michell truss optimization written in Mathematica. *Struct. Multidiscip. Optim.* **2011**, *43*, 181–190. [CrossRef]
388. Sigmund, O. EML webinar overview: Topology Optimization—Status and Perspectives. *Extrem. Mech. Lett.* **2020**, *39*, 100855. [CrossRef]
389. Spaeth, A.B. Editorial. *Arch. Sci. Rev.* **2020**, *63*, 103–104. [CrossRef]
390. Brown, N.C.; Mueller, C.T. Design for structural and energy performance of long span buildings using geometric multi-objective optimization. *Energy Build.* **2016**, *127*, 748–761. [CrossRef]
391. Kazakis, G.; Kanellopoulos, I.; Sotiropoulos, S.; Lagaros, N.D. Topology optimization aided structural design: Interpretation, computational aspects and 3D printing. *Heliyon* **2017**, *3*, e00431. [CrossRef]
392. Cicconi, P.; Germani, M.; Bondi, S.; Zuliani, A.; Cagnacci, E. A Design Methodology to Support the Optimization of Steel Structures. *Procedia CIRP* **2016**, *50*, 58–64. [CrossRef]
393. Fischer, A.W.; Guest, J.K.; Schafer, B.W. Novel Building Diaphragm Layouts Generated through Topology Optimization. *Ce/Pap.* **2019**, *3*, 505–510. [CrossRef]
394. Weldeyesus, A.G.; Stolpe, M. Free material optimization for laminated plates and shells. *Struct. Multidiscip. Optim.* **2016**, *53*, 1335–1347. [CrossRef]
395. Nan, B.; Bai, Y.; Wu, Y. Multi-objective optimization of spatially truss structures based on node movement. *Appl. Sci.* **2020**, *10*, 1964. [CrossRef]
396. Changizi, N.; Jalalpour, M. Stress-Based Topology Optimization of Steel-Frame Structures Using Members with Standard Cross Sections: Gradient-Based Approach. *J. Struct. Eng.* **2017**, *143*, 4017078. [CrossRef]
397. Weldeyesus, A.G.; Gondzio, J.; He, L.; Gilbert, M.; Shepherd, P.; Tyas, A. Truss geometry and topology optimization with global stability constraints. *Struct. Multidiscip. Optim.* **2020**, *62*, 1721–1737. [CrossRef]
398. Larsen, S.D.; Sigmund, O.; Groen, J.P. Optimal truss and frame design from projected homogenization-based topology optimization. *Struct. Multidiscip. Optim.* **2018**, *57*, 1461–1474. [CrossRef]
399. Kaveh, A.; Mahdavi, V.R. Colliding bodies optimization for size and topology optimization of truss structures. *Struct. Eng. Mech.* **2015**, *53*, 847–865. [CrossRef]
400. Kaveh, A.; Farhoudi, N. Layout optimization of braced frames using differential evolution algorithm and dolphin echolocation optimization. *Period. Polytech. Civ. Eng.* **2015**, *59*, 441–449. [CrossRef]
401. Lu, H.; Gilbert, M.; Tyas, A. Theoretically optimal bracing for pre-existing building frames. *Struct. Multidiscip. Optim.* **2018**, *58*, 677–686. [CrossRef]
402. Baradaran, M.; Madhkhan, M. Determination of Optimal Configuration for Mega Bracing Systems in Steel Frames using Genetic Algorithm. *Ksce J. Civ. Eng.* **2019**, *23*, 3616–3627. [CrossRef]
403. Hassanzadeh, A.; Gholizadeh, S. Collapse-performance-aided design optimization of steel concentrically braced frames. *Eng. Struct.* **2019**, *197*, 109411. [CrossRef]
404. Nouri, F.; Ashtari, P. Weight and topology optimization of outrigger-braced tall steel structures subjected to the wind loading using GA. *Wind Struct. Int. J.* **2015**, *20*, 489–508. [CrossRef]
405. Beghini, L.L.; Beghini, A.; Baker, W.F.; Paulino, G.H. Integrated Discrete/Continuum Topology Optimization Framework for Stiffness or Global Stability of High-Rise Buildings. *J. Struct. Eng.* **2015**, *141*, 4014207. [CrossRef]
406. Angelucci, G.; Mollaioli, F.; Alshawa, O. Evaluation of optimal lateral resisting systems for tall buildings subject to horizontal loads. *Procedia Manuf.* **2020**, *44*, 457–464. [CrossRef]

407. Angelucci, G.; Spence, S.M.J.; Mollaioli, F. An integrated topology optimization framework for three-dimensional domains using shell elements. *Struct. Des. Tall Spec. Build.* **2020**, 1–17. [CrossRef]
408. Zakian, P.; Kaveh, A. Topology optimization of shear wall structures under seismic loading. *Earthq. Eng. Eng. Vib.* **2020**, *19*, 105–116. [CrossRef]
409. Kaveh, A.; Ghafari, M.H.; Gholipour, Y. Optimal seismic design of 3D steel moment frames: Different ductility types. *Struct. Multidiscip. Optim.* **2017**, *56*, 1353–1368. [CrossRef]
410. Qiao, S.; Han, X.; Zhou, K.; Ji, J. Seismic analysis of steel structure with brace configuration using topology optimization. *Steel Compos. Struct.* **2016**, *21*, 501–515. [CrossRef]
411. Ribeiro, T.; Rigueiro, C.; Borges, L.; Sousa, A. A comprehensive method for fatigue life evaluation and extension in the context of predictive maintenance for fixed ocean structures. *Appl. Ocean. Res.* **2020**. [CrossRef]
412. Natarajan, A.; Stolpe, M.; Njomo Wandji, W. Structural optimization based design of jacket type sub-structures for 10MW offshore wind turbines. *Ocean. Eng.* **2019**, *172*, 629–640. [CrossRef]
413. Sandal, K.; Verbart, A.; Stolpe, M. Conceptual jacket design by structural optimization. *Wind Energy* **2018**, *21*, 1423–1434. [CrossRef]
414. Sandal, K.; Latini, C.; Zania, V.; Stolpe, M. Integrated optimal design of jackets and foundations. *Mar. Struct.* **2018**, *61*, 398–418. [CrossRef]
415. Savsani, V.; Dave, P.; Raja, B.D.; Patel, V. Topology optimization of an offshore jacket structure considering aerodynamic, hydrodynamic and structural forces. *Eng. Comput.* **2020**. [CrossRef]
416. Cicconi, P.; Castorani, V.; Germani, M.; Mandolini, M.; Vita, A. A multi-objective sequential method for manufacturing cost and structural optimization of modular steel towers. *Eng. Comput.* **2020**, *36*, 475–497. [CrossRef]
417. Kaveh, A.; Rezaei, M.; Shiravand, M.R. Optimal design of nonlinear large-scale suspendome using cascade optimization. *Int. J. Sp. Struct.* **2018**, *33*, 3–18. [CrossRef]
418. Baandrup, M.; Sigmund, O.; Polk, H.; Aage, N. Closing the gap towards super-long suspension bridges using computational morphogenesis. *Nat. Commun.* **2020**, *11*, 1–7. [CrossRef] [PubMed]
419. Kristiansen, H.; Poulios, K.; Aage, N. Topology optimization for compliance and contact pressure distribution in structural problems with friction. *Comput. Methods Appl. Mech. Eng.* **2020**, *364*, 112915. [CrossRef]
420. Rodrigues, T.A.; Duarte, V.; Miranda, R.M.; Santos, T.G.; Oliveira, J.P. Current status and perspectives on wire and arc additive manufacturing (WAAM). *Materials* **2019**, *12*, 1121. [CrossRef] [PubMed]
421. Tankova, T.; da Silva, L.S. Robotics and Additive Manufacturing in the Construction Industry. *Curr. Robot. Rep.* **2020**, *1*, 13–18. [CrossRef]
422. Lange, J.; Feucht, T.; Erven, M. 3D printing with steel: Additive Manufacturing for connections and structures. *Steel Constr.* **2020**, *13*, 144–153. [CrossRef]
423. Herzog, D.; Seyda, V.; Wycisk, E.; Emmelmann, C. Additive manufacturing of metals. *Acta Mater.* **2016**. [CrossRef]
424. Leach, R.K.; Bourell, D.; Carmignato, S.; Donmez, A.; Senin, N.; Dewulf, W. Geometrical metrology for metal additive manufacturing. *CIRP Ann.* **2019**, *68*, 677–700. [CrossRef]
425. DebRoy, T.; Wei, H.L.; Zuback, J.S.; Mukherjee, T.; Elmer, J.W.; Milewski, J.O.; Beese, A.M.; Wilson-Heid, A.; De, A.; Zhang, W. Additive manufacturing of metallic components–Process, structure and properties. *Prog. Mater. Sci.* **2018**, *92*, 112–224. [CrossRef]
426. Davila Delgado, J.M.; Oyedele, L.; Ajayi, A.; Akanbi, L.; Akinade, O.; Bilal, M.; Owolabi, H. Robotics and automated systems in construction: Understanding industry-specific challenges for adoption. *J. Build. Eng.* **2019**, *26*, 100868. [CrossRef]
427. Seifi, H.; Rezaee Javan, A.; Lin, X.; Xie, Y.M. An innovative and inexpensive experimental setup for testing connections in gridshell structures. *Eng. Struct.* **2020**, *207*, 110257. [CrossRef]
428. Wang, L.; Du, W.; He, P.; Yang, M. Topology Optimization and 3D Printing of Three-Branch Joints in Treelike Structures. *J. Struct. Eng.* **2020**, *146*, 4019167. [CrossRef]
429. Kanyilmaz, A.; Berto, F. Robustness-oriented topology optimization for steel tubular joints mimicking bamboo structures. *Mater. Des. Process. Commun.* **2019**, *1*, e43. [CrossRef]
430. Kanyilmaz, A.; Berto, F.; Paoletti, I.; Caringal, R.J.; Mora, S. Nature-inspired optimization of tubular joints for metal 3D printing. *Struct. Multidiscip. Optim.* **2020**. [CrossRef]
431. Alberdi, R.; Murren, P.; Khandelwal, K. Connection topology optimization of steel moment frames using metaheuristic algorithms. *Eng. Struct.* **2015**, *100*, 276–292. [CrossRef]
432. Moghadasi, A.; Held, A.; Seifried, R. Modeling of Revolute Joints in Topology Optimization of Flexible Multibody Systems. *J. Comput. Nonlinear Dyn.* **2017**, *12*, 1–8. [CrossRef]
433. Wang, J.; Zhu, J.; Hou, J.; Wang, C.; Zhang, W. Lightweight design of a bolt-flange sealing structure based on topology optimization. *Struct. Multidiscip. Optim.* **2020**. [CrossRef]
434. Ambrozkiewicz, O.; Kriegesmann, B. Simultaneous topology and fastener layout optimization of assemblies considering joint failure. *Int. J. Numer. Methods Eng.* **2020**, 1–26. [CrossRef]
435. Wang, X.; Hu, P.; Kang, Z. Layout optimization of continuum structures embedded with movable components and holes simultaneously. *Struct. Multidiscip. Optim.* **2020**, *61*, 555–573. [CrossRef]
436. Kang, S.W.; Kim, S.I.; Kim, Y.Y. Topology optimization of planar linkage systems involving general joint types. *Mech. Mach. Theory* **2016**, *104*, 130–160. [CrossRef]

437. Neves, M.M.; Rodrigues, H.; Guedes, M. Generalized topology criterion design of structures with a buckling load. *Struct. Optim.* **1995**, *10*, 71–78. [CrossRef]
438. Townsend, S.; Kim, H.A. A level set topology optimization method for the buckling of shell structures. *Struct. Multidiscip. Optim.* **2019**, *60*, 1783–1800. [CrossRef]
439. Clausen, A.; Aage, N.; Sigmund, O. Exploiting Additive Manufacturing Infill in Topology Optimization for Improved Buckling Load. *Engineering* **2016**, *2*, 250–257. [CrossRef]
440. Doan, Q.H.; Lee, D. Optimum topology design of multi-material structures with non-spurious buckling constraints. *Adv. Eng. Softw.* **2017**, *114*, 110–120. [CrossRef]
441. Thomsen, C.R.; Wang, F.; Sigmund, O. Buckling strength topology optimization of 2D periodic materials based on linearized bifurcation analysis. *Comput. Methods Appl. Mech. Eng.* **2018**, *339*, 115–136. [CrossRef]
442. Ferrari, F.; Sigmund, O. Towards solving large-scale topology optimization problems with buckling constraints at the cost of linear analyses. *Comput. Methods Appl. Mech. Eng.* **2020**, *363*, 112911. [CrossRef]
443. Pedersen, P.; Pedersen, N.L. Discussion on Problems in Buckling Analysis of a Continua. In Proceedings of the 30th Nordic Seminar on Computational Mechanics (NSCM-30), DTU Mechanical Engineering, Lyngby, Denmark, 25–27 October 2017.
444. Pedersen, P.; Pedersen, N. Discussion on Buckling Load Optimization for Continuum Models Subjected to Eccentric Loads. In Proceedings of the 6th International Conference on Engineering Optimization, Lisbon, Portugal, 17–19 September 2018.
445. Wang, F.; Sigmund, O. Numerical investigation of stiffness and buckling response of simple and optimized infill structures. *Struct. Multidiscip. Optim.* **2020**, *61*, 2629–2639. [CrossRef]
446. Pedersen, N.L.; Pedersen, P. Buckling load optimization for 2D continuum models, with alternative formulation for buckling load estimation. *Struct. Multidiscip. Optim.* **2018**, *58*, 2163–2172. [CrossRef]
447. Pedersen, N.L.; Pedersen, P. Local analytical sensitivity analysis for design of continua with optimized 3D buckling behavior. *Struct. Multidiscip. Optim.* **2018**, *57*, 293–304. [CrossRef]
448. Tugilimana, A.; Filomeno Coelho, R.; Thrall, A.P. Including global stability in truss layout optimization for the conceptual design of large-scale applications. *Struct. Multidiscip. Optim.* **2018**, *57*, 1213–1232. [CrossRef]
449. Xu, X.; Wang, Y.; Luo, Y.; Hu, D. Topology Optimization of Tensegrity Structures Considering Buckling Constraints. *J. Struct. Eng.* **2018**, *144*, 4018173. [CrossRef]
450. Xu, X.; Wang, Y.; Luo, Y. An improved multi-objective topology optimization approach for tensegrity structures. *Adv. Struct. Eng.* **2018**, *21*, 59–70. [CrossRef]
451. Zhao, Z.; Wu, J.; Liu, H.; Liang, B.; Zhang, N. Shape optimization of reticulated shells with constraints on member instabilities. *Eng. Optim.* **2019**, *51*, 1463–1479. [CrossRef]
452. Massaroppi, E.; Zampaolo, T.C.; Abambres, M.; Ribeiro, T.P. Collapse of i-section tapered beam-columns in medium-span steel frames: Finite element model validation and parameters influence evaluation. *Lat. Am. J. Solids Struct.* **2020**. [CrossRef]
453. Farahmand-Tabar, S.; Ashtari, P. Simultaneous size and topology optimization of 3D outrigger-braced tall buildings with inclined belt truss using genetic algorithm. *Struct. Des. Tall Spec. Build.* **2020**, *29*, 1–15. [CrossRef]
454. da Silva, G.A.; Cardoso, E.L.; Beck, A.T. Non-probabilistic robust continuum topology optimization with stress constraints. *Struct. Multidiscip. Optim.* **2019**, *59*, 1181–1197. [CrossRef]
455. Simões, L.; Templeman, A. Entropy-based Synthesis of Pretensioned Cable Net Structures. *Eng. Opt.* **1989**, *1*, 121–140.
456. Marler, R.T.; Arora, J.S. Survey of multi-objective optimization methods for engineering. *Struct. Multidiscip. Optim.* **2004**, *26*, 369–395. [CrossRef]
457. Luo, Z.; Chen, L.P.; Yang, J.; Zhang, Y.Q.; Abdel-Malek, K. Fuzzy tolerance multilevel approach for structural topology optimization. *Comput. Struct.* **2006**, *84*, 127–140. [CrossRef]
458. Li, H.; Gao, L.; Li, P. Topology optimization of structures under multiple loading cases with a new compliance-volume product. *Eng. Optim.* **2014**, *46*, 725–744. [CrossRef]
459. Pintér, E.; Lengyel, A.; Lógó, J. Structural topology optimization with stress constraint considering loading uncertainties. *Period. Polytech. Civ. Eng.* **2015**, *59*, 559–565. [CrossRef]
460. Wang, D.; Gao, W. Robust topology optimization under multiple independent uncertainties of loading positions. *Int. J. Numer. Methods Eng.* **2020**, *121*, 4944–4970. [CrossRef]
461. Csébfalvi, A. A new compliance-function-shape-oriented robust approach for volume-constrained continuous topology optimization with uncertain loading directions. *Period. Polytech. Civ. Eng.* **2018**, *62*, 219–225. [CrossRef]
462. Csébfalvi, A. Robust topology optimization: A new algorithm for volume-constrained expected compliance minimization with probabilistic loading directions using exact analytical objective and gradient. *Period. Polytech. Civ. Eng.* **2017**, *61*, 154–163. [CrossRef]
463. Chan, Y.C.; Shintani, K.; Chen, W. Robust topology optimization of multi-material lattice structures under material and load uncertainties. *Front. Mech. Eng.* **2019**, *14*, 141–152. [CrossRef]
464. Nishino, T.; Kato, J. Robust topology optimization based on finite strain considering uncertain loading conditions. *Int. J. Numer. Methods Eng.* **2020**. [CrossRef]
465. Yi, G.; Sui, Y. TIMP method for topology optimization of plate structures with displacement constraints under multiple loading cases. *Struct. Multidiscip. Optim.* **2016**, *53*, 1185–1196. [CrossRef]

466. Tang, J.; Xie, Y.M.; Felicetti, P. Topology optimization of building structures considering wind loading. *Appl. Mech. Mater.* **2012**, *166–169*, 405–408. [CrossRef]
467. Lógó, J.; Balogh, B.; Pintér, E. Topology optimization considering multiple loading. *Comput. Struct.* **2018**, *207*, 233–244. [CrossRef]
468. Alkalla, M.G.; Helal, M.; Fouly, A. Revolutionary Superposition Layout Method for Topology Optimization of Non-Concurrent Multi-load Models: Connecting-Rod Case Study. *Int. J. Numer. Methods Eng.* **2020**. [CrossRef]
469. Tsavdaridis, K.D.; Nicolaou, A.; Mistry, A.D.; Efthymiou, E. Topology optimisation of lattice telecommunication tower and performance-based design considering wind and ice loads. *Structures* **2020**, *27*, 2379–2399. [CrossRef]
470. Silva, A.; Santos, L.; Ribeiro, T.; Castro, J.M. Improved Seismic Design of Concentrically X-Braced Steel Frames to Eurocode 8. *J. Earthq. Eng.* **2018**. [CrossRef]
471. Ribeiro, T.; Sousa, A. Methods for conceptual and preliminary seismic design of buildings with steel structure. *Av. En Cienc. E Ing.* **2019**, *11*. [CrossRef]
472. Amir, O.; Mass, Y. Topology optimization for staged construction. *Struct. Multidiscip. Optim.* **2018**, *57*, 1679–1694. [CrossRef]
473. Sobótka, M. Shape optimization of flexible soil-steel culverts taking non-stationary loads into account. *Structures* **2020**, *23*, 612–620. [CrossRef]
474. Bos, F.P.; Lucas, S.S.; Wolfs, R.J.M.; Salet, T.A.M. *Second RILEM on Concrete and Conference International Digital Fabrication*; Springer: Berlin/Heidelberg, Germany, 2020; Volume 28, ISBN 9783030499150.
475. Vantyghem, G.; Boel, V.; De Corte, W. *Compliance, Stress-Based and Multi-physics Topology Optimization for 3D-Printed Concrete Structures*; Springer International Publishing: Berlin/Heidelberg, Germany, 2019; Volume 1, ISBN 9783319995199.
476. Ritter, W. Die Bauweise Hennebique. *Schweiz. Bauztg.* **1899**, *33*, 59–61.
477. Schlaich, J.; Schaefer, K.; Jennewein, M. Toward a Consistent Design of Structural Concrete. *PCI J.* **1987**. [CrossRef]
478. Zhou, L.Y.; He, Z.Q.; Liu, Z. Investigation of optimal layout of ties in STM developed by topology optimization. *Struct. Concr.* **2016**, *17*, 175–182. [CrossRef]
479. Yang, Z.; Zhou, K.; Qiao, S. Topology optimization of reinforced concrete structure using composite truss-like model. *Struct. Eng. Mech.* **2018**, *67*, 79–85. [CrossRef]
480. Jewett, J.L.; Carstensen, J.V. Experimental investigation of strut-and-tie layouts in deep RC beams designed with hybrid bi-linear topology optimization. *Eng. Struct.* **2019**, *197*, 109322. [CrossRef]
481. Xia, Y.; Langelaar, M.; Hendriks, M.A.N. A critical evaluation of topology optimization results for strut-and-tie modeling of reinforced concrete. *Comput. Civ. Infrastruct. Eng.* **2020**, *35*, 850–869. [CrossRef]
482. Pastore, T.; Mercuri, V.; Menna, C.; Asprone, D.; Festa, P.; Auricchio, F. Topology optimization of stress-constrained structural elements using risk-factor approach. *Comput. Struct.* **2019**, *224*, 106104. [CrossRef]
483. Qiao, S.; Han, X.; Zhou, K. Bracing configuration and seismic performance of reinforced concrete frame with brace. *Struct. Des. Tall Spec. Build.* **2017**, *26*, 1–14. [CrossRef]
484. Venini, P.; Ceresa, P. A rational H∞-norm–based approach for the optimal design of seismically excited reinforced concrete frames. *Earthq. Eng. Struct. Dyn.* **2018**, *47*, 1522–1543. [CrossRef]
485. Camacho, V.T.; Horta, N.; Lopes, M.; Oliveira, C.S. Optimizing earthquake design of reinforced concrete bridge infrastructures based on evolutionary computation techniques. *Struct. Multidiscip. Optim.* **2020**, *61*, 1087–1105. [CrossRef]
486. Amir, O.; Shakour, E. Simultaneous shape and topology optimization of prestressed concrete beams. *Struct. Multidiscip. Optim.* **2018**, *57*, 1831–1843. [CrossRef]
487. Wu, J.; Wu, L. Revised Level Set-Based Method for Topology Optimization and Its Applications in Bridge Construction. *Open Civ. Eng. J.* **2017**, *11*, 153–166. [CrossRef]
488. Zhang, W.; Zhu, J.; Gao, T. *Topology Optimization in Engineering Structure Design*; ISTE Press Ltd.: London, UK, 2016; ISBN 9788578110796.
489. Calabrese, M.; Primo, T.; Del Prete, A. Lattice structures integration with conventional topology optimization. In Proceedings of the AIP Conference; 2017; Volume 1896.
490. Carneiro, P.M.C.; Gamboa, P. Structural analysis of wing ribs obtained by additive manufacturing. *Rapid Prototyp. J.* **2019**, *25*, 708–720. [CrossRef]
491. SHI, G.; GUAN, C.; QUAN, D.; WU, D.; TANG, L.; GAO, T. An aerospace bracket designed by thermo-elastic topology optimization and manufactured by additive manufacturing. *Chin. J. Aeronaut.* **2020**, *33*, 1252–1259. [CrossRef]
492. Willner, R.; Lender, S.; Ihl, A.; Wilsnack, C.; Gruber, S.; Brandão, A.; Pambaguian, L.; Riede, M.; López, E.; Brueckner, F.; et al. Potential and challenges of additive manufacturing for topology optimized spacecraft structures. *J. Laser Appl.* **2020**, *32*, 032012. [CrossRef]
493. Mantovani, S.; Barbieri, S.G.; Giacopini, M.; Croce, A.; Sola, A.; Bassoli, E. Synergy between topology optimization and additive manufacturing in the automotive field. *Proc. Inst. Mech. Eng. Part. B J. Eng. Manuf.* **2020**. [CrossRef]
494. van de Ven, E.; Maas, R.; Ayas, C.; Langelaar, M.; van Keulen, F. Overhang control based on front propagation in 3D topology optimization for additive manufacturing. *Comput. Methods Appl. Mech. Eng.* **2020**, *369*, 113169. [CrossRef]
495. Mass, Y.; Amir, O. Topology optimization for additive manufacturing: Accounting for overhang limitations using a virtual skeleton. *Addit. Manuf.* **2017**, *18*, 58–73. [CrossRef]
496. Topaç, M.M.; Karaca, M.; Aksoy, B.; Deryal, U.; BïLal, L. Lightweight design of a rear axle connection bracket for a heavy commercial vehicle by using topology optimisation: A case study. *Mechanika* **2020**, *26*, 64–72. [CrossRef]

497. Kumar, A.; Sharma, S. Development of Methodology for Full Bus Body Optimisation and Strengthening by Numerical Simulation. *SAE Tech. Pap.* **2017**. [CrossRef]
498. Mantovani, S.; Campo, G.A.; Giacalone, M. Steering column support topology optimization including lattice structure for metal additive manufacturing. *Proc. Inst. Mech. Eng. Part. C J. Mech. Eng. Sci.* **2020**. [CrossRef]
499. Li, C.; Kim, I.Y. Topology, size and shape optimization of an automotive cross car beam. *Proc. Inst. Mech. Eng. Part. D J. Automob. Eng.* **2015**, *229*, 1361–1378. [CrossRef]
500. Tamijani, A.Y.; Velasco, S.P.; Alacoque, L. Topological and morphological design of additively-manufacturable spatially-varying periodic cellular solids. *Mater. Des.* **2020**, *196*, 109155. [CrossRef]
501. Tromme, E.; Kawamoto, A.; Guest, J.K. Topology optimization based on reduction methods with applications to multiscale design and additive manufacturing. *Front. Mech. Eng.* **2020**, *15*, 151–165. [CrossRef]
502. Sigmund, O. Design of material structures using topology optimization. *Dep. Solid Mech.* **1994**.
503. de Lima, C.R.; Paulino, G.H. Auxetic structure design using compliant mechanisms: A topology optimization approach with polygonal finite elements. *Adv. Eng. Softw.* **2019**, *129*, 69–80. [CrossRef]
504. Wang, C.; Gu, X.; Zhu, J.; Zhou, H.; Li, S.; Zhang, W. Concurrent design of hierarchical structures with three-dimensional parameterized lattice microstructures for additive manufacturing. *Struct. Multidiscip. Optim.* **2020**, *61*, 869–894. [CrossRef]
505. Zhang, J.; Yanagimoto, J. Topology optimization of microlattice dome with enhanced stiffness and energy absorption for additive manufacturing. *Compos. Struct.* **2021**, *255*, 112889. [CrossRef]
506. Duan, S.; Xi, L.; Wen, W.; Fang, D. Mechanical performance of topology-optimized 3D lattice materials manufactured via selective laser sintering. *Compos. Struct.* **2020**, *238*. [CrossRef]
507. Deng, H.; Hinnebusch, S.; To, A.C. Topology optimization design of stretchable metamaterials with Bézier skeleton explicit density (BSED) representation algorithm. *Comput. Methods Appl. Mech. Eng.* **2020**, *366*, 113093. [CrossRef]
508. Collet, M.; Noël, L.; Bruggi, M.; Duysinx, P. Topology optimization for microstructural design under stress constraints. *Struct. Multidiscip. Optim.* **2018**, *58*, 2677–2695. [CrossRef]
509. Kang, Z.; Wu, C.; Luo, Y.; Li, M. Robust topology optimization of multi-material structures considering uncertain graded interface. *Compos. Struct.* **2019**, *208*, 395–406. [CrossRef]
510. Bluhm, G.L.; Sigmund, O.; Wang, F.; Poulios, K. Nonlinear compressive stability of hyperelastic 2D lattices at finite volume fractions. *J. Mech. Phys. Solids* **2020**, *137*, 103851. [CrossRef]
511. Huang, J.; Chen, Q.; Jiang, H.; Zou, B.; Li, L.; Liu, J.; Yu, H. A survey of design methods for material extrusion polymer 3D printing. *Virtual Phys. Prototyp.* **2020**, *15*, 148–162. [CrossRef]
512. El Jai, M.; Saidou, N.; Zineddine, M.; Bachiri, H. Mathematical design and preliminary mechanical analysis of the new lattice structure: "GE-SEZ*" structure processed by ABS polymer and FDM technology. *Prog. Addit. Manuf.* **2020**. [CrossRef]
513. Mirzendehdel, A.M.; Suresh, K. A Pareto-Optimal Approach to Multimaterial Topology Optimization. *J. Mech. Des. Trans. Asme* **2015**, *137*, 1–12. [CrossRef]
514. Chin, T.W.; Leader, M.K.; Kennedy, G.J. A scalable framework for large-scale 3D multimaterial topology optimization with octree-based mesh adaptation. *Adv. Eng. Softw.* **2019**, *135*, 102682. [CrossRef]
515. Valente, M.; Sambucci, M.; Sibai, A.; Musacchi, E. Multi-physics analysis for rubber-cement applications in building and architectural fields: A preliminary analysis. *Sustainability* **2020**, *12*, 5993. [CrossRef]
516. Chen, J.; Xu, Y.; Gao, Y. Topology optimization of metal and carbon fiber reinforced plastic (CFRP) laminated battery-hanging structure. *Polymers* **2020**, *12*, 2495. [CrossRef]
517. Bykerk, L.; Liu, D.; Waldron, K. A topology optimisation based design of a compliant gripper for grasping objects with irregular shapes. In Proceedings of the IEEE/ASME International Conference on Advanced Intelligent Mechatronics (AIM), Banff, AB, Canada, 12–15 July 2016; pp. 383–388.
518. da Silva, G.A.; Beck, A.T.; Sigmund, O. Topology optimization of compliant mechanisms with stress constraints and manufacturing error robustness. *Comput. Methods Appl. Mech. Eng.* **2019**, *354*, 397–421. [CrossRef]
519. Suresh, S.; Lindström, S.B.; Thore, C.J.; Klarbring, A. Topology optimization for transversely isotropic materials with high-cycle fatigue as a constraint. *Struct. Multidiscip. Optim.* **2020**. [CrossRef]
520. Rigo, P.; Caprace, J.-D.; Sekulski, Z.; Bayatfar, A.; Echeverry, S. Structural Design Optimization—Tools and Methodologies. In *A Holistic Approach to Ship Design: Optimisation of Ship Design and Operation for Life Cycle*; Papanikolaou, A., Ed.; Springer International Publishing: Berlin/Heidelberg, Germany, 2018; Volume 1, ISBN 9783030028107.
521. Zhu, J.; Li, Y.; Wang, F.; Zhang, W. Shape preserving design of thermo-elastic structures considering geometrical nonlinearity. *Struct. Multidiscip. Optim.* **2020**, *61*, 1787–1804. [CrossRef]
522. Cheng, L.; Liu, J.; Liang, X.; To, A.C. Coupling lattice structure topology optimization with design-dependent feature evolution for additive manufactured heat conduction design. *Comput. Methods Appl. Mech. Eng.* **2018**, *332*, 408–439. [CrossRef]
523. Perumal, V.I.; Najafi, A.R.; Kontsos, A. A novel digital design approach for metal additive manufacturing to address local thermal effects. *Designs* **2020**, *4*, 41. [CrossRef]
524. Conlan-Smith, C.; Ramos-García, N.; Sigmund, O.; Andreasen, C.S. Aerodynamic shape optimization of aircraft wings using panel methods. *AIAA J.* **2020**, *58*, 3765–3776. [CrossRef]
525. Pollini, N.; Sigmund, O.; Andreasen, C.S.; Alexandersen, J. A "poor man's" approach for high-resolution three-dimensional topology design for natural convection problems. *Adv. Eng. Softw.* **2020**, *140*, 102736. [CrossRef]

526. Lim, Y.E.; Kim, N.H.; Choi, H.J.; Park, K. Design for additive manufacturing of customized cast with porous shell structures. *J. Mech. Sci. Technol.* **2017**, *31*, 5477–5483. [CrossRef]
527. Jiang, L.; Chen, S.; Sadasivan, C.; Jiao, X. Structural topology optimization for generative design of personalized aneurysm implants: Design, additive manufacturing, and experimental validation. In Proceedings of the IEEE Healthcare Innovations and Point of Care Technologies, HI-POCT 2017, Bethesda, MD, USA, 6–8 November 2017; pp. 9–13. [CrossRef]
528. Chen, X.J.; Hu, J.L.; Zhou, Q.L.; Politis, C.; Sun, Y. An automatic optimization method for minimizing supporting structures in additive manufacturing. *Adv. Manuf.* **2020**, *8*, 49–58. [CrossRef]
529. Reintjes, C.; Lorenz, U. *Bridging Mixed Integer Linear Programming for Truss Topology Optimization and Additive Manufacturing*; Springer: Berlin/Heidelberg, Germany, 2020; ISBN 0123456789.
530. Gao, W.; Zhang, Y.; Ramanujan, D.; Ramani, K.; Chen, Y.; Williams, C.B.; Wang, C.C.L.; Shin, Y.C.; Zhang, S.; Zavattieri, P.D. The status, challenges, and future of additive manufacturing in engineering. *Cad. Comput. Aided Des.* **2015**. [CrossRef]
531. Standard ASTM F2792-12, *Standard Terminology for Additive Manufacturing Technologies*; ASTM International: West Conshohocken, PA, USA, 2012.
532. ISO/ASTM, *ISO/ASTM 52900: Additive manufacturing-General principles-Terminology*; International Standard, 2015.
533. Berrio Bernal, J.D.; Silva, E.C.N.; Montealegre Rubio, W. Characterization of effective Young's modulus for Fused Deposition Modeling manufactured topology optimization designs. *Int. J. Adv. Manuf. Technol.* **2019**, *103*, 2879–2892. [CrossRef]
534. Takezawa, A.; Koizumi, Y.; Kobashi, M. High-stiffness and strength porous maraging steel via topology optimization and selective laser melting. *Addit. Manuf.* **2017**, *18*, 194–202. [CrossRef]
535. Grossmann, A.; Weis, P.; Clemen, C.; Mittelstedt, C. Optimization and re-design of a metallic riveting tool for additive manufacturing—A case study. *Addit. Manuf.* **2020**, *31*. [CrossRef]
536. Pellens, J.; Lombaert, G.; Michiels, M.; Craeghs, T.; Schevenels, M. Topology optimization of support structure layout in metal-based additive manufacturing accounting for thermal deformations. *Struct. Multidiscip. Optim.* **2020**, *61*, 2291–2303. [CrossRef]
537. Nirish, M.; Rajendra, R. Suitability of metal additive manufacturing processes for part topology optimization–A comparative study. *Mater. Today Proc.* **2020**, *27*, 1601–1607. [CrossRef]
538. Brant, A.; Sundaram, M. A Novel Electrochemical Micro Additive Manufacturing Method of Overhanging Metal Parts without Reliance on Support Structures. *Procedia Manuf.* **2016**, *5*, 928–943. [CrossRef]
539. Hirtler, M.; Jedynak, A.; Sydow, B.; Sviridov, A.; Bambach, M. Investigation of microstructure and hardness of a rib geometry produced by metal forming and wire-arc additive manufacturing. *Matec Web Conf.* **2018**, *190*, 1–6. [CrossRef]
540. Seabra, M.; Azevedo, J.; Araújo, A.; Reis, L.; Pinto, E.; Alves, N.; Santos, R.; Pedro Mortágua, J. Selective laser melting (SLM) and topology optimization for lighter aerospace componentes. *Procedia Struct. Integr.* **2016**, *1*, 289–296. [CrossRef]
541. Gebisa, A.W.; Lemu, H.G. Design for manufacturing to design for Additive Manufacturing: Analysis of implications for design optimality and product sustainability. *Procedia Manuf.* **2017**, *13*, 724–731. [CrossRef]
542. Valjak, F.; Bojčetić, N.; Lukić, M. Design for additive manufacturing: Mapping of product functions. In Proceedings of the International Design Conference (DESIGN), Dubrovnik, Croatia, 21–24 May 2018; Volume 3, pp. 1369–1380.
543. Chen, W.; Zheng, X.; Liu, S. Finite-element-mesh based method for modeling and optimization of lattice structures for additive manufacturing. *Materials* **2018**, *11*, 2073. [CrossRef]
544. Doagou-Rad, S.; Jensen, J.S.; Islam, A.; Mishnaevsky, L. Multiscale molecular dynamics-FE modeling of polymeric nanocomposites reinforced with carbon nanotubes and graphene. *Compos. Struct.* **2019**, *217*, 27–36. [CrossRef]
545. Jung, J.; Jeong, C.H.; Jensen, J.S. Spectrally smooth and spatially uniform sound radiation from a thin plate structure using band gaps. *J. Sound Vib.* **2020**, *471*, 115187. [CrossRef]
546. Ren, S.; Galjaard, S. Modelling BehaviourTopology Optimisation for Steel Structural Design with Additive Manufacturing. In *Modelling Behaviour*; Springer: Cham, Switzerland, 2015.
547. Pellens, J.; Lombaert, G.; Lazarov, B.; Schevenels, M. Combined length scale and overhang angle control in minimum compliance topology optimization for additive manufacturing. *Struct. Multidiscip. Optim.* **2019**, *59*, 2005–2022. [CrossRef]
548. Zhang, P.; Liu, J.; To, A.C. Role of anisotropic properties on topology optimization of additive manufactured load bearing structures. *Scr. Mater.* **2017**, *135*, 148–152. [CrossRef]
549. Wang, C.; Qian, X. Simultaneous optimization of build orientation and topology for additive manufacturing. *Addit. Manuf.* **2020**, *34*, 101246. [CrossRef]
550. Sigmund, O.; Clausen, A.; Groen, J.P.; Wu, J. Topology optimization of structures and infill for additive manufacturing. In Proceedings of the Simulation for Additive Manufacturing, Munich, Germany, 11–13 October 2017; pp. 4–6.
551. Senck, S.; Happl, M.; Reiter, M.; Scheerer, M.; Kendel, M.; Glinz, J.; Kastner, J. Additive manufacturing and non-destructive testing of topology-optimised aluminium components. *Nondestruct. Test. Eval.* **2020**, *35*, 315–327. [CrossRef]
552. Rankouhi, B.; Bertsch, K.M.; Meric de Bellefon, G.; Thevamaran, M.; Thoma, D.J.; Suresh, K. Experimental validation and microstructure characterization of topology optimized, additively manufactured SS316L components. *Mater. Sci. Eng. A* **2020**, *776*, 139050. [CrossRef]
553. Mirzendehdel, A.M.; Suresh, K. Multi-Material Topology Optimization for Additive Manufacturing. In Proceedings of the ASME 2015 International Design Engineering Technical Conferences & Computers and Information in Engineering Conference, Boston, MA, USA, 2–5 August 2015; Volume 46268, pp. 1–11.

554. Gaynor, A.T.; Guest, J.K. Topology optimization considering overhang constraints: Eliminating sacrificial support material in additive manufacturing through design. *Struct. Multidiscip. Optim.* **2016**, *54*, 1157–1172. [CrossRef]
555. Langelaar, M. Topology optimization of 3D self-supporting structures for additive manufacturing. *Addit. Manuf.* **2016**, *12*, 60–70. [CrossRef]
556. Ranjan, R.; Samant, R.; Anand, S. Integration of Design for Manufacturing Methods with Topology Optimization in Additive Manufacturing. *J. Manuf. Sci. Eng. Trans. Asme* **2017**, *139*, 1–14. [CrossRef]
557. Booth, J.W.; Alperovich, J.; Chawla, P.; Ma, J.; Reid, T.N.; Ramani, K. The design for additive manufacturing worksheet. *J. Mech. Des. Trans. Asme* **2017**, *139*. [CrossRef]
558. Wu, J.; Clausen, A.; Sigmund, O. Minimum compliance topology optimization of shell–infill composites for additive manufacturing. *Comput. Methods Appl. Mech. Eng.* **2017**, *326*, 358–375. [CrossRef]
559. Walton, D.; Moztarzadeh, H. Design and Development of an Additive Manufactured Component by Topology Optimisation. *Procedia Cirp* **2017**, *60*, 205–210. [CrossRef]
560. Tammas-Williams, S.; Todd, I. Design for additive manufacturing with site-specific properties in metals and alloys. *Scr. Mater.* **2017**, *135*, 105–110. [CrossRef]
561. Wang, X.; Zhang, C.; Liu, T. A Topology Optimization Algorithm Based on the Overhang Sensitivity Analysis for Additive Manufacturing. *Iop Conf. Ser. Mater. Sci. Eng.* **2018**, *382*. [CrossRef]
562. Zhang, W.; Zhou, L. Topology optimization of self-supporting structures with polygon features for additive manufacturing. *Comput. Methods Appl. Mech. Eng.* **2018**, *334*, 56–78. [CrossRef]
563. Mhapsekar, K.; McConaha, M.; Anand, S. Additive Manufacturing Constraints in Topology Optimization for Improved Manufacturability. *J. Manuf. Sci. Eng. Trans. Asme* **2018**, *140*. [CrossRef]
564. Mezzadri, F.; Bouriakov, V.; Qian, X. Topology optimization of self-supporting support structures for additive manufacturing. *Addit. Manuf.* **2018**, *21*, 666–682. [CrossRef]
565. Kuo, Y.H.; Cheng, C.C.; Lin, Y.S.; San, C.H. Support structure design in additive manufacturing based on topology optimization. *Struct. Multidiscip. Optim.* **2018**, *57*, 183–195. [CrossRef]
566. Vogiatzis, P.; Ma, M.; Chen, S.; Gu, X.D. Computational design and additive manufacturing of periodic conformal metasurfaces by synthesizing topology optimization with conformal mapping. *Comput. Methods Appl. Mech. Eng.* **2018**, *328*, 477–497. [CrossRef]
567. Steuben, J.C.; Iliopoulos, A.P.; Michopoulos, J.G. Multiscale topology optimization for additively manufactured objects. *J. Comput. Inf. Sci. Eng.* **2018**, *18*. [CrossRef]
568. Weiss, B.M.; Ganter, M.A.; Hamel, J.M.; Storti, D.W. Data-driven additive manufacturing constraints for topology optimization. In Proceedings of the ASME International Design Engineering Technical Conferences; 2018; pp. 1–10. [CrossRef]
569. Allaire, G.; Jakabain, L. *Taking into Account Thermal Residual Stresses in Topology Optimization of Structures Built by Additive Manufacturing*; 2018; Volume 28, ISBN 0218202518500.
570. Garaigordobil, A.; Ansola, R.; Veguería, E.; Fernandez, I. Overhang constraint for topology optimization of self-supported compliant mechanisms considering additive manufacturing. *CAD Comput. Aided Des.* **2019**, *109*, 33–48. [CrossRef]
571. Barroqueiro, B.; Andrade-Campos, A.; Valente, R.A.F. Designing self supported SLM structures via topology optimization. *J. Manuf. Mater. Process.* **2019**, *3*, 68. [CrossRef]
572. Wang, L.; Xia, H.; Yang, Y.; Cai, Y.; Qiu, Z. A novel approach of reliability-based topology optimization for continuum structures under interval uncertainties. *Rapid Prototyp. J.* **2019**, *25*, 1455–1474. [CrossRef]
573. Langelaar, M. Integrated component-support topology optimization for additive manufacturing with post-machining. *Rapid Prototyp. J.* **2019**, *25*, 255–265. [CrossRef]
574. Orlov, A.; Masaylo, D.; Polozov, I.; Ji, P. Designing of topology optimized parts for additive manufacturing. *Key Eng. Mater.* **2019**, *822*, 526–533. [CrossRef]
575. Liu, J.; Yu, H. Self-Support Topology Optimization with Horizontal Overhangs for Additive Manufacturing. *J. Manuf. Sci. Eng. Trans. Asme* **2020**, *142*. [CrossRef]
576. Luo, Y.; Sigmund, O.; Li, Q.; Liu, S. Additive manufacturing oriented topology optimization of structures with self-supported enclosed voids. *Comput. Methods Appl. Mech. Eng.* **2020**, *372*, 113385. [CrossRef]
577. Dalpadulo, E.; Gherardini, F.; Pini, F.; Leali, F. Integration of topology optimisation and design variants selection for additive manufacturing-based systematic product redesign. *Appl. Sci.* **2020**, *10*, 7841. [CrossRef]
578. Liu, J.; Zheng, Y.; Ma, Y.; Qureshi, A.; Ahmad, R. A Topology Optimization Method for Hybrid Subtractive–Additive Remanufacturing. *Int. J. Precis. Eng. Manuf. Green Technol.* **2020**, *7*, 939–953. [CrossRef]
579. Wang, W.; Munro, D.; Wang, C.C.L.; van Keulen, F.; Wu, J. Space-time topology optimization for additive manufacturing: Concurrent optimization of structural layout and fabrication sequence. *Struct. Multidiscip. Optim.* **2020**, *61*. [CrossRef]
580. Crispo, L.; Kim, I.Y. Assembly level topology optimization towards a part consolidation algorithm for additive manufacturing. In *AIAA Scitech 2020 Forum*; 2020; pp. 1–9. [CrossRef]
581. Fritz, K.; Kim, I.Y. Simultaneous topology and build orientation optimization for minimization of additive manufacturing cost and time. *Int. J. Numer. Methods Eng.* **2020**, *121*, 3442–3481. [CrossRef]
582. Zhang, K.; Cheng, G. Three-dimensional high resolution topology optimization considering additive manufacturing constraints. *Addit. Manuf.* **2020**, *35*, 101224. [CrossRef]

583. Liu, B.; Jiang, C.; Li, G.; Huang, X. Topology optimization of structures considering local material uncertainties in additive manufacturing. *Comput. Methods Appl. Mech. Eng.* **2020**, *360*, 112786. [CrossRef]
584. Kim, G.-W.; Park, Y.-I.; Park, K. Topology Optimization And Additive Manufacturing Of Automotive Component By Coupling Kinetic And Structural Analyses. *Int. J. Automot. Technol.* **2020**. [CrossRef]
585. Li, S.; Yuan, S.; Zhu, J.; Wang, C.; Li, J.; Zhang, W. Additive manufacturing-driven design optimization: Building direction and structural topology. *Addit. Manuf.* **2020**, *36*, 101406. [CrossRef]
586. Li, H.; Gao, L.; Li, H.; Tong, H. Spatial-varying multi-phase infill design using density-based topology optimization. *Comput. Methods Appl. Mech. Eng.* **2020**, *372*, 113354. [CrossRef]
587. Wang, C.; Xu, B.; Meng, Q.; Rong, J.; Zhao, Y. Numerical performance of Poisson method for restricting enclosed voids in topology optimization. *Comput. Struct.* **2020**, *239*, 106337. [CrossRef]
588. Aliyi, A.M.; Lemu, H.G. Case study on topology optimized design for additive manufacturing. In *IOP Conference Series: Materials Science and Engineering*; IOP Publishing: Bristol, UK, 2019; Volume 659.
589. Fu, Y.F.; Rolfe, B.; Chiu, L.N.S.; Wang, Y.; Huang, X.; Ghabraie, K. Design and experimental validation of self-supporting topologies for additive manufacturing. *Virtual Phys. Prototyp.* **2019**, *14*, 382–394. [CrossRef]
590. Dinar, M.; Rosen, D.W. A design for additive manufacturing ontology. *J. Comput. Inf. Sci. Eng.* **2017**, *17*. [CrossRef]
591. Acar, E.; Du, J.; Kim, Y.Y.; Saka, M.P.; Sigmund, O.; Silva, E.C.N. Special issue for the 13th world congress on structural and multidisciplinary optimization—editorial note. *Struct. Multidiscip. Optim.* **2020**, *61*, 2225–2226. [CrossRef]
592. Sangree, R.; Carstensen, J.V.; Gaynor, A.T.; Zhu, M.; Guest, J.K. Topology Optimization as a Teaching Tool for Undergraduate Education in Structural Engineering. In Proceedings of the Structures Congress 2015, Portland, OR, USA, 23–25 April 2015.
593. Fu, Y.F. Recent advances and future trends in exploring Pareto-optimal topologies and additive manufacturing oriented topology optimization. *Math. Biosci. Eng.* **2020**, *17*, 4631–4656. [CrossRef] [PubMed]
594. Meng, L.; Zhang, W.; Quan, D.; Shi, G.; Tang, L.; Hou, Y.; Breitkopf, P.; Zhu, J.; Gao, T. From Topology Optimization Design to Additive Manufacturing: Today's Success and Tomorrow's Roadmap. *Arch. Comput. Methods Eng.* **2020**, *27*, 805–830. [CrossRef]
595. Liu, J.; Gaynor, A.T.; Chen, S.; Kang, Z.; Suresh, K.; Takezawa, A.; Li, L.; Kato, J.; Tang, J.; Wang, C.C.L.; et al. Current and future trends in topology optimization for additive manufacturing. *Struct. Multidiscip. Optim.* **2018**, *57*, 2457–2483. [CrossRef]
596. Mendez-Dominguez, E.; Fortuny-Guasch, J. *Optimization Techniques for MIMO Radar Antenna Systems*; European Commission Joint Research Center, 2008.
597. Suresh, S.; Lindström, S.B.; Thore, C.J.; Torstenfelt, B.; Klarbring, A. Topology optimization using a continuous-time high-cycle fatigue model. *Struct. Multidiscip. Optim.* **2020**, *61*, 1011–1025. [CrossRef]
598. Kahlin, M.; Ansell, H.; Basu, D.; Kerwin, A.; Newton, L.; Smith, B.; Moverare, J.J. Improved fatigue strength of additively manufactured Ti6Al4V by surface post processing. *Int. J. Fatigue* **2020**, *134*, 105497. [CrossRef]
599. Livermore Software Technology Corporation EP2251805A2-Improved Topology Optimization for Designing Engineering Product 2010. Available online: https://worldwide.espacenet.com/patent/search/family/042668333/publication/EP2251805A2?q=EP2251805A2 (accessed on 21 February 2021).
600. Tushar, G.; Willem, J. RouxLivermore Software Technology Corporation—Topology Optimization for Designing Engineering Product. U.S. Patent 8126684, 2012.
601. Livermore Software Technology Corporation—Structural Topology Optimization Using Numerical Derivatives. U.S. Patent 0160078161, 2016.
602. Willem, J.R. Livermore Software Technology Corporation—Enhanced Global Design Variables Used in Structural Topology Optimization of a Product in an Impact Event. U.S. Patent 0170255724, 2020.
603. Bonner, D.L.; Pedersen, C.B.W. Dassault Systemes EP3502931A1—Designing a part by topology optimization. U.S. Patent Application No 16/232,802, 2019.
604. Bonner, D.L.; Pedersen, C.B.W. Dassault Systemes US20190197210A1—Designing A Part By Topology Optimization. U.S. Patent Application No 16/232,802, 2019.
605. Schmidt, M.H.; Del Castillo, A.O. Dassault Systemes EP3647973A1—Designing a mechanical part with topology optimization. U.S. Patent Application No 16/673,649, 2020.
606. Schmidt, M.H.; Del Castillo, A.O. Dassault Systemes JP2020071887A—Designing a mechanical part with topology optimization. U.S. Patent Application No 16/673,649, 2020.
607. Autodesk—Topology Optimization For Subtractive Manufacturing Techniques. U.S. Patent 2018/0349531 A1, 2018.
608. Zhou, M.; Fleury, R. Altair Engineering US10354024 B2—Failsafe Topology Optimization. U.S. Patent 10,354,024, 2019.
609. Wang, C.; Liao, C.; Chen, P.Y.; Luo, T.L. Industrial Technology Research Institute US20160140269—Structural Topology Optimization Design Method. U.S. Patent Application No 14/583,471, 2016.
610. Taggart, D.; Dewhurst, P.; Nair, A. US8335668—Systems and methods for finite element based topology optimization. U.S. Patent 8,335,668, 2012.
611. Dalian Univesity of Technology—Structural Topology Optimization Method Based on Material-Field Reduction Series Expansion. U.S. Patent WO2020215533A1, 2020.
612. South China University of Technology Guangdong—Topology Optimization Design Method for Flexible Hinge. U.S. Patent EP3285189A1, 2018.
613. Networks, A. EP3292657A1—Multi-Layer Network Topology Optimization. U.S. Patent Application No 15/572,460, 2016.

614. Aria Networks US0180139130—Multi-Layer Network Topology Optimization. U.S. Patent 0180139130, 2018.
615. University of Michigan. IHI Corporation US0170161405—Topology Optimization Using Reduced Length Boundaries On Structure Segments Of Different Thicknesses. U.S. Patent Application No 15/368,225, 201, 2017.
616. The Hong Kong University of Science and Technology. US0200134918—Methods of High-Definition Cellular Level Set in B-Splines for Modeling and Topology Optimization of Three-Dimensional Cellular Structures. U.S. Patent Application No 16/665,369, 2020.
617. Wisconsin Alumni Research Foundation—Support Structure Constrained Topology Optimization for Additive Manufacturing. U.S. Patent 010613496, 2020.
618. Caterpillar—Stress-Based Topology Optimization Method and Tool. U.S. Patent 0100274537, 2010.
619. Caterpillar—Fatigue-Based Topology Optimization Method and Tool. U.S. Patent 0140156229, 2014.
620. Nomura, T.; Saitou, K.; Zhou, Y. Toyota WO2019152596A1—Methods for Topology Optimization Using a Membership Variable. U.S. Patent Application No 16/163,950, 2019.
621. Toyota—Methods for Combinatorial Constraint in Topology Optimization Using Shape Transformation. U.S. Patent 20190236221, 2019.
622. Chakravarty, R.R.; Xu., W.S. GM Global Technology Operations US20150103698—System and Method for Topology Optimization with a Plurality of Materials. U.S. Patent Application No 14/051,097, 2015.
623. MRL. Materials Resources US0200180228—Microstructure-Based Topology Optimization for Structural Components Made by Additive Manufacturing. U.S. Patent Application No 16/697,713, 2020.
624. Armstrong, N. Freespace Composites—Manufacturing System Using Topology Optimization Design Software, Novel Three-Dimensional. U.S. Patent 0150239178, 2015.
625. Armstrong, N. Freespace Composites US9789652—Manufacturing System Using Topology Optimization Design Software, Novel Three-Dimensional Printing Mechanisms and Structural Composite Materials. U.S. Patent 9,789,652, 2017.
626. Thales Alenia Space Italia, S.p.A.—Adaptive Topology Optimization for Additive Layer Manufacturing. U.S. Patent Ep3545443a1, 2019.
627. Siemens—Method for Structure Preserving Topology Optimization of Lattice Structures for Additive Manufacturing. U.S. Patent WO2015106021A1, 2015.
628. Musuvathy, S.R.; Arisoy, E. Siemens US009789651—Method for Structure Preserving Topology Optimization of Lattice Structures for Additive Manufacturing. U.S. Patent 9,789,651, 2017.
629. Siemens—Topology Optimization with Design-Dependent Loads and Boundary Conditions for Multi-Physics Applications. U.S. Patent WO2019178199A1, 2019.
630. Siemens—System for Machine Learning-Based Acceleration of a Topology Optimization Process. U.S. Patent WO2020160099A1, 2020.
631. Siemens—Topology Optimization of Thermoelastic Structures for an Additive Manufacturing Process. U.S. Patent WO2020159812A1, 2020.
632. Cohn, M.Z.; Dinovitzer, A.S. Application of Structural Optimization. *J. Struct. Eng.* **1994**, *120*, 617–650. [CrossRef]

*applied sciences*

MDPI

*Article*

# Guidelines for Topology Optimization as Concept Design Tool and Their Application for the Mechanical Design of the Inner Frame to Support an Ancient Bronze Statue

**Abas Ahmad, Michele Bici * and Francesca Campana**

Department of Mechanical and Aerospace Engineering, Sapienza Università di Roma, Via Eudossiana 18, 00184 Rome, Italy; abas.ahmad@uniroma1.it (A.A.); francesca.campana@uniroma1.it (F.C.)
* Correspondence: michele.bici@uniroma1.it

**Abstract:** For the past few decades, topology optimization (TO) has been used as a structural design optimization tool. With the passage of time, this kind of usage of TO has been extended to many application fields and branches, thanks to a better understanding of how manufacturing constraints can achieve a practical design solution. In addition, the advent of additive manufacturing and its subsequent advancements have further increased the applications of TO, raising the chance of competitive manufacturing. Design for additive manufacturing has also promoted the adoption of TO as a concept design tool of structural components. Nevertheless, the most frequent applications are related to lightweight design with or without design for assembly. A general approach to integrate TO in concept designs is still missing. This paper aims to close this gap by proposing guidelines to translate design requirements into TO inputs and to include topology and structural concerns at the early stage of design activity. Guidelines have been applied for the concept design of an inner supporting frame of an ancient bronze statue, with several constraints related to different general design requirements, i.e., lightweight design, minimum displacement, and protection of the statue's structural weak zones to preserve its structural integrity. Starting from the critical analysis of the list of requirements, a set of concepts is defined through the application of TO with different set-ups (loads, boundary conditions, design and non-design space) and ranked by the main requirements. Finally, a validation of the proposed approach is discussed comparing the achieved results with the ones carried out through a standard iterative concept design.

**Keywords:** topology optimization; lightweight design; design methodology; restoration of ancient statues

**Citation:** Ahmad, A.; Bici, M.; Campana, F. Guidelines for Topology Optimization as Concept Design Tool and Their Application for the Mechanical Design of the Inner Frame to Support an Ancient Bronze Statue. *Appl. Sci.* **2021**, *11*, 7834. https://doi.org/10.3390/app11177834

Academic Editors: Marco Mandolini, Patrick Pradel and Paolo Cicconi

Received: 30 June 2021
Accepted: 18 August 2021
Published: 25 August 2021



## 1. Introduction

Topology optimization (TO) naturally gives a wider set of possibilities and a larger degree of freedom with regard to the design space availability, as compared to size and shape optimization. Due to this, TO has been the most widely accepted and well-established structural optimization tool [1]. To improve the performance of TO, passing through a natural process of scientific progress, several approaches have been established, such as the homogenization method [2,3], the SIMP method [4,5], the level set method [6,7], the MMC/MMV method [8,9], and some evolutionary and heuristic methods [10–12]. For the last few decades, TO has been implemented effectively as a conceptual design tool to achieve the desired mechanical performance with feasible manufacturing [1,13,14]. The desired mechanical performance is achieved with the help of a specific objective function (e.g., compliance minimization, maximization of fundamental frequency, mass reduction, etc.) and suitable design constraints (e.g., displacement, fatigue, stress, etc.). Similarly, the feasible manufacturing is obtained through several manufacturing constraints (e.g., draw, extrusion, member size, overhang size, symmetry, etc.). Before the advent of additive manufacturing (AM), manufacturing of the topologically optimized designs was challenging task, sometimes not totally achievable, due to complex geometrical features and hidden cavities. Though several manufacturing constraints could have been considered in order to facilitate conventional manufacturing, a process of remodeling of the optimized design was required for achieving a feasible manufacturing. This procedure resulted not only in an increase of the final mass of the piece, but also in an increase of the design processing time. Therefore, the advent of additive manufacturing and its subsequent advancements further increased the applications of topology optimization with regard to a feasible manufacturing of complex geometries without any remodeling (or, in a few cases, with little remodeling operations connected to specific AM constraints).

TO presents numerous remarkable practical applications such as the design of structures with multi-objectives [15], multi-scale [16–18] and multi-material [19,20], design of functionally graded structures [21,22], design of energy absorption structures to improve resistance against crash failures [23,24], design of periodic structures to improve stability against buckling failure [25], design of architectural structures [26,27], production of sustainable design [28], etc. Similarly, TO provides several other worthwhile applications in conjunction with other optimization tools such as TO with lattice structures to improve resistance against buckling failure [29,30], and designing efficient heat dissipation structures [31], TO with shape optimization [32,33] for better shape and maximum weight reduction, TO with generative design [34,35] to achieve better structural performance, or with tools able to select design variants for optimizing both product and process features [36], etc. Furthermore, researchers are currently working on TO in order to produce some state-of-the-art designs regarding structures with negative Poisson ratio [37], structures with negative thermal expansion [38], structures with natural convection [39], load-supporting structures for civil engineering [40,41], etc. On the other hand, there are still some open issues evolving around TO. These issues are mainly concerned with the aesthetic problem [42]; additive manufacturing problems related to large overhanging members [43–46], the remodeling of the optimized structure due to poor shape geometry [47]; stress-related problems [48–50]; the application of manufacturing constraints [14]; the optimization of multi-scale and multi-material structures [16–20], etc.

Conceptual design is a fundamental milestone at the initial phase of any product development process, helping the early assessment of optimal design solution with regard to other crucial factors such as manufacturing, time-to-market, performance, testing and cost [1,13,14]. Therefore, the designers always come up with several design solutions in the concept design phase and after a meticulous evaluation, select the best one. According to literature [51,52], around 80% of the product quality/cost ratio can be determined by the end of the initial design phase and therefore, the initial design phase plays a vital role. To bolster the initial design phase, numerous design tools have been developed to expedite the design process. In addition, the latest technological advancements in the field of computer, programming, artificial intelligence, CAD tools (i.e., integrated in many commercial software) and numerical solvers have further enhanced the design process with regard to fast computations and complex simulations [53–55]. For the last few decades, topology optimization (TO) has been extensively studied together with generative design (GD) as superior conceptual design methodologies [1,13,14,42,56]. Despite this, TO is usually seen as a specific structural optimization tool instead of a way to define concepts. On the contrary, GD has the advantage of simultaneously considering many materials and manufacturing methods for a single simulation, which reduces the simulation time. In addition, GD provides many design solutions as a combination of different structures and materials in a single simulation, while TO provides a single design solution. Furthermore, TO can implement manufacturing constraints to drive the final shape to a feasible manufacturing, while GD uses manufacturing methods and provides much better optimized shape for manufacturing, and therefore requires little or no remodeling. On the other hand, TO is useful in higher mass reduction and fast computation. However, both TO and GD have numerous applications inside automotive, aerospace, civil, and medical engineering [1,13,14,42,56–61].

This paper investigates TO as a concept design tool in mechanical engineering, providing guidelines to assess its input from a list of requirements so that preliminary sketches may be provided. The paper aims to extend its assessment from the methodological point of view to foster its adoption in the creative definition of innovative components, constrained by mechanical criteria, at the early stage of design. In Section 2, the methodological part is provided in terms of mathematical formulation and computational set-up (Section 2.1) of the proposed TO approach, and the design workflow and guidelines (Section 2.2). Then, in Section 3, a case study is provided to show an example of how carrying out the requirements translations into TO input conditions and, in Section 4, results are discussed to compare it with a standard design workflow. Finally, in Section 5, conclusions are outlined.

## 2. Methods

TO may reduce design efforts to review concepts of the geometrical models of the parts in accordance with the targets defined in the list of requirements. In mechanical design it can also support preliminary assessment of assemblies if proper guidelines are defined to assist the work. In this section, we propose guidelines to apply TO at the early concept design stage, so that, through a proper definition of the geometric domain of the parts/components, effective topologies may be derived while respecting design intents. Since guidelines are usually tailored on specific context, in Section 2.1, the mathematical formulation and computation set-up of the methodology we propose to use are provided. Then, in Section 2.2, the design workflow and guidelines of the proposed approach are described.

### 2.1. Mathematical Formulation and Computation Set-Up

Density-based topology optimization is formulated through the finite element method (FEM). Element density acts as a design variable, so that an optimal distribution of elements layout may be provided within the chosen design space.

Design space is approximated by a finite element analysis (FEA) mesh. FEA determines the displacement and stress gradients that help TO in the identification of elements that do not affect the optimization problem (mass reduction, compliance minimization, frequency response, etc.). During each iteration, sensitivity analysis is performed by the solver to evaluate the effect of the element density variable on the objective function and, subsequently, the element densities are updated in the design space for the next iteration and the same process is continued until a final convergence is obtained. Mathematically, for convergence, the derivative of the objective function with respect to the design variable, the element density, is studied. Basically, in density-based TO, the elements are assigned a density value from 0 to 1, where 0 represents a void, 1 represents a solid and values from 0 to 1 represent intermediate densities. In order to enforce the solution to the solids and voids discrete combination, density-based TO exploits the SIMP (solid isotropic material penalization) approach to eliminate the elements with intermediate densities with the help of power law (as shown below):

$$k_i = (\rho_i)^P k_o \tag{1}$$

where "$i$" is the number of elements in the design space, $i = 1, 2, 3, \ldots, L$; "$\rho_i$" is the density of the $i$th element, "$P$" is the penalization factor in the range $1 < P \leq 3$; "$k_0$" and "$k_i$" are the stiffnesses of $i$th element before and after penalization, respectively. TO with SIMP approach has been declared the most implemented and developed approach with respect to math simplicity and computational efficiency. As per numerical experimentations, a penalty factor of value 3 has been found the most suitable one for penalization purpose.

Distinction between design and non-design space allows an easy set-up of the conceptual geometric domain, assuming as non-design space regions constrained by functional interfaces (e.g., displacement constraints, contact surfaces, other assigned boundary conditions). The optimization problem solved by SIMP is a minimization problem related to design space volume fraction or compliance ($C$), or in another words, maximization of the stiffness of the structure that can be defined for the overall domain or only for the design space. According to compliance theory, compliance is inversely proportional to the stiffness of the structure. The global compliance is the sum of the element's elastic characteristics or strain energies (as shown below).

$$C = \frac{1}{2}FU^T = \frac{1}{2}\sum_{i=1}^{L} f_i\, u_i{}^T \quad (2)$$

where "$F$" is the external load vector, "$U$" is the global displacement vector and "$L$" is the overall number of finite elements.

According to Hook's Law $F = KU$, therefore $C = \frac{1}{2}F^2K^{-1}$ hence $\frac{1}{2}$ and $F$ are constants, therefore $C\ \alpha\ K^{-1}$ where $K$ is the global stiffness matrix. Using the SIMP approach $K$ is computed as:

$$K = \sum_{i=1}^{L} k_i = \sum_{i=1}^{L} (\rho_i)^P k_o \quad (3)$$

where "$\rho_i$" is in the range $\eta \leq \rho_i \leq 1$ and $i = 1,\ 2,\ 3 \ldots\ldots\ldots, L$; and $\eta$ is a very small number but greater than zero to prevent the global matrix from being singular and so ensure the FEA stability.

Different constraints may be set (e.g., mass, load, displacement in specific ranges) and also weighted customized objective functions may be defined. Similarly, TO may be subjected to manufacturing constraints to achieve feasible manufacturing shapes (e.g., direction of protrusion, forging shapes, etc.) [1]. From Equation (2), the general formulation of SIMP for the compliance minimization problem becomes:

$$Minimize\ C = \frac{1}{2}\sum_{i=1}^{L} f_i\, u_i{}^T = \frac{1}{2}\sum_{i=1}^{L} f_i{}^2\left((\rho_i)^P k_o\right)^{-1} \quad (4)$$

Subject to :

1. $F = KU$; To ensure static equilibrium;
2. $G = \sum_{j=1}^{M} g_j \leq 0$; Design Constraints, $j = 1,2,\ldots\ldots\ldots, M$;
3. $H = \sum_{k=1}^{N} h_k \leq 0$; Manufacturing Constraints, $k = 1,2,\ldots\ldots\ldots, N$;
4. $\eta \leq \rho_i \leq 1$ and $i = 1,2,3\ldots\ldots\ldots, L$.

The above objective function is used for a single loadstep. When multiple loadsteps occur, the minimization is carried out on the weighted compliance ($W_c$), defined as:

$$Minimize\ (W_c) = \sum_{q=1}^{R} W_q\, C_q{}^T = \frac{1}{2}\sum_{q=1}^{R} W_q\, f_q\, u_q{}^T \quad (5)$$

where "$W_q$" is the specific weight of each loadstep, "$C_q$" is the compliance of individual loadstep and "$R$" is the overall number of total loadsteps.

### 2.2. Design Workflow and Guidelines

Common applications of TO start from the CAD model of the preliminary design of the component. Here, its approach is discussed to foster innovative geometric parts with respect to mechanical targets that may derive from the list of requirements defined for mechanical engineering design.

A standard TO process is performed according to Figure 1. As an optimization step during preliminary or executive design, its definition is strongly connected to a preliminary CAD model of the component and to the FEA modeling accuracy. To gain the proper accuracy, material properties and physical approximation of load conditions must be properly defined (e.g., contact friction and/or constraints compliance if present). Manufacturing constraints are not included directly in the optimization loop, under the assumption that during TO, the designer is also looking for disruptive solutions (despite the manufacturing constraints). Nevertheless, in accordance with an integrated product-process approach, manufacturing constraints may collapse in the design constraints step.

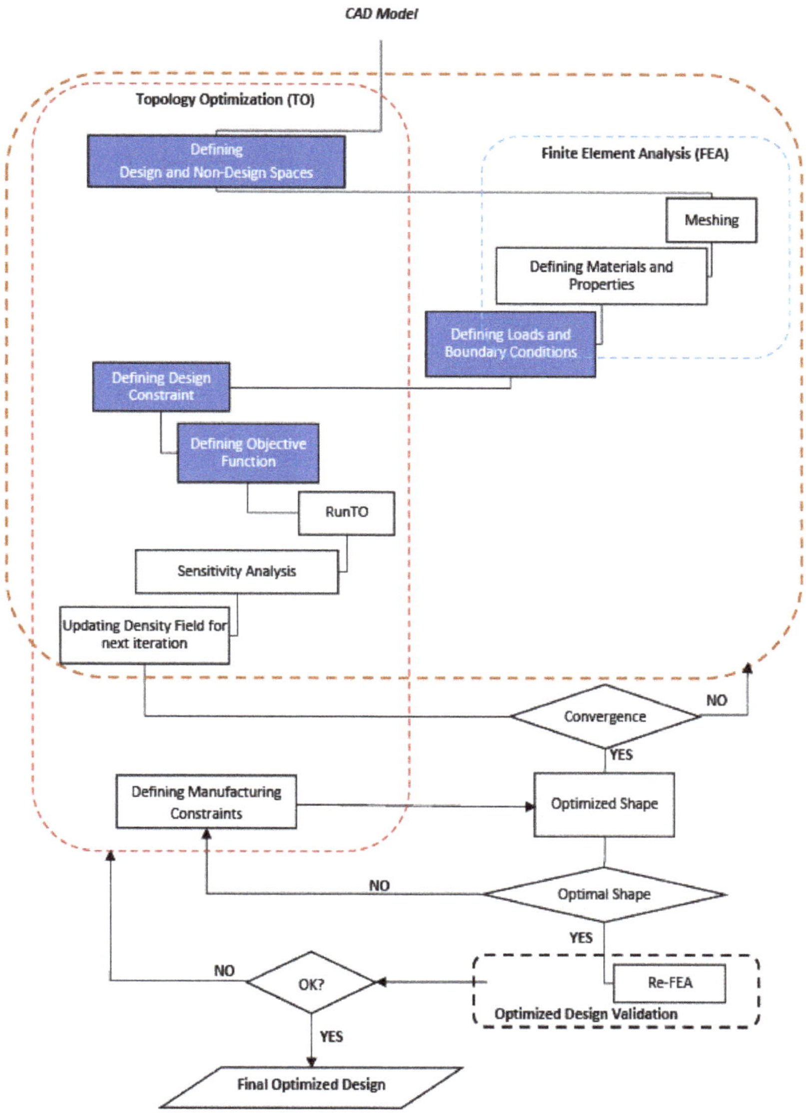


**Figure 1.** Classical TO Workflow.

In concept design, the final CAD model of the component/structure must be the output of the workflow, not the input. The input model of the TO must represent a CAD of the design and non-design space. Design space may be the maximum lengths possible with the subtraction of volumes and areas that are functional interfaces for loads and boundary conditions (displacements, fluxes, velocity, etc.). Design and non-design spaces are defined as enveloping defeatured volumes avoiding any influence given by designers' choices [14]. Through these considerations, the modified workflow is proposed in Figure 2.

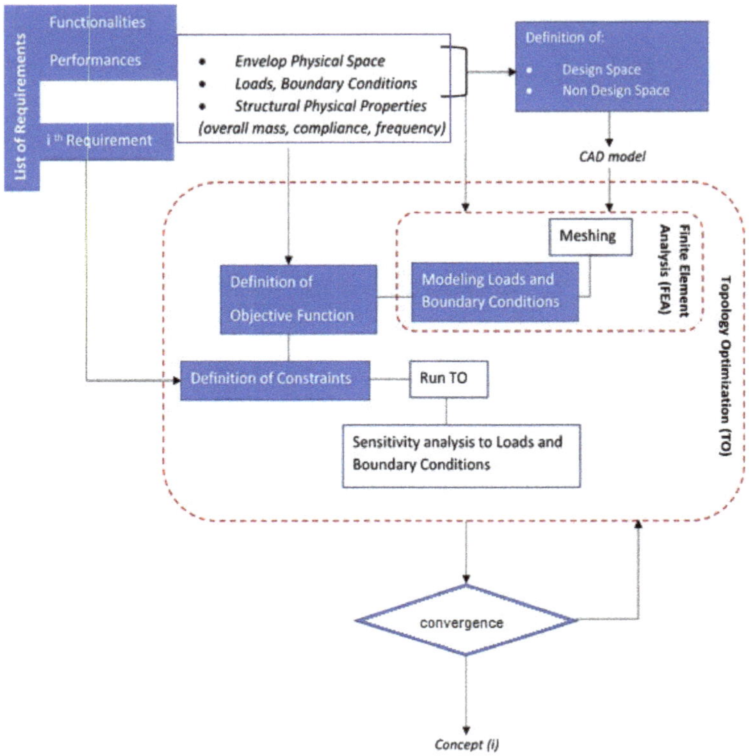


**Figure 2.** The adopted TO workflow in concept design.

In this case, the preliminary FEA step has the role of helping the translation of the performances into TO inputs (mainly loads and boundary constraints). From this, basic guidelines that may provide benefits to the concept design workflow are:

- a set of requirements (e.g., aesthetics, ergonomics, transportability, etc.) may constrain the final envelop of the shapes. The definition of design and non-design spaces operatively forces this kind of evaluation, which can be multi-criteria (thus taking into account together many requirements) or not. In this case, requirements may help to define geometrical variants in terms of design space.
- The need for simulating the structural behaviour of the system by FEA requires an anticipated evaluation of the target values, enforcing the physical feasibility of the concept evaluation.
- Looking for optimal topology in assemblies, a clear model of the operative conditions must be defined, so that a preliminary understanding of the worst kinematic/dynamic conditions can be defined. Envelop volumes coupled with multibody analysis such as the equivalent static load method may help the assessing of the problems of flexible multibody assemblies [62].

## 3. Case Study

The case study concerns the application of TO as a concept design tool for an inner frame of a virtually approximated model of an ancient bronze statue named Vittoria Alata. It explores how to obtain TO optimization requirements from general design requirements (DR) and how to derive design and non-design Space (DS and NDS) and, through them, obtain concepts.

The original statue was recently restored and moved to a new exhibition space in the ancient Capitolium of Brescia (Italy). Restoration also included a new inner frame, described in [63]. The present application assumes the requirements assessed in [63] as the starting point of the concept design with TO and the actual result with its standard design workflow as a reference for its validation.

The statue is composed of 5 parts: a central body, left and right arm, left and right wings (Figure 3).

**Figure 3.** Case study—ancient bronze statue (Vittoria Alata): (**a**) The Vittoria Alata di Brescia (© Foto StudioRapuzzi Brescia); (**b**) Application of loads in their respective COGs.

In Figure 3b, the $P_a$, $P_b$, and $P_{cc}$ are describing the effect of the loads as concentrated masses of the wings, arms, and central body at their respective COGs (center of gravity).

### *3.1. Design Requirements and the Definition of Design Space and Non-Design Space*

Requirements of the design problem [63] are summarized in Table 1, both as design requirements (DR) and topological optimization requirements (OR).

The maximum envelope of the DS is the volume inside the statue, without taking into account the head, which is considered bulk (during the restoration process, the ancient filling has been removed except for the head filling, judged able to be safely maintained). DS is the volume where the inner frame can be placed. The statue itself represents part of the non-design space where loads act (the weight of central body, wings, and arms) and where interfaces between the inner frame and bronze are possible. Non-design space has been modeled starting from a 3D acquisition of the outer surface of the statue, made by structured light. Non-design space also includes the connection interface with the ground as prescribed by DR6 (→OR6.1) and a central volume preliminary assumed as a circular beam to accomplish DR4 and 5 (→OR4.1 and 5.1). The central beam's orientation is constrained by the smallest section, highlighted in Figure 4a as S1, and by the necessity of correct alignment during assembly, since it will be the first element to be inserted during the mounting procedure. Its position can be defined using a pocket already present in the inner filling of the head (the only ancient filling maintained) and other reference contacts at the opposite side (e.g., at Sa or S1 of Figure 4a).

**Table 1.** Design Requirements declared by the restorers and their translation to optimization requirements.

| Design Requirements (DR) | | TO Optimization Requirements (OR) |
|---|---|---|
| 1 Preserve structural integrity of the ancient material | → | 1.1 Maximize the safety margin for local stress (at least equal to 2)<br>1.2 Avoid interfaces on thin or weak areas |
| 2 Limit relative displacement at the neck | → | 2.1 Reduce the head displacement<br>2.2 Minimize bending compliance in the front and lateral planes of the statue |
| 3 Reduce stress at the right part of the hip | → | 3.1 Minimize bending compliance in the front and lateral planes of the statue |
| 4 Do not apply loads on the base of the skirt | → | 4.1 Suspended directly on the inner frame |
| 5 Avoid contact among wings and back of the central body | → | 5.1 Suspended directly on the inner frame |
| 6 Connect the statue as a rigid body to the anti-seismic basement | → | 6.1 Transfer loads to the ground for the static balance<br>6.2 Reduce relative compliance between central body and arms and wings, making the system as a rigid body |
| 7 Guarantee the position of the arms and the wings in the respect of the central body | → | 7.1 Wings and arms must be connected directly to the central beam through upper openings of the central body |
| 8 Allow assembly and disassembly through the limited areas of the openings i.e., at the base, around the arms and wings connection areas | → | 8.1 Reduction of the design space at the minimum section of the statue to guarantee maneuvering<br>8.2 Possible manufacturing constraints related to the adoption of a wireframe structures made of beams and mold/machined parts at the interfaces with bronze |

These reasonings, deduced as target values of DR4, DR5, and DR8, define the functional role of the central beam as part of the non-design space (NDS) and fix its position and section length. Solidity, as derived in OR6.2, asks for constrained sections; that means that they must move together without relative rotation in respect to the inner frame along the horizontal plane. Figure 4a shows where these sections may be localized due to the accessibility for assembly (Sa, Sb, and Sc of Figure 4a).

Minimum condition suggests highlighting three points for Sa and Sc that may be contact points with the statue. Sb has been introduced as a contrast point with the statue with respect to contact at the leg. BCs at Sa are defined on the central beam, to simulate the connection with the base. In these points, boundary constraints (BCs) are defined to lock relative motion among DS and NDS as shown in Figure 4b. BCs are also selected to reproduce the support scheme on the left leg and right shoulder. This means locking the DOF in z direction at the center of the thigh and shoulder, as depicted by P1 and P2 in Figure 4b.

### 3.2. Design Requirements and the Definition of Boundary Constraints and Load Conditions

Load conditions, to be applied, are stress-strain conditions derived from the static analysis of the central body and the effects due to the assembly of arms and wings on the inner frame, as described in Figure 3b. Wings and arms introduce a bending moment in the lateral plane of the statue that passes through its left knee (P1 in Figure 4a). From the design requirements point of view, the central body stress must be maintained in the same conditions as before the restoration, characterized by the presence of a filling of resin able to minimize relative rotation between the inner frame and the statue. This filling has been removed for material preservation requirements, hence the necessity of a new inner frame. To define the worst reference condition derived from the filling removal, a preliminary FEA on the central body was done. Figure 4c shows von Mises equivalent stress superimposed to the magnified plot of the relative displacements from the undeformed shape (grey mesh), as derived from the gravity load, without imposing any filling inside. The gravity load of the central body defines a bending effect, in the frontal plane, that overstresses the base of the neck, where a visible crack is present. This defines a major constraint on the compliance of the structure that must reduce the final displacement of the head (OR2.1, 2.2 and 3.1).

Areas useful for suspending the central body with a simple contact interface are the left leg (above the knee) and the right shoulder (P2 and P1 in Figure 4a,b). These have been derived after a deep phase of material analysis, with several specialists involved (chemists, physicists, restorers, engineers); only these areas have been validated as stiff and safe enough to be used for support interfaces, according to DR1 that determines OR1.2: "Avoid interfaces on thin or weak areas". According to OR1.1, maximum equivalent stress at the contact interface must be at least half the value of 80 MPa, considering a safety factor equal to 2. This value is consistent with the aim of achieving a final condition able to preserve ancient material, assuming its yielding value in the range of [80 ÷ 150] MPa.

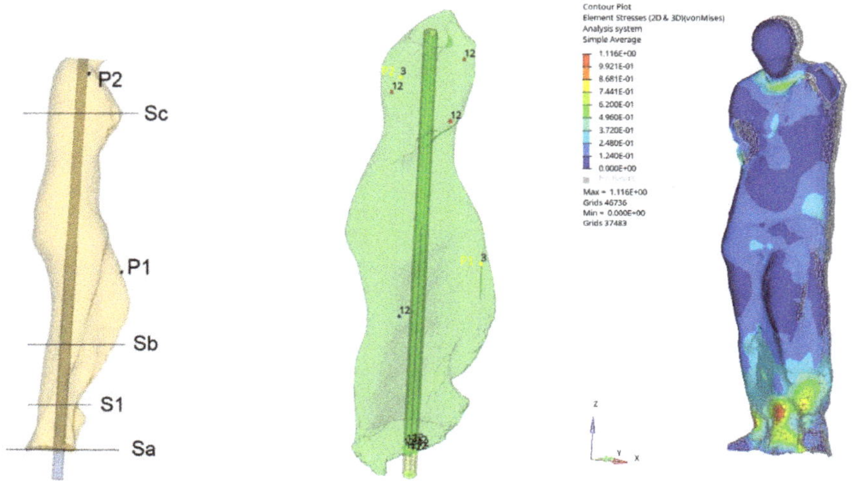


(a) (b) (c)

**Figure 4.** (**a**) Lateral view of DS and central beam NDS; (**b**) Boundary conditions (numbers represent locked DOFs: 1 = DOF locked in x direction, 2 = DOF locked in y, 3 = DOF locked in z). (**c**) von Mises equivalent stress (MPa).

### 3.3. Topological Optimization Set-Up

As previously described, the first guess concept assumes, as design space (DS), a bulk 3D filling of the statue, surrounding the central beam from the feet up to the base of the neck. Equivalent loads of the central body are applied at the bottom of the central beam. Rigid links are used to transfer force and moment. Table 2 summarizes design intent in terms of TO problem set-up and numerical constraints as derived by the DRs.

**Table 2.** TO condition derived from DRs (for boundary conditions see Figure 4b).

| | | |
|---|---|---|
| Loads | Central Body Equivalent loads @ basement | Force: (0, 0, 2111) N<br>Moment: (−53,740, 25,310, 0) Nm |
| | Arms and wings Equivalent loads @ basement | Force: (0, 0, 866) N<br>Moment: (203,600, 65,050, −56,890) Nm |
| Stress-Strain Constraints | Max displacement at neck<br>Maximum stress at hip<br>Maximum stress at neck | <0.9 mm<br>Less than 40 MPa<br>Less than 40 MPa |
| Mass Constraint | Volume Fraction (%) | Less than 10% |
| Objective Function | Central Body compliance | To be minimized |

The first guess concept minimizes compliance of the design space, looking for a volume fraction of the DS less than 0.1. The adoption of this extreme threshold derives from the assembly requirements DR8 and OR8.2, which highlight a wireframe structure as preferred.

For what concerns TO solver options, using Optistruct, we set up the usage of SIMP approach with a dual optimization algorithm for achieving convergence, in addition to any limit in terms of minimum member size. This choice represents an advisable set-up for concept level optimization, especially when many design variables are involved.

## 4. Results and Discussion

Load effects on the final structure were investigated as single effects of central body, wings and arms (designated CB, W, and A, respectively) and overall load, so that their single and interaction effects on the final topology of the concepts may be seen. The presence of BC at Sa has been highlighted through the code Sa. The sensitivity to its effect has been introduced to understand its relevance in the final reaction, considering that it may be redundant to S1.

Figures 5–8 show an overall view of the results in terms of density distribution. Red areas represent density equal to 1. Admissible domains are in the range of $0.5 \div 1$. They are shown in back and front view to highlight the distribution of density (on the left of each figure). On the right of the figures, the final material distribution is highlighted by section cuts (in blue), made along two concurrent/intersecting planes that pass through the central beam and P1 and P2, respectively. Doing so, the loading path between the interface with statue and the central beam may be seen.

CB case shows that load is well supported by areas around P1 and P2 as shown by the section cuts of the related cases, reported in Figures 5b and 6b. As reported in Figures 5 and 6, the massive part of the optimized resulting structure is connected to the central beam at the hip height, slightly upwards of the position of the original center of mass of the CB.

In the CBWA case, this bulk part is reduced. The left leg increases its relevance due to the wings, which produce bending tension along the plane passing in P2. Due to this occurrence, the role of the area around the shoulder becomes less relevant with respect to the CB load case (as reported in Figures 7 and 8).

Loadcase CBWA_Sa (Figure 8) reveals that the adoption of the in-plane BC at Sa avoids mass concentration at the bottom. The major problem of this solution is related to a reduced capability of connecting the structure to the central beam. From Figure 8b, reported also as Figure 9a, it can be seen that the connection of the obtained structure from shoulder area (P2) to the central beam is made by low density elements ($0.5 \div 0.6$) and is really partial in the section cut through P2 and the central beam, as highlighted by the arrow in Figure 9a.

Figure 9b better displays it, showing three section cuts (cyan, blue in transversal planes similar but not equal to the one passing through P2, and yellow passing through P1). In greater detail, the structure connects P2 to branches and thus to the central beam near the hip, as highlighted by the arrow in Figure 9b. This structure development may be seen as a consequence of the simplification made by applying the WA load condition.

Figure 10 shows the variant of CBWA_Sa, changing the application of the WA load to a more realistic section for the assembly condition. Blue sections highlight contact with the central beam. The low-density parts, previously present in the bottom (as in the blue section cut of Figure 9b) are not present in this case. As reported on the right part of Figure 10, the final mass distribution with a density threshold of 0.67 shows the bulkiest solution on the upper part to contrast the WA effects, correctly increasing the stiffness of that portion.

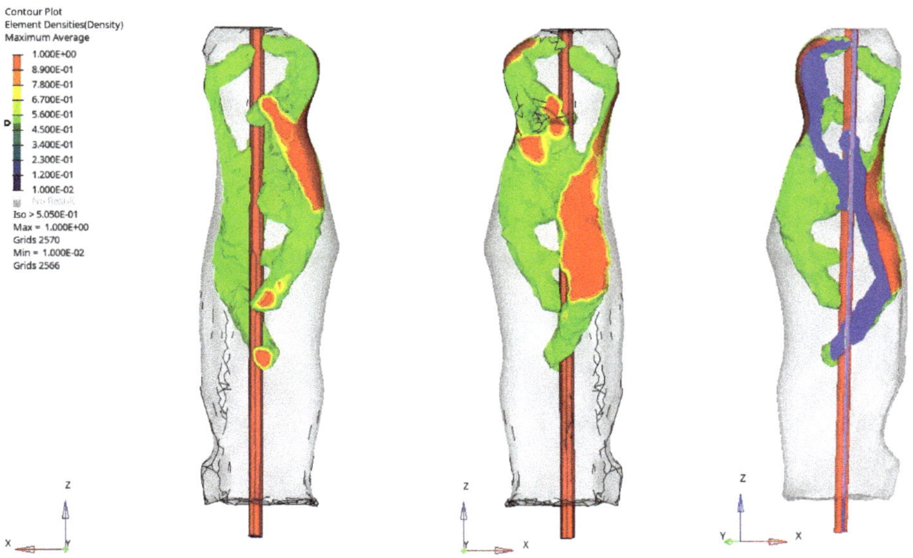


(a) (b)

**Figure 5.** CB case: density distribution (**a**) Back and front View) (**b**) Two intersecting sectional views (in blue).

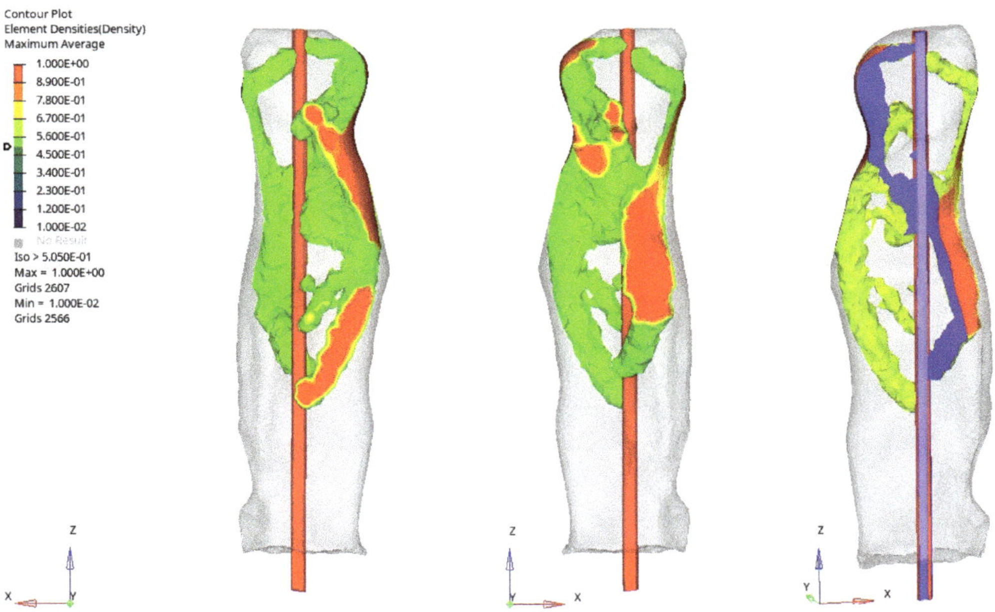


(a) (b)

**Figure 6.** CB_Sa case: density distribution (**a**) Back and front View) (**b**) Two intersecting sectional views (in blue).

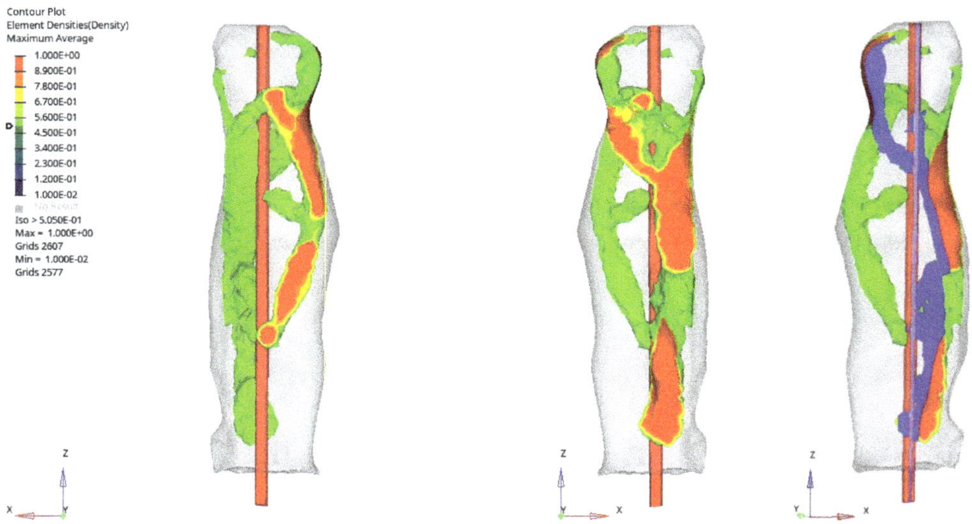


(a) (b)

**Figure 7.** CBWA case: density distribution (**a**) Back and front View) (**b**) Two intersecting sectional views (in blue).

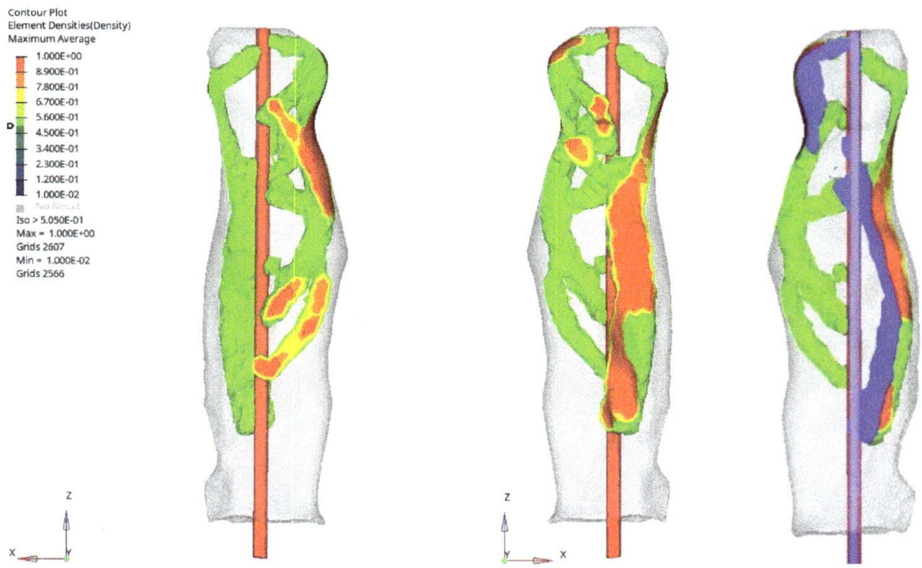


(a) (b)

**Figure 8.** CBWA_Sa case: density distribution (**a**) Back and front View) (**b**) Two intersecting sectional views (in blue).

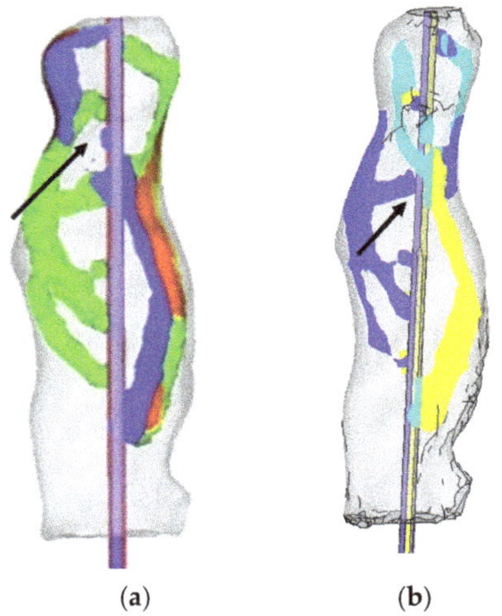

**Figure 9.** (**a**) Density distribution due to CBWA_Sa (sectional View); (**b**) Integrated sectional view.

The absence of material at the opposite part of the legs with respect to CB load conditions may be seen as a minor effect of that area regarding the CB load, confirmed by the density map of Figure 8.

The assembly constraints impose the definition of a wireframe structure and the statue preservation imposes contacts only at a few points, as defined through the assigned boundary conditions. This suggests a refinement of the DS, limiting the volumes accessible from the arm openings and at the left leg above the knee. Figure 11a shows the solution in case of CB_Sa condition. Figure 11b is related to CBWA_Sa.

Finally, a comparison of these results has been made with the actual engineered solution implemented through a classical design approach (Figure 11c), as referred in [63]. The double supported interface at P1 in the engineered solution corresponds to the final TO results (Figure 11), as well as in the slope of the support beam. The right interface, corresponding to the right armpit, reports elements in a lower range of densities (under 0.65) so it cannot be considered as fundamental. A more relevant effect of this interface may be present in applying arms as separate loads since it increases bending in the front plane. This highlights a limit of the applied load scheme not completely correct to capture the local behaviour induced by the actual type of load. Nevertheless, it is considered as being of the second order since the computation problem has been highlighted heuristically.

In terms of maximum displacement, all the solutions present a value less than 1 mm. Von Mises equivalent stress at P1 and P2 are less than 10 MPa. Considering that actual interfaces are wider than a single contact point, the effects on the bronze are in the allowable range. The final stress-strain performances achieved by all the variants and the computational efforts spent for the analysis (less than one week without considering the mesh set-up of the NDS related to the statue, which was more expensive, but also necessary for the classical design workflow) confirm, also in this case, that TO may support the direct accomplishment of qualitative definition of topologies and their quantitative evaluation in terms of DR targets [64]. This provides benefits that shorten the time to market through an early verification of mechanical requirements during concept design stage.

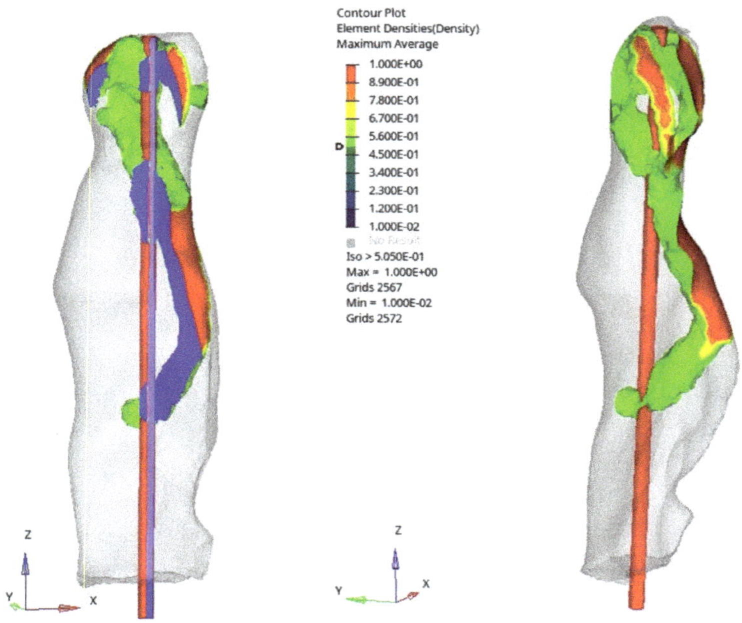


(a) (b)

**Figure 10.** Design variant: (**a**) section cut (in blue); (**b**) density distribution.

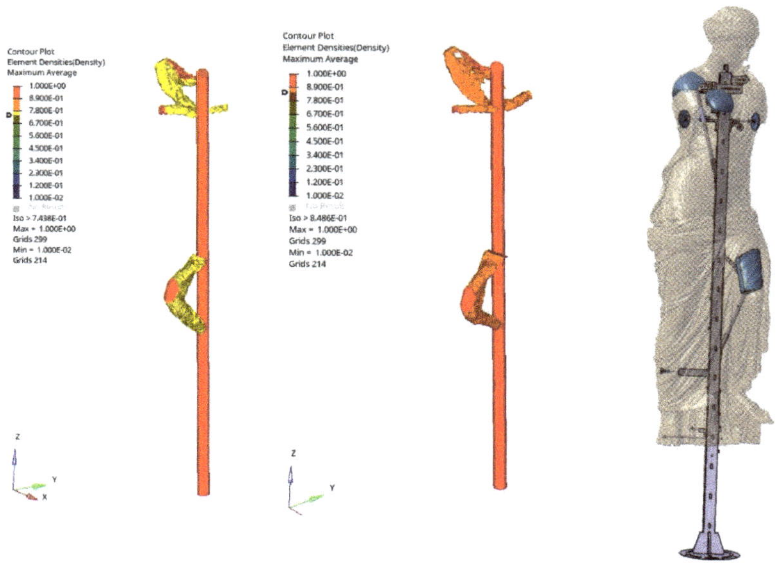

(a) (b) (c)

**Figure 11.** (**a**) Solution obtained through TO due to CB_sa; (**b**) Solution obtained through TO due to CBWA_sa; (**c**) Engineered solution obtained through a classical design workflow.

## 5. Conclusions

This paper explores how TO may support concept design to generate sketches able to accomplish general design purposes related to the definition of functional interfaces and structural requirements. A general workflow able to convert the targets of a list of requirements into TO input prescription has been provided, so that the geometric topology of the concept may be defined consistently with DR target earlier, during sketching concepts, shortening time to market in a step that is crucial for the quality/cost ratio.

The general workflow has been applied to a test case related to the development of an inner frame of an ancient statue. Two variants were analyzed, and one of them was selected and compared to the one achieved and engineered successfully with a classic design workflow. The first variant is achieved starting from the most general and widest design space, in accordance with the requirements. It represents an ideal solution, impractical due to assembly problems but effective for understanding the relevance of the statue-frame interface areas. The second variant, related to a DS that has been reduced to improve maneuverability, confirms once again the capability of TO in obtaining results aligned to "good engineering practice". Finally, the comparison with the actual engineering solution confirms the valuable result of the second variant.

Besides the benefit related to time to market, TO, as applied in this paper, may close the gap between experts and young designers, supporting preliminary target evaluations together with topology conception. Concept design, in fact, is strictly related to unpredictable factors external to the complexity of the design problem itself, due to its creative nature [65,66]. This reduces the ability of younger designers to speed up the structural optimization during the concept design, so that proper tools must be promoted to avoid this limit.

**Author Contributions:** Conceptualization, F.C. and M.B.; methodology, F.C., M.B. and A.A.; software, A.A.; validation, F.C. and M.B.; formal analysis, F.C., M.B. and A.A.; investigation, M.B. and A.A.; resources, M.B. and A.A.; data curation, M.B. and A.A.; writing—original draft preparation, F.C. and A.A.; writing—review and editing, M.B.; visualization, F.C., M.B. and A.A.; supervision, F.C.; project administration, F.C.; funding acquisition, F.C. All authors have read and agreed to the published version of the manuscript.

**Funding:** This research was funded by Fondazione Brescia Musei as a part of the multidisciplinary research entitled: "Studio e progettazione di supporti per la ricomposizione e l'esposizione di bronzi antichi di grandi dimensioni, con particolare riferimento alla Vittoria Alata di Brescia—Study and design of supports for the recomposition and exposition of ancient bronzes of large dimensions, with particular reference to the Vittoria Alata of Brescia".

**Institutional Review Board Statement:** Not applicable.

**Informed Consent Statement:** Not applicable.

**Data Availability Statement:** Not applicable.

**Acknowledgments:** Authors want to acknowledge Fondazione Brescia Musei and especially Francesca Morandini and Stefano Karadjov. They want also to thank Opificio delle Pietre Dure di Firenze and its restorers and specialists, in particular Annalena Brini, Stefano Casu, Sveta Gennai, Anna Patera, Elisa Pucci. The last, but not least, acknowledgements are for Capoferri SpA, in the persons of Sergio Capoferri and Riccardo Guarnieri.

**Conflicts of Interest:** The authors declare no conflict of interest.

## References

1. Ahmad, A.; Raza, M.A.; Campana, F. Simulation Based Topology Optimization Assessment with Manufacturing Constraints. In Proceedings of the 2020 17th International Bhurban Conference on Applied Sciences and Technology (IBCAST), Islamabad, Pakistan, 14–18 January 2020; pp. 174–182.
2. Beghini, L.L.; Beghini, A.; Katz, N.; Baker, W.F.; Paulino, G.H. Connecting architecture and engineering through structural topology optimization. *Eng. Struct.* **2014**, *59*, 716–726. [CrossRef]

3. Suzuki, K.; Kikuchi, N. A homogenization method for shape and topology optimization. *Comput. Methods Appl. Mech. Eng.* **1991**, *93*, 291–318. [CrossRef]
4. Bendsoe, M.P.; Sigmund, O. *Topology Optimization: Theory, Methods, and Applications*; Springer Science & Business Media: Berlin/Heidelberg, Germany, 2013.
5. Rozvany, G.I.; Zhou, M.; Birker, T. Generalized shape optimization without homogenization. *Struct. Optim.* **1992**, *4*, 250–252. [CrossRef]
6. Yulin, M.; Xiaoming, W. A level set method for structural topology optimization and its applications. *Adv. Eng. Softw.* **2004**, *35*, 415–441. [CrossRef]
7. Allaire, G.; Jouve, F.; Toader, A.M. A level-set method for shape optimization. *Comptes Rendus Math.* **2002**, *334*, 1125–1130. [CrossRef]
8. Guo, X.; Zhang, W.; Zhong, W. Doing topology optimization explicitly and geometrically—A new moving morphable components-based framework. *J. Appl. Mech.* **2014**, *81*, 081009. [CrossRef]
9. Zhang, W.; Chen, J.; Zhu, X.; Zhou, J.; Xue, D.; Lei, X.; Guo, X. Explicit three-dimensional topology optimization via Moving Morphable Void (MMV) approach. *Comput. Methods Appl. Mech. Eng.* **2017**, *322*, 590–614. [CrossRef]
10. Khan, S.A.; Engelbrecht, A.P. A fuzzy particle swarm optimization algorithm for computer communication network topology design. *Appl. Intell.* **2012**, *36*, 161–177. [CrossRef]
11. Blank, L.; Garcke, H.; Sarbu, L.; Srisupattarawanit, T.; Styles, V.; Voigt, A. Phase-field approaches to structural topology optimization. In *Constrained Optimization and Optimal Control for Partial Differential Equations*; Springer: Berlin/Heidelberg, Germany, 2012; pp. 245–256.
12. Watanabe, K.; Suga, T.; Kitabatake, S. Topology optimization based on the on/off method for synchronous motor. *IEEE Trans. Magn.* **2017**, *54*, 1–4. [CrossRef]
13. Vlah, D.; Žavbi, R.; Vukašinović, N. Evaluation of topology optimization and generative design tools as support for conceptual design. In *Proceedings of the Design Society: DESIGN Conference*; Cambridge University Press: Cambridge, UK, 2020; Volume 1, pp. 451–460.
14. Bici, M.; Broggiato, G.B.; Campana, F. Topological Optimization in Concept Design: Starting approach and a validation case study. In *Advances on Mechanics, Design Engineering and Manufacturing*; Springer: Berlin/Heidelberg, Germany, 2017; pp. 289–299.
15. Guirguis, D.; Hamza, K.; Aly, M.; Hegazi, H.; Saitou, K. Multi-objective topology optimization of multi-component continuum structures via a Kriging-interpolated level set approach. *Struct. Multidiscip. Optim.* **2015**, *51*, 733–748. [CrossRef]
16. Sivapuram, R.; Dunning, P.D.; Kim, H.A. Simultaneous material and structural optimization by multiscale topology optimization. *Struct. Multidiscip. Optim.* **2016**, *54*, 1267–1281. [CrossRef]
17. Gao, J.; Luo, Z.; Xia, L.; Gao, L. Concurrent topology optimization of multiscale composite structures in Matlab. *Struct. Multidiscip. Optim.* **2019**, *60*, 2621–2651. [CrossRef]
18. Hoang, V.N.; Tran, P.; Vu, V.T.; Nguyen-Xuan, H. Design of lattice structures with direct multiscale topology optimization. *Compos. Struct.* **2020**, *252*, 112718. [CrossRef]
19. Zuo, W.; Saitou, K. Multi-material topology optimization using ordered SIMP interpolation. *Struct. Multidiscip. Optim.* **2017**, *55*, 477–491. [CrossRef]
20. Blasques, J.P.; Stolpe, M. Multi-material topology optimization of laminated composite beam cross sections. *Compos. Struct.* **2012**, *94*, 3278–3289. [CrossRef]
21. Radman, A.; Huang, X.; Xie, Y.M. Topology optimization of functionally graded cellular materials. *J. Mater. Sci.* **2013**, *48*, 1503–1510. [CrossRef]
22. Taheri, A.H.; Suresh, K. An isogeometric approach to topology optimization of multi-material and functionally graded structures. *Int. J. Numer. Methods Eng.* **2017**, *109*, 668–696. [CrossRef]
23. Aulig, N.; Nutwell, E.; Menzel, S.; Detwiler, D. A weight balanced multi-objective topology optimization for automotive development. In Proceedings of the 10th European LS-DYNA Conference, Würzburg, Germany, 15–17 June 2015.
24. Duddeck, F.; Hunkeler, S.; Lozano, P.; Wehrle, E.; Zeng, D. Topology optimization for crashworthiness of thin-walled structures under axial impact using hybrid cellular automata. *Struct. Multidiscip. Optim.* **2016**, *54*, 415–428. [CrossRef]
25. Neves, M.M.; Sigmund, O.; Bendsøe, M.P. Topology optimization of periodic microstructures with a penalization of highly localized buckling modes. *Int. J. Numer. Methods Eng.* **2002**, *54*, 809–834. [CrossRef]
26. Osanov, M.; Guest, J.K. Topology optimization for architected materials design. *Annu. Rev. Mater. Res.* **2016**, *46*, 211–233. [CrossRef]
27. Kingman, J.; Tsavdaridis, K.D.; Toropov, V.V. Applications of topology optimization in structural engineering. In Proceedings of the Civil Engineering for Sustainability and Resilience International Conference (CESARE), Leeds, UK, 24–27 April 2014.
28. Ahmad, A.; Campana, F.; Bici, M. Application of Topology Optimization to Reduce Automotive Exhaust Emissions. *SAE Int. J. Sustain. Transp. Energy Environ. Policy* **2021**, accepted.
29. Yi, B.; Zhou, Y.; Yoon, G.H.; Saitou, K. Topology optimization of functionally-graded lattice structures with buckling constraints. *Comput. Methods Appl. Mech. Eng.* **2019**, *354*, 593–619. [CrossRef]
30. Cheng, L.; Bai, J.; To, A.C. Functionally graded lattice structure topology optimization for the design of additive manufactured components with stress constraints. *Comput. Methods Appl. Mech. Eng.* **2019**, *344*, 334–359. [CrossRef]

31. Cheng, L.; Liu, J.; Liang, X.; To, A.C. Coupling lattice structure topology optimization with design-dependent feature evolution for additive manufactured heat conduction design. *Comput. Methods Appl. Mech. Eng.* **2018**, *332*, 408–439. [CrossRef]
32. Karadere, G.; Düzcan, Y.; Yıldız, A.R. Light-weight design of automobile suspension components using topology and shape optimization techniques. *Mater. Test.* **2020**, *62*, 454–464. [CrossRef]
33. Liu, T.; Wang, S.; Li, B.; Gao, L. A level-set-based topology and shape optimization method for continuum structure under geometric constraints. *Struct. Multidiscip. Optim.* **2014**, *50*, 253–273. [CrossRef]
34. Tyflopoulos, E.; Tollnes, F.D.; Steinert, M.; Olsen, A. State of the art of generative design and topology optimization and potential research needs. In Proceedings of the NordDesign 2018, Linköping, Sweden, 14–17 August 2018.
35. Chen, X.A.; Tao, Y.; Wang, G.; Kang, R.; Grossman, T.; Coros, S.; Hudson, S.E. Forte: User-driven generative design. In Proceedings of the 2018 CHI Conference on Human Factors in Computing Systems, Montréal, QC, Canada, 21–26 April 2018; pp. 1–12.
36. Dalpadulo, E.; Gherardini, F.; Pini, F.; Leali, F. Integration of topology optimisation and design variants selection for additive manufacturing-based systematic product redesign. *Appl. Sci.* **2020**, *10*, 7841. [CrossRef]
37. Vogiatzis, P.; Chen, S.; Wang, X.; Li, T.; Wang, L. Topology optimization of multi-material negative Poisson's ratio metamaterials using a reconciled level set method. *Comput. Aided Des.* **2017**, *83*, 15–32. [CrossRef]
38. Takezawa, A.; Kobashi, M. Design methodology for porous composites with tunable thermal expansion produced by multi-material topology optimization and additive manufacturing. *Compos. Part. B Eng.* **2017**, *131*, 21–29. [CrossRef]
39. Joo, Y.; Lee, I.; Kim, S.J. Topology optimization of heat sinks in natural convection considering the effect of shape-dependent heat transfer coefficient. *Int. J. Heat Mass Transf.* **2017**, *109*, 123–133. [CrossRef]
40. Kingman, J.J.; Tsavdaridis, K.; Toropov, V. Applications of topology optimisation in structural engineering: High-rise buildings & steel components. *Jordan J. Civ. Eng.* **2015**, *9*, 335–357.
41. Gaynor, A.T.; Guest, J.K.; Moen, C.D. Reinforced concrete force visualization and design using bilinear truss-continuum topology optimization. *J. Struct. Eng.* **2013**, *139*, 607–618. [CrossRef]
42. Oh, S.; Jung, Y.; Kim, S.; Lee, I.; Kang, N. Deep generative design: Integration of topology optimization and generative models. *J. Mech. Des.* **2019**, *141*, 111405. [CrossRef]
43. Liu, J.; Gaynor, A.T.; Chen, S.; Kang, Z.; Suresh, K.; Takezawa, A.; To, A.C. Current and future trends in topology optimization for additive manufacturing. *Struct. Multidiscip. Optim.* **2018**, *57*, 2457–2483. [CrossRef]
44. Zhang, K.; Cheng, G.; Xu, L. Topology optimization considering overhang constraint in additive manufacturing. *Comput. Struct.* **2019**, *212*, 86–100. [CrossRef]
45. Garaigordobil, A.; Ansola, R.; Santamaría, J.; De Bustos, I.F. A new overhang constraint for topology optimization of self-supporting structures in additive manufacturing. *Struct. Multidiscip. Optim.* **2018**, *58*, 2003–2017. [CrossRef]
46. Gaynor, A.T.; Guest, J.K. Topology optimization considering overhang constraints: Eliminating sacrificial support material in additive manufacturing through design. *Struct. Multidiscip. Optim.* **2016**, *54*, 1157–1172. [CrossRef]
47. Ali Banijamali, S.M.; Oftadeh, R.; Nazarian, A.; Goebel, R.; Vaziri, A.; Nayeb-Hashemi, H. Effects of different loading patterns on the trabecular bone morphology of the proximal femur using adaptive bone remodeling. *J. Biomech. Eng.* **2015**, *137*, 011011. [CrossRef]
48. Xu, S.; Liu, J.; Zou, B.; Li, Q.; Ma, Y. Stress constrained multi-material topology optimization with the ordered SIMP method. *Comput. Methods Appl. Mech. Eng.* **2021**, *373*, 113453. [CrossRef]
49. Lee, E.; James, K.A.; Martins, J.R. Stress-constrained topology optimization with design-dependent loading. *Struct. Multidiscip. Optim.* **2012**, *46*, 647–661. [CrossRef]
50. Holmberg, E.; Torstenfelt, B.; Klarbring, A. Stress constrained topology optimization. *Struct. Multidiscip. Optim.* **2013**, *48*, 33–47. [CrossRef]
51. Wang, L.; Shen, W.; Xie, H.; Neelamkavil, J.; Pardasani, A. Collaborative conceptual design—State of the art and future trends. *Comput. Aided Des.* **2002**, *34*, 981–996. [CrossRef]
52. Anderson, D.M. *Design for Manufacturability: How to Use Concurrent Engineering to Rapidly Develop Low-Cost, High-Quality Products for Lean Production*; CRC Press: Boca Raton, FL, USA, 2020.
53. Wisthoff, A.; Ferrero, V.; Huynh, T.; DuPont, B. Quantifying the impact of sustainable product design decisions in the early design phase through machine learning. In *ASME 2016 International Design Engineering Technical Conferences and Computers and Information in Engineering Conference*; American Society of Mechanical Engineers Digital Collection: New York, NY, USA, 2016.
54. Zeballos, L.J.; Méndez, C.A.; Povoa, A.P.B. Mixed-integer linear programming approach for product design for life-cycle profit. *Comput. Ind. Eng.* **2019**, *137*, 106079. [CrossRef]
55. Pan, Z.; Wang, X.; Teng, R.; Cao, X. Computer-aided design-while-engineering technology in top-down modeling of mechanical product. *Comput. Ind.* **2016**, *75*, 151–161. [CrossRef]
56. Wu, J.; Quian, X.; Wang, M.Y. Advances in generative design. *Comput. Aided Des.* **2019**, *116*, 102733. [CrossRef]
57. Li, C.; Kim, I.Y. Multi-material topology optimization for automotive design problems. *Proc. Inst. Mech. Eng. Part. D J. Automob. Eng.* **2018**, *232*, 1950–1969. [CrossRef]
58. Zhu, J.H.; Zhang, W.H.; Xia, L. Topology optimization in aircraft and aerospace structures design. *Arch. Comput. Methods Eng.* **2016**, *23*, 595–622. [CrossRef]
59. Chang, C.L.; Chen, C.S.; Huang, C.H.; Hsu, M.L. Finite element analysis of the dental implant using a topology optimization method. *Med. Eng. Phys.* **2012**, *34*, 999–1008. [CrossRef]

60. Feist, S.; Barreto, G.; Ferreira, B.; Leitao, A. Portable Generative Design for Building Information Modelling. In Proceedings of the 21st International Conference on Computer Aided Architectural Design Research in Asia (CAADRIA 2016 Conference), Melbourne, Australia, 30 March–2 April 2016.
61. Bagassi, S.; Lucchi, F.; De Crescenzio, F.; Persiani, F. Generative design: Advanced design optimization processes for aeronautical applications. In Proceedings of the 30th Congress of the International Council of the Aeronautical Sciences, Daejeon, Korea, 25–30 September 2016.
62. Yang, Z.-J.; Chen, X.; Kelly, R. A topological optimization method for flexible multi-body dynamic system using epsilon algorithm. *Struct. Eng. Mech.* **2011**, *37*, 475–487. [CrossRef]
63. Bici, M.; Brini, A.; Campana, F.; Capoferri, S.; Guarnieri, R.; Morandini, F.; Patera, A. Design of the new inner frame for the Vittoria Alata di Brescia: How engineering design may support ancient bronze restoration. *Lect. Notes Mech. Eng. Proc. ADM2021 Int. Conf.* **2021**, accepted.
64. Blösch-Paidosh, A.; Shea, K. Design Heuristics for Additive Manufacturing Validated Through a User Study. *ASME J. Mech. Des.* **2019**, *141*, 041101. [CrossRef]
65. Fu, K.K.; Yang, M.C.; Wood, K.L. Design principles: Literature review, analysis, and future directions. *J. Mech. Des. Trans. ASME* **2016**, *138*, 101103. [CrossRef]
66. Yilmaz, S.; Seifert, C.M. Creativity through design heuristics: A case study of expert product design. *Des. Stud.* **2011**, *32*, 384–415. [CrossRef]

applied sciences

MDPI

*Article*

# Smooth Design of 3D Self-Supporting Topologies Using Additive Manufacturing Filter and SEMDOT

**Yun-Fei Fu [1,*], Kazem Ghabraie [1], Bernard Rolfe [1], Yanan Wang [1] and Louis N. S. Chiu [2]**

1 School of Engineering, Deakin University, Waurn Ponds, VIC 3217, Australia; k.ghabraie@deakin.edu.au (K.G.); bernard.rolfe@deakin.edu.au (B.R.); yanan.wang@deakin.edu.au (Y.W.)
2 Department of Materials Science and Engineering, Monash University, Clayton, VIC 3800, Australia; louis.chiu@monash.edu
* Correspondence: fuyunf@deakin.edu.au; Tel.: +61-470-273-301

**Abstract:** The smooth design of self-supporting topologies has attracted great attention in the design for additive manufacturing (DfAM) field as it cannot only enhance the manufacturability of optimized designs but can obtain light-weight designs that satisfy specific performance requirements. This paper integrates Langelaar's AM filter into the Smooth-Edged Material Distribution for Optimizing Topology (SEMDOT) algorithm—a new element-based topology optimization method capable of forming smooth boundaries—to obtain print-ready designs without introducing post-processing methods for smoothing boundaries before fabrication and adding extra support structures during fabrication. The effects of different build orientations and critical overhang angles on self-supporting topologies are demonstrated by solving several compliance minimization (stiffness maximization) problems. In addition, a typical compliant mechanism design problem—the force inverter design—is solved to further demonstrate the effectiveness of the combination between SEMDOT and Langelaar's AM filter.

**Keywords:** SEMDOT; Langelaar's AM filter; print-ready design

**Citation:** Fu, Y.-F.; Ghabraie, K.; Rolfe, B.; Wang, Y.; Chiu, L.N.S. Smooth Design of 3D Self-Supporting Topologies Using Additive Manufacturing Filter and SEMDOT. *Appl. Sci.* **2021**, *11*, 238. https://doi.org/10.3390/app11010238

Received: 2 December 2020
Accepted: 24 December 2020
Published: 29 December 2020



## 1. Introduction

Design for additive manufacturing (DfAM) represents a range of design methods through which performance and/or other key considerations such as manufacturability, reliability and cost can be optimized subject to the capabilities of additive manufacturing (AM) technologies [1–4]. In recent years, topology optimization for AM has become one of the most important branches in DfAM, as topology optimization has great potential to fully exploit the significant benefits provided by the increased design freedom offered by AM [5,6].

As traditional element-based algorithms such as solid isotropic material with penalization (SIMP), rational material with penalization (RAMP) and bi-directional evolutionary structural optimization (BESO) will inevitably form non-smooth boundaries, post-processing or redesign methods have to be used to obtain accurate boundary information for the purpose of engineering applications [7–9], meaning that extra efforts have to be made after topology optimization. Given the significance of accurate boundary representation, some element-based algorithms that are capable of forming smooth boundaries such as multiresolution topology optimization (MTO) methods [10–12], elemental volume fraction-based methods [13–16] and a method using floating projection [17,18] have been developed in recent years.

Optimized topologies are fabricated layer-by-layer and have to be sufficiently supported to prevent the component distortion caused by high bending stresses in some AM technologies such as fused deposition modeling (FDM) and selective laser melting (SLM). Even though introducing support structures to topological designs can resolve this issue, more materials need to be wasted, and extra efforts have to be made to remove support structures after manufacturing [19–21]. Therefore, optimizing self-supporting topologies directly is an effective way to mitigate overhang angle limitations, save materials and avoid additional effort. Even though there are a great number of investigations regarding 2D self-supporting topologies [22–24], these studies have more theoretical value than application value compared to the investigations on 3D self-supporting designs. Studies on 3D self-supporting design have attracted scholarly attention in recent years. The early work in this field was Langelaar's gradient-based AM filter [25], followed by an improved method capable of executing simultaneous topology optimization and support structures considering metal AM constraints and post-print machining requirements [26]. Han et al. [27] proposed hybrid additive–subtractive manufacturing constraints for 3D continuum structures based on BESO. Mezzadri and Qian [28] proposed a second-order measure of boundary oscillations for overhang control to enhance the manufacturability of designs in AM, and a 3D cantilever beam case was tested. Zhao et al. [29] formulated the self-supporting requirements as an explicit quadratic continuous constraint, and 3D benchmark problems were tested. Zhang and Chen [30] integrated the proposed constraint that was imposed on the overhang angle, directional-dependent overhang angle and horizontal minimum length scale into the MTOP method to conduct the smooth design of 3D self-supporting topologies. van de Ven et al. [31] presented a 3D front propagation-based overhang filter through which printable topologies can be obtained. Most recently, Bi et al. [32] developed a new layer-wise geometric self-supporting constraint for 3D continuum structures based on BESO, and an AM experiment was conducted to validate the effectiveness of the proposed constraint by printing a hinge frame used in the aerospace field.

The authors of this study merely focused on the smooth design of 2D self-supporting topologies through the combination of the Smooth-Edged Material Distribution for Optimizing Topology (SEMDOT) algorithm and Langelaar's AM filter in previous works [33,34]. Most recently, the authors presented a simple 3D self-supporting case using the Optimality Criteria (OC) optimizer in [35]. In this paper, the smooth design of 3D self-supporting topologies is shown in detail.

The remainder of this paper is structured as follows. Section 2 mathematically demonstrates the combination of SEMDOT and Langelaar's AM filter. Section 3 presents some numerical experiments to demonstrate the effectiveness and generality of this combination. Finally, conclusions are summarized in Section 4.

## 2. Integrating Langelaar's AM Filter into SEMDOT

### 2.1. Problem Statement

This study uses SEMDOT as an optimization platform to form smooth topologies and Langelaar's AM filter to obtain 3D self-supporting designs. SEMDOT is a new topology optimization platform that was developed by the authors [16], and its effectiveness has been thoroughly validated. For more details on the advantages of SEMDOT compared to some existing algorithms that are capable of generating smooth boundaries, see [15,16]. Langelaar's AM filter is a well-established restriction method that can be used with gradient-based topology optimization algorithms, and it has become one of the most extensively used methods for generating self-supporting designs in this community. In SEMDOT, the elemental volume fraction used in the Finite Element Analysis (FEA) model is defined as

$$X_e = \frac{1}{N}\sum_{g=1}^{N}\rho_{e,g} \tag{1}$$

where $N$ is the total number of grid points in each element and $\rho_{e,g}$ is the density of the $g$th grid point in the $e$th element.

The filtering scheme selected in this study is

$$\tilde{X}_e = \frac{\sum_{l=1}^{N_e} \omega_{el} X_l}{\sum_{l=1}^{N_e} \omega_{el}} \tag{2}$$

where $\tilde{X}_e$ is the filtered elemental volume fraction, $N_e$ is the neighborhood set of elements within the filter domain for the $e$th element that is a circle centered at the centroid of this element with a predefined filter radius $r_{\min}$, and $\omega_{el}$ is a linear weight factor defined as

$$\omega_{el} = \max(0, r_{\min} - \Delta(e, l)) \tag{3}$$

where $\Delta(e, l)$ is the center-to-center distance of the $l$th element within the filter domain to the $e$th element.

Filtered elemental volume fractions are updated for the next round of FEA by summing up the grid points for each element:

$$\tilde{X}_e^{new} = \frac{1}{N} \sum_{g=1}^{N} \rho_{e,g}^{new} \tag{4}$$

where $\rho_{e,g}^{new}$ is the density of the grid point obtained by the Heaviside smooth function.

The relationship between $\tilde{X}_e$ and $\tilde{X}_e^{new}$ can be established as

$$\delta_e = \tilde{X}_e^{new} - \tilde{X}_e \tag{5}$$

where $\delta_e$ is the deviation of the $e$th element before and after the update. A detailed discussion of this relationship is presented in [16].

The basic idea of Langelaar's AM filter is that the value of the element to be supported should not be less than the maximum value of the elements in the supporting region. Figure 1 schematically demonstrates the method of generating self-supporting designs, where the blue element is the element to be supported and the five green elements ($S_{(i,j,k)}$) form the supporting region. Mathematical expressions of the 3D Langelaar's AM filter [25] are given by

$$\xi_{(i,j,k)} = \min(\tilde{\mathbf{X}}_{(i,j,k)}, \Xi_{(i,j,k)}) \quad \text{with} \tag{6}$$

$$\Xi_{(i,j,k)} = \max(\xi_{(i-1,j,k-1)}, \xi_{(i,j-1,k-1)}, \xi_{(i,j,k-1)}, \xi_{(i,j+1,k-1)}, \xi_{(i+1,j,k-1)}) \tag{7}$$

where $\boldsymbol{\xi}$ is the vector of printed elemental volume fractions, $\tilde{\mathbf{X}}$ is the vector of filtered elemental volume fractions and $\Xi$ is the vector of the maximum printed elemental volume fractions.

According to [5,25], the following approximation is made for gradient-based optimization:

$$\xi = \frac{1}{2}\left(\tilde{\mathbf{X}} + \Xi - \left((\tilde{\mathbf{X}} - \Xi)^2 + \varepsilon\right)^{\frac{1}{2}} + \sqrt{\varepsilon}\right) \tag{8}$$

where $\varepsilon$ is the parameter that controls the accuracy of the approximation.

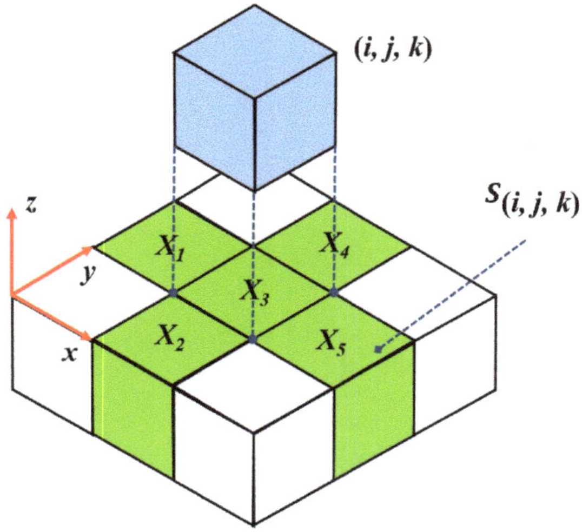


**Figure 1.** Illustration of the 3D Langelaar's additive manufacturing (AM) filter [25].

The softmax function is used to calculate the maximum value of the elements in the supporting region instead of the P-Q max function in origianl Langelaar's AM filter [36]. This is because the P-Q max function will generally cause a higher absolute error (the difference between real maximum and maximum smooth approximation) than the softmax function. The softmax function is expressed as defined by Barroqueiro et al. [36]:

$$\text{smax}(\xi_k) = \Xi(\xi_k) = \frac{\sum\limits_{k=1}^{n_s} \xi_k e^{P\xi_k}}{\sum\limits_{k=1}^{n_s} e^{P\xi_k}} \tag{9}$$

where $\xi_k$ is the printed elemental volume fraction in the supporting region relevant to the considered element, $P$ is the parameter that controls the smoothness of the approximation ($10^{-4}$), and $n_s$ is the number of elements in the supporting region (for 3D cases, $n_s$ = 5).

## 2.2. Topology Optimization Problems

Two typical topology optimization problems: compliance minimization and compliant mechanism design problems are considered in this study. The mathematical statement of compliance minimization design is

$$\begin{aligned} \min : C(X_e) &= \mathbf{f}^{\mathbf{T}}\mathbf{u} \\ \text{subject to} : \mathbf{K}(X_e)\mathbf{u} &= \mathbf{f} \\ &\frac{\sum\limits_{e=1}^{M} X_e V_e}{\sum\limits_{e=1}^{M} V_e} - V^* \leq 0 \\ &0 < \rho_{\min} \leq X_e \leq 1;\ e = 1,2,\cdots,M \end{aligned} \tag{10}$$

where $C$ is compliance; $\mathbf{f}$ and $\mathbf{u}$ are the global force and displacement vectors, respectively; $\mathbf{K}$ is the global stiffness matrix; $V_e$ is the volume of the $e$th element; $V^*$ is the prescribed value of the allowable volume; $M$ is the total number of elements; and $\rho_{\min}$ is a small artificial parameter, 0.001.

The compliant mechanism design problem is stated as

$$\begin{aligned}
&\min : C(X_e) = -\mathbf{L}^{\mathrm{T}}\mathbf{u} = -u_{out} \\
&\text{subject to} : \mathbf{K}(X_e)\mathbf{u} = \mathbf{f}_{in} \\
&\qquad \frac{\sum_{e=1}^{M} X_e V_e}{\sum_{e=1}^{M} V_e} - V^* \leq 0 \\
&\qquad 0 < \rho_{\min} \leq X_e \leq 1;\ e = 1, 2, \cdots, M
\end{aligned} \tag{11}$$

where $\mathbf{L}$ is a unit length vector with zeros at all degrees of freedom except at the output point, where it is one; $u_{out}$ is the output port displacement; and $\mathbf{f}_{in}$ is the input force vector.

*2.3. Sensitivity Analysis*

Sensitivity analysis measures the effectiveness of changing the elemental volume fractions on the objective function in order to determine the search direction [37,38]. The sensitivity of $C_p(\boldsymbol{\xi})$ with respect to $\mathbf{X}$ (the vector of elemental volume fractions) is calculated with the chain rule as

$$\frac{\partial C_p(\boldsymbol{\xi})}{\partial \mathbf{X}} = \frac{\partial C_p(\boldsymbol{\xi})}{\partial \boldsymbol{\xi}} \frac{\partial \boldsymbol{\xi}}{\partial \tilde{\mathbf{X}}^{\mathrm{new}}} \frac{\partial \tilde{\mathbf{X}}^{\mathrm{new}}}{\partial \tilde{\mathbf{X}}} \frac{\partial \tilde{\mathbf{X}}}{\partial \mathbf{X}} \tag{12}$$

Given the discussions of the relationship between $\tilde{X}_e$ and $\tilde{X}_e^{new}$ in [16], it can be assumed that $\partial \boldsymbol{\xi}^{new} / \partial \boldsymbol{\xi} = 1$ in this paper for simplicity. The sensitivity of $C_p(\boldsymbol{\xi})$ with respect to $\tilde{\mathbf{X}}$ can be calculated based on the derivation of adjoint sensitivity analysis as [5]:

$$\frac{\partial C_p(\boldsymbol{\xi})}{\partial \tilde{\mathbf{X}}_j} = \boldsymbol{\lambda}_j^{\mathrm{T}} \frac{\partial \boldsymbol{\xi}_j}{\partial \tilde{\mathbf{X}}_j} \tag{13}$$

where $\boldsymbol{\lambda}_j$ is the adjoint field, defined as [5]

$$\begin{aligned}
\boldsymbol{\lambda}_j^{\mathrm{T}} &= \frac{\partial C_p(\boldsymbol{\xi})}{\partial \boldsymbol{\xi}_j} + \boldsymbol{\lambda}_{j+1}^{\mathrm{T}} \frac{\partial \boldsymbol{\lambda}_{j+1}}{\partial \boldsymbol{\lambda}_j}, \quad \text{for} \quad 1 <= j < n_i \\
\boldsymbol{\lambda}_{n_i}^{\mathrm{T}} &= \frac{\partial C_p(\boldsymbol{\xi})}{\partial \boldsymbol{\xi}_{n_i}}
\end{aligned} \tag{14}$$

The remaining derivatives required for executing the sensitivity analysis are calculated as [36]

$$\frac{\partial \boldsymbol{\xi}}{\partial \tilde{\mathbf{X}}} = \frac{1}{2}\left(1 - (\tilde{\mathbf{X}} - \Xi)\left((\tilde{\mathbf{X}} - \Xi)^2 + \varepsilon\right)^{-\frac{1}{2}}\right) \tag{15}$$

$$\frac{\partial \boldsymbol{\xi}}{\partial \Xi} = \frac{1}{2}\left(1 + (\tilde{\mathbf{X}} - \Xi)\left((\tilde{\mathbf{X}} - \Xi)^2 + \varepsilon\right)^{-\frac{1}{2}}\right) \tag{16}$$

$$\frac{\partial \Xi_{ki}}{\partial \boldsymbol{\xi}_k} = \frac{e^{P\xi_k}}{\sum_{ki=1}^{n_s} e^{P\xi_{ki}}}[1 + P(\xi_k - \Xi)] \tag{17}$$

For more details on the combination of SEMDOT and Langelaar's AM filter, see the work presented in [33,34]. As the authors' previous works regarding smooth self-supporting topologies [33,34] were based on the very initial version of SEMDOT, we therefore also refer readers to the paper about the latest version of SEMDOT [16] for further reference.

## 3. Numerical Experiments

The 3D Matlab code of SEMDOT was developed based on the codes presented by the authors [16] and Liu and Tovar [39], and the 3D Matlab code of Langelaar's AM filter was developed based on the 2D code provided by Langelaar [5]. Several benchmark optimization problems are solved to validate the effectiveness of the combination of SEMDOT and Langelaar's AM filter. Parameters in the SEMDOT algorithm are set as presented in [15]. An isotropic linear elastic material model is assumed with a Young's modulus of 1 MPa and Poisson's ratio of 0.3 for all examples. A grid with $5 \times 5 \times 5$ points in each element is used to maintain a proper balance between the smoothness of boundaries and computational cost. Unless otherwise stated, the prescribed value of the allowable volume $V^*$ is set to 0.3, meaning that 30% of materials within the design domain will be retained; the filter radius $r_{\min}$ is set to 1.5 time elements width ($r_{\min}$ = 1.5), a dimensionless size; and the critical overhang angle is set to 45° in Langelaar's AM filter. In addition, the maximum number of iterations is set to 500. The method of moving asymptotes (MMA) proposed by Svanberg [40] is used as the optimizer. Default parameters in MMA are adopted for compliance minimization design, and the move limit is set to 0.05 for compliant mechanism design to improve the convergence stability.

### 3.1. Different Build Orientations

A deep cantilever beam shown in Figure 2 is considered to demonstrate the influences of different build orientations on compliance, the number of iterations and topologies. In AM, the build orientation is the accumulating orientation of materials when building the part [41]. As illustrated in Figure 2, the left end is fixed and a unit vertical load ($F_y$ = 1 N—Figure 2) is imposed in the middle of the lower edge on the free end. In this study, three build orientations—$x_{\max}$, $y_{\min}$ and $z_{\max}$—are considered (see Figure 2). The design domain is discretized by a $60 \times 40 \times 30$ finite element mesh.

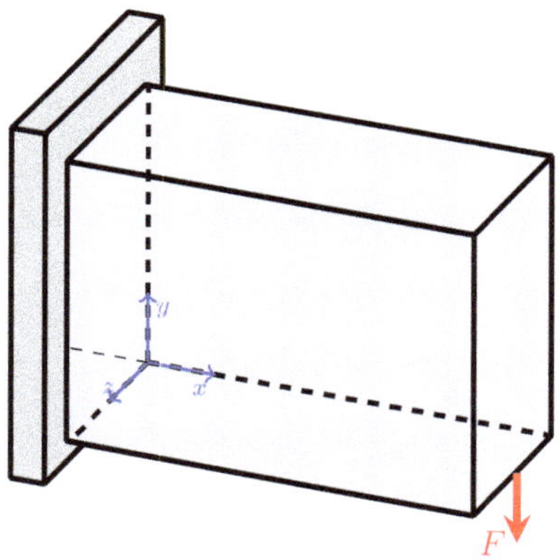

**Figure 2.** Design domain of deep cantilever beam.

Figure 3a shows the reference design case, which converges at 5.226 J after 88 iterations; the optimized topology with the build orientation $x_{\max}$ converges at 5.292 J after 201 iterations; the optimized topology with the build orientation $y_{\min}$ converges at 5.374 J after 165 iterations; and the optimized topology with the build orientation $z_{\max}$ converges at 5.378 J after 139 iterations. It is noted that the reference case is the topological design obtained without using Langelaar's AM filter (i.e., general topology optimization), and J (N·mm) denotes the multiplication of the force (N) by displacement (mm). Performance sacrifices of 1.25 %, 2.83 %, and 2.91 % are made to form self-supporting designs for build orientations $x_{\max}$, $y_{\min}$ and $z_{\max}$, respectively. Even though the self-supporting design in the build orientation $y_{\min}$ (Figure 3d) is the most similar to the reference design (Figure 3b), the self-supporting design in the build orientation $x_{\max}$ (Figure 3c) is the closest to the reference design in terms of performance. The self-supporting design in the build orientation $z_{max}$ (Figure 3e) obtains the worst performance, as great changes have to be made for this specific direction.

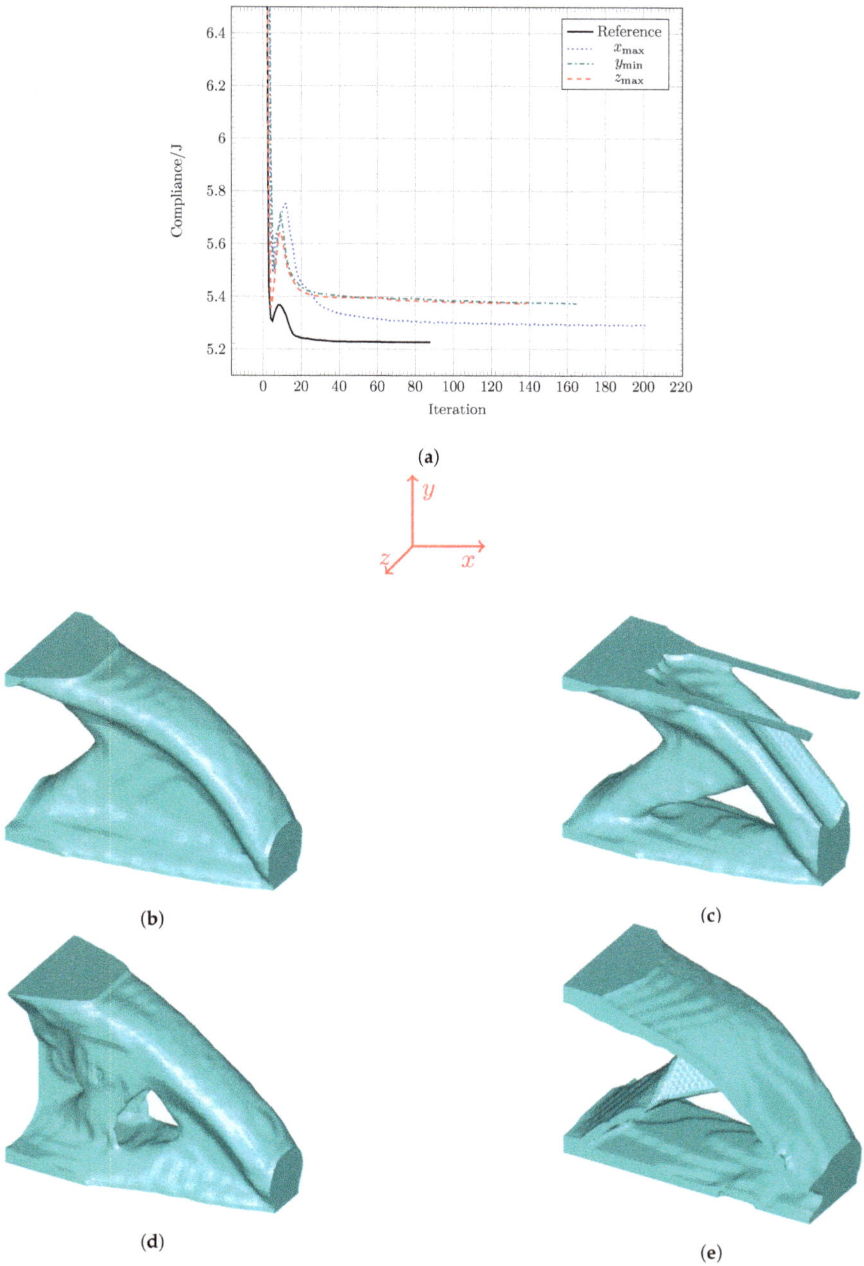


**Figure 3.** Convergence processes and optimized topologies with different build orientations: (**a**) convergence processes, (**b**) reference, (**c**) $x_{max}$, (**d**) $y_{min}$, (**e**) $z_{max}$.

### 3.2. Different Target Volume Fractions

A simply supported deep beam, as shown in Figure 4, is used to investigate the effects of different target volume fractions on compliance, the number of iterations and topologies. As illustrated in Figure 4, four bottom corners are prevented, and a unit vertical load ($F_y$ = 1 N—Figure 4) is applied in the middle of the top surface. Six different target volume fractions (i.e., 0.15, 0.2, 0.3, 0.4, 0.5 and 0.6) are considered in this study. The design domain is discretized by a $60 \times 40 \times 20$ finite element mesh.

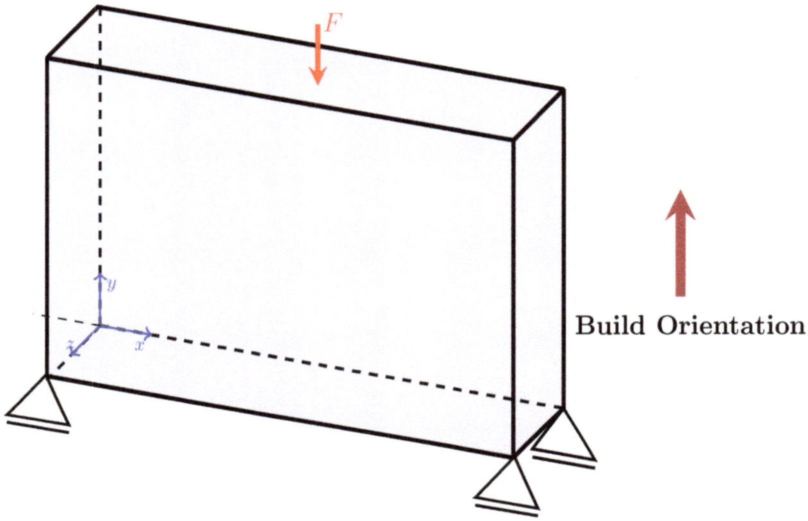


**Figure 4.** Design domain of a simply supported deep beam.

Figure 5a shows that with an increase in the target volume fraction, the performance of optimized topologies gradually imroves, and there are some difficulties in convergence when small target volume fractions ($V^* = 0.15$ and $V^* = 0.2$) are considered. Optimized topologies with different target volume fractions are shown in Figure 5b–g. There is a thin feature in the middle of the optimized topology, as shown in Figure 5b. The value of the element at the bottom of this thin feature is small, whereas elements above this bottom element have greater values, which violates the basic strategy of Langelaar's AM filter. The formation of this thin feature is because of the absolute error caused by the softmax function (mentioned in Section 2.1).

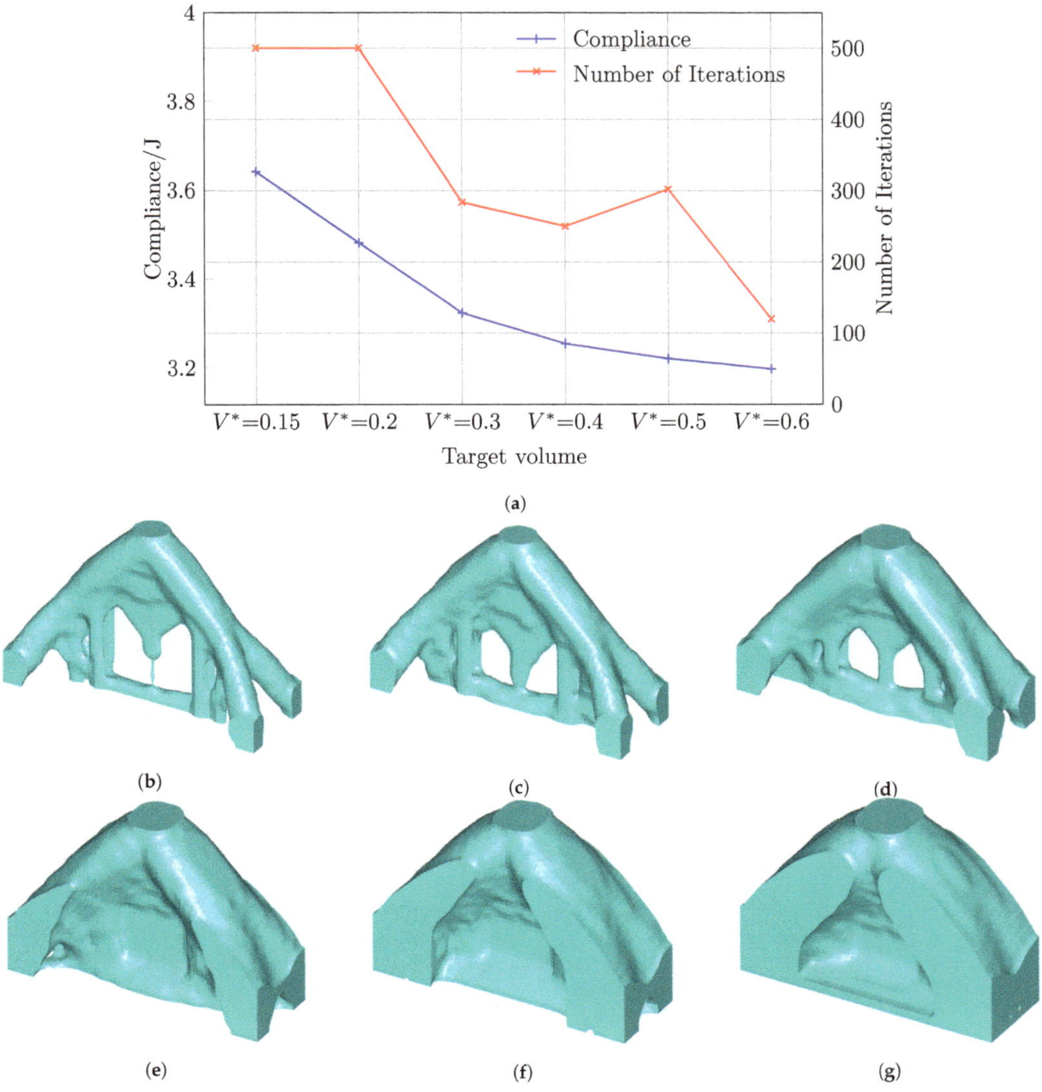


**Figure 5.** Compliance, number of iterations and optimized topologies with different target volume fractions: (**a**) compliance and number of iterations, (**b**) $V^* = 0.15$, (**c**) $V^* = 0.2$, (**d**) $V^* = 0.3$, (**e**) $V^* = 0.4$, (**f**) $V^* = 0.5$, (**g**) $V^* = 0.6$.

The first solution to solve this specific thin feature issue is to introduce the weighting factor of $w^*$ to the location of element $X_3$ in the supporting region (see Figure 1). Equation (7) is therefore rewritten as

$$\Xi_{(i,j,k)} = \max(\xi_{(i-1,j,k-1)}, \xi_{(i,j-1,k-1)}, w^*\xi_{(i,j,k-1)}, \xi_{(i,j+1,k-1)}, \xi_{(i+1,j,k-1)}) \quad (18)$$

The related equations are accordingly changed.

The reference design (Figure 6a) converges at 3.621 J after 86 iterations. For the rest of the self-supporting designs (Figure 6b–d), the maximum number of iterations (500) is reached. Among self-supporting designs, the use of a higher weighting factor $w^*$ will generally contribute to relatively improved performance. More importantly, the thin feature issue in Figure 5b has been solved by introducing the weighting factor $w^*$, as illustrated in Figure 6b–d.

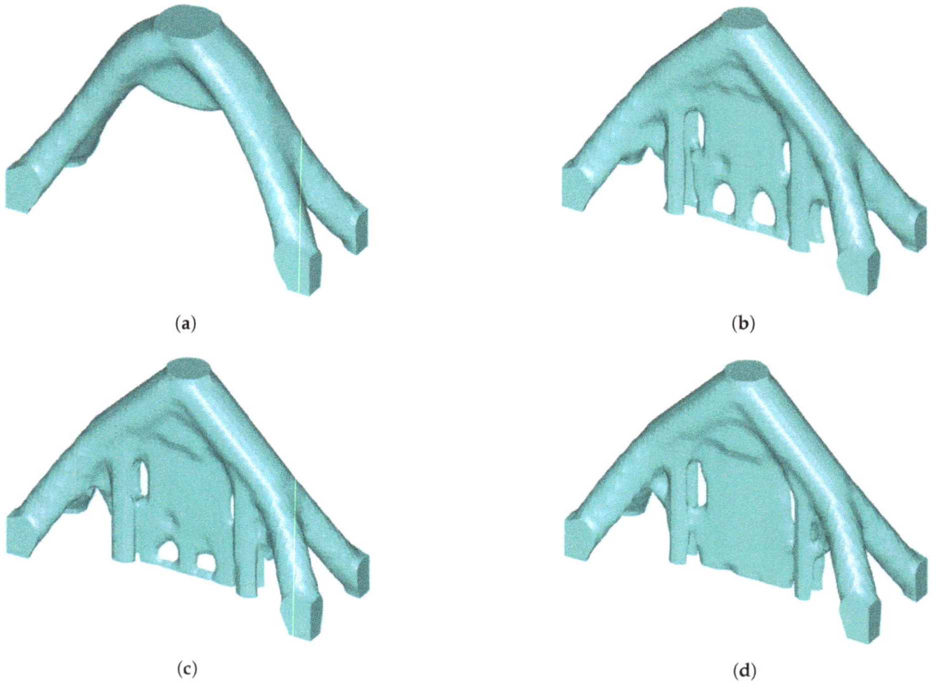

**Figure 6.** Optimized topologies with different weighting factors $w^*$: (**a**) reference, $C$ = 3.621 J. (**b**) $w^* = 1.01$, $C$ = 3.665 J. (**c**) $w^* = 1.015$, $C$ = 3.662 J. (**d**) $w^* = 1.02$, $C$ = 3.659 J.

Another solution is to use a fine mesh, as it can provide more design freedom. The mesh size of $90 \times 60 \times 30$ is used as an example. As the element size is scaled with a certain ratio, the filter radius is scaled with the same ratio to make sure its absolute value remains constant. Figure 7a shows that the self-supporting design case with the fine mesh for $V^* = 0.15$ converges at 3.484 J after reaching the maximum number of iterations (500), and the topological boundary error gradually decreases to almost 0% when the converged topology is obtained. The performance of the fine mesh case is 4.32% better than that of the optimized topology in Figure 5b (3.641 J). Figure 7b shows a much better self-supporting design than that in Figure 5b in terms of manufacturability. There is almost no structural resemblance between coarse mesh (Figure 5b) and fine mesh cases (Figure 7b) as Langelaar's AM filter is highly mesh-dependent. Compared to the strategy of introducing the weighting factor $w^*$, using the fine mesh is a more general way to solve the thin feature issue.

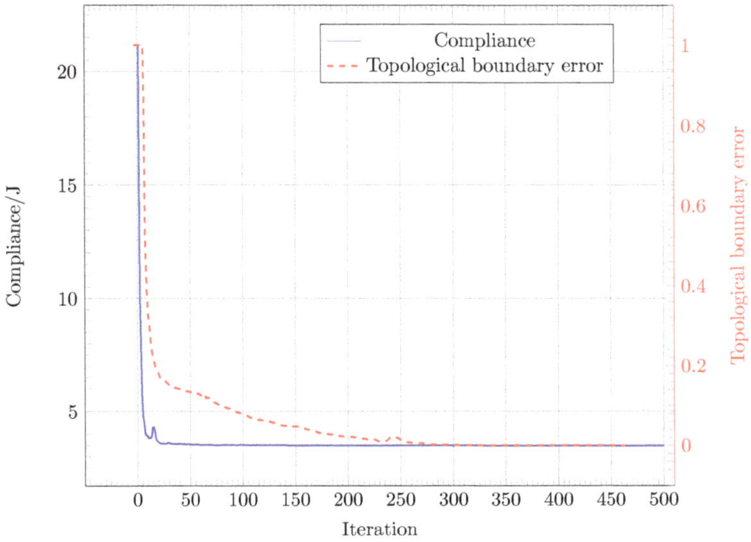


(**a**)

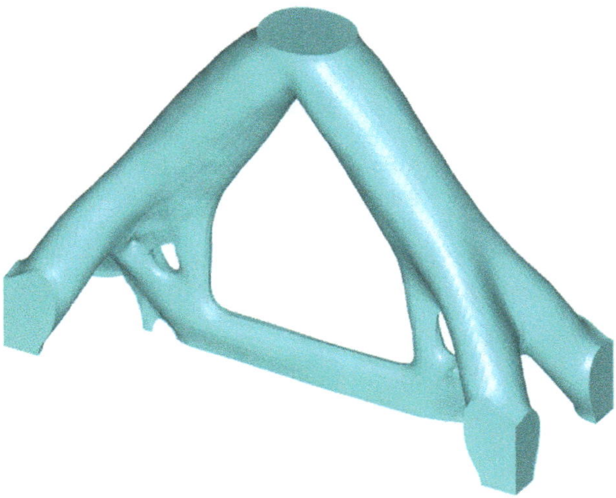

(**b**)

**Figure 7.** Convergence process and optimized topology with a fine mesh for a target volume fraction of 0.15: (**a**) convergence process, (**b**) Optimized topology.

## 3.3. Different Critical Overhang Angles

When Langelaar's AM filter is integrated into traditional element-based algorithms such as SIMP, the critical overhang angle is fixed at 45° because of the defined element layout (see Figure 1). As intermediate elements can be cut during the formation of smooth boundaries in SEMDOT, the possibility of exploring different critical overhang angles is therefore provided by the combination of Langelaar's AM filter and SEMDOT. To obtain self-supporting topologies with different critical overhang angles, the weighting factor of $w_s = 1/\tan\alpha$ is introduced to the locations of elements $X_1$, $X_2$, $X_4$, and $X_5$ in the supporting region (refer to Figure 1) where $\alpha$ is the critical overhang angle. Equation (7) is therefore rewritten as

$$\Xi_{(i,j,k)} = \max(w_s\xi_{(i-1,j,k-1)}, w_s\xi_{(i,j-1,k-1)}, \xi_{(i,j,k-1)}, w_s\xi_{(i,j+1,k-1)}, w_s\xi_{(i+1,j,k-1)}) \quad (19)$$

The related equations are accordingly changed.

Following Section 3.2, a simply supported deep beam with a finite element mesh of $30 \times 20 \times 20$ is used to explore self-supporting topologies with different critical overhang angles (i.e., 30°, 45° and 60°). The objective function value is multiplied by a scale factor of 100.

Figure 8a shows that the reference design case converges at 326.4 J after 83 iterations, and self-supporting design cases with different critical overhang angles of 30°, 45°, and 60° converge at 328.4 J after 133 iterations, 328.7 J after 109 iterations and 332.2 J after 261 iterations, respectively. Generally, a higher critical overhang angle will cause worse performance as it provides less design freedom. Based on the results presented in Figure 8a, the critical overhang angle of 45° requires the least number of iterations to converge among self-supporting design cases. The optimized reference topology is shown in Figure 8b, and optimized topologies with different critical overhang angles are shown in Figure 8c–e.

It should be acknowledged that the mentioned strategy of exploring different critical overhang angles is not suitable for all test cases. For some specific cases, numerical issues may occur during the optimization process, resulting in improper self-supporting designs.

### 3.4. Force Inverter Design

The force inverter design problem is demonstrated in Figure 9 where an input load is defined in the positive direction, and symmetric constraints are applied on top and side faces (two pink faces). In the case of force inverter design, the goal is to maximize the negative horizontal output displacement. The design domain is discretized by a $40 \times 20 \times 5$ finite element mesh.

Figure 10a shows that the reference design case converges at 1.195 mm after 131 iterations, and the self-supporting design case converges at 1.105 mm after 287 iterations. That is, a performance sacrifice of 7.53% is made to form the self-supporting topology suitable for AM. Compared to the reference topology (Figure 10b), self-supporting features are observed in the optimized topology with Langelaar's AM filter (Figure 10c).

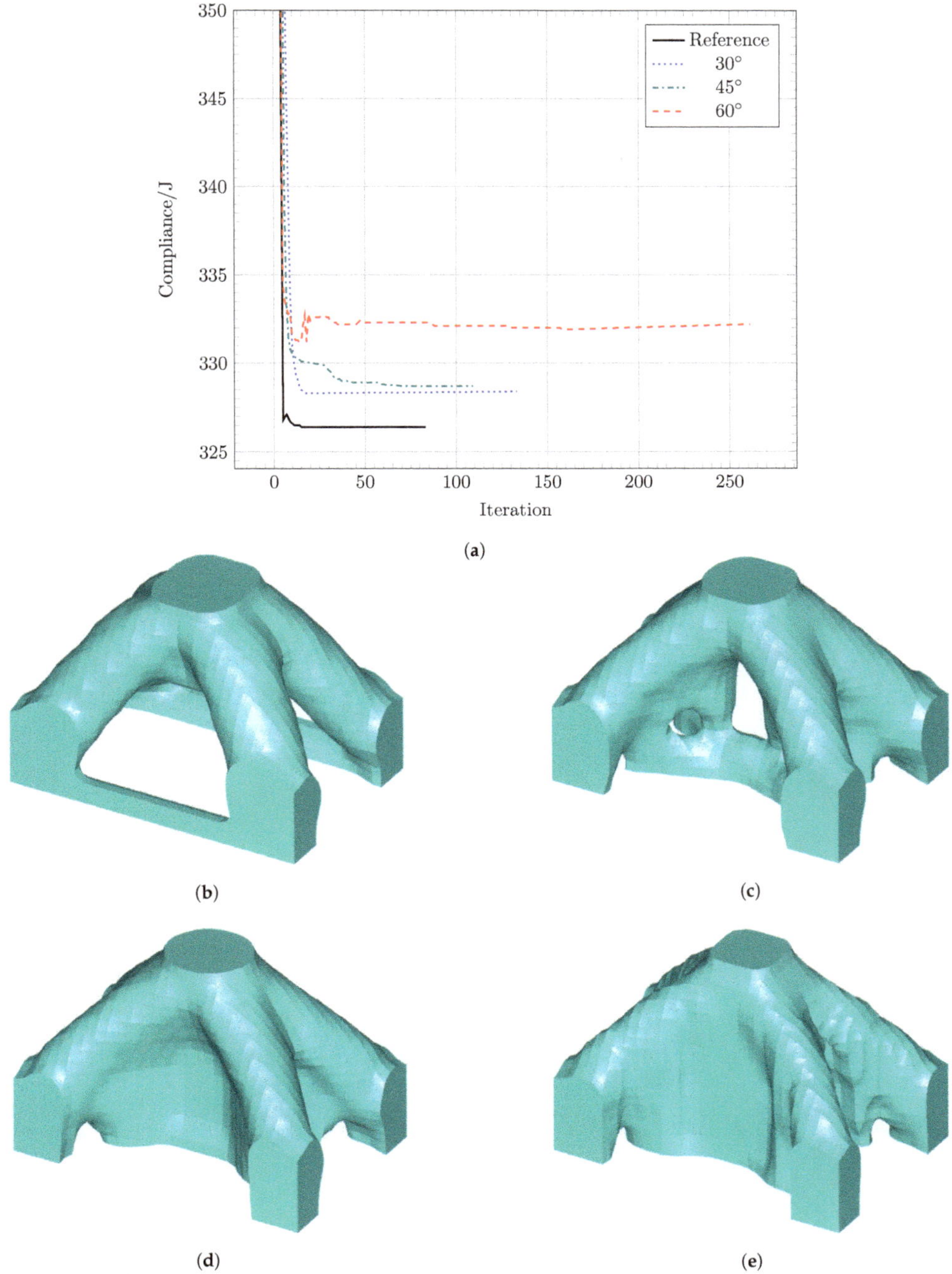


**Figure 8.** Convergence processes and optimized topologies with different critical overhang angles: (**a**) convergence processes, (**b**) reference, (**c**) 30°, (**d**) 45°, (**e**) 60°.

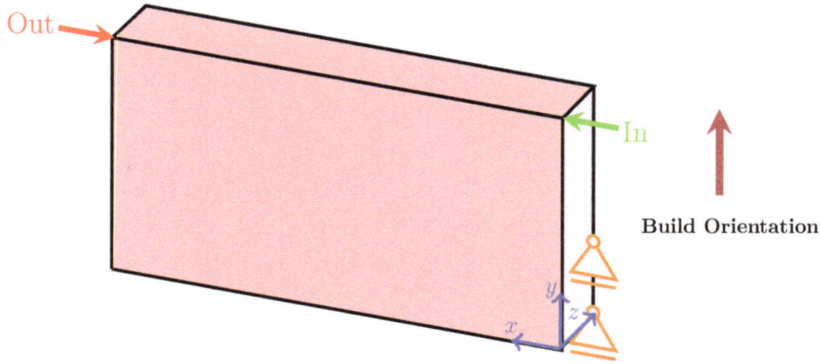


**Figure 9.** Design domain of force inverter design.

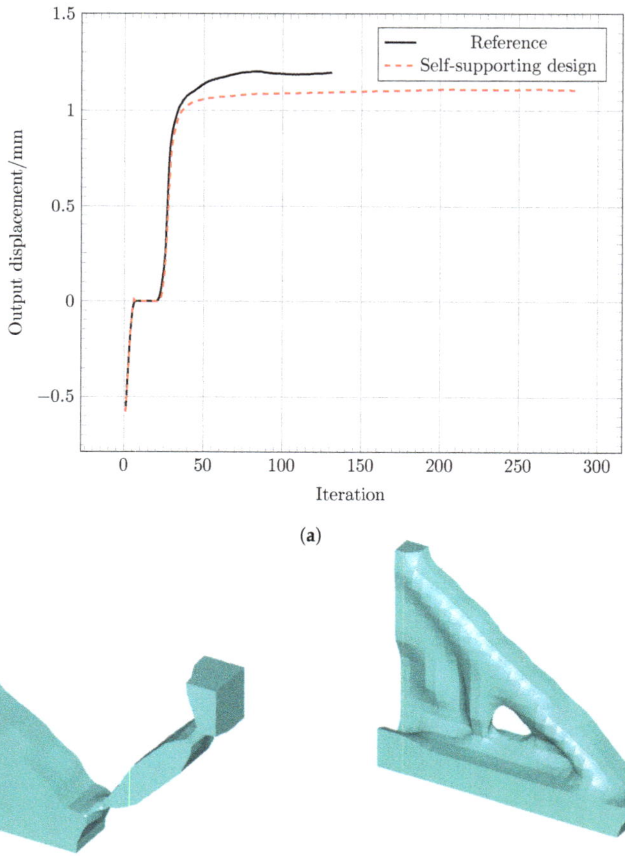


(a)

(b)

(c)

**Figure 10.** Convergence processes and optimized topologies solving the force inverter design case: (**a**) convergence processes, (**b**) reference, (**c**) self-supporting design.

The compliant mechanism design problem itself is more challenging than the compliance minimization design problem in terms of sensitivity analysis. The problems of self-supporting design for a compliant mechanism in this study demand more accurate sensitivity analysis than general compliant mechanism design problems. Therefore, the successful formation of the self-supporting topology for compliant mechanism design (Figure 10c) further validates the effectiveness of the sensitivity analysis method assumed in SEMDOT. More details on the mentioned sensitivity analysis method can be found in [16].

Based on the results presented from Sections 3.1–3.4, it can be concluded that the generation of a self-supporting design needs a longer convergence process than its corresponding reference design. The similar phenomenon was also observed by the authors when 2D self-supporting topologies were explored [33,34].

## 4. Conclusions

This paper integrates Langelaar's AM filter into SEMDOT—a newly developed element-based algorithm capable of forming smooth boundaries—to implement the smooth design of 3D self-supporting topologies. The obtained conclusions are summarized as follows:

- The integration of Langelaar's AM filter in SEMDOT is capable of forming converged 3D self-supporting topologies with smooth boundary representation.
- As Langelaar's AM filter is mesh-dependent, the fine mesh is recommended to form proper self-supporting designs when the target volume fraction is small.
- Different critical overhang angles can be explored using the combination of Langelaar's AM filter and SEMDOT.
- The effectiveness of the sensitivity analysis method adopted in SEMDOT is further validated using a 3D compliant mechanism design problem for self-supporting design.

As overhang angle and length constraints interact to determine buildability, the overhang length constraint should be considered in terms of obtaining self-supporting designs in the future. In addition, the experimental validation of the 3D self-supporting topologies obtained using the combination of Langelaar's AM filter and SEMDOT is required.

**Author Contributions:** Conceptualization, Y.-F.F., K.G., B.R., Y.W. and L.N.S.C.; methodology, Y.-F.F., K.G., B.R., Y.W. and L.N.S.C.; software, Y.-F.F.; validation, K.G., B.R., Y.W. and L.N.S.C.; formal analysis, Y.-F.F.; investigation, Y.-F.F.; resources, B.R.; writing—original draft preparation, Y.-F.F.; writing—review and editing, K.G., B.R., Y.W. and L.N.S.C.; visualization, Y.-F.F.; supervision, K.G., B.R., Y.W. and L.N.S.C.; project administration, K.G. All authors have read and agreed to the published version of the manuscript.

**Funding:** This research received no external funding.

**Institutional Review Board Statement:** Not applicable.

**Informed Consent Statement:** Not applicable.

**Data Availability Statement:** The data presented in this study are available on request from the corresponding author. The data are not publicly available due to privacy restriction.

**Acknowledgments:** The authors would like to thank Krister Svanberg for providing the MMA optimizer code. Yun-Fei Fu would like to thank Matthijs Langelaar, Delft University of Technology, for his invaluable guidance on his 3D AM filter, and Bruno Barroqueiro, University of Aveiro, for his help with the softmax approximation function.

**Conflicts of Interest:** The authors declare no conflict of interest.

## References

1. Thompson, M.K.; Moroni, G.; Vaneker, T.; Fadel, G.; Campbell, R.I.; Gibson, I.; Bernard, A.; Schulz, J.; Graf, P.; Ahuja, B.; et al. Design for additive manufacturing: Trends, opportunities, considerations, and constraints. *CIRP Ann.* **2016**, *65*, 737–760. [CrossRef]
2. Oh, Y.; Zhou, C.; Behdad, S. Part decomposition and assembly-based (Re) design for additive manufacturing: A review. *Addit. Manuf.* **2018**, *22*, 230–242. [CrossRef]

3. Izadi, M.; Farzaneh, A.; Mohammed, M.; Gibson, I.; Rolfe, B. A review of laser engineered net shaping (LENS) build and process parameters of metallic parts. *Rapid Prototyp. J.* **2020**, *26*, 2228–2236. [CrossRef]
4. Jiang, J.C.; Xiong, Y.; Zhang, Z.Y.; Rosen, D.W. Machine learning integrated design for additive manufacturing. *J. Intell. Manuf.* **2020**, 1–14.
5. Langelaar, M. An additive manufacturing filter for topology optimization of print-ready designs. *Struct. Multidiscip. Optim.* **2017**, *55*, 871–883. [CrossRef]
6. Langelaar, M. Combined optimization of part topology, support structure layout and build orientation for additive manufacturing. *Struct. Multidiscip. Optim.* **2018**, *57*, 1985–2004. [CrossRef]
7. Ghabraie, K.; Xie, Y.M.; Huang, X.D.; Ren, G. Shape and reinforcement optimization of underground tunnels. *J. Comput. Sci. Technol.* **2010**, *4*, 51–63. [CrossRef]
8. Ghabraie, K.; Chan, R.; Huang, X.D.; Xie, Y.M. Shape optimization of metallic yielding devices for passive mitigation of seismic energy. *Eng. Struct.* **2010**, *32*, 2258–2267. [CrossRef]
9. Liu, D.D.; Chiu, L.N.S.; Davies, C.; Yan, W. A post-processing method to remove stress singularity and minimize local stress concentration for topology optimized designs. *Adv. Eng. Softw.* **2020**, *145*, 102815. [CrossRef]
10. Nguyen, T.H.; Paulino, G.H.; Song, J.; Le, C.H. A computational paradigm for multiresolution topology optimization (MTOP). *Struct. Multidiscip. Optim.* **2010**, *41*, 525–539. [CrossRef]
11. Nguyen, T.H.; Paulino, G.H.; Song, J.; Le, C.H. Improving multiresolution topology optimization via multiple discretizations. *Int. J. Numer. Methods Eng.* **2012**, *92*, 507–530. [CrossRef]
12. Nguyen, T.H.; Le, C.H.; Hajjar, J.F. Topology optimization using the p-version of the finite element method. *Struct. Multidiscip. Optim.* **2017**, *56*, 571–586. [CrossRef]
13. Da, D.C.; Xia, L.; Li, G.Y.; Huang, X.D. Evolutionary topology optimization of continuum structures with smooth boundary representation. *Struct. Multidiscip. Optim.* **2018**, *57*, 2143–2159. [CrossRef]
14. Fu, Y.F.; Rolfe, B.; Chiu, N.S.L.; Wang, Y.N.; Huang, X.D.; Ghabraie, K. Topology Optimization of Continuum Structures using Smooth Boundary. In Proceedings of the 13th World Congress on Structural and Multidisciplinary Optimization-Continued, Beijing, China, 20–24 May 2019; pp. 448–454.
15. Fu, Y.F.; Rolfe, B.; Chiu, N.S.L.; Wang, Y.N.; Huang, X.D.; Ghabraie, K. Smooth topological design of 3D continuum structures using elemental volume fractions. *Comput. Struct.* **2020**, *231*, 106213. [CrossRef]
16. Fu, Y.F.; Rolfe, B.; Chiu, N.S.L.; Wang, Y.N.; Huang, X.D.; Ghabraie, K. SEMDOT: Smooth-edged material distribution for optimizing topology algorithm. *Adv. Eng. Softw.* **2020**, *150*, 102921. [CrossRef]
17. Huang, X.D. Smooth topological design of structures using the floating projection. *Eng. Struct.* **2020**, *208*, 110330. [CrossRef]
18. Huang, X.D. On smooth or 0/1 designs of the fixed-mesh element-based topology optimization. *Adv. Eng. Softw.* **2021**, *151*, 102942. [CrossRef]
19. Jiang, J.C.; Xu, X.; Stringer, J. Support structures for additive manufacturing: A review. *J. Manuf. Mater. Process.* **2018**, *2*, 64. [CrossRef]
20. Jiang, J.C.; Fu, Y.F. A short survey of sustainable material extrusion additive manufacturing. *Aust. J. Mech. Eng.* **2020**, 1–10. doi:10.1080/14484846.2020.1825045. [CrossRef]
21. Jiang, J.C. A novel fabrication strategy for additive manufacturing processes. *J. Clean. Prod.* **2020**, *272*, 122916. [CrossRef]
22. Liu, J.K.; Gaynor, A.T.; Chen, S.K.; Kang, Z.; Suresh, K.; Takezawa, A.; Li, L.; Kato, J.; Tang, J.Y.; Wang, C.C.; et al. Current and future trends in topology optimization for additive manufacturing. *Struct. Multidiscip. Optim.* **2018**, *57*, 2457–2483. [CrossRef]
23. Meng, L.; Zhang, W.; Quan, D.; Shi, G.; Tang, L.; Hou, Y.; Breitkopf, P.; Zhu, J.; Gao, T. From topology optimization design to additive manufacturing: Today's success and tomorrow's roadmap. *Arch. Comput. Methods Eng.* **2020**, *27*, 805–830. [CrossRef]
24. Fu, Y.F. Recent advances and future trends in exploring Pareto-optimal topologies and additive manufacturing oriented topology optimization. *Math. Biosci. Eng.* **2020**, *17*, 4631–4656. [CrossRef] [PubMed]
25. Langelaar, M. Topology optimization of 3D self-supporting structures for additive manufacturing. *Addit. Manuf.* **2016**, *12*, 60–70. [CrossRef]
26. Langelaar, M. Integrated component-support topology optimization for additive manufacturing with post-machining. *Rapid Prototyp. J.* **2019**, *25*, 255–265. [CrossRef]
27. Han, Y.S.; Xu, B.; Zhao, L.; Xie, Y.M. Topology optimization of continuum structures under hybrid additive-subtractive manufacturing constraints. *Struct. Multidiscip. Optim.* **2019**, *60*, 2571–2595. [CrossRef]
28. Mezzadri, F.; Qian, X. A second-order measure of boundary oscillations for overhang control in topology optimization. *J. Comput. Phys.* **2020**, *410*, 109365. [CrossRef]
29. Zhao, D.Y.; Li, M.; Liu, Y.S. A novel application framework for self-supporting topology optimization. *Vis. Comput.* **2020**, 1–16. [CrossRef]
30. Zhang, K.Q.; Chen, G.D. Three-dimensional high resolution topology optimization considering additive manufacturing constraints. *Addit. Manuf.* **2020**, *35*, 101224. [CrossRef]
31. van de Ven, E.; Maas, R.; Ayas, C.; Langelaar, M.; van Keulen, F. Overhang control based on front propagation in 3D topology optimization for additive manufacturing. *Comput. Methods Appl. Mech. Eng.* **2020**, *369*, 113169. [CrossRef]
32. Bi, M.H.; Tran, P.; Xie, Y.M. Topology optimization of 3D continuum structures under geometric self-supporting constraint. *Addit. Manuf.* **2020**, *36*, 101422.

33. Fu, Y.F.; Rolfe, B.; Chiu, L.N.S.; Wang, Y.N.; Huang, X.D.; Ghabraie, K. Design and experimental validation of self-supporting topologies for additive manufacturing. *Virtual Phys. Prototyp.* **2019**, *14*, 382–394. [CrossRef]
34. Fu, Y.F.; Rolfe, B.; Chiu, L.N.S.; Wang, Y.N.; Huang, X.D.; Ghabraie, K. Parametric studies and manufacturability experiments on smooth self-supporting topologies. *Virtual Phys. Prototyp.* **2020**, *15*, 22–34. [CrossRef]
35. Fu, Y.F.; Ghabraie, K.; Rolfe, B.; Wang, Y.N.; Chiu, L.N.S.; Huang, X.D. Optimizing 3D Self-Supporting Topologies for Additive Manufacturing. In Proceedings of the 12th International Conference on Computer Modeling and Simulation, Brisbane, Australia, 22–24 June 2020; pp. 220–223.
36. Barroqueiro, B.; Andrade-Campos, A.; Valente, R.A.F. Designing self supported SLM structures via topology optimization. *J. Manuf. Mater. Process.* **2019**, *3*, 68.
37. Ghabraie, K. The ESO method revisited. *Struct. Multidiscip. Optim.* **2015**, *51*, 1211–1222. [CrossRef]
38. Ghabraie, K. An improved soft-kill BESO algorithm for optimal distribution of single or multiple material phases. *Struct. Multidiscip. Optim.* **2015**, *52*, 773–790. [CrossRef]
39. Liu, K.; Tovar, A. An efficient 3D topology optimization code written in Matlab. *Struct. Multidiscip. Optim.* **2014**, *50*, 1175–1196. [CrossRef]
40. Svanberg, K. The method of moving asymptotes – A new method for structural optimization. *Int. J. Numer. Methods Eng.* **1987**, *24*, 359–373. [CrossRef]
41. Qin, Y.C.; Qi, Q.F.; Scott, P.J.; Jiang, X.Q. Determination of optimal build orientation for additive manufacturing using Muirhead mean and prioritised average operators. *J. Intell. Manuf.* **2019**, *30*, 3015–3034. [CrossRef]

*applied sciences*

MDPI

*Review*

# Geometrical Degrees of Freedom for Cellular Structures Generation: A New Classification Paradigm

**Ken M. Nsiempba [1,*], Marc Wang [1] and Mihaela Vlasea [1,2]**

[1] Multi-Scale Additive Manufacturing Lab, University of Waterloo, Waterloo, ON N2L 3G1, Canada; m384wang@uwaterloo.ca (M.W.); mihaela.vlasea@uwaterloo.ca (M.V.)
[2] Department Mechanical and Mechatronics Engineering, University of Waterloo, Waterloo, ON N2L 3G1, Canada
* Correspondence: kmnsiemp@uwaterloo.ca

**Abstract:** Cellular structures (CSs) have been used extensively in recent years, as they offer a unique range of design freedoms. They can be deployed to create parts that can be lightweight by introducing controlled porous features, while still retaining or improving their mechanical, thermal, or even vibrational properties. Recent advancements in additive manufacturing (AM) technologies have helped to increase the feasibility and adoption of cellular structures. The layer-by-layer manufacturing approach offered by AM is ideal for fabricating CSs, with the cost of such parts being largely independent of complexity. There is a growing body of literature concerning CSs made via AM; this presents an opportunity to review the state-of-the-art in this domain and to showcase opportunities in design and manufacturing. This review will propose a novel way of classifying cellular structures by isolating their Geometrical Degrees of Freedom (GDoFs) and will explore the recent innovations in additively manufactured CSs. Based on the present work, the design inputs that are common in CSs generation will be highlighted. Furthermore, the work explores examples of how design inputs have been used to drive the design domain through various case studies. Finally, the review will highlight the manufacturability limitations of CSs in AM.

**Keywords:** cellular structures; additive manufacturing; cellular design; hierarchical structures; lattice structures

**Citation:** Nsiempba, K.M.; Wang, M.; Vlasea, M. Geometrical Degrees of Freedom for Cellular Structures Generation: A New Classification Paradigm. *Appl. Sci.* **2021**, *11*, 3845. https://doi.org/10.3390/app11093845

Academic Editors: Marco Mandolini, Patrick Pradel and Paolo Cicconi

Received: 08 March 2021
Accepted: 21 April 2021
Published: 23 April 2021



## 1. Introduction

Cellular structures (CSs) are hierarchical materials that are composed of repetitive unit cells. CSs are an intrinsic part of nature; examples are found in bones, wood and seashells. Such materials are known to impart a balance between weight and strength. Over the past decades, CSs have gained a growing focus in design for industrial and biomedical applications. Such applications vary from structural components [1–3], where the stiffness and strength of a part can be controlled based on external data, to vibrational [4,5], where a well-selected CS can be used as a shock absorber in a car, as well as thermal [6,7] where CSs may be used to enhance heat transfer, or in mass transport in the case of scaffolds for implants [8].

The growing popularity of CSs correlates positively with the increasing popularity and adoption of AM processes, where a design is printed layer per layer. As such, AM fabrication enables users to achieve higher design complexity than traditional manufacturing methods [9,10]. For example, using AM technologies, porosity can be varied non-uniformly across the volume of a product to tailor one side of a part to be stiffer than the other using AM [11], without necessarily increasing the manufacturing cost. Using the layer-by-layer AM principle, it is possible to build micro-architected materials with controlled properties [12]. This review paper presents recent research advancements in CSs; more precisely, this work reviews how the advent of AM and Design for Additive Manufacturing (DfAM) have expanded the opportunities for manufacturing CSs and highlights the limitations and opportunities in fabricating such structures. As such, the focus of this work is twofold:

1. To review the design and geometrical considerations for CS and their classification based on unit cell feature properties.
2. To highlight the manufacturability limitations and opportunities in DfAM with CSs.

In recent years, a few other articles have focused on reviewing one specific or more aspects of AM cellular structures [13–19]. Other reviews have focused on specific types of porous structures, such as biodegradable structures for porous implants [20–22], Functionally Gradient Material (FGM) or Gradient Lattice Structures (GLS) [23,24]. Furthermore, some reviews have touched to some extent on all the aspects CSs [25]. This review will bring forth a new classification paradigm, in addition to highlighting the issues that are specifically related to additively manufactured CSs.

## 2. Design and Geometrical Consideration

### 2.1. Cellular Structure Design Classification

There are numerous classifications of CS that have already been established, such as the work by Tang and Zhao [26]. The basis for the classification, presented in the present work, is inspired by these existing classifications and attempts to further harmonize the available literature in this space. Tang and Zhao [26] have divided CSs into three structural design classes: foam structures, two-dimensional (2D) CSs referring to extruded 2D cells and three-dimensional (3D) CSs referring to structures that periodically repeat in 3D. They then described an emerging classification for 3D CSs, which is based on how ordered the CSs are, ranging from randomized to pseudo-periodic and periodic structures. The classification for this review work will consider both randomized and ordered classifications. Other classifications, such as the one by Hadi et al. [27], rely on the different types of design variables for CSs: patterns (referring to the characteristics of the minimum CS unit replicated), surface limits (referring to the size of the shape of the CS boundary), progressivity (referring to the variation of the CS unit cell's thickness) and conformity (referring to whether or not the CS unit cells are varied according to strict parameters). Many advances in the design of CSs have shown that there is a growing body of potential unit cells, with a push towards implicit modeling for the design of CSs. Therefore, in the present review, the CSs will not be classified based on their unit cells. In addition, some reviewers such as Tamburrino et al. [19] have divided CSs stochastic structures into open cells and closed cell foams. That concept also has its limitations, as there are examples of triply periodic minimal structures with a varying volume fraction, which can be seen as open on one end and completely closed on the other. Hence, it is more important to isolate the sources of morphology variation inside a CS rather than trying to classify them in different distinct, non-intersecting groups.

The present review challenges the current paradigm, which aims at grouping CSs into multiple distinct and non-overlapping groups. Instead, the current approach identifies the main geometric degrees of freedom (GDoF) that come into play when designing a CS.

It is important to first define some concepts that will recur throughout the article. The design space refers to the volume that contains the CS (the object being lightweight), and the Representative Volume Element (RVE) refers to the smallest element that is replicated throughout the design space. There are multiple types of RVEs and some of the most famous will be discussed later in Section 2.2. The RVE can vary throughout the design space in many ways: it can be transformed (scaled or rotated), its volume fraction can be changed and its topology/type can also be changed. The RVE can be modified to optimize the design space as a response to multiple factors: based on the boundary of the design space (or the surface of the design), based on the loads that are applied to the CS and based on random or external factors (such as trying to replicate the density of bones). All of those have been considered when creating Table 1.

**Table 1.** Cellular structure classification based on geometrical degrees of freedom and representative examples.

| | Change of RVE Morphology | Volume Fraction Change of RVE | Scaling & Rotation of RVE |
|---|---|---|---|
| Boundary induced variation | [28] | [29] | [29–36] |
| Load induced variation | [37] | [30,37–49] | [34,38–40,50–53] |
| Random and external sources induced variations | [54–56] | [55,57,58] | [33,55,59–64] |
| No variation | [4,6,11,65–79] | | |

The RVEs inside the CS may also vary based on both the loads and the boundary conditions, whereas for other CSs, their RVEs may be transformed and changed based on imposed volume fractions. Thus, certain case studies can appear in multiple entries on the table. This is the case for the work of Tang et al. [30], wherein the authors have generated a strut-based CS that considers the boundary of the design space where the strut-based RVE are then modified towards the surface of the part. However, the thickness of the struts is varied based on the load applied to the design space. That paper, therefore, appears in two locations in the CSs classification table. It is important to notice that several design paths can lead to the same design result. The purpose of the summary table is thus not to show that every entry is unique, rather to showcase all the design freedoms the designer has access to.

Another important thing to note is that it is extremely hard to create a classification that is exclusive; no matter the classification, there will always be a CS that can belong to multiple entries in any classification table. For example, the RVE can be changed in multiple ways and for multiple reasons. There are also varying degrees of randomness that can be attributed to a CS—some CS arrangement can be random in some locations and ordered in others. Table 1 shows the GDoFs that go into making a CS. In addition, the GDoFs that were selected to create the table did not include the type of RVE. With the advent of implicit modeling, there are more and more RVEs that emerge as design solutions, making a classification based on RVEs challenging to upkeep. The next sections will focus on some of the most popular and emerging RVEs in literature.

### 2.2. *Overview of Representative Volume Elements*

The smallest replicated unit or the representative volume element (RVE) alludes to the smallest unit that is replicated throughout the design domain or in a sub-region of a design domain; such RVEs are also referred to as unit cells. Below are the different existing types of RVEs or unit cells.

#### 2.2.1. Strut-Based

Struts-based unit cells are RVEs with a specific arrangement of struts. Such examples of unit cells are presented in Figure 1. An emerging type of investigated strut-based unit cells are the ones used in bi-auxetic structures, from the Greek word "auxetos" meaning "that may increase". Bi-auxetic structures are structures with a negative Poisson's ratio [80], which means that when stretched in one direction, the structure expands in all other directions, whereas when compressed, the opposite is expected to happen. Auxetic materials have traditionally been fabricated through the use of foams; with the advent of AM such structures can be better tailored based on the intended application. Auxetic materials can be found in a few applications such as shock and sound-absorbing materials for vehicles and aircraft [80]. Other works, such as by Queheillalt and Wadley [81], have studied strut-based unit cells with hollow struts; hollow struts are not widely used as of now in the AM domain. The manufacturing of hollow struts still poses a problem in AM processes due to challenges in removing trapped powder or support materials within the struts.

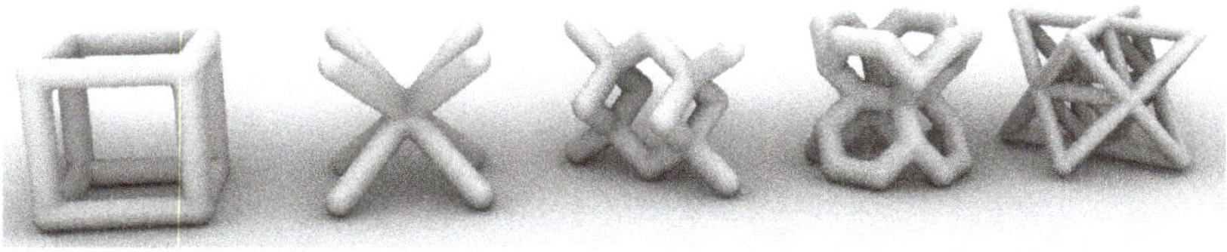

**Figure 1.** Strut-based unit cells. From left to right: grid, X (or BCC), diamond, vintiles, octet. Made using IntraLattice [82].

Sometimes, the strut-based unit cell can just be one strut. For the classification in this review, when a CS is made of a truss network but does not contain any clear strut-based unit cell (an arrangement of strut replicated throughout the design space) then the RVE is considered to be a strut. An example is given by Smith et al. [39], whereby they describe a design methodology based on layout optimization where nodes distributed across the design space are interconnected by potential members. Those final members of the structures are chosen to minimize the total structural volume. Following this step, the size optimization algorithm uses the updated layout as the ground structure. Finally, in the last step, every member is resized to ensure that none of the members buckle [39].

### 2.2.2. Extruded 2D Cells

Generally, 2D cellular structures can be described as CSs, where the repeated units are extrusions of polygons. The type of polygon that is extruded defines the type of extruded 2D CSs. Common extruded 2D CSs include Kagome, square, honeycomb [83], and triangular, as can be seen in Figure 2. The recent work from Ongaro offers an in-depth review of common extruded 2D CSs, as well as their mechanical simulation [69].

Furthermore, as it is shown by Hu and Wang [84], auxetic structures can also be made from extruded 2D cells. However, in that case, the expansion/contraction can only be done in the 2D plane perpendicular to the direction of the extrusion.

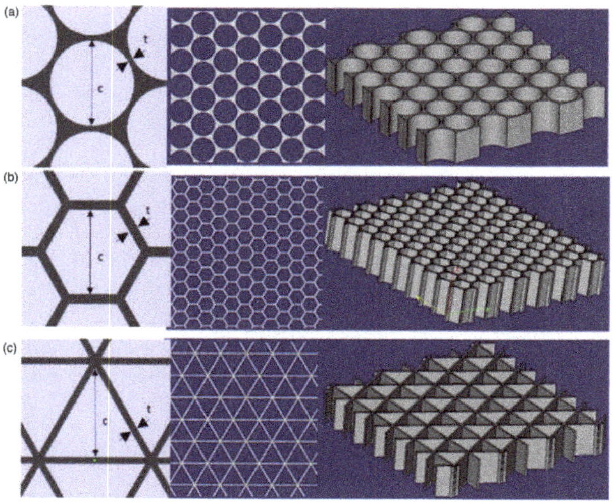


**Figure 2.** Extruded 2D unit cells with visualizations capturing a three-unit cell, a top view, and an isometric view of each configuration: (**a**) circular (c = 9.8 mm, t = 0.6 mm); (**b**) hexagonal (c = 3.125 mm, t = 0.6 mm); and (**c**) triangular (c = 9.14 mm, t = 0.6 mm) [85].

### 2.2.3. Triply Periodic Minimal Surfaces

Minimal surface areas have been researched for more than 200 years. A minimal surface area is a surface with the smallest area bounded by a contour [86]. Triply Periodic

Minimal Surfaces (TPMS) are minimal surfaces in all three x, y and z directions.They are defined by an implicit function such as $f(x,y,z) = t$ describing the locus of points at which the function takes the value $t$. To generate a solid structure or to describe a volume, the equation is replaced by an inequality, for instance $f(x,y,z) < t$. The function is used to define one unit cell, and the unit cell is replicated throughout the structure. The constant $t$ can be adapted regionally to allow different volume fractions [13] (see Figure 3).

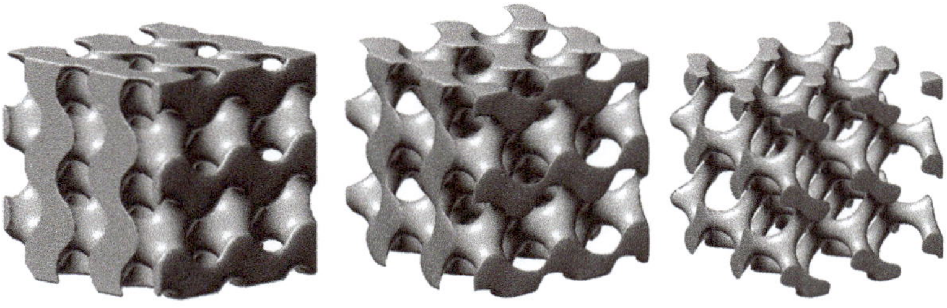

**Figure 3.** Gyroid structure for varying threshold, t = 0, t = 0.5 and t = 1.

More specifically, a change in $t$ will result in a different volume fraction. Since TPMSs are defined by functions, the transition between two unit cells is smooth, which means the structure has fewer stress concentrations. Hao et al. [87] have described an image-based algorithm and an implicit function to properly define a unit cell. Their unit cell is easy to discretize using finite element and it also minimizes the use of overhangs. Figure 4 shows a list of typical TPMS cells. Sometimes, there are simiarities between strut-based unit cells and TPMS. For example, Zhao et al. [88] have highlighted that controlling the volume fraction of the gyroid structure, one can get something close to the BCC lattice structure.

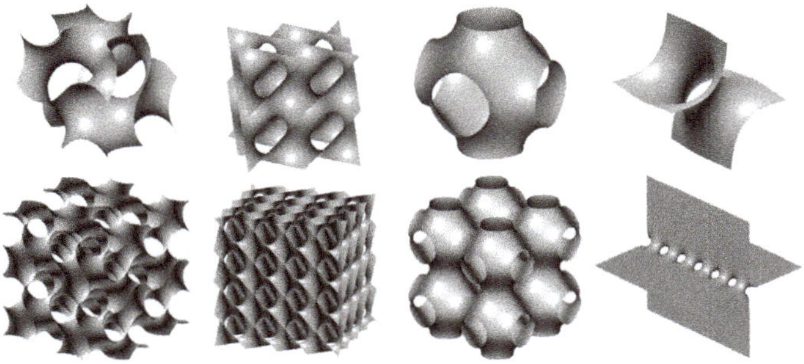

**Figure 4.** Examples of TPMS; from left to right—Schoen's Gyroid, Schwarz's Primitive, Shwarz's Diamond and Sherk's first surface. In each case the bottom picture shows the results when the cells are duplicated.

2.2.4. Unit Cells Obtained through Topology Optimization or Other Numerical Methods

According to Coelho et al. [73], TPMS-based RVEs are site-dependent, and it might be difficult for TPMS-based unit cells to balance scaffold elasticity and permeability. In contrast, Topologically Optimized Microstructures (TOMs) are microstructures that have been topologically optimized and smoothed; such microstructures can be deployed as the RVEs for CSs. Coelho et al. [73] suggest that such TOMs are advantageous over TPMS as they allow for anisotropic optimal design (Figure 5). Another way to obtain a TOM is described by Hollister [89]. The article shows one way to create libraries of unit cells to allow hierarchical design. There are two ways to construct those boundaries, either by image-based design approaches, or using CAD software. In the study, they rely heavily on imaging techniques to create the scaffold architecture with 3D anatomic defects. Globally, their process goes as follows: the microstructure is optimized for maximum permeability and the effective modulus must match the human bone tissue. Subsequently, they use imaging techniques to design the scaffold exterior. Following this, the global anatomic and architecture design are integrated. They then use AM to fabricate the design [89]. Similarly to Coelho et al. [73], Takezawa et al. [74] have used topology optimization to optimize the microstructures of parts.

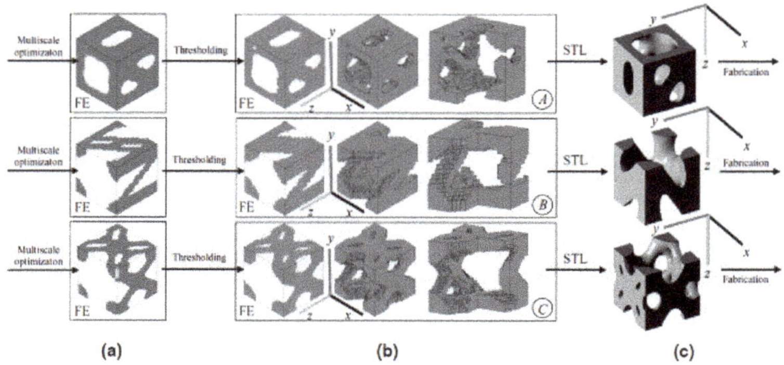


**Figure 5.** Unit cell topologies A, B and C from design to fabrication: (**a**) solutions as they were obtained via topology optimization on the top of the finite element mesh (20 × 20 × 20 or 30 × 30 × 30 of 8-node hexahedral isoparametric elements); (**b**) microstructures after thresholding, 0–1 design, solid phase and cut views; (**c**) conversion into STL format. Figure by Coelho et al. [73].

### 2.2.5. Origami-Inspired Materials

An emerging class of RVEs are based on origami-inspired materials. Such designs are comprised of stacked layers of material [90]. Figure 6 gives an example of how they are parametrized.

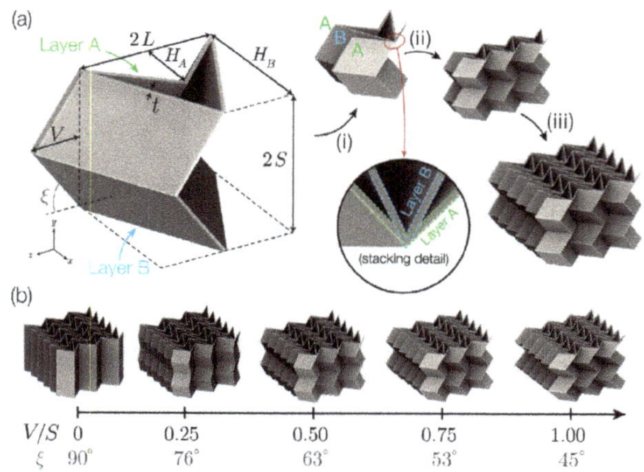


**Figure 6.** (**a**) Parameterisation of the unit cell of the stacked Miura-ori design, and (i)–(iii) the patterning of this unit cell to construct the cellular solid. A and B denote adjacent Miura-ori layers in the stack. (**b**) Variation of the stacked origami structure with increasing fold angles $V/S$ . Figure by Harris and McShane [90].

### 2.2.6. Void RVE

In some cases, the RVE is not the solid that is being stacked to create CSs, rather the RVE constitutes the solid material that is being removed to create the CSs. Some have discussed the need to remove random spheres in order to generate the CS [91]; this concept will also be discussed in further detail in Section 2.5.

2.2.7. Further thoughts on RVEs and CSs

It is important to mention that not all porous structures are CSs. CSs are obtained by the repetitiveness of an RVE, whether it is a solid RVE or a void RVE. Some examples of porous structures that are not considered to be CSs are structures that may result from topology optimization or other mathematical operations. Examples of other porous structures that are not CSs include Spinodoids [92] and structures that are the result of the thresholding of Gaussian fields Hyman and Winter [93]. Furthermore, many randomized structures seem to use Voronoi cells. A Voronoi cell can be observed in different ways. The RVE might be an edge of the cell, in which case the RVE is rotated and scaled throughout the design space. In some other instances, the RVE can be a Voronoi cell, which would mean that an RVE is defined by its number of vertices; in that case, since there are many polygons (in 2D) with different dimensions but the same number of vertices, the RVEs are varied (in type) but are also scaled or rotated.

### 2.3. RVE Variation Methods

2.3.1. RVE Morphology Variation

A CSs can vary in terms of the type of RVE (unit cell) present throughout the design space. The first column of Table 1 is dedicated to illustrating cases where there are RVE morphology variations. For example, Yang et al. [55] have proposed a method to vary the morphology of the CSs based on the location within the design space (see Figure 7). The authors were able to ensure a continuous transition between morphologies using control points. Varying the morphologies of the RVE based on the location can pave the way toward mimicking natural structures such as tissues by increasing the degrees of freedom required to locally control the properties of the material.

Jin et al. [56] have presented an interesting way of changing the morphology by superimposing two TPMSs. Varying the volume fraction of one TPMS based on the location is like changing the morphology of the resulting structure.

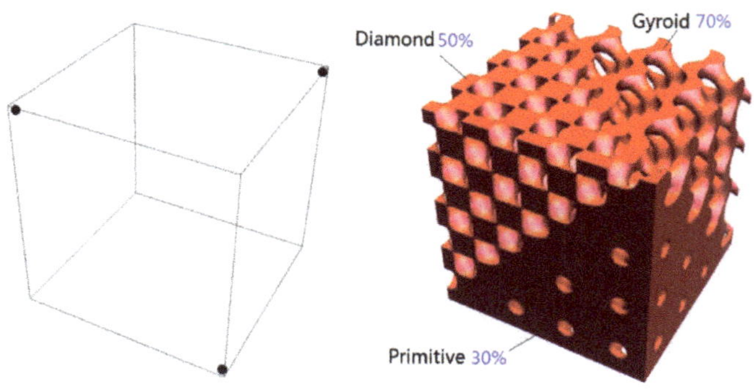


(a) (b)

**Figure 7.** RVE gradient in terms of morphology types and porosities. (**a**) Three control points. (**b**) The 3D structure integrates type Primitive of 30% porosity, type Diamond of 50% porosity, and type Gyroid of 70% porosity [55].

2.3.2. RVE Volume Fraction Variation

Volume fraction can also be varied to create porous Functionally Graded Materials (FGMs). In the case of a strut-based unit cell, volume fraction refers to the relative thickness of the struts. Figure 8 shows the difference between a structure with a constant volume fraction and a gradient volume fraction. The second column of Table 1 is dedicated to highlighting examples of research where the volume fraction of the RVE is changed throughout the CS to result in non-homogeneous properties.

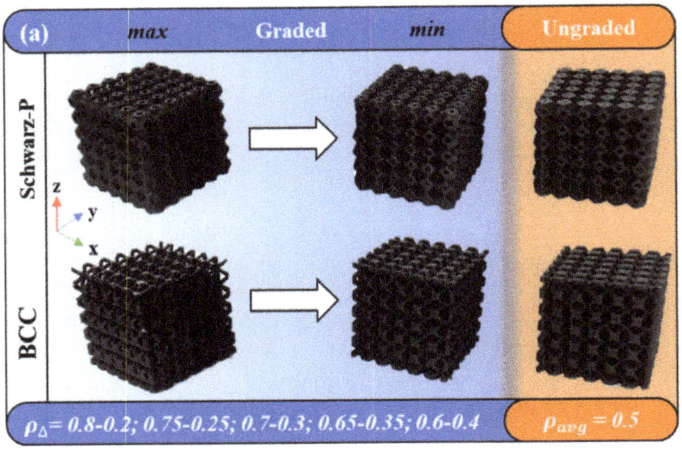


**Figure 8.** RVE gradient in terms of unit cell density grading across the CS [29]—$\rho_\Delta$ is the difference between the maximum and minimum densities, $\rho_{avg}$ is the average density.

Numerous papers have studied how to map the variation in the volume fraction (or conversely density) of CSs based on the gray scale gradient obtained from structural topology optimization algorithms. In such efforts, the typical goal is to ensure that the volume fraction of the RVE in the volume neighborhood maps to the topology optimization results in that region [11,30,51,94,95]. One such example is in the work by Alzahrani et al. [11], whereby they propose a method where the relative density is adjusted using the strut size in the structure. Similarly, Plocher and Panesar [29] have constructed multiple examples of CSs for which they have varied the density in a linear manner based on topologically optimized design spaces.

### 2.3.3. RVE Transformations

The third column of Table 1 is dedicated to cases where the volume fraction of the RVE undergoes scaling or rotation mathematical transformations. An example is shown in Figure 9. In some cases, RVEs can undergo scaling, rotation or any other type of geometric transformation, while still being recognizable. Plocher and Panesar [29] have also constructed multiple CSs for which they have varied the scale. There are numerous cases that have been included in that column that might seem a bit more subtle. For example, in many of those case studies, the unit cell that is being scaled and rotated throughout the design space is simply a strut [40,50].

The next subsections will tackle the second question—"why or according to what factor does the RVE vary throughout the design space?".

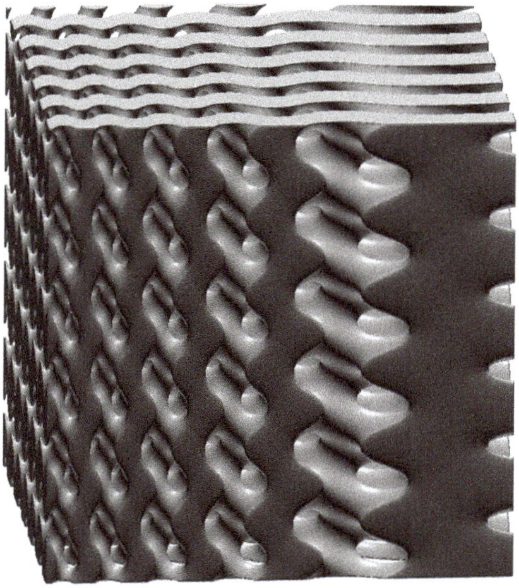

**Figure 9.** RVE transformation of a gyroid structure with scaled unit cells.

## 2.4. RVE Variation Factors

### 2.4.1. RVE Variation Based Boundary Information

Sometimes, the CS unit cells can vary based on the boundary of the design space; this is the case for conformal CSs. For conformal CSs, Tang et al.'s approach consists of using conformal CSs to create a lightweight structure [30]. There are different advantages in using conformal CSs, most importantly, the non-uniformity can potentially increase the strength of the CS.

Figure 10 shows the difference between a uniform and a conformal strut arrangement for a lattice structure. Other authors such as Nguyen et al. [31] have written about conformal CSs; they suggested a novel method of generating conformal CSs by offsetting the surfaces of the design space and dividing the result into tetrahedra. The same method had been used by Engelbrecht et al. [96]. Another interesting application of conformal CSs structures was showcased by Brennan-Craddock et al. [97]. The team used CSs for body protection and investigated their ability to absorb energy from the impact. The conclusion was that foams are suitable for energy absorption use. They defined four types of what they call conformal methods. In their framework, only the swept and the meshed structures correspond to the definition of conformal CSs. This framework is also used in this paper. As the name "conformal" implies, in the trimmed method the struts are cut at the boundaries. The "swept method" requires two surfaces, often parallel, in which unit cells are swept in the normal direction between the two surfaces. The mesh method is more robust as it does not require any specific surface configuration, nor does it require them to be parallel. The main disadvantage of the trimmed method is that unit cells near the boundary are weakened. To remedy that, the same authors have suggested wrapping a skin around the design space. Furthermore, Melpal [98] discussed a conformal CS technique in his PhD thesis; the technique consists of slicing design spaces (in STL format), taking some of the resulting intersection points and then generating conformal structures by offsetting the original design space [98]. An accurate example is given by Yang et al. [28]. Another example of the volume fraction varied between two boundaries is in the work by Plocher and Panesar [29]; a custom example of that case was made in Figure 11.

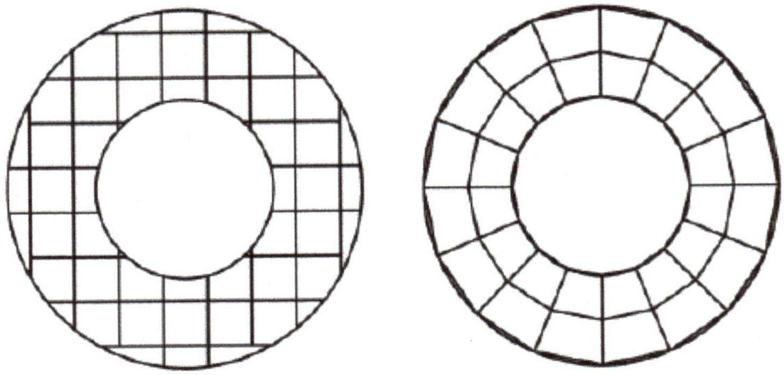

**Figure 10.** Representative volume element variation based boundary information—illustration of the difference between uniform and conformal lattice structure [99].

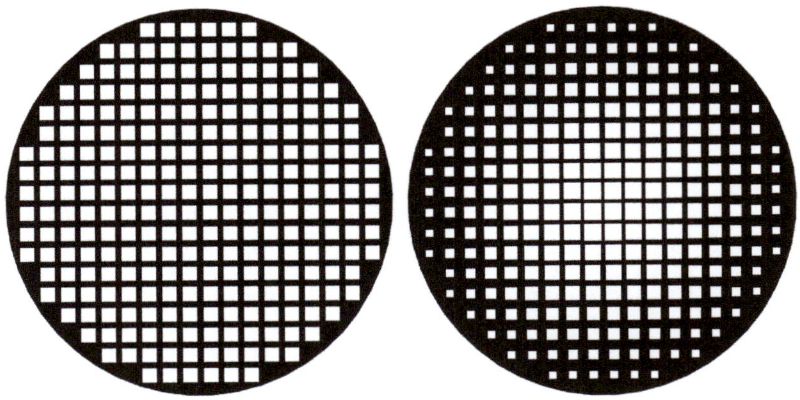

(a) (b)

**Figure 11.** Representative volume element variation based on boundary information—illustration of the difference between a cellular structure cell sizes are (**a**) not influenced by the boundary and (**b**) a cellular structure in which the cells' size vary based on their distance to the boundary.

2.4.2. RVE Variation Based on Load Conditions

In the context of CSs, the designer typically deploys such architectures towards structural performance customization. One example is the use of CSs in the frame of structural topology optimization to balance lightweight and compliance requirements (see Figure 12).

When it comes to "load adaptive" cellular structures, the shape of the cellular structure wireframe is based on a response to loads applied on the surface. Chen [45] explains how space warping can be used to properly distribute the load. For example, in order to maximize strength, the CS will be deformed using a warping function. Another example is given by Reinhart and Teufelhart [50]. The authors developed a method that uses stress fields in order to create CSs. Robbins et al. [100] described an example of a multi-step optimization procedure where geometry is topologically optimized and then discretized using a hexahedral mesh. The mesh is then populated with unit cells.

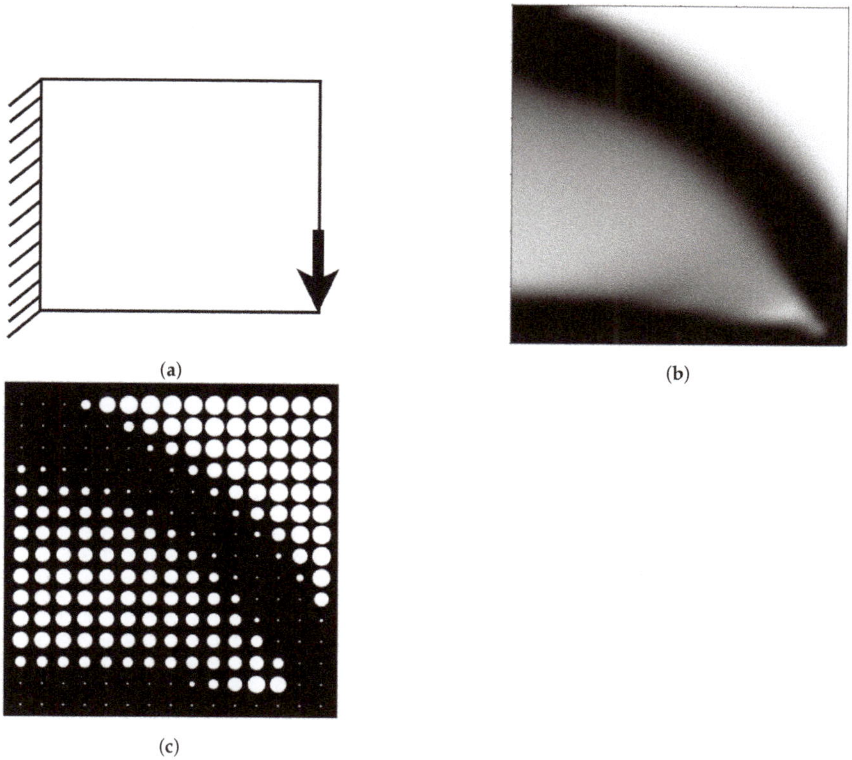


**Figure 12.** Representative volume element variation based on structural load conditions starting with (**a**) design space with boundary condition(s), followed by (**b**) topology optimization results, then finally (**c**) generated cellular structure.

A similar concept can be applied in other types of "loads". Another type of assumed "load" is permittivity for dielectric structures, as mentioned by Larimore et al. [47]. The idea is to create dielectric structures with spatially varying electromagnetic properties via AM. The unit RVE structures obtained by the researchers are made based on a curve wrapped within a rectangle. The curve is wrapped into multiple cells, such that when they are combined, they are able to change the permittivity of the cells by changing how dense the curve is wrapped at each of the cell's locations (see Figure 13).

An example of stiffness maximization using lattice structure is given by Alzahrani et al. [11]. In their paper, they have redesigned a micro aerial vehicle, used for surveillance in hazardous areas. They have done so using a relative density mapping. Their technique consists of the following steps. First, they analyze the boundary conditions as well as the loads acting upon the vehicle. Then, they topologically optimize the part via the SIMP method. Afterward, they select the type of microstructure required. Finally, they fill each element on the part with a unit cell. The density of the element is what dictates the size of the unit cell struts. Furthermore, a minimum strut thickness can be imposed in order to properly fill every element. This addresses the problem of discontinuity in topologically optimized parts caused partially by elements with zero density.

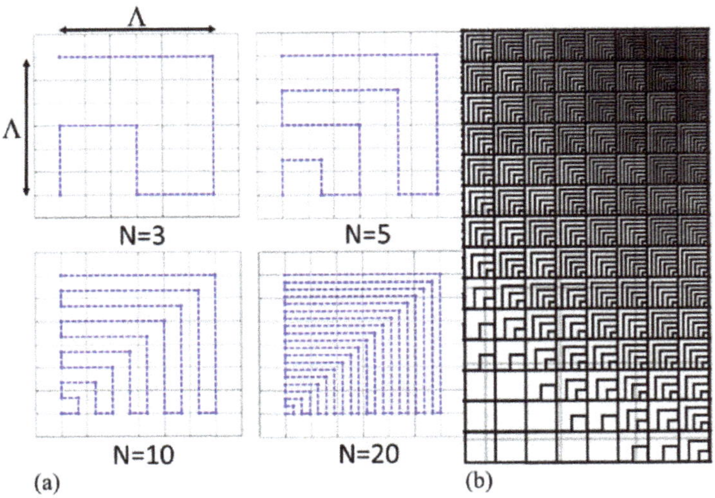


**Figure 13.** Representative volume element variation based on permittivity conditions. (**a**) Illustration showing examples of the space-filling curve geometry used to create spatially varying permittivity distributions. By varying the order N, the user can tune the local volume fraction of printed material and thus the local effective permittivity. (**b**) Illustration showing how the unit cell geometries can be connected end-to-end resulting in a single continuous curve with spatially varying effective EM properties [47].

The CS parameters (or implicitly the local RVE properties) can be changed in order to meet a specific optimization objective. One of those objectives is thermal applications. In the article written by Seepersad et al. [7], the authors show an example of a two-stage optimization of a CS. The first optimization is meant to reach certain customized structural properties and the second optimization to improve heat transfer without affecting the structural properties of the part. To do a multifunctional optimization, the typical force equation was replaced with the following: $K \cdot D = F + G$, Where $K$ is the global stiffness matrix for $N$ elements stiffness matrices frame elements, $D$ is the vector of global displacements, $F$ is the vector of applied nodal loads and $G$ is the vector of loads that account for thermal heating. The result has revealed itself to be promising toward structural and thermal multi-objective design optimization.

The article by Liu et al. [37] shows a rare example of varying the RVE (unit cells) using topology optimization results. The team has first constructed a unit-cell library in which they have analyzed the connectivity of the cell as well as some of their mathematical properties. In parallel, they have topologically optimized a structure and they have filled each element with a unit cell based on the density of the element. The volume fraction of the struts can then be optimized as well.

#### 2.4.3. RVE Variation Based on Random and External Sources

Random and external sources are grouped together because they both have little to do with the geometry of the part, nor with what the part is going through. Many definitions have been given to randomized CSs; however, in this article, the following definition is given: a CS is called randomized if a designer goes through the process of designing a specific CS twice with the same inputs without any guarantee that the resulting structure will be exactly the same.

When generating a randomized CS, one might not be able to predict the exact layout of the material; however, it can be possible to approximately predict the stiffness of a randomized CS. This can be done through the global or localized control of porosity or conversely, density. An example can be shown in Figure 14. There are many applications of randomized CSs within the medical field, one of them is described by Fantini et al. [101]. The authors have developed a method to design biomimetic scaffolds using a Voronoi diagram. In the article, they describe an algorithm that takes as inputs the information from the scan of a patient's bone, the desired porosity, and a target pore size. Then, based on the scattered points within a volume, the proposed algorithm creates Voronoi cells. The edges of the Voronoi cells are then thickened. The resulting CS will mimic the structure of the human bones. Their process is randomized because when generating the points or the seeds of the Voronoi cells, the user does not precisely control the location of those points but they can control the distribution or density of those points inside the design space. Similar work is done in Brackett et al. [51], where they use an error distribution method to map a Voronoi diagram to a grayscale image. Almonti et al. [54] have also proposed an approach for a completely customized structure using the Voronoi tessellation of a specific region to realize cellular structure typical of metal foams. In their case, the random change affects the topology of the cells.

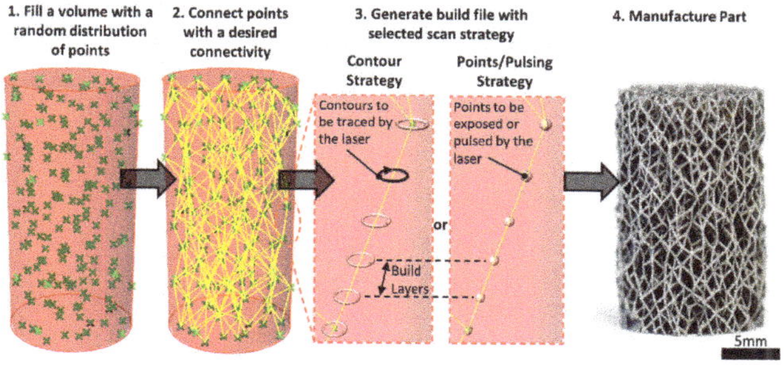


**Figure 14.** Manufacturing workflow from specimen design to build file generation to manufacture. Note: 1–3 are simplified for clarity and not to scale [63].

In the work by Savio et al. [46], the authors have classified cellular structures for AM in the biomedical field; they have classified the conformal lattice in the pseudo-random category and Voronoi in the random category in a few instances. A Voronoi diagram is not necessarily a randomized CS, it all depends on how the seeds are generated, if the seeds have a fixed position then the Voronoi diagram will always be the same; hence, it cannot always be classified as random. The main difference with Voronoi diagrams is that the design variables are the seeds' position and it is easier to generate random points than to directly generate randomly placed unit cells. In the latter case, one has to ensure connectivity, which is an issue that does not occur when working directly with a Voronoi diagram. According to Martínez et al. [102], the absence of a regular structure (for example, the stacking of more complex strut-based unit cells) affords for a simple approach to grade the foam geometry. To create randomized shapes, the vertices of existing unit cells can be varied in a random fashion, as is the case for Zhang et al. [61]'s work. A similar method is captured in the work of Reis et al. [62]. They create a network of struts that it subtracted from an original design space. This shows that sometimes, the unit cells are not the solid, but the voids.

Furthermore, the variations in a CS are not in response to a load nor to a boundary, but rather to an external source such as the CT scan of a part. For example, in the work by Cadman et al. [58], the authors have examined the microstructure of cuttlebone (bones of cuttle fish) and they have extracted properties such as the local volume fractions to try and reproduce the properties of the cuttlebone. Other sources that can induce a variation of unit cells often involved CT scans. The field of biomimicking contains a number of such examples.

As alluded to previously, the authors acknowledge that they can be multiple factors combined that result in RVE variations throughout a CS. For example, in the work by Tang and Zhao [103], the authors first used the boundary of the shapes to design a tetrahedral mesh and then, based on the stress within the part, they have been able to dictate the thickness of a strut. Similarly, the work by Martínez et al. [102] illustrates CS based on both random and load response. In this work, points (seeds) are scattered randomly, but the density of points follows a probability that is related to the level of localized stress within the parts.

Lim et al. [34] have used a tetrahedral mesh from FEA. Tetrahedral mesh vary based on the boundary of the structure. However, the authors have changed the way the density of strut is based on the stress level. Tetrahedral mesh is an example of boundary-induced cell transformation because the arrangement of the strut (the RVE in that case) is affected by the boundary of the shape.

### 2.5. CS Generation

This section will go over how one can generate and store CS digital information. There are many types of CS generation methods. However, only three of them will be discussed here: voxels, Constructive Solid Geometry (CSG), and implicit modeling.

#### 2.5.1. CS Generation via Voxel-Based Approaches

Generally, CSs are built by populating a design space with existing cells, as seen in Figure 1. Voxels are a relatively new concept, but an intuitive manner of representing solids. Each solid is created by adding small cubes in 3D (or voxels). The main disadvantage of this method is the computational cost, discretizing a shape accurately requires a high number of voxels [104].

As an example, in the work by Aremu et al. [104], the authors use voxels to generate CSs. The target geometry is first voxelized, as shown in Figure 15, then a grid was mapped onto the domain and trimmed, as shown in Figure 16. The main advantage of this method is that trimming and filling gaps becomes trivial when using a voxel representation. Unfortunately, the voxel representation cannot exactly represent any curved surface; therefore, staircases arise around such features. This problem can be minimized by using smaller voxels (i.e., higher resolution); however, this increases the computational cost. Another issue is that voxel representation is not well supported in the current CAD software and AM ecosystem. Most CAD software or AM build environments only accept geometry using Boundary REPresentation (BREP) or CSG. Unfortunately, the conversion from voxel to standard CAD file format nulls out the computational cost advantage of using voxels. When converting from voxel to mesh, the resulting STL file is extremely large as well [105].

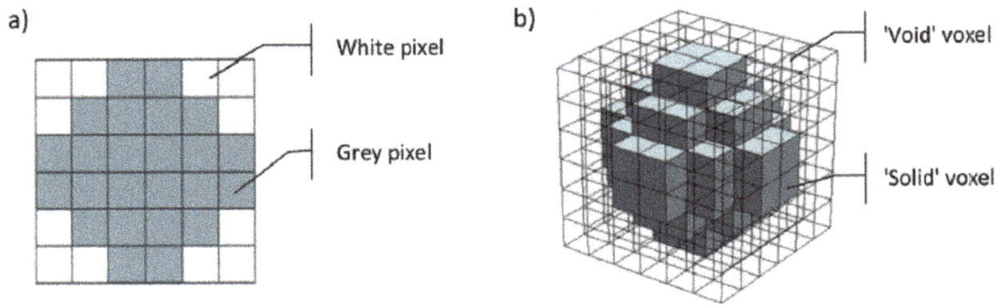


**Figure 15.** Illustration of cellular structure generation via voxel-based approaches. (**a**) Low-resolution pixel image of a circle. (**b**) Similarly low-resolution voxel model of a sphere [104].

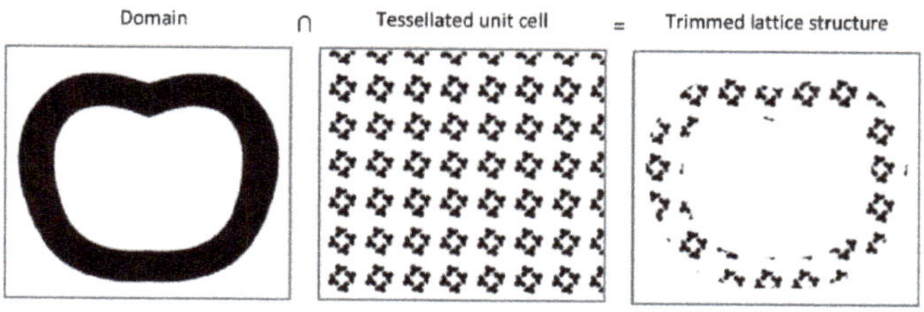


**Figure 16.** Generation of a trimmed structure slice—analogous to a Boolean intersection [104].

2.5.2. CS Generation via Constructive Solid Geometry

Constructive solid geometry (CSG) uses a sequential approach to represent a complex solid. CSG performs boolean operations on a set of primitives (cones, prisms, spheres) and stores the operations in a tree, as seen in Figure 17. A common method to represent a lattice structure using CSG is by creating CSs via representing the lines as cylinders and the nodes as spheres [4] and then joining them. Many problems arise from this method. First, the "join" operation can take a long time; this is because the only primitives that can be joined at once are the ones that are in contact. The second problem is the lack of smooth transition at the junction of the primitives. Eventually, when the structure is printed, the rough transition may result in stress concentrations. One simple way that has been found to overcome this challenge is to represent each node as a sphere where all the cylinder tips lie inside, as illustrated in Figure 18. Consequently, a boolean operation can be done to merge all cylinders and spheres into an RVE or a collection of RVEs into a more complex CS. Another way is to calculate the intersection between the cylinders and trim them such that they are flush with each other and then fill any gaps, as illustrated in Figure 19 [105,106].

In an effort to tackle the issue of nodes connection, Goel and Anand [42] have suggested using an additional unit cell. Overall, their methodology goes as follows: they perform topology optimization, then they use a mapping function, they then populate the space with Functional Gradient Lattice (FGL). The interesting part of their work is how they connect the different unit cells; they create a different unit cell to take care of the connection between the regular unit cells. That connection is made using B-spline principles, which comes in handy when the initial unit cells have different diameters.

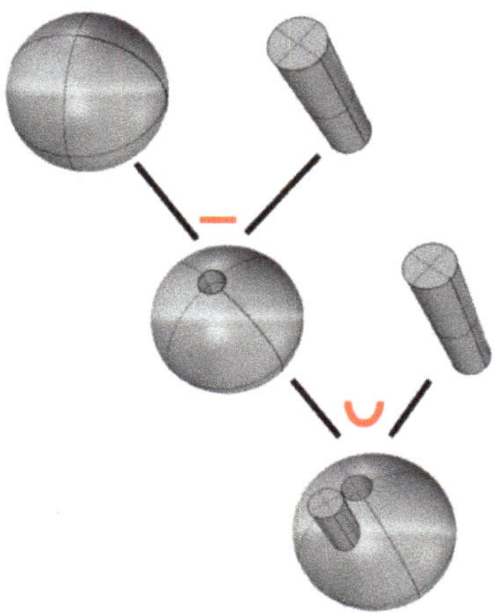

**Figure 17.** Example of a CSG tree: The "U" symbol stands for union and the "-" symbol stands for substraction.

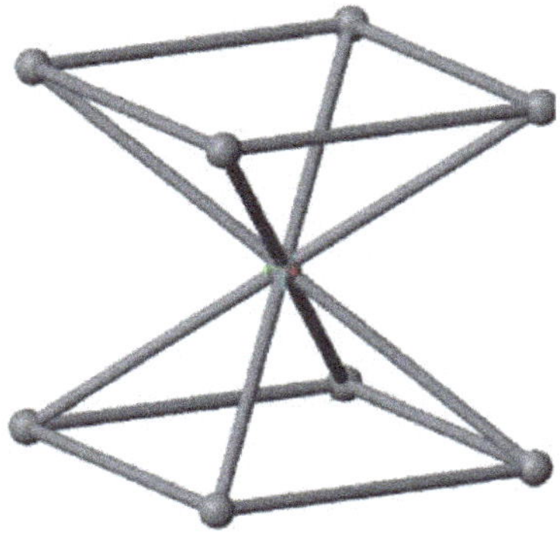

**Figure 18.** Cellular structure generation via CSG where spheres represent nodes and cylinders represent struts [4].

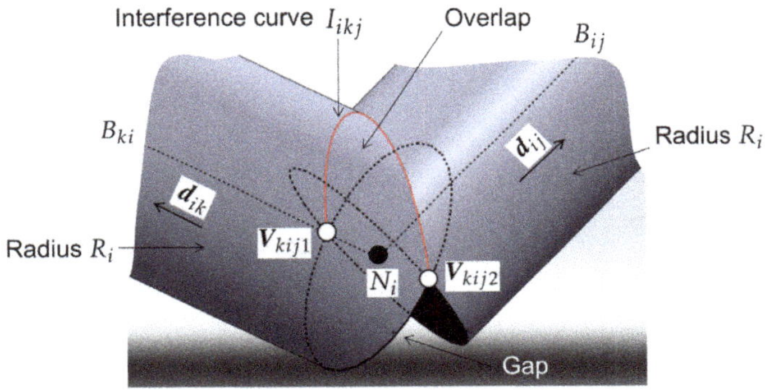


(**a**) Interference around the extremities of two connected beams

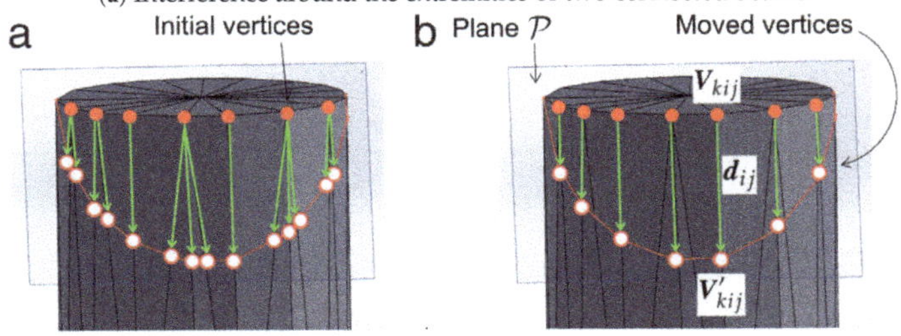


(**b**) Displacement of vertices

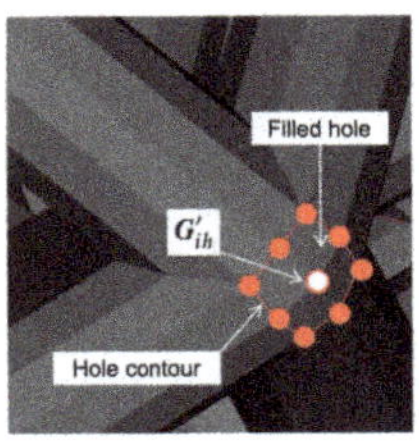


(**c**) Hole Filling

**Figure 19.** Cellular structure generation via CSG—resolving beam intersection using lattice structure lightweight triangulation algorithm [105].

The work of Zeinalabedini et al. [91] presents a method similar to Aremu et al. [104], where the authors resorted to subtracting a set of randomly spaced spheres from the design space to obtain the CS, as seen in Figure 20. In this methodology, the user can control the size of the spheres, and thus the overall porosity, as well as the overlap between spheres to control the degree of pore necking and inter-connectivity. In this particular case, the RVE is the void itself. Uhlířová and Pabst [107] have discussed a similar technique. There are two disadvantages to this approach. Firstly, there is a high computational cost of the boolean difference because the spheres are not necessarily overlapping, the number of subtractions that need to be done can be as high as the total number of spheres in the set. Secondly, the other drawback is the potential failure of such an operation; depending on the software used, subtracting two intersecting solids may not be possible. The main advantage of this is that it uses existing functions available in most CAD software; therefore, it is easy to implement for users with low experience.

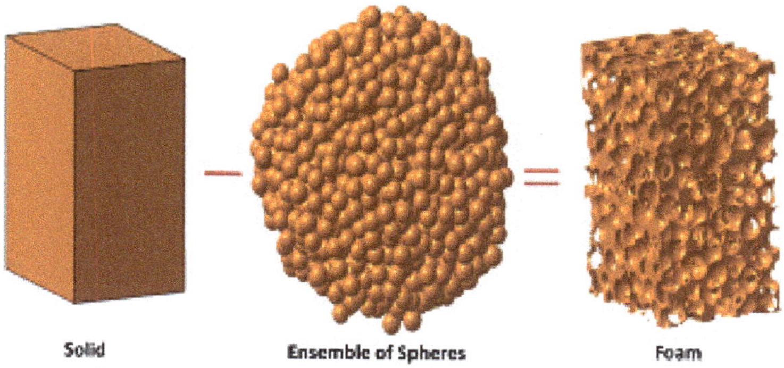


**Figure 20.** Cellular structure generation via CSG—foam structure generation [91].

Presently, most CAD software uses a combination of BREP and CSG [108] to capture the complexity of the CS design space and to visualize such structures. The BREP format describes geometry as a combination of vertices, edges, and faces [108], while in CSG, it is done through union, intersection, and difference of primitive solids (e.g., squares, spheres) [108].

2.5.3. CS Generation via Implicit Modeling and Mesh Data Structures

Solids can be represented by the volume enclosed within their representative surfaces. Such surfaces can be seen as the boundary between void and solid. To represent void and solid domains, numbers can be attributed to points within the design space volume, where the values of the points that are below a threshold value are seen as void while the ones above the threshold value are seen as solid. Two-dimensional examples of this concept can be seen in Figure 21). This procedure is known as level sets or implicit modeling. In order to create the contours in 3D, many algorithms can be used; one of the most famous is the marching cube algorithm [109]. This algorithm discretizes the surface with triangles and stores the coordinate of each vertex as a mesh. This process leads to meshes that can be exported as Standard Tessellation Language (STL) files, which can be used to visualize structures.

One advantage of this method is that it allows a smooth transition between the primitives without the need for fillets, as seen in Figure 22. The other advantage of such a method is that the computational cost is mainly affected by the density of the points' grid. The number of operations done on the grid to change the value at each point also affects the computational cost; however, some could argue that an intricate CS will require a highly dense grid, which will greatly increase the computational cost.

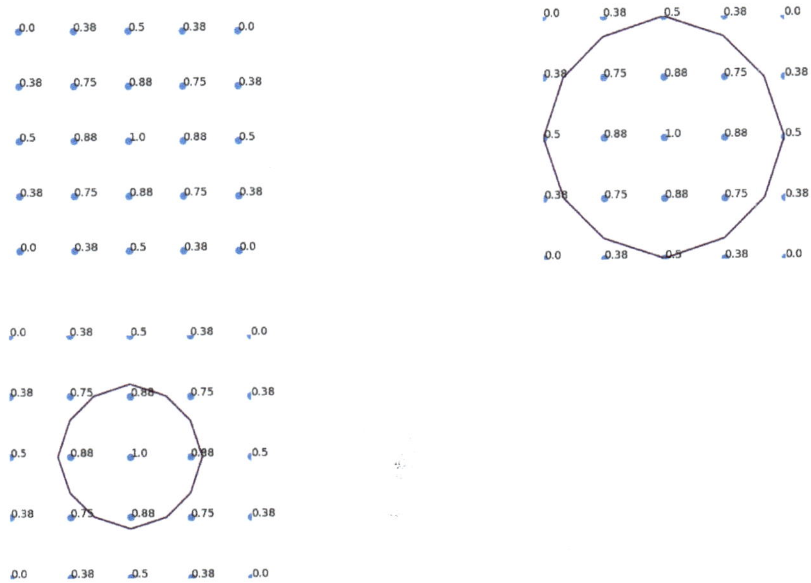


**Figure 21.** The images above show a 2D version of grid points and some of their isocontours. From left to right: 2D grid points with their respective values, isocontour at 0.5, and isocontour at 0.8.

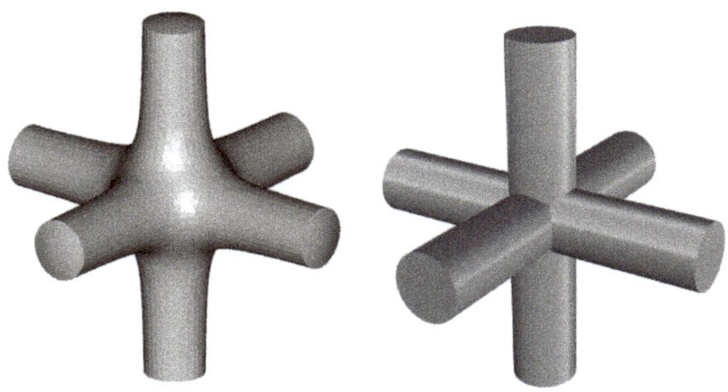

**Figure 22.** Intersection of cylinders using implicit modeling (**left**) and CSG (**right**).

Researchers such as Tang et al. [110] have done some work towards helping to improve solid modeling of lattice structures. To do so, they have designed a hybrid approach, where the method first starts by designing and generating the lattice frame, then implicit functions are used to thicken the frame. The final structure is then voxelized such that it can be directly used for AM. Similarly, strut-based unit cells can be thickened using implicit functions. An example of this generation method is given by Intralattice [82], a plug-in for the Rhinoceros3D CAD software. In this platform, the unit cell is first defined and then replicated in the design space periodically or pseudo-periodically. Finally, the resulting wireframes are thickened, as seen in Figure 23.

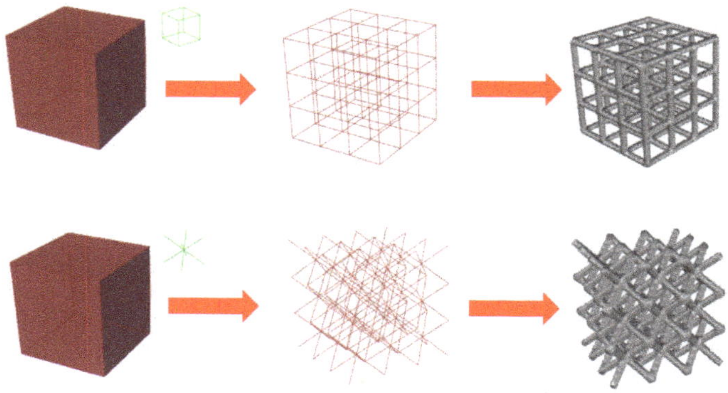

**Figure 23.** Cellular structure generation via implicit modeling—intralattice lattice generation framework [82]. From right to left: design space (unit cell in green), lattice wireframe, lattice mesh.

Triply Periodic Minimized Structures (TPMS) are examples of structures that are implemented using implicit modeling. Some algorithms, such as highlighted by Hsieh and Valdevit [111] have used implicit functions to represent triply periodic minimum surfaces. In the article by Kumar et al. [92], structures were also built using implicit modeling. In the article by Yang et al. [55], an example is shown of the transition that can be made between the primitives. Furthermore, as topology optimization can also be done using level sets; it is only a matter of time before TOMs can be rendered using implicit functions. In other occasions, meshes that already exist can be modified using known mesh operations. In Savio et al. [46]'s work, an FE model is first built based on loads and the surface is subdivided using computer graphic techniques. Researchers such as Stadlbauer et al. [112] have worked on an approach to interactively generate cellular structures on existing meshes. With surface meshes as inputs, they can decompose their meshes into hollow cellular structures that act as a skeleton on the surface and a set of thin shells. Liang et al. [113] have given an example of a lattice structure that is built using implicit functions. The values of the grid points are given based on the distance to the closest point on the lattice core (skeleton).

For digital storage and handling of the resulting geometry, the de facto standard in AM is the STL file. With the advent of multi-material printers and the increasing complexity of additive manufactured parts, the STL format is starting to become limiting, mainly in its inability to store other types of information (such as materials, colors, copyrights), in addition to geometry data. New file formats (3MF and AMF) were created to address such shortcomings. Both 3MF and AMF are XML (extensible markup language)-based, which enables them to store other information in addition to geometry. In the 3MF and AMF file, geometry is still stored as a mesh. In AMF, a porous structure can be specified as a material. On the other hand, 3MF now supports the representation of CSs as a set of beam elements. Since file formats store geometry information as a surface, most of the algorithms described in the previous sections aim at obtaining a mesh-based representation of CSs. Here lies the challenge in generating cellular structures; as the complexity of CSs increases, the complexity of the mesh needed to describe it increases exponentially [27]. This drives the computation time and file size upward when processing and generating CSs. The complexity of CS also pushes CAD technology to the limit. Representing large numbers of features using BREP and CSG becomes impractical [104]. Hence, there is a continued interest in representing and generating CSs either as beam elements, implicit models or voxels.

## 3. Highlights of Manufacturability Challenges of Cellular Structures

One of the main challenges in AM is that the manufacturability of components is at the intersection of the design space, material properties and process physics, where the interactions are not fully understood. Furthermore, the effect of the process parameters on the manufacturability of lattices depends on the AM technique used. This section presents the set of design limitations relevant to each technology.

The parameters that affect manufacturability can be divided into two categories: the lattice design parameters and parameters related to the AM process. For the lattice design, this includes the size of the RVEs (unit cells) and the primitive properties (the strut angles and diameters), which is typically be dictated based on the resolution of the AM technology (design guidelines) and experience with each material and machine (design rules). For AM parameters, each technology has a specific set of inputs; such brief examples are the scanning speed, in the case of powder bed processes, or the extrusion speed in the case of material extrusion processes. In the context of this work, manufacturability refers to the capacity to achieve three things: dimensional fidelity, low surface roughness, and low porosity defects. It is important to have a firm understanding of those elements to be able to determine how the mechanical properties, among others, will be affected. Table 2 summarizes the various design constraints with example works across the more common AM fabrication techniques [114]. Mechanical properties will not be explored in this context since such properties are dependent on the material used.

One design limitation is the need to remove unused material from inside the lattice structure. For the powder bed AM process such as Laser Powder Bed Fusion (LPBF) or Selective Laser Melting (SLM), Selective Laser Sintering (SLS), Binder Jetting (BJ), and Electron Beam (EBM) powder bed fusion, the lattice inner voids need to be large enough to allow powder to flow freely and to exit the volume during de-powdering. Adam and Zimmer [115] mention a minimum size requirement for SLM and SLS, although such requirement is likely dependent on the powder size distribution and material rheological properties. For stereolithography (SLA), the unused resin flows freely out and only requires an exit opening if the lattice is fully enclosed. For Fused Deposition Modeling (FDM) and Material Jetting (MJ), internal material removal requirements are largely driven by the need to remove support structures via mechanical or chemical processes, if such support structures are present.

If the lattice structures contain overhangs as defined by a technology class-specific threshold angle, the overhanging features need to be supported through the generation of expendable support structures. This can be mitigated either by changing the lattice design, or by modifying the build orientation of the product. For multi-material capable systems like FDM or MJ, support structures can be printed out of dissolvable materials. For FDM, all overhang needs to be supported due to gravity. Adam and Zimmer [115] and Qattawi et al. [116] illustrate a few design rules for FDM. Support structures are also necessary for LPBF and EBM. For LPBF, support structures act as thermal dissipation structures, to prevent warping and material vaporization due to the accumulation of residual stress and thermal insulating properties of powders [117], respectively. Adam and Zimmer [115] and Kranz et al. [118] show a few design guidelines regarding overhangs and slopes for LPBF. For EBM, support structures are technically only necessary for anchoring components to the build plate, as the powder around the part is sintered by the process into a powder cake. The powder cake provides enough mechanical support and heat dissipation properties for the object [117]. For SLS and BJ, support structures are generally not necessary for overhangs, as the powder provides enough gravity support [115,117]. Furthermore, such processes do not experience excessive thermal gradients. For SLA, although not robustly documented in literature, design guidelines by 3D Hubs and Materialise specify that support structures are only needed to anchor the object to the build plate and ensure print object continuity in respect to the build plate [119,120].

Table 2. The common process design constraints for manufacturability of CS.

| Typical RVE Design Constraints | LPBF | SLS | SLA | FDM | BJ | EBM | MJ |
|---|---|---|---|---|---|---|---|
| Need for dimensional fidelity | x | x | x | x | x | x | x |
| Need for material removal | x | x | - | - | x | x | - |
| Need for support overhang structures | x | - | - | x | - | - | x |
| Need for avoidance of pore defects | x | x | - | x | x | x | - |

Another design limitation stems from the material–process–design relationship, where the minimum feature size of an RVE correlates with the resulting porous defects in the CS. For LPBF and EBM, the dimension of the energy source, the powder size and the minimum desired feature size in an RVE, as well as the surrounding RVE neighbors result in a complex set of melt pool and heat transfer phenomena. Such interplay can often result in pore defects; the mechanism of pore formation, the classification of pores, and strategies to mitigate such defects in powder bed fusion processes have been reviewed and summarized by Echeta et al. [121] and Sola and Nouri [122]. For BJ technologies, the scale of the interaction between the layer resolution, the liquid binder droplets, and the powder morphology and size can pose limitations in terms of the lattice structure geometric and green density quality [123]; such qualities are also impacted by the sintering and densification process. For SLA, pore defects are only mentioned when printing ceramic-loaded polymers into complex-shaped objects, where the pores are mainly created by the sintering post-processing steps rather than the AM process itself [124]. As for FDM, the minimum feature size of an RVE and the nozzle diameter play a role in pore defect generation. The RVE inner features may be inadequately filled due to discontinuity in the toolpath and gaps between the extruded plastic filament [125]. For MJ, pore defects are not mentioned [126] except in the context of polymer-loaded ceramic production [127].

## 4. Conclusions

The advent of AM has expanded the horizons of cellular structure design. In this paper, a new cellular structures classification paradigm has been developed. The design methods for cellular structures have been classified by identifying all the geometric degrees of freedom that are involved in the construction of cellular structures. The classification was brought forth by reviewing the numerous case studies, with representative works highlighted. Moreover, cellular structure generation and CAD manipulation techniques have also been reviewed. The review explored the different design constraints imposed by AM processes on the quality of cellular structures. Finally, manufacturability issues that occur when printing CSs are also highlighted. In the future, more work is needed to develop new simulation methods and design tools for CSs, to also analyze the printing parameters that have an effect on the dimensional accuracy and the properties (mechanical, thermal, etc.) of CSs, to then establish CS design rules that should be implemented at the design stage.

**Author Contributions:** Cellular structure design classification, overview of representative volume elements, RVE variation methods, RVE variation factors, K.M.N.; CS generation, K.M.N., M.W.; highlights of manufacturability challenges of cellular structures, M.W.; conceptualization, supervision and correction, M.V. All authors have read and agreed to the published version of the manuscript.

**Funding:** The authors appreciate the funding support from the Federal Economic Development Agency for Southern Ontario (FedDev Ontario) grant number 814654.

**Institutional Review Board Statement:** Not applicable.

**Informed Consent Statement:** Not applicable.

**Data Availability Statement:** Not applicable.

**Conflicts of Interest:** The authors declare no conflict of interest.

## Abbreviations

Abbreviations

The following abbreviations are used in this manuscript:

| | |
|---|---|
| AM | Additive manufacturing |
| BEM | Boundary element method |
| BESO | Bi-directional evolutionary structural optimization |
| BJ | Binder Jetting |
| CAD | Computer Aided Design |
| CpTi | Commercially pure titanium |
| CS | Cellular structure |
| CSG | Constructive solid geometry |
| EBM | Electron beam melting |
| ESO | Evolutionary structural optimization |
| FDM | Fused deposition modeling |
| FEA | Finite element analysis |
| FFF | Fused filament fabrication |
| FGM | Functionally graded material |
| GDoF | Geometrical degrees of freedom |
| GLS | Gradient lattice structures |
| HIP | Hot isostatic pressure |
| LPBF | Laser powder bed fusion |
| MJ | Material Jetting |
| SIMP | Solid isotropic material with penalization method |
| SLA | Stereolithography |
| SLS | Selective laser sintering |
| STL | Standard Tessellation Language |

## References


1. Huynh, L.; Rotella, J.; Sangid, M.D. Fatigue behavior of IN718 microtrusses produced via additive manufacturing. *Mater. Des.* **2016**, *105*, 278–289. [CrossRef]
2. Ozdemir, Z.; Hernandez-Nava, E.; Tyas, A.; Warren, J.A.; Fay, S.D.; Goodall, R.; Todd, I.; Askes, H. Energy absorption in lattice structures in dynamics: Experiments. *Int. J. Impact Eng.* **2016**, *89*, 49–61. [CrossRef]
3. Shamvedi, D.; McCarthy, O.J.; Donoghue, E.O.; Danilenkoff, C.; OLeary, P.; Raghavendra, R. 3D Metal printed heat sinks with longitudinally varying lattice structure sizes using direct metal laser sintering. *Virtual Phys. Prototyp.* **2018**, *13*, 301–310. [CrossRef]
4. Syam, W.P.; Jianwei, W.; Zhao, B.; Maskery, I.; Elmadih, W.; Leach, R. Design and analysis of strut-based lattice structures for vibration isolation. *Precis. Eng.* **2018**, *52*, 494–506. [CrossRef]
5. Zhang, X.; Xia, Y.; Wang, J.; Yang, Z.; Tu, C.; Wang, W. Medial axis tree—An internal supporting structure for 3D printing. *Comput. Aided Geom. Des.* **2015**, *35*, 149–162. [CrossRef]
6. Mazur, M.; Leary, M.; Sun, S.; Vcelka, M.; Shidid, D.; Brandt, M. Deformation and failure behaviour of Ti-6Al-4V lattice structures manufactured by selective laser melting (SLM). *Int. J. Adv. Manuf. Technol.* **2016**, *84*, 1391–1411. [CrossRef]
7. Seepersad, C.C.; Allen, J.K.; McDowell, D.L.; Mistree, F. Multifunctional topology design of cellular material structures. *J. Mech. Des.* **2008**, *130*, 031404. [CrossRef]
8. Chu, C.; Graf, G.; Rosen, D.W. Design for additive manufacturing of cellular structures. *Comput. Aided Des. Appl.* **2008**, *5*, 686–696. [CrossRef]
9. Wang, B.; Wu, L.; Ma, L.; Sun, Y.; Du, S. Mechanical behavior of the sandwich structures with carbon fiber-reinforced pyramidal lattice truss core. *Mater. Des. (1980–2015)* **2010**, *31*, 2659–2663. [CrossRef]
10. Kooistra, G.W.; Wadley, H.N. Lattice truss structures from expanded metal sheet. *Mater. Des.* **2007**, *28*, 507–514. [CrossRef]
11. Alzahrani, M.; Choi, S.K.; Rosen, D.W. Design of truss-like cellular structures using relative density mapping method. *Mater. Des.* **2015**, *85*, 349–360. [CrossRef]
12. Wang, Y.; Chen, F.; Wang, M.Y. Concurrent design with connectable graded microstructures. *Comput. Methods Appl. Mech. Eng.* **2017**, *317*, 84–101. [CrossRef]
13. Panesar, A.; Abdi, M.; Hickman, D.; Ashcroft, I. Strategies for functionally graded lattice structures derived using topology optimisation for additive manufacturing. *Addit. Manuf.* **2018**, *19*, 81–94. [CrossRef]
14. Dong, G.; Tang, Y.; Zhao, Y.F. A survey of modeling of lattice structures fabricated by additive manufacturing. *J. Mech. Des.* **2017**, *139*, 100906. [CrossRef]

15. Helou, M.; Kara, S. Design, analysis and manufacturing of lattice structures: an overview. *Int. J. Comput. Integr. Manuf.* **2018**, *31*, 243–261. [CrossRef]
16. Schaedler, T.A.; Carter, W.B. Architected cellular materials. *Annu. Rev. Mater. Res.* **2016**, *46*, 187–210. [CrossRef]
17. Plocher, J.; Panesar, A. Review on design and structural optimisation in additive manufacturing: Towards next-generation lightweight structures. *Mater. Des.* **2019**, *183*, 108164. [CrossRef]
18. Nagesha, B.; Dhinakaran, V.; Shree, M.V.; Kumar, K.M.; Chalawadi, D.; Sathish, T. Review on characterization and impacts of the lattice structure in additive manufacturing. *Mater. Today Proc.* **2020**, *21*, 916–919. [CrossRef]
19. Tamburrino, F.; Graziosi, S.; Bordegoni, M. The design process of additively manufactured mesoscale lattice structures: A review. *J. Comput. Inf. Sci. Eng.* **2018**, *18*. [CrossRef]
20. Qin, Y.; Wen, P.; Guo, H.; Xia, D.; Zheng, Y.; Jauer, L.; Poprawe, R.; Voshage, M.; Schleifenbaum, J.H. Additive manufacturing of biodegradable metals: Current research status and future perspectives. *Acta Biomater.* **2019**, *98*, 3–22. [CrossRef]
21. du Plessis, A.; Broeckhoven, C.; Yadroitsava, I.; Yadroitsev, I.; Hands, C.H.; Kunju, R.; Bhate, D. Beautiful and functional: A review of biomimetic design in additive manufacturing. *Addit. Manuf.* **2019**, *27*, 408–427. [CrossRef]
22. Yuan, L.; Ding, S.; Wen, C. Additive manufacturing technology for porous metal implant applications and triple minimal surface structures: A review. *Bioact. Mater.* **2019**, *4*, 56–70. [CrossRef] [PubMed]
23. Zhang, C.; Chen, F.; Huang, Z.; Jia, M.; Chen, G.; Ye, Y.; Lin, Y.; Liu, W.; Chen, B.; Shen, Q.; et al. Additive manufacturing of functionally graded materials: A review. *Mater. Sci. Eng. A* **2019**, *764*, 138209. [CrossRef]
24. Seharing, A.; Azman, A.H.; Abdullah, S. A review on integration of lightweight gradient lattice structures in additive manufacturing parts. *Adv. Mech. Eng.* **2020**, *12*, 1687814020916951. [CrossRef]
25. Nazir, A.; Abate, K.M.; Kumar, A.; Jeng, J.Y. A state-of-the-art review on types, design, optimization, and additive manufacturing of cellular structures. *Int. J. Adv. Manuf. Technol.* **2019**, *104*, 3489–3510. [CrossRef]
26. Tang, Y.; Zhao, Y.F. A survey of the design methods for additive manufacturing to improve functional performance. *Rapid Prototyp. J.* **2016**, *22*, 569–590. [CrossRef]
27. Hadi, A.; Vignat, F.; Villeneuve, F. Design configurations and creation of lattice structures for metallic additive manufacturing. In Proceedings of the 14ème Colloque National AIP PRIMECA, La Plagne, France, 31 March–2 April 2015.
28. Yang, N.; Quan, Z.; Zhang, D.; Tian, Y. Multi-morphology transition hybridization CAD design of minimal surface porous structures for use in tissue engineering. *Comput. Aided Des.* **2014**, *56*, 11–21. [CrossRef]
29. Plocher, J.; Panesar, A. Effect of density and unit cell size grading on the stiffness and energy absorption of short fibre-reinforced functionally graded lattice structures. *Addit. Manuf.* **2020**, 101171. [CrossRef]
30. Tang, Y.; Yang, S.; Zhao, Y.F. Design Method for Conformal Lattice-Skin Structure Fabricated by AM Technologies. In Proceedings of the ASME 2016 International Design Engineering Technical Conferences and Computers and Information in Engineering Conference, Charlotte, NC, USA, 21–24 August 2016; p. V01AT02A037.
31. Nguyen, J.; Park, S.I.; Rosen, D.W.; Folgar, L.; Williams, J. Conformal lattice structure design and fabrication. In Proceedings of the Solid Freeform Fabrication Symposium (SFF), Austin, TX, USA, 6–8 August 2012; pp. 6–8.
32. Terriault, P.; Brailovski, V. Modeling and simulation of large, conformal, porosity-graded and lightweight lattice structures made by additive manufacturing. *Finite Elem. Anal. Des.* **2018**, *138*, 1–11. [CrossRef]
33. Wang, Y.; Zhang, L.; Daynes, S.; Zhang, H.; Feih, S.; Wang, M.Y. Design of graded lattice structure with optimized mesostructures for additive manufacturing. *Mater. Des.* **2018**, *142*, 114–123. [CrossRef]
34. Lim, Y.E.; Park, J.H.; Park, K. Automatic design of 3D conformal lightweight structures based on a tetrahedral mesh. *Int. J. Precis. Eng. Manuf. Green Technol.* **2018**, *5*, 499–506. [CrossRef]
35. Guo, Y.; Liu, K.; Yu, Z. Tetrahedron-Based Porous Scaffold Design for 3D Printing. *Designs* **2019**, *3*, 16. [CrossRef]
36. Ambu, R.; Morabito, A.E. Porous scaffold design based on minimal surfaces: Development and assessment of variable architectures. *Symmetry* **2018**, *10*, 361. [CrossRef]
37. Liu, Y.; Zhuo, S.; Xiao, Y.; Zheng, G.; Dong, G.; Zhao, Y.F. Rapid modeling and design optimization of multi-topology lattice structure based on unit-cell library. *J. Mech. Des.* **2020**, *142*. [CrossRef]
38. Daynes, S.; Feih, S.; Lu, W.F.; Wei, J. Optimisation of functionally graded lattice structures using isostatic lines. *Mater. Des.* **2017**, *127*, 215–223. [CrossRef]
39. Smith, C.J.; Gilbert, M.; Todd, I.; Derguti, F. Application of layout optimization to the design of additively manufactured metallic components. *Struct. Multidiscip. Optim.* **2016**, *54*, 1297–1313. [CrossRef]
40. Teufelhart, S.; Reinhart, G. Optimization of strut diameters in lattice structures. In Proceedings of the 23th Annual Solid Freeform Fabrication Symposium, Austin, TX, USA, 6–8 August 2012; pp. 719–733.
41. Cheng, L.; Liu, J.; To, A.C. Concurrent lattice infill with feature evolution optimization for additive manufactured heat conduction design. *Struct. Multidiscip. Optim.* **2018**, 1–25. [CrossRef]
42. Goel, A.; Anand, S. Design of functionally graded lattice structures using B-splines for additive manufacturing. *Procedia Manuf.* **2019**, *34*, 655–665. [CrossRef]
43. Zhang, P.; Toman, J.; Yu, Y.; Biyikli, E.; Kirca, M.; Chmielus, M.; To, A.C. Efficient design-optimization of variable-density hexagonal cellular structure by additive manufacturing: theory and validation. *J. Manuf. Sci. Eng.* **2015**, *137*, 021004. [CrossRef]
44. Nguyen, C.H.P.; Kim, Y.; Choi, Y. Design for additive manufacturing of functionally graded lattice structures: A design method with process induced anisotropy consideration. *Int. J. Precis. Eng. Manuf. Green Technol.* **2019**, 1–17. [CrossRef]

45. Chen, Y. 3D texture mapping for rapid manufacturing. *Comput. Aided Des. Appl.* **2007**, *4*, 761–771. [CrossRef]
46. Savio, G.; Meneghello, R.; Concheri, G. Geometric modeling of lattice structures for additive manufacturing. *Rapid Prototyp. J.* **2018**. [CrossRef]
47. Larimore, Z.; Jensen, S.; Parsons, P.; Good, B.; Smith, K.; Mirotznik, M. Use of space-filling curves for additive manufacturing of three dimensionally varying graded dielectric structures using fused deposition modeling. *Addit. Manuf.* **2017**, *15*, 48–56. [CrossRef]
48. Song, G.H.; Jing, S.K.; Zhao, F.L.; Wang, Y.D.; Xing, H.; Qie, L.F. Design of Lattice Structures Using Local Relative Density Mapping Method. *Chin. J. Mech. Eng.* **2018**, *31*, 89. [CrossRef]
49. Li, D.; Liao, W.; Dai, N.; Dong, G.; Tang, Y.; Xie, Y.M. Optimal design and modeling of gyroid-based functionally graded cellular structures for additive manufacturing. *Comput. Aided Des.* **2018**, *104*, 87–99. [CrossRef]
50. Reinhart, G.; Teufelhart, S. Optimization of mechanical loaded lattice structures by orientating their struts along the flux of force. *Procedia CIRP* **2013**, *12*, 175–180. [CrossRef]
51. Brackett, D.; Ashcroft, I.; Wildman, R.; Hague, R.J. An error diffusion based method to generate functionally graded cellular structures. *Comput. Struct.* **2014**, *138*, 102–111. [CrossRef]
52. Opgenoord, M.M.; Willcox, K.E. Design for additive manufacturing: cellular structures in early-stage aerospace design. *Struct. Multidiscip. Optim.* **2019**, *60*, 411–428. [CrossRef]
53. Rodríguez-Montaño, Ó.L.; Cortés-Rodríguez, C.J.; Naddeo, F.; Uva, A.E.; Fiorentino, M.; Naddeo, A.; Cappetti, N.; Gattullo, M.; Monno, G.; Boccaccio, A. Irregular Load Adapted Scaffold Optimization: A Computational Framework Based on Mechanobiological Criteria. *ACS Biomater. Sci. Eng.* **2019**, *5*, 5392–5411. [CrossRef]
54. Almonti, D.; Baiocco, G.; Tagliaferri, V.; Ucciardello, N. Design and Mechanical Characterization of Voronoi Structures Manufactured by Indirect Additive Manufacturing. *Materials* **2020**, *13*, 1085. [CrossRef] [PubMed]
55. Yang, N.; Du, C.f.; Wang, S.; Yang, Y.; Zhang, C. Mathematically defined gradient porous materials. *Mater. Lett.* **2016**, *173*, 136–140. [CrossRef]
56. Jin, Y.; Kong, H.; Zhou, X.; Li, G.; Du, J. Design and Characterization of Sheet-Based Gyroid Porous Structures with Bioinspired Functional Gradients. *Materials* **2020**, *13*, 3844. [CrossRef] [PubMed]
57. Niknam, H.; Akbarzadeh, A. Graded lattice structures: Simultaneous enhancement in stiffness and energy absorption. *Mater. Des.* **2020**, *196*, 109129. [CrossRef]
58. Cadman, J.; Zhou, S.; Chen, Y.; Li, W.; Appleyard, R.; Li, Q. Characterization of cuttlebone for a biomimetic design of cellular structures. *Acta Mech. Sin.* **2010**, *26*, 27–35. [CrossRef]
59. Mullen, L.; Stamp, R.C.; Fox, P.; Jones, E.; Ngo, C.; Sutcliffe, C.J. Selective laser melting: A unit cell approach for the manufacture of porous, titanium, bone in-growth constructs, suitable for orthopedic applications. II. Randomized structures. *J. Biomed. Mater. Res. Part Appl. Biomater. Off. J. Soc. Biomater. Jpn. Soc. Biomater. Aust. Soc. Biomater. Korean Soc. Biomater.* **2010**, *92*, 178–188. [CrossRef]
60. Liu, F.; Mao, Z.; Zhang, P.; Zhang, D.Z.; Jiang, J.; Ma, Z. Functionally graded porous scaffolds in multiple patterns: new design method, physical and mechanical properties. *Mater. Des.* **2018**, *160*, 849–860. [CrossRef]
61. Zhang, Z.; Yuan, L.; Lee, P.D.; Jones, E.; Jones, J.R. Modeling of time dependent localized flow shear stress and its impact on cellular growth within additive manufactured titanium implants. *J. Biomed. Mater. Res. Part Appl. Biomater.* **2014**, *102*, 1689–1699. [CrossRef]
62. Reis, S.; Vasconcelos, V.; Leite, M.; Vasconcelos, W. Development of a Computer Application to Simulate Porous Structures. *Mater. Res.* **2002**, *5*, 275–279. [CrossRef]
63. Ghouse, S.; Babu, S.; Nai, K.; Hooper, P.A.; Jeffers, J.R. The influence of laser parameters, scanning strategies and material on the fatigue strength of a stochastic porous structure. *Addit. Manuf.* **2018**, *22*, 290–301. [CrossRef]
64. Liu, F.; Ran, Q.; Zhao, M.; Zhang, T.; Zhang, D.Z.; Su, Z. Additively manufactured continuous cell-size gradient porous scaffolds: pore characteristics, mechanical properties and biological responses in vitro. *Materials* **2020**, *13*, 2589. [CrossRef]
65. Cuan-Urquizo, E.; Martínez-Magallanes, M.; Crespo-Sánchez, S.E.; Gómez-Espinosa, A.; Olvera-Silva, O.; Roman-Flores, A. Additive manufacturing and mechanical properties of lattice-curved structures. *Rapid Prototyp. J.* **2019**. [CrossRef]
66. Yang, L.; Harrysson, O.; Cormier, D.; West, H.; Gong, H.; Stucker, B. Additive manufacturing of metal cellular structures: Design and fabrication. *JOM* **2015**, *67*, 608–615. [CrossRef]
67. Beyer, C.; Figueroa, D. Design and analysis of lattice structures for additive manufacturing. *J. Manuf. Sci. Eng.* **2016**, *138*, 121014. [CrossRef]
68. du Plessis, A.; Yadroitsava, I.; Yadroitsev, I. Ti6Al4V lightweight lattice structures manufactured by laser powder bed fusion for load-bearing applications. *Opt. Laser Technol.* **2018**, *108*, 521–528. [CrossRef]
69. Ongaro, F. Estimation of the effective properties of two-dimensional cellular materials: A review. *Theor. Appl. Mech. Lett.* **2018**, *8*, 209–230. [CrossRef]
70. Rehme, O.; Emmelmann, C. Selective laser melting of honeycombs with negative Poissonś ratio. *J. Laser Micro Nanoeng* **2009**, *4*, 128–134. [CrossRef]
71. Yan, C.; Hao, L.; Hussein, A.; Young, P. Ti–6Al–4V triply periodic minimal surface structures for bone implants fabricated via selective laser melting. *J. Mech. Behav. Biomed. Mater.* **2015**, *51*, 61–73. [CrossRef] [PubMed]

72. Yan, C.; Hao, L.; Hussein, A.; Young, P.; Raymont, D. Advanced lightweight 316L stainless steel cellular lattice structures fabricated via selective laser melting. *Mater. Des.* **2014**, *55*, 533–541. [CrossRef]
73. Coelho, P.G.; Hollister, S.J.; Flanagan, C.L.; Fernandes, P.R. Bioresorbable scaffolds for bone tissue engineering: optimal design, fabrication, mechanical testing and scale-size effects analysis. *Med. Eng. Phys.* **2015**, *37*, 287–296. [CrossRef]
74. Takezawa, A.; Koizumi, Y.; Kobashi, M. High-stiffness and strength porous maraging steel via topology optimization and selective laser melting. *Addit. Manuf.* **2017**, *18*, 194–202. [CrossRef]
75. Sercombe, T.B.; Xu, X.; Challis, V.; Green, R.; Yue, S.; Zhang, Z.; Lee, P.D. Failure modes in high strength and stiffness to weight scaffolds produced by Selective Laser Melting. *Mater. Des.* **2015**, *67*, 501–508. [CrossRef]
76. Limmahakhun, S.; Oloyede, A.; Sitthiseripratip, K.; Xiao, Y.; Yan, C. 3D-printed cellular structures for bone biomimetic implants. *Addit. Manuf.* **2017**, *15*, 93–101. [CrossRef]
77. Maskery, I.; Aremu, A.; Parry, L.; Wildman, R.; Tuck, C.; Ashcroft, I. Effective design and simulation of surface-based lattice structures featuring volume fraction and cell type grading. *Mater. Des.* **2018**, *155*, 220–232. [CrossRef]
78. Yu, S.; Sun, J.; Bai, J. Investigation of functionally graded TPMS structures fabricated by additive manufacturing. *Mater. Des.* **2019**, *182*, 108021. [CrossRef]
79. Kladovasilakis, N.; Tsongas, K.; Tzetzis, D. Mechanical and FEA-Assisted Characterization of Fused Filament Fabricated Triply Periodic Minimal Surface Structures. *J. Compos. Sci.* **2021**, *5*, 58. [CrossRef]
80. Evans, K.E. Auxetic polymers: A new range of materials. *Endeavour* **1991**, *15*, 170–174. [CrossRef]
81. Queheillalt, D.T.; Wadley, H.N. Pyramidal lattice truss structures with hollow trusses. *Mater. Sci. Eng. A* **2005**, *397*, 132–137. [CrossRef]
82. Kurtz, A. IntraLattice. Available online: http://intralattice.com/case_studies/ (accessed on 20 April 2021).
83. Rahman, H.; Yarali, E.; Zolfagharian, A.; Serjouei, A.; Bodaghi, M. Energy Absorption and Mechanical Performance of Functionally Graded Soft–Hard Lattice Structures. *Materials* **2021**, *14*, 1366. [CrossRef]
84. Hu, J.; Wang, B. Enhanced fatigue performance of auxetic honeycomb/substrate structures under thermal cycling. *Int. J. Mech. Sci.* **2021**, 106432. [CrossRef]
85. Yap, Y.L.; Yeong, W.Y. Shape recovery effect of 3D printed polymeric honeycomb: This paper studies the elastic behaviour of different honeycomb structures produced by PolyJet technology. *Virtual Phys. Prototyp.* **2015**, *10*, 91–99. [CrossRef]
86. Karcher, H.; Polthier, K. Construction of triply periodic minimal surfaces. *Philos. Trans. R. Soc. London. Ser. A Math. Phys. Eng. Sci.* **1996**, *354*, 2077–2104.
87. Hao, L.; Raymont, D.; Yan, C.; Hussein, A.; Young, P. Design and additive manufacturing of cellular lattice structures. In *The International Conference on Advanced Research in Virtual and Rapid Prototyping (VRAP)*; Taylor & Francis Group: Leiria, Portugal, 2011; pp. 249–254.
88. Zhao, M.; Liu, F.; Fu, G.; Zhang, D.Z.; Zhang, T.; Zhou, H. Improved mechanical properties and energy absorption of BCC lattice structures with triply periodic minimal surfaces fabricated by SLM. *Materials* **2018**, *11*, 2411. [CrossRef]
89. Hollister, S.J. Porous scaffold design for tissue engineering. *Nat. Mater.* **2005**, *4*, 518. [CrossRef] [PubMed]
90. Harris, J.; McShane, G. Metallic stacked origami cellular materials: Additive manufacturing, properties, and modelling. *Int. J. Solids Struct.* **2020**, *185*, 448–466. [CrossRef]
91. Zeinalabedini, H.; Yildiz, Y.O.; Zhang, P.; Laux, K.; Kirca, M.; To, A.C. Homogenization of additive manufactured polymeric foams with spherical cells. *Addit. Manuf.* **2016**, *12*, 274–281. [CrossRef]
92. Kumar, S.; Tan, S.; Zheng, L.; Kochmann, D.M. Inverse-designed spinodoid metamaterials. *NPJ Comput. Mater.* **2020**, *6*, 1–10. [CrossRef]
93. Hyman, J.D.; Winter, C.L. Stochastic generation of explicit pore structures by thresholding Gaussian random fields. *J. Comput. Phys.* **2014**, *277*, 16–31. [CrossRef]
94. Cheng, L.; Zhang, P.; Biyikli, E.; Bai, J.; Pilz, S.; To, A.C. Integration of topology optimization with efficient design of additive manufactured cellular structures. In Proceedings of the Solid Freeform Fabrication Symposium, Austin, TX, USA, 10–12 August 2015; pp. 1370–1377.
95. Zhao, J.; Zhang, M.; Zhu, Y.; Li, X.; Wang, L.; Hu, J. A novel optimization design method of additive manufacturing oriented porous structures and experimental validation. *Mater. Des.* **2019**, *163*, 107550. [CrossRef]
96. Engelbrecht, S.; Folgar, L.; Rosen, D.W.; Schulberger, G.; Williams, J. Cellular structures for optimal performance. In Proceedings of the SFF Symposium, Austin, TX, USA, 3–5 August 2009; pp. 831–842.
97. Brennan-Craddock, J.; Brackett, D.; Wildman, R.; Hague, R. The design of impact absorbing structures for additive manufacture. *J. Physics Conf. Ser. IOP Publ.* **2012**, *382*, 012042. [CrossRef]
98. Melpal, G.R. Conformal Lattice Structures in Additive Manufacturing (AM). Ph.D. Thesis, University of Cincinnati, Cincinnati, OH, USA, 2018.
99. Tang, Y.; Kurtz, A.; Zhao, Y.F. Bidirectional Evolutionary Structural Optimization (BESO) based design method for lattice structure to be fabricated by additive manufacturing. *Comput. Aided Des.* **2015**, *69*, 91–101. [CrossRef]
100. Robbins, J.; Owen, S.; Clark, B.; Voth, T. An efficient and scalable approach for generating topologically optimized cellular structures for additive manufacturing. *Addit. Manuf.* **2016**, *12*, 296–304. [CrossRef]
101. Fantini, M.; Curto, M.; De Crescenzio, F. A method to design biomimetic scaffolds for bone tissue engineering based on Voronoi lattices. *Virtual Phys. Prototyp.* **2016**, *11*, 77–90. [CrossRef]

102. Martínez, J.; Dumas, J.; Lefebvre, S. Procedural voronoi foams for additive manufacturing. *ACM Trans. Graph. (TOG)* **2016**, *35*, 1–12. [CrossRef]
103. Tang, Y.; Zhao, Y.F. Design method for lattice-skin structure fabricated by additive manufacturing. In Proceedings of the ASME 2014 International Mechanical Engineering Congress and Exposition, Montreal, QC, Canada, 14–20 November 2014; p. V02BT02A030.
104. Aremu, A.; Brennan-Craddock, J.; Panesar, A.; Ashcroft, I.; Hague, R.J.; Wildman, R.D.; Tuck, C. A voxel-based method of constructing and skinning conformal and functionally graded lattice structures suitable for additive manufacturing. *Addit. Manuf.* **2017**, *13*, 1–13. [CrossRef]
105. Chougrani, L.; Pernot, J.P.; Véron, P.; Abed, S. Lattice structure lightweight triangulation for additive manufacturing. *Comput.-Aided Des.* **2017**, *90*, 95–104. [CrossRef]
106. McMillan, M.; Jurg, M.; Leary, M.; Brandt, M. Programmatic lattice generation for additive manufacture. *Procedia Technol.* **2015**, *20*, 178–184. [CrossRef]
107. Uhlířová, T.; Pabst, W. Poisson's ratio of porous and cellular materials with randomly distributed isometric pores or cells. *J. Am. Ceram. Soc.* **2020**. [CrossRef]
108. Hoffmann, C.M. Solid Modeling. In *Handbook of Discrete and Computational Geometry*; Goodman, J.E., O'Rourke, J., Eds.; Chapman | & Hall/CRC: Boca Raton, FL, USA, 2004; Volume 56, pp. 1257–1278.
109. Lorensen, W.E.; Cline, H.E. Marching cubes: A high resolution 3D surface construction algorithm. In Proceedings of the ACM Siggraph Computer Graphics, Anaheim, CA, USA, 27–31 July 1987; Volume 21, pp. 163–169.
110. Tang, Y.; Dong, G.; Zhao, Y.F. A hybrid geometric modeling method for lattice structures fabricated by additive manufacturing. *Int. J. Adv. Manuf. Technol.* **2019**, *102*, 4011–4030. [CrossRef]
111. Hsieh, M.T.; Valdevit, L. Minisurf–A minimal surface generator for finite element modeling and additive manufacturing. *Softw. Impacts* **2020**, *6*, 100026. [CrossRef]
112. Stadlbauer, P.; Mlakar, D.; Seidel, H.P.; Steinberger, M.; Zayer, R. Interactive Modeling of Cellular Structures on Surfaces with Application to Additive Manufacturing. In *Computer Graphics Forum*; Wiley Online Library: Hoboken, NJ, USA, 2020; Volume 39, pp. 277–289.
113. Liang, Y.; Zhao, F.; Yoo, D.J.; Zheng, B. Design of conformal lattice structures using the volumetric distance field based on parametric solid models. *Rapid Prototyp. J.* **2020**. [CrossRef]
114. I.S.O. *ASTM52900-15, Standard Terminology for Additive Manufacturing—General Principles—Terminology*; ASTM International: West Conshohocken, PA, USA, 2015
115. Adam, G.A.; Zimmer, D. Design for Additive Manufacturing—Element transitions and aggregated structures. *CIRP J. Manuf. Sci. Technol.* **2014**, *7*, 20–28. [CrossRef]
116. Qattawi, A.; Ablat, M.A. Design consideration for additive manufacturing: fused deposition modelling. *Open J. Appl. Sci.* **2017**, *7*, 291–318.
117. Diegel, O.; Nordin, A.; Motte, D. *A Practical Guide to Design for Additive Manufacturing*; Springer: Berlin/Heidelberg, Germany, 2019.
118. Kranz, J.; Herzog, D.; Emmelmann, C. Design guidelines for laser additive manufacturing of lightweight structures in TiAl6V4. *J. Laser Appl.* **2015**, *27*, S14001. [CrossRef]
119. Armstrong, C. How to Design Parts for SLA 3D Printing. Available online: https://www.3dhubs.com/knowledge-base/how-design-parts-sla-3d-printing/ (accessed on 20 April 2021).
120. Design Guidelines—PerFORM—Stereolithography. Available online: https://www.materialise.com/en/manufacturing/materials/perform/design-guidelines (accessed on 20 April 2021).
121. Echeta, I.; Feng, X.; Dutton, B.; Leach, R.; Piano, S. Review of defects in lattice structures manufactured by powder bed fusion. *Int. J. Adv. Manuf. Technol.* **2020**, *106*, 2649–2668. [CrossRef]
122. Sola, A.; Nouri, A. Microstructural porosity in additive manufacturing: The formation and detection of pores in metal parts fabricated by powder bed fusion. *J. Adv. Manuf. Process.* **2019**, *1*, e10021. [CrossRef]
123. Mirzababaei, S.; Pasebani, S. A review on binder jet additive manufacturing of 316L stainless steel. *J. Manuf. Mater. Process.* **2019**, *3*, 82. [CrossRef]
124. An, D.; Liu, W.; Xie, Z.; Li, H.; Luo, X.; Wu, H.; Huang, M.; Liang, J.; Tian, Z.; He, R. A strategy for defects healing in 3D printed ceramic compact via cold isostatic pressing: Sintering kinetic window and microstructure evolution. *J. Am. Ceram. Soc.* **2019**, *102*, 2263–2271. [CrossRef]
125. Eiliat, H.; Urbanic, R.J. Minimizing voids for a material extrusion-based process. *Rapid Prototyp. J.* **2018**. [CrossRef]
126. Egan, P.; Wang, X.; Greutert, H.; Shea, K.; Wuertz-Kozak, K.; Ferguson, S. Mechanical and biological characterization of 3D printed lattices. *3D Print. Addit. Manuf.* **2019**, *6*, 73–81. [CrossRef]
127. Fayazfar, H.; Liravi, F.; Ali, U.; Toyserkani, E. Additive manufacturing of high loading concentration zirconia using high-speed drop-on-demand material jetting. *Int. J. Adv. Manuf. Technol.* **2020**, *109*, 2733–2746. [CrossRef]

applied sciences

MDPI

Article

# Data-Efficient Neural Network for Track Profile Modelling in Cold Spray Additive Manufacturing

**Daiki Ikeuchi [1,2,*,†], Alejandro Vargas-Uscategui [2], Xiaofeng Wu [1] and Peter C. King [2]**

1 School of Aerospace, Mechanical and Mechatronic Engineering, The University of Sydney, Sydney, NSW 2006, Australia; xiaofeng.wu@sydney.edu.au
2 Commonwealth Scientific and Industrial Research Organisation Manufacturing, Private Bag 10, Clayton, VIC 3169, Australia; alejandro.vargas@csiro.au (A.V.-U.); peter.king@csiro.au (P.C.K.)
* Correspondence: di261@cam.ac.uk; Tel.: +44-012-2376-4772
† Now at: Department of Engineering, University of Cambridge, Cambridge, CB2 1PZ, UK.

**Featured Application: This study presents a data-efficient modelling approach for a single-track profile in Cold Spray Additive Manufacturing using an artificial neural network. The approach presented in this study can be extended to modelling cases of other deposition-based additive manufacturing technologies with a high deposition rate, such as Wire and Arc Additive Manufacturing and Laser Cladding. The developed model can serve as a tool in simulation software by defining a realisable feature size at product design phases and predicting an as-fabricated product in these near-net-shaped manufacturing technologies. Hence, it allows designers to form a better idea of product design limitation and potential material waste after post-machining, as well as assessing and minimising economic and environmental impact with the aid of an appropriate toolpath planning algorithm.**

**Abstract:** Cold spray is emerging as an additive manufacturing technique, particularly advantageous when high production rate and large build sizes are in demand. To further accelerate technology's industrial maturity, the problem of geometric control must be improved, and a neural network model has emerged to predict additively manufactured geometry. However, limited data on the effect of deposition conditions on geometry growth is often problematic. Therefore, this study presents data-efficient neural network modelling of a single-track profile in cold spray additive manufacturing. Two modelling techniques harnessing prior knowledge or existing model were proposed, and both were found to be effective in achieving the data-efficient development of a neural network model. We also showed that the proposed data-efficient neural network model provided better predictive performance than the previously proposed Gaussian function model and purely data-driven neural network. The results indicate that a neural network model can outperform a widely used mathematical model with data-efficient modelling techniques and be better suited to improving geometric control in cold spray additive manufacturing.

**Keywords:** cold spray; neural network; additive manufacturing; data-efficient; model; profile; geometry; spray angle; limited data; machine learning

**Citation:** Ikeuchi, D.; Vargas-Uscategui, A.; Wu, X.; King, P.C. Data-Efficient Neural Network for Track Profile Modelling in Cold Spray Additive Manufacturing. *Appl. Sci.* **2021**, *11*, 1654. https://doi.org/10.3390/app11041654

Academic Editor: Marco Mandolini

Received: 25 January 2021
Accepted: 9 February 2021
Published: 12 February 2021



## 1. Introduction

Cold spray is a solid-state materials deposition technology that employs a supersonic gas jet to accelerate powder particles to 500–1000 m/s. Due to the particles' kinetic energy, local metallurgical bonding and mechanical interlocking are achieved without in-flight melting. This characteristic provides unique advantages that are difficult to achieve otherwise, including deposition free of melting-induced microstructure changes, the ability to handle oxygen-sensitive materials without a protective atmosphere and a high deposition rate with a narrow nozzle diameter [1–4].

Cold spray has recently been recognised to possess great potential as an alternative additive manufacturing technology and in this context is referred to as Cold Spray Additive Manufacturing (CSAM) [5–8]. This potential is particularly important when high production rates, large build sizes and repair or building on an existing structure are in demand, e.g., in aerospace industries [8,9]. The protective atmosphere-free environment allows for the fabrication of large components that are not possible with other additive manufacturing technologies, e.g., powder bed fusion, while providing a flexible selection of oxygen-sensitive powder materials [8,10,11]. These benefits have resulted in several successful demonstrations of the technology at different levels of fabrication complexity, ranging from a simple tubular structure [12], pyramidal fin array [13], to more complex parts such as topologically optimised components [14].

However, several fundamental and practical challenges need to be addressed to fully adopt the CSAM technology in commercial applications. One of these is geometric control, which is a common problem for other high production rate additive manufacturing (HPRAM) processes, including Wire and Arc Additive Manufacturing (WAAM) [15,16] and Laser Cladding (LC) [17,18]. Poor geometric control places many limitations on applying HPRAM technologies; examples include varying geometric quality, difficulty producing complex geometries, and geometry-induced property variations [7,8,19]. Hence, geometric control must be addressed to facilitate further development and commercial integration of CSAM and other HPRAM technologies.

Given the track-by-track and layer-by-layer nature of HPRAM, a high-accuracy process model based on the shape of a characteristic processing unit (e.g., single-track profile) provides a promising solution to the problem and often forms a basis for the modelling of higher geometric processing units, such as overlapping and overlayer models [20,21]. The single-track profile modelling in HPRAM was previously attempted using two distinct approaches: mathematical and data-driven modelling.

In WAAM, the symmetric single-track bead profile has been approximated using various basic mathematical function models ranging from parabolic, cosine and arcs [22,23]. These mathematical function models are often combined with another regression model to provide predictive capabilities. For example, Suryakumar et al. developed a quadratic regression model based on experimental data, computing the coefficients of the parabolic function model to describe a single-track bead profile [22]. In CSAM, a mathematical Gaussian function model is often chosen due to the mass distribution of jetted powder being assumed to be of Gaussian function profile [24,25]. Some other studies have utilised different mathematical function models, such as triangular [26,27] and trapezoidal [28]. Due to the complex processes underlying each HPRAM, there is no agreement on the choice of a single mathematical function model with simplifying assumptions, and the suitable model often depends on process conditions and their combination [23], leading to limited prediction accuracy over a wide range of process conditions using only a single mathematical functional model.

Data-driven modelling has emerged as an alternative approach due to its excellent non-linear mapping capability and increased accessibility of available software options [18,29]. Xiong et al. developed an Artificial Neural Network (ANN) model to predict the height and width of a single-track bead profile in WAAM [30]. The results were compared with those of a quadratic regression model and showed that the ANN model outperformed in the prediction of key geometric features. However, data-driven modelling has previously been limited to only predicting the height and width in HPRAM, and unlike the mathematical modelling approach, it has not been adopted to describe an entire track profile. Thus, there has been no exploration of the technique beyond symmetric single-track profiles or predicting details in track profiles. Our previous study attempted to address this issue in CSAM, focusing on the ANN modelling of a single-track profile with high morphology at normal and off-normal spray angles [31]. Our results demonstrated the potential of a data-driven modelling approach for better prediction accuracy than a mathematical counterpart, i.e., the Gaussian function model.

However, the limitation of a data-driven modelling approach was also observed in our previous study [31], namely, the necessity of a large amount of process training data to achieve a high prediction accuracy, which has also been identified recently in relevant manufacturing studies [32,33]. This data scarcity issue is associated with high experimental costs and the lack of an automated measurement system in HPRAM. Liu et al. applied a grey modelling technique for the first time in a thermal spray process in an attempt to overcome the issue, harnessing both mathematical (or white box) and data-driven (or black-box) modelling approaches [32]. Despite the reasonable prediction accuracy achieved in this study, the authors concluded that more complex and nonlinear phenomena existed and suggested further exploration of data-efficient modelling approaches to improve prediction accuracy.

Therefore, this study focuses on the prediction of a single-track profile in CSAM, at both normal and off-normal spray angles, using a data-efficient ANN (DANN) approach to demonstrate that data-driven modelling can achieve better prediction accuracy than its mathematical counterpart that has already been adopted in CSAM. Inspired by the study by Liu et al. [32], we leverage a mathematical function model as domain knowledge or the existing model at hand into the development of a DANN model. Specifically, a Gaussian function model, the model adopted elsewhere in CSAM studies, is selected with its coefficients computed by a quadratic regression model as applied in [22]. The significance of this study is four-fold: (1) the application of a data-driven modelling approach with a data-efficient focus in the prediction of a single-track profile in CSAM; (2) the comparative study among purely mathematical function, purely data-driven and data-efficient data-driven modelling approaches, in the context of HPRAM; (3) the demonstration that data-driven modelling can outperform more widely used mathematical modelling with appropriate data-efficient techniques in HPRAM; and (4) that existing models at hand can contribute to the development of a new data-driven model with better prediction accuracy without further experimentation.

## 2. Materials and Methods

An ANN is a feed-forward network model for supervised machine learning that performs the mapping of an input–output relationship based on appropriate training data. The development of an ANN with sufficient prediction accuracy depends on several pre-processing factors, including selecting appropriate input variables, quality of data and network architecture [34,35]. In this study, three experimental process variables were selected as inputs to an ANN model: spray angle, traverse speed and standoff distance, together with other input variables subsequently introduced in Section 2.2. This selection was made based on previous studies, demonstrating their influence on the geometry of a track profile in CSAM [24,36] and precise control with a robotic system [37].

A full factorial method was employed to define the experimental process variables' values in the ANN training dataset and design the set of experimental conditions. This approach was selected due to the nonlinear nature of CSAM and the affordable number of the process variables in this study. Here, three levels were considered for traverse speed and standoff distance, while four levels were adopted to effectively capture the effects of spray angle on track profiles in CSAM. The values of these process variables at each level are listed in Table 1. The minimum and maximum level values of each process variable corresponded to their operating limits to ensure the sufficient quality of track profiles. The intermediate level values were then equally placed between the values at each extreme level to maximise possible interactions between the process variables [38]. The resulting experimental design matrix in the full factorial method required 36 experimental single-track profiles for the proposed ANN modelling as a training dataset. The detailed experimental conditions of each single-track profile are summarised in Tables S1 and S2 in the Supplementary Materials.

**Table 1.** The levels of process variables in the experimental design matrix for the preparation of single-track profiles used for the training of the proposed data-efficient neural network model.

| Level | Spray Angle (°) | Traverse Speed (mm/s) | Standoff Distance (mm) |
|---|---|---|---|
| 1 | 45 | 25 | 30 |
| 2 | 60 | 100 | 40 |
| 3 | 75 | 200 | 50 |
| 4 | 90 | - | - |

### 2.1. Sample Preparation

All experimental single-track profiles were prepared using a commercial Impact Innovations (Haun, Germany) 5/11 cold spray gun guided by an ABB (Zurich, Switzerland) 4600 robot with 6 degrees of freedom, as can be seen in [39]. The gun was equipped with a long pre-chamber and an OUT1 tungsten carbide de Laval nozzle with a 6.2 mm exit diameter from Impact Innovations. The powder feedstock in this study was commercial purity grade −2 titanium from AP&C (Boisbriand, Canada) which was prepared by gas atomisation and distributed within the size of 15 to 45 μm (i.e., $D_{10}$ = 19 μm, $D_{50}$ = 34 μm and $D_{90}$ = 45 μm). The working gas was Nitrogen, preheated to 600 °C at a pressure of 5 MPa, accelerating the powder particles injected into the nozzle upstream at a feed rate of 1.9 kg/h. All spray variables and conditions were held constant during all experiments, except those listed in Table 1. A strip of commercial purity grade −2 titanium was used as a substrate with a dimension of 6 × 30 × 200 mm, having its surface processed with a milling machine from Avemax Machinery (Taichung City, Taiwan) and subsequently ground with a P120-SiC emery paper from LECO (Moenchengladbach, Germany). This surface processing was followed by cleaning with ethanol before the experiments. The fabrication of experimental single-track profiles was randomised to ensure statistically unbiased results with minimal effects of potential extraneous factors [40]. RobotStudio® software version 6.08 (ABB Robotics, Zurich, Switzerland) was used to confirm that there was sufficient distance beyond the substrate's edge to ensure that the robot's trajectory and traverse speed were stabilised before fabricating the profiles.

The geometry of each single-track profile was measured at five randomly selected locations using a LEXT OLS4000 confocal laser scanning microscope (Tokyo, Japan) and scanControl 2950–100 laser scanner from Micro-Epsilon (Ortenburg, Germany) with a z-axis measuring precision of at least 12 μm. These measurements were processed with the in-built filtering methods: flat Surface filtering in LEXT OLS4000 and average filtering with a filter size of 7 in scanControl Configuration Tool version 6.0. Additional filtering was applied with a local regression method using weighted linear least square and second-order polynomial model in MATLAB version R2018a. The five filtered track profiles were averaged to form each sample profile, as depicted in Figures S1–S3 in the Supplementary Materials, which was then considered for all modelling approaches in this study.

### 2.2. Data-Efficent Artificial Neural Network Model Design and Training

To demonstrate the effectiveness of leveraging a previous modelling attempt or existing model, a mathematical Gaussian function model, previously proposed in [24] and expressed in Equation 1, was selected and built with its free coefficients, A and σ, being predicted using a quadratic regression model as applied in [22].

$$y = \frac{A}{\sigma\sqrt{2\pi}} e^{-\frac{1}{2}\left(\frac{x}{\sigma}\right)^2}, \quad (1)$$

This selection was due to the mathematical model framework being capable of predicting an asymmetric single-track profile at off-normal angles and being often used in cold spray and CSAM [24,25,36]. For preparing the training dataset of outputs for the quadratic regression model, appropriate free coefficients were found through a Gaussian function equation curve fitting to each single-track profile shown in Figures S1–S3 of the

Supplementary Materials. The Curve Fitting Tool in MATLAB version R2018a was used with the trust-region-reflective algorithm and nonlinear least square method. The resulting free coefficients are summarised in Table S3 of the Supplementary Materials. With the experimental process parameters listed in Table 1 as inputs, the quadratic model was developed with the QR decomposition algorithm using the iterative reweighted least square method in Statistics and Machine Learning Toolbox, MATLAB version R2018a.

The proposed DANN modelling framework is shown in Figure 1, using a static ANN model for the geometric prediction of a single-track profile in CSAM. The DANN was developed to predict a polar length at a polar angle from the Tool Centre Point (TCP). A data-driven model for predicting a complete single-track profile can be developed by sampling a sufficient number of geometric points from the fabricated single-track profiles, as demonstrated by the area validation method in our previous work [31]. The polar length was sampled at 2.72° intervals around the TCP, resulting in 67 points from each single-track profile.

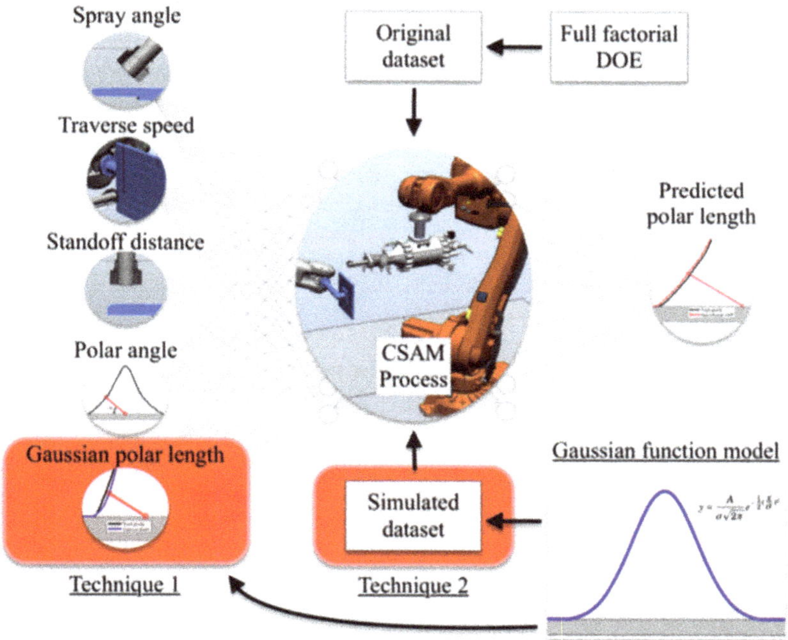


**Figure 1.** The proposed data-efficient artificial neural network modelling framework for the prediction of a single-track profile in cold spray additive manufacturing. The CSAM process part represents a digital version of the experimental equipment and setup described in Section 2.1.

Two data-efficient techniques were adopted in the proposed modelling framework, leveraging the Gaussian function model as an existing model, discussed above, to develop an ANN model. In Technique 1, the polar length approximated from the Gaussian function model was used as another input variable in addition to spray angle, traverse speed, standoff distance and polar angle. This technique explicitly leverages the partial domain knowledge of cold spray deposition, represented by the previously proposed Gaussian function model. At the same time, the DANN model learnt to compensate for the discrepancies between this knowledge and the true CSAM process, as successfully observed in other physical science fields [41,42]. Technique 2 was the augmentation of training data with a virtual input–output subset generated by the Gaussian function model. Therefore, the overall training dataset consisted of an empirical dataset prepared using the DOE method in Section 2.1 and a virtual dataset created using the Gaussian function model. The virtual dataset was generated from Gaussian function profiles using identical CSAM process parameters to those employed to make the physical test tracks and comprised of 804 training data points (i.e., 67 geometric sampling points from 12 simulated Gaussian function profiles). Both Technique 1 and 2 were independently employed in the preparation of training data and then simultaneously utilised in the development of the proposed DANN model.

For the development of the DANN model, this study iteratively changed the hidden layer architecture with the different number of hidden neurons (i.e., 1–15 neurons) per hidden layer and of hidden layers (i.e., 1–2 layers) to determine the optimal architecture using Mean Squared Error (MSE) as a performance evaluation function on an independent testing dataset. This range of hidden structures and evaluation function were selected given the limited data availability and frequent use in relevant studies respectively [29,30]. Following the 75–25 division method for the training-testing dataset, 12 single-track profiles were fabricated using the experimental methods described in Section 2.1 to form the testing dataset. The experimental process parameters for these testing profiles were randomly determined within the boundary of each parameter in Table 1 with the aid of the default random number generator in MATLAB version R2018a, summarised in Table S2 of the Supplementary Materials. Due to the limited training data, Bayesian regularised back-propagation was selected as the training method, eliminating the need for a validation dataset [43]. With this training method, hyperbolic tangent sigmoid and linear activation functions were selected for hidden and output layers, respectively, and all input and output variables were scaled to $[-1\ 1]$ for improving a training process [44]. The training of a DANN model was performed using the Deep Learning Toolbox in MATLAB version R2018a. Each architecture candidate was retrained 100 times to avoid local optima convergence due to initially allocated weights and biases. A purely data-driven ANN model was also developed for comparison in predictive performance, using the same methods presented above for the DANN model. The difference was two-fold: (1) the number of input variables was four without the approximated polar length by the Gaussian function model, and (2) only the original training dataset prepared from the experimental single-track profiles was used.

## 3. Results

The quality of the fabricated single-track profiles was validated against the cold spray and CSAM studies in our previous study [31], confirming that each process parameter's effects were consistent with previous relevant studies for the geometry of a single-track profile. Therefore, relevant and meaningful datasets could be generated from these single-track profiles that contained true representation of the CSAM process.

### 3.1. Data-Efficient Artificial Neural Network Model Validation

The Gaussian function model was built and evaluated on the coefficients taken from the testing single-track profiles (listed in Table S4 of the Supplementary Materials), showing the mean absolute error of 6.407%.

The iterative investigation of different hidden layer architectures found that the proposed DANN model, having two hidden layers with 11 and 4 hidden neurons respectively, provided the best predictive performance (i.e., [5 11 4 1]). During the training process, an MSE of $1.032 \times 10^{-4}$ was achieved on the normalised independent testing dataset. The normalised predictive results are shown in Figure 2a with the resulting Mean Absolute Percent Error (MAPE) of 1.230% and Maximum Absolute Percent Error (MXAPE) of 5.748%. These predictive performances were comparable to another study of data-efficient machine learning modelling in manufacturing, e.g., MAPE of 5.483% [32]. Figure 2b shows the developed DANN model's training process, confirming that the model was free of overfitting and underfitting and achieved the best performance at 445 epochs (or training iterations). Consequently, these results demonstrate the successful application of a data-efficient data-driven modelling approach to predict a single-track profile in CSAM.

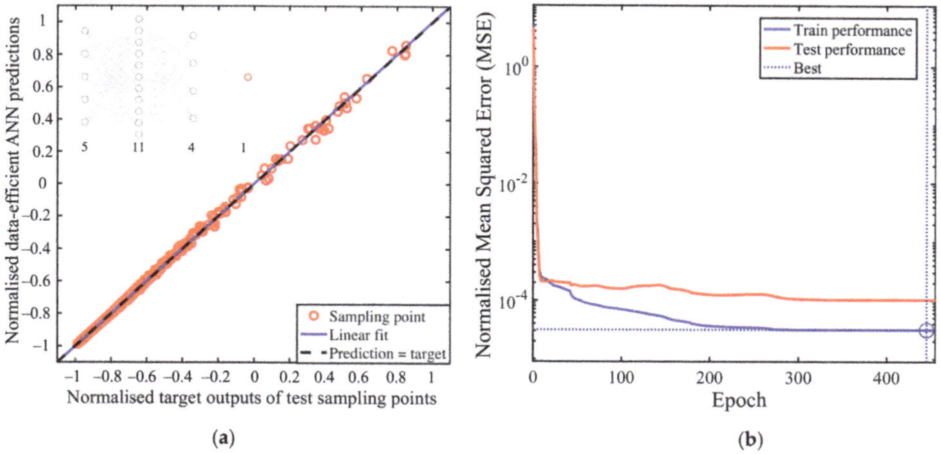


**Figure 2.** The results of the developed data-efficient neural network model with [5 11 4 1] architecture: (**a**) normalised data-efficient ANN predictions vs. target outputs (or polar lengths) with a Mean Squared Error (MSE) of $1.032 \times 10^{-4}$; (**b**) the training process of the developed data-efficient ANN model, showing no overfitting and underfitting.

### *3.2. Data-Efficient Artificial Neural Network Model Evaluation and Comparison*

The proposed DANN model, incorporating Techniques 1 and 2, was evaluated and compared with other modelling approaches, including a mathematical Gaussian function model [24] and a purely data-driven ANN model [31]. Here, the mathematical Gaussian function model was not the Gaussian function model used for the data-efficient ANN model as the existing model, but one with the optimal coefficients, listed in Table S4 of the Supplementary Materials, that were found through the curve-fitting method described in Section 2.2. The resulting model was referred to as the curve-fitted Gaussian function model and prepared to allow for the comparison of the proposed data-efficient ANN model against the best predictive performance that could be achieved using the previously proposed framework for mathematical Gaussian function modelling [24]. For the purely data-driven ANN model, the best performance was achieved with the architecture [4 5 7 1], resulting in an MSE of $3.852 \times 10^{-3}$. Furthermore, to investigate the effectiveness of each data-efficient technique, the data-efficient ANN models built using Technique 1 or Technique 2 solely were evaluated and compared. The prediction results of each model are summarised in Table 2 in absolute percent error and visually presented in Figure 3.

Compared with the purely data-driven ANN model, the data-efficient ANN model with Technique 1 or 2 alone showed better predictive performance with lower MAPEs and MXAPEs. This result indicates that both Technique 1 and 2 effectively achieved data-efficient learning and development of a data-driven ANN model. Furthermore, Technique 1 was more effective than Technique 2, with a MAPE half that of Technique 2. This result might be attributed to Technique 1 being more direct in guiding the learning process of weights and biases through the approximated target output (or polar length) than augmentation of the training dataset.

**Table 2.** Summary of the prediction results in absolute percent error for the testing single-track profiles in CSAM. The results are presented for: data-efficient ANN with the two techniques applied individually (Tech. 1 and Tech. 2), both applied (Tech. 1 + 2) also presented in Figure 2, curve-fitted Gaussian function model and purely data-driven ANN model. $R^2$ values are also listed.

| Absolute Error % | Data-efficient ANN | | | Curve-Fitted Gaussian | Purely Data-Driven ANN |
|---|---|---|---|---|---|
| | Tech. 1 | Tech. 2 | Tech. 1 + 2 | | |
| Mean | 2.060 | 4.040 | 1.230 | 1.873 | 7.174 |
| Minimum | 0.003 | 0.003 | 0.006 | 0.001 | 0.060 |
| Lower Q | 0.8147 | 1.113 | 0.3724 | 0.2682 | 2.510 |
| Median | 1.719 | 2.795 | 0.9081 | 0.8204 | 5.306 |
| Upper Q | 3.004 | 5.173 | 1.753 | 2.619 | 9.831 |
| Maximum | 9.685 | 20.78 | 5.748 | 11.83 | 33.26 |
| $R^2$ | 0.9984 | 0.9964 | 0.9988 | 0.9931 | 0.9925 |

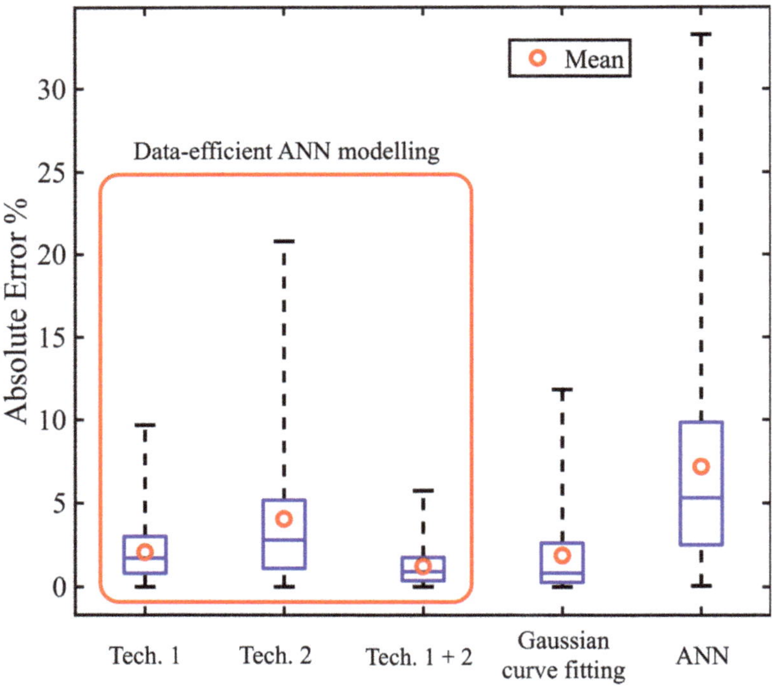


**Figure 3.** Graphical summary of the prediction results in absolute percent error for the testing single-track profiles in Cold Spray Additive Manufacturing (CSAM). The proposed data-efficient Artificial Neural Network (ANN) model is shown as Tech. 1 + 2 data-efficient model.

Hence, the proposed DANN model that combined the two data-efficient techniques achieved better predictive performance than the purely data-driven ANN model, showing that all of the prediction errors fell below the MAPE of the purely data-driven ANN model. The DANN model was also found to outperform the curved-fitted Gaussian function model with a lower MAPE and MXAPE. Notably, there was a lower number of predictions with large absolute percent errors (i.e., narrower upper quartile), as seen in Figure 3. This predictive capability became more significant when the entire single-track profile was predicted in the CSAM profiles, as presented in Figure 4.

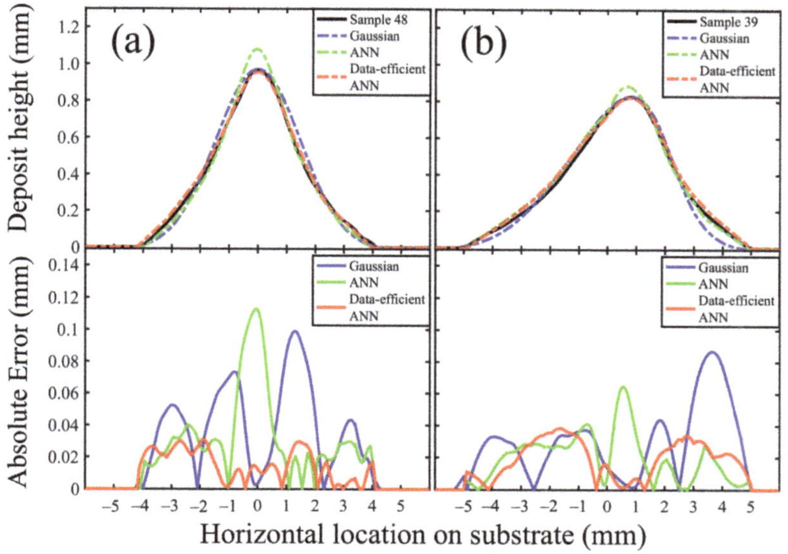


**Figure 4.** The experimental single-track profiles of the two selected profiles in the testing dataset as illustrative cases (black), plotted with the corresponding prediction results of the curve-fitted Gaussian function model (blue), purely data-driven ANN model (green) and the data-efficient ANN model (red): **(a)** Sample 48 (spray angle: 90°, traverse speed: 39 mm/s, standoff distance: 39 mm) and **(b)** Sample 39 (spray angle: 39°, traverse speed: 34 mm/s, standoff distance: 41 mm).

Figure 4 shows the single-track profile of the two selected testing samples as an illustration: (a) symmetric Sample 48 at a normal spray angle of 90° and (b) asymmetric Sample 39 at an off-normal spray angle of 39°. The prediction results of all other testing profiles are shown in Figure S4 in the Supplementary Materials. It was observed that the purely data-driven ANN model showed a higher track profile in Sample 48 and physically inconsistent predictions around the peak region in Sample 39. In contrast, the proposed DANN model outperformed in these regions. This result suggests that the lower prediction accuracy caused by the data-scarcity around profile peak regions, as also identified in our previous study [31], was overcome in this study by using the data-efficient techniques.

Compared with the curve-fitted Gaussian function model, the DANN model showed better predictive performance in both illustrative cases in Figure 4. For the symmetric Sample 48, the single-track profile was rather a triangular-shape, as previously observed in other cold spray studies [26,27], resulting in the curve-fitted Gaussian function model showing cyclic errors across the entire single-track profile. For the asymmetric Sample 39, the curve-fitted Gaussian function model showed a larger deviation on the spray-tilted side (i.e., the right end of the profile). In this particular region of single-track profiles, the particles land on the deposit closer to the normal angle, which combined with a shorter effective standoff distance results in an increased local accumulation in the deposit with improved deposition efficiency [26]. In contrast, the proposed DANN model could capture this physical phenomenon and predict significantly better at the profile end regions in Figure 4.

## 4. Conclusions

This study presented the application of a data-driven modelling approach with two techniques for leveraging the existing model at hand (i.e., the Gaussian function model as a demonstration) to achieve data-efficient learning and development of a new ANN model. The comparative study was performed for the prediction of the testing single-track profiles in CSAM; both DANN models with Technique 1 or 2 alone outperformed the purely data-driven ANN model with lower MAPE and MXAPE, demonstrating the effectiveness of the data-efficient techniques. Furthermore, the proposed DANN model, incorporating both Techniques 1 and 2, was compared against the curve-fitted Gaussian function model and found to provide better predictive performance. This result demonstrates that a data-driven modelling approach can outperform a conventionally used mathematical function model in CSAM, both at normal and off-normal spray angles, with appropriate data-efficient modelling techniques. Moreover, these techniques harnessed the existing model in developing a new data-driven ANN model without further experimentation. This result may indicate that previously built models of HPRAM can be improved by following this study's modelling strategy. In future works, we plan to incorporate the developed data-efficient ANN model into our toolpath planning algorithm to improve geometric control and achieve more complex-shaped product designs in cold spray additive manufacturing.

**Supplementary Materials:** The following are available online at https://www.mdpi.com/2076-3417/11/4/1654/s1, Figure S1: the experimental single-track profile S1–S12, Figure S2: the experimental single-track profile S13–S24, Figure S3: the experimental single-track profile S25–S36, Figure S4: the experimental single-track profile S37–S48 with the prediction of all models presented in this study, Table S1: Process parameters for the single-track profiles S1–S36 in the training dataset, Table S2: Process parameters for the single-track profiles S37–S48 in the testing dataset, Table S3: Curve-fitted coefficients of mathematical Gaussian function model for the single-track profiles S1–S36, Table S4: Curve-fitted coefficients of mathematical Gaussian function model for the single-track profiles S37–S48.

**Author Contributions:** Conceptualization, D.I.; methodology, D.I.; software, D.I.; validation, D.I.; formal analysis, D.I.; investigation, D.I.; resources, D.I., A.V.-U. and P.C.K.; data curation, D.I.; writing—original draft preparation, D.I.; writing—review and editing, D.I., A.V.-U., and P.C.K.; visualization, D.I.; supervision, X.W. and P.C.K.; project administration, D.I. and P.C.K.; funding acquisition, D.I., A.V.-U. and P.C.K. All authors have read and agreed to the published version of the manuscript.

**Funding:** This research was funded by CSIRO's Active Integrated Matter Future Science Platform (AIM–FSP) under the testbed number: WP10_TB04 and Research Training Program (International) Scholarship awarded to the first author from Department of Education, Skills and Employment, Australia.

**Informed Consent Statement:** Not applicable.

**Acknowledgments:** We would like to acknowledge and thank CSIRO Characterisation for access to LEXT OLS4000 confocal laser scanning microscope.

**Conflicts of Interest:** The authors declare no conflict of interest.

## References

1. Gärtner, F.; Stoltenhoff, T.; Schmidt, T.; Kreye, H. The cold spray process and its potential for industrial applications. *J. Therm. Spray Technol.* **2006**, *15*, 223–232. [CrossRef]
2. Karthikeyan, J. The advantages and disadvantages of the cold spray coating process. In *The Cold Spray Materials Deposition Process*; Champagne, V.K., Ed.; Elsevier: Amsterdam, The Netherlands, 2007; pp. 62–71, ISBN 978-1-84569-181-3.
3. Villafuerte, J. Current and future applications of cold spray technology. *Met. Finish.* **2010**, *108*, 37–39. [CrossRef]
4. Luo, X.-T.; Li, C.-X.; Shang, F.-L.; Yang, G.-J.; Wang, Y.-Y.; Li, C.-J. High velocity impact induced microstructure evolution during deposition of cold spray coatings: A review. *Surf. Coat. Technol.* **2014**, *254*, 11–20. [CrossRef]
5. Sova, A.; Grigoriev, S.; Okunkova, A.; Smurov, I. Potential of cold gas dynamic spray as additive manufacturing technology. *Int. J. Adv. Manuf. Technol.* **2013**, *69*, 2269–2278. [CrossRef]
6. Pathak, S.; Saha, G. Development of sustainable cold spray coatings and 3D additive manufacturing components for repair/manufacturing applications: A critical review. *Coatings* **2017**, *7*, 122. [CrossRef]
7. Li, W.; Yang, K.; Yin, S.; Yang, X.; Xu, Y.; Lupoi, R. Solid-state additive manufacturing and repairing by cold spraying: A review. *J. Mater. Sci. Technol.* **2018**, *34*, 440–457. [CrossRef]
8. Yin, S.; Cavaliere, P.; Aldwell, B.; Jenkins, R.; Liao, H.; Li, W.; Lupoi, R. Cold spray additive manufacturing and repair: Fundamentals and applications. *Addit. Manuf.* **2018**, *21*, 628–650. [CrossRef]
9. Aggour, K.S.; Gupta, V.K.; Ruscitto, D.; Ajdelsztajn, L.; Bian, X.; Brosnan, K.H.; Kumar, N.C.; Dheeradhada, V.; Hanlon, T.; Iyer, N.; et al. Artificial intelligence/machine learning in manufacturing and inspection: A GE perspective. *MRS Bull.* **2019**, *44*, 545–558. [CrossRef]

10. Mutombo, K. Research and development of Ti and Ti alloys: Past, present and future. *IOP Conf. Ser. Mater. Sci. Eng.* **2018**, *430*, 0120071–0120076. [CrossRef]
11. Titomic Titomic Kinetic FusionTM. Available online: https://www.titomic.com/titomic-kinetic-fusion.html (accessed on 22 March 2019).
12. Barnett, B.; Trexler, M.; Champagne, V. Cold sprayed refractory metals for chrome reduction in gun barrel liners. *Int. J. Refract. Met. Hard Mater.* **2015**, *53*, 139–143. [CrossRef]
13. Cormier, Y.; Dupuis, P.; Jodoin, B.; Corbeil, A. Pyramidal fin arrays performance using streamwise anisotropic materials by cold spray additive manufacturing. *J. Therm. Spray Technol.* **2016**, *25*, 170–182. [CrossRef]
14. Lynch, M.E.; Gu, W.; El-Wardany, T.; Hsu, A.; Viens, D.; Nardi, A.; Klecka, M. Design and topology/shape structural optimisation for additively manufactured cold sprayed components. *Virtual Phys. Prototyp.* **2013**, *8*, 213–231. [CrossRef]
15. Ma, G.; Zhao, G.; Li, Z.; Yang, M.; Xiao, W. Optimization strategies for robotic additive and subtractive manufacturing of large and high thin-walled aluminum structures. *Int. J. Adv. Manuf. Technol.* **2019**, *101*, 1275–1292. [CrossRef]
16. Li, Y.; Li, X.; Zhang, G.; Horváth, I.; Han, Q. Interlayer closed-loop control of forming geometries for wire and arc additive manufacturing based on fuzzy-logic inference. *J. Manuf. Process.* **2020**. [CrossRef]
17. Liu, H.; Qin, X.; Huang, S.; Jin, L.; Wang, Y.; Lei, K. Geometry characteristics prediction of single track cladding deposited by high power diode laser based on genetic algorithm and neural network. *Int. J. Precis. Eng. Manuf.* **2018**, *19*, 1061–1070. [CrossRef]
18. Gonçalves, D.A.; Stemmer, M.R.; Pereira, M. A convolutional neural network approach on bead geometry estimation for a laser cladding system. *Int. J. Adv. Manuf. Technol.* **2020**, *106*, 1811–1821. [CrossRef]
19. Frazier, W.E. Metal additive manufacturing: A review. *J. Mater. Eng. Perform.* **2014**, *23*, 1917–1928. [CrossRef]
20. Ding, D.; Pan, Z.; Cuiuri, D.; Li, H. A multi-bead overlapping model for robotic wire and arc additive manufacturing (WAAM). *Robot. Comput. Integr. Manuf.* **2015**, *31*, 101–110. [CrossRef]
21. Nenadl, O.; Kuipers, W.; Koelewijn, N.; Ocelík, V.; de Hosson, J.T.M. A versatile model for the prediction of complex geometry in 3D direct laser deposition. *Surf. Coat. Technol.* **2016**, *307*, 292–300. [CrossRef]
22. Suryakumar, S.; Karunakaran, K.P.; Bernard, A.; Chandrasekhar, U.; Raghavender, N.; Sharma, D. Weld bead modeling and process optimization in Hybrid Layered Manufacturing. *Comput. Des.* **2011**, *43*, 331–344. [CrossRef]
23. Xiong, J.; Zhang, G.; Gao, H.; Wu, L. Modeling of bead section profile and overlapping beads with experimental validation for robotic GMAW-based rapid manufacturing. *Robot. Comput. Integr. Manuf.* **2013**, *29*, 417–423. [CrossRef]
24. Chen, C.; Xie, Y.; Verdy, C.; Liao, H.; Deng, S. Modelling of coating thickness distribution and its application in offline programming software. *Surf. Coat. Technol.* **2017**, *318*, 315–325. [CrossRef]
25. Wu, H.; Xie, X.; Liu, M.; Chen, C.; Liao, H.; Zhang, Y.; Deng, S. A new approach to simulate coating thickness in cold spray. *Surf. Coat. Technol.* **2020**, *382*, 125151. [CrossRef]
26. Kotoban, D.; Grigoriev, S.; Okunkova, A.; Sova, A. Influence of a shape of single track on deposition efficiency of 316L stainless steel powder in cold spray. *Surf. Coat. Technol.* **2017**, *309*, 951–958. [CrossRef]
27. Klinkov, S.V.; Kosarev, V.F.; Shikalov, V.S. Influence of nozzle velocity and powder feed rate on the coating mass and deposition efficiency in cold spraying. *Surf. Coat. Technol.* **2019**, *367*, 231–243. [CrossRef]
28. Zhu, W.; Zhang, X.; Zhang, M.; Tian, X.; Li, D. Integral numerical modeling of the deposition profile of a cold spraying process as an additive manufacturing technology. *Prog. Addit. Manuf.* **2019**, *4*, 357–370. [CrossRef]
29. Deng, J.; Xu, Y.; Zuo, Z.; Hou, Z.; Chen, S. Bead geometry prediction for multi-layer and multi-bead wire and arc additive manufacturing based on XGBoost. In *Transactions on Intelligent Welding Manufacturing*; Chen, S., Zhang, Y., Feng, Z., Eds.; Springer: Singapore, 2019; pp. 125–135. ISBN 978-981-13-8667-1.
30. Xiong, J.; Zhang, G.; Hu, J.; Wu, L. Bead geometry prediction for robotic GMAW-based rapid manufacturing through a neural network and a second-order regression analysis. *J. Intell. Manuf.* **2014**, *25*, 157–163. [CrossRef]
31. Ikeuchi, D.; Vargas-Uscategui, A.; Wu, X.; King, P.C. Neural network modelling of track profile in cold spray additive manufacturing. *Materials* **2019**, *12*, 2827. [CrossRef]
32. Liu, M.; Zhang, Y.; Dong, W.; Yu, Z.; Liu, S.; Gomes, S.; Liao, H.; Deng, S. Grey modeling for thermal spray processing parameter analysis. *Grey Syst. Theory Appl.* **2020**, *10*, 265–279. [CrossRef]
33. Olleak, A.; Xi, Z. Calibration and validation framework for selective laser melting process based on multi-fidelity models and limited experiment data. *J. Mech. Des.* **2020**, *142*, 1–13. [CrossRef]
34. Shafi, I.; Ahmad, J.; Shah, S.I.; Kashif, F.M. Impact of varying neurons and hidden layers in neural network architecture for a time frequency application. In Proceedings of the 2006 IEEE International Multitopic Conference, Islamabad, Pakistan, 23–24 December 2006; pp. 188–193.
35. May, R.; Dandy, G.; Maier, H. Review of input variable selection methods for artificial neural networks. In *Artificial Neural Networks—Methodological Advances and Biomedical Applications*; InTech: London, UK, 2011; pp. 19–44. ISBN 9789533072432.
36. Cai, Z.; Deng, S.; Liao, H.; Zeng, C.; Montavon, G. The effect of spray distance and scanning step on the coating thickness uniformity in cold spray process. *J. Therm. Spray Technol.* **2014**, *23*, 354–362. [CrossRef]
37. Evjemo, L.D.; Moe, S.; Gravdahl, J.T.; Roulet-Dubonnet, O.; Gellein, L.T.; Brøtan, V. Additive manufacturing by robot manipulator: An overview of the state-of-the-art and proof-of-concept results. In Proceedings of the IEEE International Conference on Emerging Technologies and Factory Automation ETFA, Limassol, Cyprus, 12–15 September 2017; Volume 99, pp. 1–8.

38. Noriega, A.; Blanco, D.; Alvarez, B.J.; Garcia, A. Dimensional accuracy improvement of FDM square cross-section parts using artificial neural networks and an optimization algorithm. *Int. J. Adv. Manuf. Technol.* **2013**, *69*, 2301–2313. [CrossRef]
39. Green, A. Cold Spray Heats Up. Available online: https://blog.csiro.au/cold-spray-heats-up/ (accessed on 4 February 2021).
40. Krishaniah, K.; Shahabudeen, P. Fundamentals of experimental design. In *Applied Design of Experiments and Taguchi Methods*; PHI Learning Pvt. Ltd.: New Delhi, India, 2012; pp. 22–48. ISBN 978-81-203-4527-0.
41. Karpatne, A.; Watkins, W.; Read, J.; Kumar, V. Physics-guided Neural Networks (PGNN): An application in lake temperature modeling. *arXiv* **2017**, arXiv:1710.11431.
42. Deist, T.M.; Patti, A.; Wang, Z.; Krane, D.; Sorenson, T.; Craft, D. Simulation-assisted machine learning. *Bioinformatics* **2019**, *35*, 4072–4080. [CrossRef]
43. Burden, F.; Winkler, D. Bayesian regularization of neural networks. In *Methods in Molecular Biology (Clifton, N.J.)*; Springer: Berlin, Germany, 2008; Volume 458, pp. 25–44.
44. Haykin, S. *Neural Networks and Learning Machines*, 3rd ed.; Pearson Education, Inc.: London, UK, 2009; ISBN 0131471392.

applied sciences

MDPI

*Article*

# Design for Additive Manufacturing: Tool Review and a Case Study

**Daniel Moreno Nieto [1,]* and Daniel Moreno Sánchez [2]**

[1] Departamento de Ingeniería Mecánica y Diseño Industrial, Escuela Superior de Ingeniería, Universidad de Cádiz, Puerto Real, 11510 Cádiz, Spain

[2] Departamento de Ciencia de los Materiales e Ingeniería Metalúrgica y Química Inorgánica, Facultad de Ciencias, IMEYMAT, Campus Río San Pedro, Universidad de Cádiz, PuertoReal, 11510 Cádiz, Spain; danielmoreno.sanchez@uca.es

* Correspondence: daniel.moreno@uca.es; Tel.: +34-676923332

**Abstract:** This paper aims to collect in a structured manner different computer-aided engineering (CAE) tools especially developed for additive manufacturing (AM) that maximize the capabilities of this technology regarding product development. The flexibility of the AM process allows the manufacture of highly complex shapes that are not possible to produce by any other existing technology. This fact enables the use of some existing design tools like topology optimization that has already existed for decades and is used in limited cases, together with other novel developments like lattice design tools. These two technologies or design approaches demand a highly flexible manufacturing system to be applied and could not be used before, due to the conventional industrial process limitations. In this paper, these technologies will be described and combined together with other generic or specific design tools, introducing the study case of an additive manufactured mechanical design of a bicycle stem.

**Keywords:** additive manufacturing; industrial design; fused deposition modeling; parametric design; industrial design; CAD computer aided design; topology optimization; lattice design

**Citation:** Moreno Nieto, D.; Moreno Sánchez, D. Design for Additive Manufacturing: Tool Review and a Case Study. *Appl. Sci.* **2021**, *11*, 1571. https://doi.org/10.3390/app11041571

Academic Editor: Marco Mandolini

Received: 11 January 2021
Accepted: 3 February 2021
Published: 9 February 2021



## 1. Introduction

Additive manufacturing (AM) is a set of manufacturing processes that consist in the generation of three-dimensional models from digital files, using different equipment and technologies that build objects layer by layer [1–4]. These technologies are reaching the age of maturity due to their effective incorporation in different industries, besides the pioneering aerospace and automotive industries, such as consumer products. There are many advantages that additive manufacturing technologies offer to designers and engineers for the development of new parts and products.

Advantages, the freedom of constructed geometries, which allows the elaboration of complex shapes that would not be possible by other technologies [5], the reduction of parts in the assemblies, by simplifying them or allowing the printing of the assembled parts [6–8], the capability of locating the necessary properties in specific areas or developing variable properties in the geometries of the parts and the possibility of manufacturing optimized structures that modify their properties according to the scale to which they refer, from the micro to the macro scale [9,10], are the most remarkable.

Through these advantages, additive manufacturing in the current state of development is increasingly finding its place in industry, while understanding that it is not a substitute technology, but rather complementary to current production technologies, in cases where some of the characteristics mentioned above are required. That is, it will take its place within the production processes and coexisting technologies, while having special importance in parts with highly complex geometries, in small series and in customized products.

The limitations of these processes have also been identified: the geometric characteristics (tolerances in the definition of wall thicknesses, holes, rounding, cantilevers, bridges and angles) and the anisotropy of the process that affects the mechanical properties of printed parts [11,12]. Those inherent to the execution of the process, such as the orientation of the part or the definition of the supports, deposition rates, finishes or production times, are decisive for the correct definition of the geometries [13–15], as well as other factors such as scale limitations, the absence of specific design and simulation tools and economic considerations. All these factors are related to the execution of the design for additive manufacturing [16,17].

Currently, new design prescriptions and standards are appearing, providing the designer a way to mitigate the limitations of the technologies and increase the functional performance of the final fabricated parts, known as design for additive manufacturing (DfAM) [18–22]. Some aspects to be considered and that are directly related with the manufacturing process are the disposition of holes that enable the evacuation of the residual material in the powder bed or liquid resin material or the deposition orientation in fused deposition modeling (FDM), for example.

In this context, in recent years, there has been an important advance in the field of specific engineering technologies for additive manufacturing, with the continuous appearance of numerous tools that consider the materials and their properties as variables, allowing for the prediction of their behavior and thus optimizing designs for manufacture. Prior to the description of these tools, we will review in a general way the main stages of the work process for the design and additive manufacturing of parts.

## 2. The Additive Manufacturing Workflow

The approach to the additive manufacturing processes establishes an orderly workflow, so that by relying on different tools, processing files are generated for subsequent printing and post-processing. There are five fundamental stages in the additive manufacturing process, which are listed below and shown in a diagram in Figure 1.

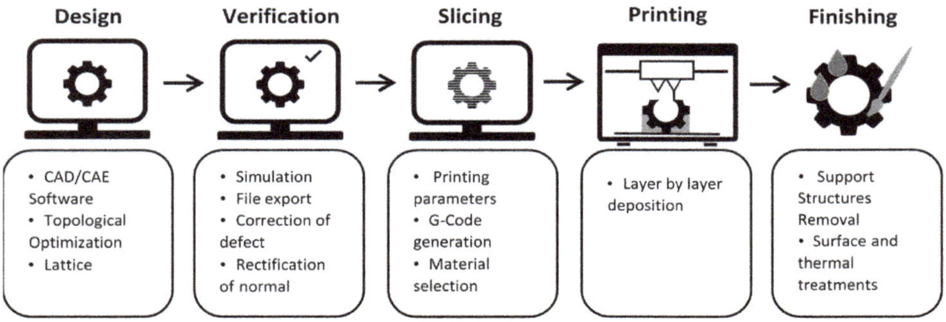


**Figure 1.** Additive manufacturing (AM) workflow. Note: CAD (Computer-Assisted Design); CAE (Computer-Aided-Engineering).

(a) Design: For the development of the parts and products that will be manufactured by additive processes, the first stage consists of the generation of three-dimensional designs by means of computer-assisted design tools (CAD). The characteristics of the processes allow the generation of complex geometries, however, not all shapes are always viable and they are closely linked to the technology used. The construction of 3D models can be done with conventional solid and surface modeling programs; likewise, due to the flexibility of the process, the use of advanced modeling tools, such as polygonal meshes or NURBS surfaces, is very common.

(b) Verification: Once the design process is completed, the next step is to export the files to the 3D printing standards. The most common file format is STL, which consists of a triangulated surface mesh that defines the complete geometry [23]. The export of files is done for their correct inclusion in the lamination tools or slicers. However, a previous step of surface verification is recommended, since sometimes discontinuities or incorrect orientations of the normals that define the surface triangles could be produced.

(c) Slicing: The lamination programs or slicers previously mentioned are the tools that generate the machine code with the characteristics of the process and the parameters, as well as the trajectories. It is important to define the correct parameters of the process according to the characteristics of the design and materials. The main parameters shared by most processes are: layer height, manufacturing speed, temperature and percentage of filling. The result of this process definition is the machine path files, which are transferred to the printing equipment for process activation.

(d) 3D printing: This is the physical process in which the materialization of the parts using the 3D printers in question is done. Its characteristics vary depending on the technology selected. A common feature in all the technologies is that this is a long process and usually requires post-processing compared with traditional manufacturing processes such as injection molding.

(e) Post-processing: Depending on the characteristics of the process, more or less intervention is necessary to achieve the specific finish and properties of the printed parts. Tasks common to all technologies are the removal of parts from the printing surface, the removal of supports and mechanical processes of surface finishing, as well as heat and surface treatments, if necessary.

## 3. Additive Manufacturing Design Tools

In order to take advantage of and mitigate the limitations of these AM processes, different tools have appeared in recent years that maximize the potential of the constructive capabilities offered by additive manufacturing technologies. These programs consist of engineering and CAD platforms such as mesostructured design and optimization programs, along with process management and simulation solutions.

One of the main objectives set in the development of this work is to configure an updated workflow, ordered and structured to optimize processes from the integration and exchange of files, as well as to maximize the potential of the possibilities of forming that AM technologies present, supported by these specific tools.

In Figure 2 is presented different CAD and computer-aided engineering (CAE) tools currently available for the AM design process. Some of these programs previously existed on the market and have been adapted to these technologies, as well as other new specific tools that are appearing as the development and implementation of them advance.

Within the software tools, in addition to the traditional CAD programs, such as parametric design programs, e.g., Catia and Solidworks (Dassault Systèmes, Vélizy-Villacoublay, France), Inventor (Autodesk, San Rafael, CA, USA) or NX (Siemens, Munich, Germany), we have to consider expanding the range of tools. Thanks to the design freedom offered by layer-by-layer construction in the AM technologies, which allows us to shape virtually any geometry, we must add other surface design tools, which include the above programs, and other new ones such as Rhino (Robert McNeel & Associates, Seattle, WA, USA) or Alias and 3D Studio (Autodesk, San Rafael, CA, USA). AM allows us to create objects with organic shapes and geometric complexity without a great impact on the cost or complexity of manufacturing. This freedom already existed in digital design for animation or video games, while CAD software traditionally associated with mechanical design had limited capabilities. Although in the latest versions of the abovementioned CAD software, the possibility of making complex geometries more easily has already been implemented, certain specialized software is still required. These programs usually work with 3D design tools based on NURBS or editable polygonal mesh that allow a freer and more organic design without limitations when generating geometries or parts.

**Figure 2.** Tools that can be used in the AM process.

Another trend in AM is the use of optimization software, based on developments that have had more than 20 years of use in the industry, however, they have never had an extensive use, because the solutions they provide, in geometric terms, are always very complex and traditional production processes for machining or casting did not meet the requirements for proper production. However, with the emergence of AM technologies and their layer-by-layer production systems, these geometries resulting from different optimizations, whether volumetric, shape or topological, as schematized in Figure 3, are perfectly achievable. The optimization tools determine formal solutions to different requirements through the application of numerical models in iterative processes, based on different calculation algorithms of shape, volume or load [24], trying to obtain ideal solutions with respect to the geometries of the parts.

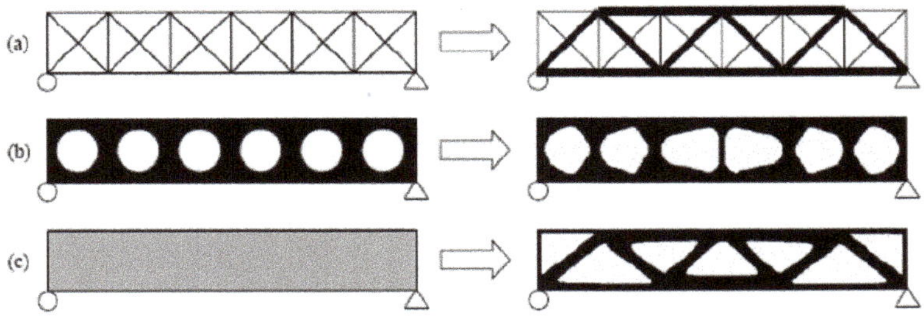


**Figure 3.** Simple beam parametric optimization (**a**) of shape (**b**) and topology (**c**). Adapted from [1].

Among the most widespread tools in the processes of AM, we must highlight the topological optimization (TO). In this type of tool, starting from some established objectives (usually minimizing mass or maximizing rigidity) and some requirements of load and restrictions of design, an orientated solution adapted to the defined objectives is obtained. The solutions provided by this tool serve as a reference shape of the part designs. Some commercial solutions are TOSCA, from Dassault Systems, a tool integrated in the Generative Design module of Catia V.6, multi-materials such as Paramatters (Paramatters, Ventura, CA, USA), Inspire based on Optistruct (Altair, Troy, NY, USA), Pareto (Sciart Software, Madison, WI, USA) or the Limitstate Form (Limitstate, Sheffield, UK). There are also processes defined in traditional CAE tools, such as finite element calculation software Ansys (Ansys Inc., Canonsburg, PA, USA), Abaqus (Dassault Systèmes, Vélizy-Villacoublay, France), Comsol (Comsol Group, Stockholm, Sweden) or Nastran (Autodesk, San Rafael, CA, USA), HyperWorks (Altair, Troy, NY, USA), which usually incorporate the simulation of the optimized parts for verification of the modifications, although they do not consider the anisotropic behavior of the printed parts. We differentiate in this section the topological optimization tools, continuous and discrete, differentiated according to the type of algorithm and the way of locating the load vectors. The first is based on continuous volumes and the second on systems of beams interconnected to the points that are subjected to maximum stress.

The most used algorithms for the optimization of these structures are the solid isotropic microstructure with penalty (SIMP) algorithm and the bidirectional evolutionary structural optimization (BESO) algorithm [24]. The established algorithms for topological optimization focus on conventional manufacturing systems and can be used in additive manufacturing if the constraints, usually volumetric and strain energy, are minimized.

Currently, TO is used in different sectors, such as wind [25] and automotive sectors [26] or the redesign of casting parts [27]. Additionally, one of the latest trends in TO is the use of a topology optimization algorithm while considering the space–time of the process. This concept has been well implemented in the bridge created by MX3D [28]. The process considered the evolution of the structure to self-support the weight of two robotic arms during the manufacturing, using wire arc additive manufacturing (WAAM) technology.

With the same objective of minimizing weight and trying to maintain mechanical properties, other programs, called mesostructure, trusses, lattices or lattice design tools are appearing, mainly oriented to powder bed technologies, which allow for modifying solids by generating lattices or trusses with different geometric characteristics that maximize properties, reducing weight significantly. These tools present methods for the definition of these mesostructures, controlling their densities, and are even able to define the mass center of the pieces. Programs like Netfabb (Autodesk, San Rafael, CA, USA), 3-Matic (Materialise, Lovaina, Belgium) or Simpleware (Synopsys, San Jose, CA, USA), or the renewed nTop suite (Ntopology, New York, NY, USA), usually incorporate the simulation of the optimized parts for verification of the modifications, again without considering the laminar character of the process and the anisotropy of the parts.

The identification of problems prior to 3D printing using simulation tools is a key factor for the correct implementation of these technologies. The deformation of the parts by thermomechanical processes or warping, together with the accumulation of tensions, inform us about the stability and success of the printing, with these being the main bottlenecks that block the wide expansion of these technologies. The advances in the materials and the design will allow the best control of these processes. However, the correct simulation of the 3D printing processes will help in the early detection of possible errors, with notable savings in materials and time [29].

The simulation of printing processes is a novel process that is still in its early stages of development. There is further development within metal AM technologies, especially for powder bed fusion processes, and not for direct energy deposition or powder bed fusion processes, notably the tools Simufact™, Ansys or Netfabb. However, in March 2017, the first commercial tool for the simulation of polymer materials processes appeared, Digimat (e-Xstream). Until then, the only existing developments were for metals. By means of these simulations, we can calculate the geometric deformations of the printed parts, as well as the residual stresses, once the printing process is completed.

These process simulations allow us to obtain different types of information: (i) they allow optimizing the process parameters to minimize the accumulated tensions and the deformations of the pieces, during and after printing; (ii) they allow us to determine the most suitable printing position; (iii) they allow us to determine the characteristics of the material that better adapt to the printing process; or (iv) they allow the optimization of the support structures.

All these processes bring the solution closer to the optimum solution as the parameters are modified. These tools also allow for integration with other finite element calculation systems, which will make it possible to predict the behavior of printed parts subjected to the loads and contour restrictions of the parts in service, taking into account the deformations and stresses of the parts once they have been produced.

The pre-production tools include different solutions that go through the revision and repair of the 3D printing files obtained from the three-dimensional model, generally of the STL type, where the polygons define the 3D model, ensuring closed surfaces and that the normal is correctly oriented. Examples of programs in this line are MeshMixer (Autodesk, San Rafael, CA, USA) MeshFix and MeshLab (Visual Computer Lab, Pisa, Italy); some are private and others are free pieces of software that are available and executable online. On the other hand, following the same line of free and private software, there are tools that help to manage production by optimizing the use of printing volume or by defining support structures, where necessary, that facilitate subsequent post-processing and space utilization. Examples are Netfabb, Magics or the application for powder bed fusion AM by Delmia (Dassault Systèmes, Vélizy-Villacoublay, France).

At this point, we cannot leave aside essential tools in the process of AM. These are the slicers or software that transform the geometric models in the machine code. There are free and private software tools; we highlight Cura (Ultimaker, Utretch, The Netherlands), Slicer3D (Harvard, US) and Simplify 3D (Cincinnati, OH, USA), and the main difference is the control possibilities, with these last ones offering more control options and the recreation of the 3D printing process in virtual media. This functionality helps in the early detection of printing errors. In these applications, it is possible to carry out an exhaustive control of the position, size and orientation of the piece as well as a multitude of process control variables, from the most essential ones, such as layer height, temperature and printing speed, to other more specific ones, such as the speed and retraction of the material in fused filament fabrication (FFF) processes.

In Table 1, a summary of the described tools that cover different needs and design approaches to exploit the capabilities of AM throughout the design, simulation and pre-production processes is described. This list intends to give an overview to different stakeholders and other researchers to understand the capabilities and different metrics of the tools, in order to facilitate the decision-making process of selecting a specific tool that could match their specific need. In this table, specific software descriptions are given with the main function of every tool, its relation to AM, whether it is directly related or not, the characteristics of usability, the complexity of the tool and price.

**Table 1.** List of remarkable software used in additive manufacturing design processes.

| Scheme | Oriented to AM | Main Functions Related to AM | Ease of Use | Cost * |
|---|---|---|---|---|
| SolidWorks/Inventor/ PTC Creo/ | Not specifically but useful | Solid and surface parametric modeling and basic FEA | Easy | Medium |
| NX/Fusion360 | Yes | Solid and surface modeling/Basic simulation | Easy | Low/Free student version |
| Free CAD | Not specifically but useful | Solid and surface modeling | Medium | Free |
| Catia | Yes | Solid and surface modeling | Easy | High |
| Rhinoceros 3D | Not specifically but useful | Surface and freeform modeling | Easy | Low |
| Blender | Not specifically but useful | Organic modeling and animation | Medium | Free |
| 3D Max | Not specifically but useful | Organic modeling and animation | Medium | Medium/Free student version |

**Table 1.** *Cont.*

| Scheme | Oriented to AM | Main Functions Related to AM | Ease of Use | Cost * |
|---|---|---|---|---|
| Inspire | Yes | Optimization/Lattice/Simulation/Printing setup | Easy | Medium/Free student version |
| Abaqus, HyperWorks, Ansys | Not specifically but useful | Simulations and TO | Complex | High |
| Ntop | Yes | Basic modeling/Lattice/TO/Simulation/Printing setup | Complex | Medium/Free student version |
| 3D Matic | Yes | Modeling/Lattice/Mesh editor/Topology optimization | Medium | Medium |
| Netfabb | Yes | Lattice/Mesh editor | Medium | High |
| Pareto | Yes | TO | Low | Medium |
| Tosca | Yes | TO and printing setup | Medium | High |
| Paramatters | Yes | TO | Low | Low |
| Digimat AM | Yes | Process simulation (polymers) | Low | Medium |
| Simufact | Yes | Process simulation (metals) | High | High |
| Alphastar | Yes | FEA/Process simulation | Medium | High |
| Cura/Prusa Slicer | Yes | Process condition definition and G-code generation | Low | Free |
| Simplify3D | Yes | Process condition definition and G-code generation | Medium | Low |
| Meshmixer | Yes | Mesh edition | Low | Free |
| Magics | Yes | Mesh edition, process conditions definition and G-code generation | Low | High |

* Cost is an estimation from the academic environment based on a comparative analysis. Suppliers give specific quotations depending on final use. They can change depending on whether the client belongs to industry, academia or a research center. Note: TO (Topological Optimization).

## 4. Results and Discussion. Case Study: A Bicycle Stem

Next, the case study design of a bicycle stem will be presented. This part is exposed to different loads and is optimized with two different strategies that will create a more efficient design. The part will undergo a topology optimization process and a lattice approach redesign. The goal of both processes is to reduce the part's weight while preserving the defined mechanical requirements.

There are several publications that show the design process with advanced tools for additive manufacturing. The application of topological optimization in different aeronautical supports [30] or in classical mechanical elements such as beams or hooks is very common [31,32]. Different studies show the possibilities of lattice structures regarding their behavior and mechanical properties. However, there are not many specific applications or study cases that illustrate the use of these tools for specific cases in terms of their procedure of use and execution with the combined use of these design tools [33].

The bicycle stem has maximum dimensions of 95 mm, and is loaded with to two main loads of 1500 N that simulate the weight of the user, increasing up to 3000 N to consider possible jumps or overloads, as shown in Figure 4. The bicycle stem also contains screws that are simulated with a load of 100 N in every hole due to the screw pressure and the pressure direction simulates the tightness of the screws. It also has a restriction of movement in all degrees of freedom due to the connection to the bicycle frame. The material selected for the manufacture and redesign of the part is polyamide 6 or nylon, due to its mechanical strength, hardness, rigidity and good toughness, in addition to a strong mechanical damping capacity and good resistance to fatigue and wear.

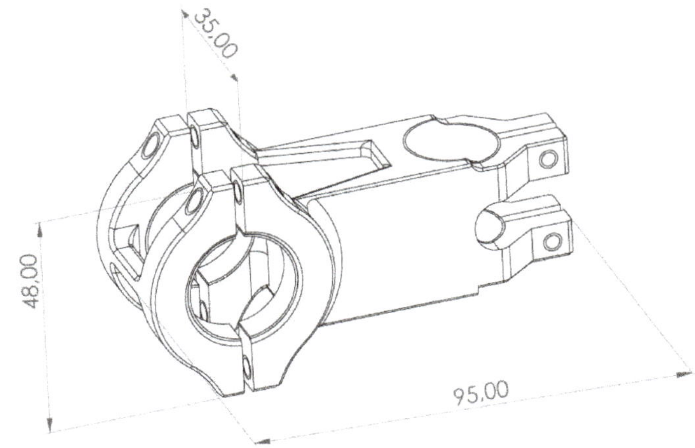


**Figure 4.** Main dimensions of the bicycle stem designed in the software SolidWorks™.

The initial design was developed with SolidWorks, a parametric design tool, for the problem specifications. Through simple sketching and basic modeling operations, we can configure the geometry. This will be the starting point for the redesign of the part. The file was exported in STEP format, which allows its import into the Inspire™ tool of Altair™. This topology optimization tool is intuitive and useful for novel users and works with the same calculation engine as the HyperWorks tool from the same company, as well as the OptiStruct software solver. The first action for a TO approach consists of assigning the material from the library to the imported pieces and, as we have mentioned, it will be polyamide 6. It is important to define within the assembly which volumes will be used for optimization and which will not. We will refer to the optimizable volumes where the material removal happens "design spaces" and those that cannot be altered "non-design spaces", as illustrated in Figure 5. The volume of the design spaces will be increased as much as possible to allow the solver to identify the most suitable solution.

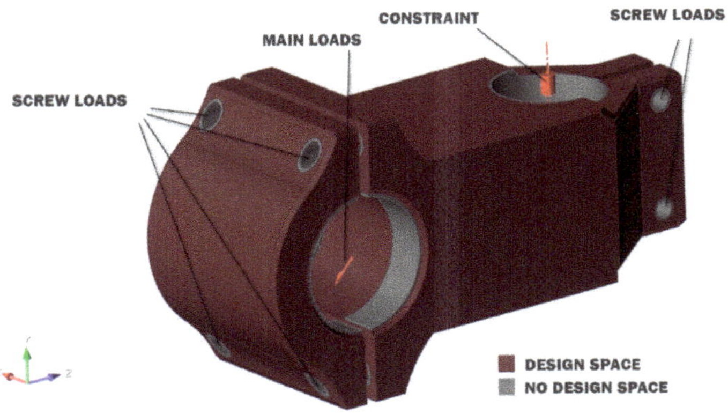


**Figure 5.** Definition of loads and constraints, as well as design and non-design spaces in the Inspire™ tool for the modified stem geometry.

The first step is to perform a static stress calculation according to the defined loads to verify that the part meets the objectives and, above all, that the parts have room for improvement for subsequent redesign. We make a stress simulation and observe the safety coefficient of each of the parts of the piece, as illustrated in Figure 6.

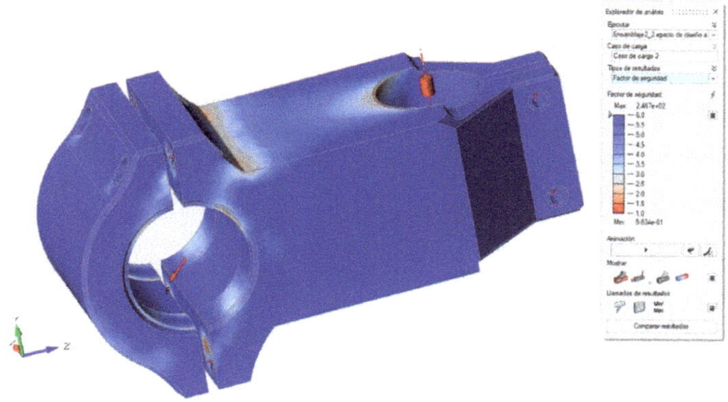

**Figure 6.** Result of the finite elements analysis expressed as safety factors of the part using the Inspire™ tool.

The result clearly expresses that the part is susceptible to optimization since the values of the safety factor greatly exceed a factor of 1.2 that we accept as the admissible limit. Once the need to reduce the mass of this part has been determined, we define a topological optimization that maintains the mentioned safety factor of 1.2, minimizing the mass and maximizing the rigidity of the part. The result will be a poorly defined organic geometry that we will reconstruct using NURBS surfaces that help us to generate a continuous and well-defined surface, as illustrated in Figure 7. This result will also be analyzed by a static calculation to validate the adopted design solution.

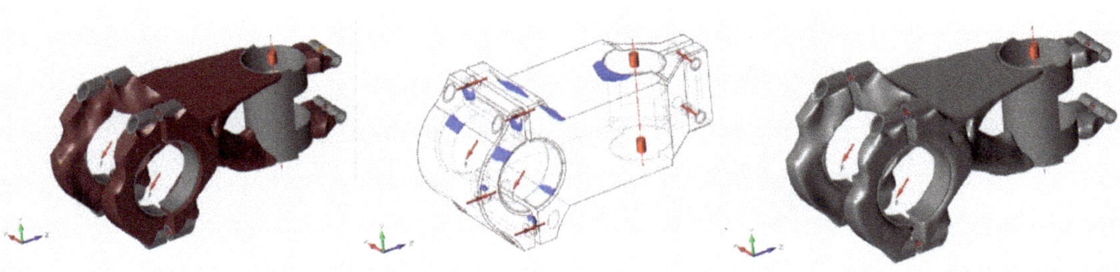

**Figure 7.** Result of the topological optimization on the left, the middle image shows the highest stress concentration areas and, on the right, the smoothed surface using NURBS, carried out with Inspire software.

With this approach to the redesign of the piece, a notable decrease in weight has been achieved, going from 46.36 g to 24.93 g, which mean a reduction of 53.8% in the optimized piece without losing functionality and according to the design specifications defined at the beginning. Again, this solution is validated by means of finite element analysis (FEA) tools. This solution not only will reduce costs and production times but can also mean important savings in energy consumption with the use of the redesigned part.

This optimized model is intended to take a further step towards weight reduction. Using the lattice tools that have already been mentioned and, more precisely, Ntop software, we intend to further reduce the weight by replacing the continuous material of the part of the piece that can be optimized by generating a microstructure of trusses without losing mechanical properties and maintaining its resistance. It is important to emphasize that the tools used do not allow the validation of the results by means of calculation by finite elements. The simulation strategy should be based on a multiscale homogenization based on the analysis of a representative volume element. Therefore, the result presented below is merely illustrative of the redesign process for mesostructure gratings.

The import/export of files is again done with STEP files. The lattice structure will be applied in the main bodies in a combination of a shell operation and the internal lattices. Ntop software offers a wide range of structures to be applied, which are different in their geometry and are structured by continuous beams or walls, which can be hexagonal, tetrahedral, cubic or random, among others, with different configurations of each of their definitions. In this case, we apply a specific structure, octec, that is an octahedral structure with only eight beams. The result is a 2 mm shell that covers the entire part together with a wired structure infill that will allow, depending on the load requirements, a certain thickness that can also be variable. This allows us to modify the density of the piece by region and even define the position of the center of mass of the part.

The steps followed to obtain the lattice structure, illustrated in Figure 8, are (i) determination of the optimization area, in this case the central body of the bicycle stem. (ii) Definition of the shell thickness and type and size of the structure. (iii) Modification of the wire structure to ensure that there is connection to the parts, and that there are no open structures that reduce their mechanical characteristics. (iv) Definition of the thickness of the cellular structure or lattice, in this case, 1 mm in diameter for each segment. (v) The Boolean union of all the parts that will provide a continuous piece capable of being printed. (vi) Finally, the exportation of the new part in a printable file, such as STL.

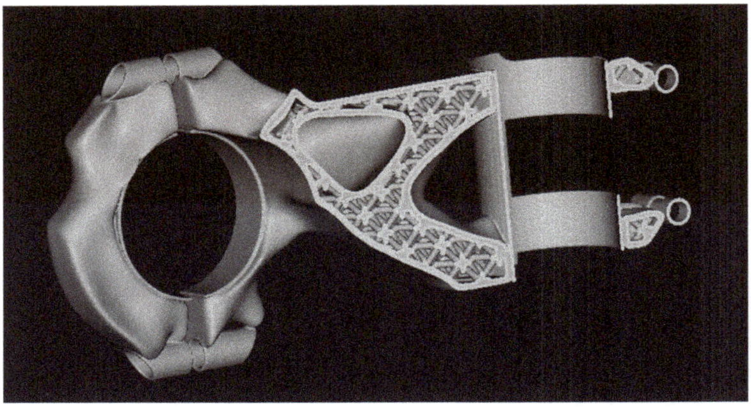

**Figure 8.** Section view of the result of the redesign using lattice-type tools with Ntop platform software.

The result, as observed in Figure 7, presents a topology-optimized geometry that has reduced weight due to the application of the octec lattice, while maintaining the performance of the product, with lower weight, production time and cost.

## 5. Conclusions

This article presents an introduction to design for additive manufacturing, illustrating the main advantages and disadvantages of these technologies along with the different stages that make up the process. The main part consists of a review and explanation of the different tools involved in each stage of the process, focusing on additive manufacturing design. In this way, different computer-aided design tools are presented. Some of these tools are well known in the scientific and engineering community, while other, less traditional tools, are new and exclusive for design oriented to additive manufacture. These tools, in their application, will allow us to extract the maximal capacities of these processes, when it is a question of obtaining robust and light design for pieces that need reduced masses. Likewise, this article illustrates the redesign process oriented towards the additive manufacture of a characteristic part of mechanical design: a bicycle stem. In this process, significant improvements in mass reduction are achieved through the application of topological and mesostructural optimization tools, illustrating their use step by step.

**Author Contributions:** Both authors, D.M.N. and D.M.S. participated in conceptualization, methodology, software, validation, formal analysis, investigation, resources, data curation, writing—original draft preparation, writing—review and editing, visualization, supervision, project administration and funding acquisition. All authors have read and agreed to the published version of the manuscript.

**Funding:** This research received no external funding.

**Institutional Review Board Statement:** The study was conducted according to the guidelines of the Declaration of Helsinki, and approved by the Institutional Review Board.

**Informed Consent Statement:** Informed consent was obtained from all subjects involved in the study.

**Data Availability Statement:** No new data were created or analyzed in this study. Data sharing is not applicable to this article.

**Acknowledgments:** The authors would like to acknowledge the support of Junta de Andalucía (IN-NANOMAT research group, TEP946) and the department of mechanical engineering and industrial design at the Engineering School from Cadiz University.

**Conflicts of Interest:** The authors declare no conflict of interest.

## References


1. Mueller, B. Additive Manufacturing Technologies – Rapid Prototyping to Direct Digital Manufacturing. In *Assembly Automation*; Springer: Berlin, Germany, 2012; Volume 32.
2. Niaki, M.K.; Nonino, F. The Management of Additive Manufacturing. Springer International Publishing: Cham, Switzerland, 2018; Volume 93.
3. Gibson, I.; Rosen, D.W.; Stucker, B. Additive Manufacturing Technologies. Springer: New York, NY, USA, 2015.
4. Gardan, J. Additive manufacturing technologies: State of the art and trends. *Addit. Manuf. Handb. Prod. Dev. Def. Ind.* **2017**, *7543*, 149–168. [CrossRef]
5. Turner, B.N.; Strong, R.; Gold, S.A. A review of melt extrusion additive manufacturing processes: I. *Process. Des. Modeling. Rapid Prototyp. J.* **2014**, *20*, 192–204. [CrossRef]
6. Sossou, G.; Demoly, F.; Montavon, G.; Gomes, S. An additive manufacturing oriented design approach to mechanical assemblies. *J. Comput. Des. Eng.* **2018**, *5*, 3–18. [CrossRef]
7. Liou, F.; Newkirk, J.; Fan, Z.; Sparks, T.; Chen, X.; Fletcher, K. *Multiscale and Multiphysics Modeling of Additive Manufacturing of Advanced Materials*; NASA: Rolla, MO, USA, 2015.
8. Vaezi, M.; Chianrabutra, S.; Mellor, B.; Yang, S. Multiple material additive manufacturing—Part 1: A review: This review paper covers a decade of research on multiple material additive manufacturing technologies which can produce complex geometry parts with different materials. *Virtual Phys. Prototyp.* **2013**, *8*, 19–50. [CrossRef]
9. Boyard, N. Design for Additive Manufacturing—DFAM. *J. Mater. Process. Technol.* **2014**, *12*, 616–622. [CrossRef]
10. Beyer, C.; Figueroa, D. Design and Analysis of Lattice Structures for Additive Manufacturing. *J. Manuf. Sci. Eng.* **2016**, *138*, 121014. [CrossRef]
11. Fernandez-Vicente, M.; Canyada, M.; Conejero, A. Identifying limitations for design for manufacturing with desktop FFF 3D printers. *Int. J. Rapid Manuf.* **2015**, *5*, 116–128. [CrossRef]
12. Rodríguez, J.F.; Thomas, J.P.; Renaud, J.E. Mechanical behavior of acrylonitrile butadiene styrene fused deposition materials modeling. *Rapid Prototyp. J.* **2003**, *9*, 219–230. [CrossRef]
13. JRodrıguez, F.; Thomas, J.P.; Renaud, J.E. Design of Fused-Deposition ABS Components for Stiffness and Strength. *J. Mech. Des.* **2003**, *125*, 545. [CrossRef]
14. Ahn, S.; Montero, M.; Odell, D.; Roundy, S.; Wright, P.K. Anisotropic material properties of fused deposition modeling ABS. *Rapid Prototyp. J.* **2002**, *8*, 248–257. [CrossRef]
15. Smyth, C. Functional Design for 3D Printing. Createspace: South Carolina, SC, USA, 2013.
16. Hopkinson, N.; Dicknes, P. Analysis of rapid manufacturing using layer manufacturing processes for production. *Proc. Inst. Mech. Eng. Part. C J. Mech. Eng. Sci.* **2003**, *217*, 31–39. [CrossRef]
17. Busachi, A.; Erkoyuncu, J.; Colegrove, P.; Martina, F.; Watts, C.; Drake, R. CIRP Journal of Manufacturing Science and Technology A review of Additive Manufacturing technology and Cost Estimation techniques for the defence sector. *CIRP J. Manuf. Sci. Technol.* **2017**, *19*, 1–12. [CrossRef]
18. RedEye, S. *Design for Additive Manufacturability: FDM Basics*; Stratasys Technical Report; ULTEM: Eden Prairie, MIN, USA, 2014.
19. Steele, K.; Ahrentzen, S. Design Guidelines. *Home Autism* **2017**, 81–170. [CrossRef]
20. London, T. Design for additive manufacturing. *Addit. Manuf. Technol.* **2015**, *17*. [CrossRef]
21. Diegel, O.; Nordin, A.; Motte, D. *A Practical Guide to Design for Additive Manufacturing*; Springer: Singapore, 2019.

22. Ben Redwood, B.G.; Filemon, S. *The 3D Printing Handbook: Technologies, Design and Applications*; Coers and Roest: Arnhem, The Netherlands, 2017.
23. *Interational Organization for Standarization/American Society for Testing Material Additive Manufacturing—General Principles—Terminology*; Technical Committee ISO/TC 261 Additive Manufacturing; Vernier: Geneva, Switzerland, 2015.
24. Aremu, A.; Ashcroft, I.; Hague, R.; Wildman, R.; Tuck, C. Suitability of SIMP and BESO topology optimization algorithms for additive manufacture. In Proceedings of the 21st Annual International Solid Freeform Fabrication Symposium—An Additive Manufacturing Conference, SFF, Austin, TX, USA, 8–10 August 2010; Volume 2010, pp. 679–692.
25. Saeed, N.; Long, K.; Rehman, A. A Review of Structural Optimization Techniques for Wind Turbines. In Proceedings of the 3rd International Conference on Computing, Mathematics and Engineering Technologies (iCoMET), Sukkur, Pakistan, 29–30 January 2020. [CrossRef]
26. Li, C.; Kim, I.Y.; Jeswiet, J. Conceptual and detailed design of an automotive engine cradle by using topology, shape, and size optimization. *Struct. Multidiscip. Optim.* **2015**, *51*, 547–564. [CrossRef]
27. Harzheim, L.; Graf, G. A review of optimization of cast parts using topology optimization: III-Topology optimization with manufacturing constraints. *Struct. Multidiscip. Optim.* **2006**, *31*, 388–399. [CrossRef]
28. Wang, W.; Munro, D.; Wang, C.C.L.; van Keulen, F.; Wu, J. Space-time topology optimization for additive manufacturing. *Struct. Multidiscip. Optim.* **2020**, *61*, 1–18. [CrossRef]
29. Talagani, M.R. Numerical Simulation of Big Area Additive Manufacturing (3D Printing) of a Full Size Car. *SAMPE J.* **2015**, *51*, 27–36.
30. Brackett, D.; Ashcroft, I.; Hague, R. Topology optimization for additive manufacturing. In Proceedings of the 21st Annual International Solid Freeform Fabrication Symposium—An Additive Manufacturing Conference, SFF, Austin, TX, USA, 8–10 August 2010; Volume 2010, pp. 348–362. [CrossRef]
31. Liu, J.; Cheng, Z.; Ma, Y. Product design-optimization integration via associative optimization feature modeling. *Adv. Eng. Inform.* **2016**, *30*, 713–727. [CrossRef]
32. Rezaie, R.; Badrossamay, M.; Ghaie, A.; Moosavi, H. Topology optimization for fused deposition modeling process. *Procedia CIRP* **2013**, *6*, 521–526. [CrossRef]
33. Primo, T.; Calabrese, M.; del Prete, A.; Anglani, A. Additive manufacturing integration with topology optimization methodology for innovative product design. *Int. J. Adv. Manuf. Technol.* **2017**, *93*, 467–479. [CrossRef]

*applied sciences*

MDPI

*Article*

# A Finite Element Method to Predict the Mechanical Behavior of a Pre-Structured Material Manufactured by Fused Filament Fabrication in 3D Printing

**Marouene Zouaoui [1,2,*], Julien Gardan [1,2], Pascal Lafon [2], Ali Makke [1], Carl Labergere [2] and Naman Recho [1,3]**

1 ERMESS, EPF-Engineering School, 2 Rue Fernand Sastre, 10430 Rosières-prés-Troyes, France; Julien.Gardan@epf.fr (J.G.); almak_21@hotmail.com (A.M.); Naman.Recho@epf.fr (N.R.)
2 Institut Charles Delaunay, LASMIS, UTT, UMR CNRS 6281, 12 Rue Marie Curie, 10010 Troyes, France; pascal.lafon@utt.fr (P.L.); carl.labergere@utt.fr (C.L.)
3 Institut Pascal CNRS-UMR 6602, Université Clermont Auvergne, PB 10448, 63000 Clermont-Ferrand, France
* Correspondence: marouene.zouaoui@utt.fr; Tel.: +33-(0)3-25-71-76-00

**Abstract:** In this paper, a numerical method is proposed to simulate the mechanical behavior of a new polymeric pre-structured material manufactured by fused filament fabrication (FFF), where the filaments are oriented along the principal stress directions. The model implements optimized filament orientations, obtained from the G code by assigning materials references in mesh elements. The Gauss points are later configured with the physical behavior while considering a homogeneous solid structure. The objective of this study is to identify the elastoplastic behavior. Therefore, tensile tests were conducted with different filament orientations. The results show that using appropriate material constants is efficient in describing the built anisotropy and incorporating the air gap volume fraction. The suggested method is proved very efficient in implementing multiplex G code orientations. The elastic behavior of the pre-structured material is quasi-isotropic. However, the anisotropy was observed at the yield point and the ultimate stress. Using the Hill criterion coupled with an experimental tabular law of the plastic flow turns out to be suitable for predicting the response of various specimens.

**Keywords:** pre-structured material; additive manufacturing; finite element modeling; mechanical behavior; fracture mechanics; Acrylonitrile Butadiene Styrene; 3D printing; smart material

**Citation:** Zouaoui, M.; Gardan, J.; Lafon, P.; Makke, A.; Labergere, C.; Recho, N. A Finite Element Method to Predict the Mechanical Behavior of a Pre-Structured Material Manufactured by Fused Filament Fabrication in 3D Printing. *Appl. Sci.* **2021**, *11*, 5075. https://doi.org/10.3390/app11115075

Academic Editor: Namhun Kim

Received: 4 May 2021
Accepted: 27 May 2021
Published: 30 May 2021



## 1. Introduction

With the recent development in manufacturing techniques, smart materials become more effective at answering the requirement of highly advanced industrial applications. Those materials have the ability to react to external stimulations and can adapt their behavior according to the environment. This research constitutes a relatively recent area, which has emerged from the prospect of producing topologically optimized and complex shapes using additive manufacturing. In [1], Gardan defined a class of smart materials applied to additive manufacturing (AM) as follows: "advanced structured materials, which are based on a static definition of complex shapes or a material's combination to achieve one or more properties that respond to a predefined functionality".

The concept of designing advanced structures by AM materials is widely applied to enhance structural performance and manufacture custom materials with specific functionality. For instance, Li et al. [2] worked on a toughening method of fracture properties using 3D printed topological patterns on a sample surface which effectively delayed crack initiation and deflected crack propagation. In a like manner, Madugula et al. [3] proposed an optimization method that uses continuous extrusion in 3D printing. It consists of an iterative procedure using re-meshing and analysis by finite element simulations. The proposed approach was applied to enhance the strength of a four-point bending beam made with Polylactic Acid PLA material.

A new stress-based criterion for filament deposition in FFF FDM 3D printing has been suggested and tested [4]. The work reported herein is based on reproducing principal stress orientations in the sample to define the filament's trajectory deposition in the manufacturing process. It was inspired by the reinforcement principle of composite materials where the fibers are oriented toward the in-plane tensile stress. Experimental studies have reported that this "enhanced" deposition method tested on an Acrylonitrile Butadiene Styrene (ABS) material generates an improvement by over 20% of fracture toughness and leads to a ductile-like fracture behavior [5–7]. In the same spirit, Kim et al. [8] developed a customized 3D printing tool-path for carbon fiber reinforced polymer where the carbon fibers reproduce the principal stress directions. This tool was used to print a spur gear geometry leading to a considerable improvement in the gear stiffness. The previously described deposition method has been proposed to surmount the problems caused by classical deposition methods such as layers alternation +45°/−45° or 0°/90° raster angle orientation. In those configurations, a locally heterogeneous structure is created because of the weld lines and the residual air gaps between the deposed filaments. These trajectories do not necessarily respect the use requirements of the part since a misplaced weld line, considered as a weakness, can lead to the reduction of the structure's strength.

Materials fabricated by fused filament fabrication (FFF) are supposed to exhibit strong anisotropy due to the filaments' orientation and the presence of an air gap [9]. Therefore, modeling their behavior will require a deep understanding of local phenomena occurring while loading the material. Rankouhi et al. [10] reveal the effect of the air-gaps size on the fracture morphology of the specimens announcing that larger air-gaps can cause inter-raster fusion bonds to fail which results in a more discretized surface area. In the study by Roy Xu et al. [11], a dual-notch void model was used to provide an explanation of the anisotropic tensile strengths of a 3D printed polymer. Most recently [12] Allum et al. demonstrated that the interface between layers in 3D printed PLA material has isotropic strength equivalent to that of the bulk filament. This study suggested that the anisotropy is caused by the welding lines' geometries and a localized strain represented by a stress concentration. It opposes the assumption of the weaker behavior of weld lines. The previously mentioned studies stand as proof of the complexity of 3D printed parts behavior and reveal many issues to model such materials.

Numerous studies investigate the effect of raster orientation on the mechanical behavior of 3D printed parts through experimentation [13–19]. To investigate the orthotropic elastic properties of 3D printed parts, Biswas et al. [20] developed computational models based on microstructures of 3D printed ABS using micromechanics of a representative volume element (RVE). In a similar fashion, Nasirov et al. [21] employed asymptotic homogenization as a tool for predicting the mechanical properties of fused filament fabricated thermoplastics with line infill and grid infill structures. Both Ghandriz et al. [22] and Li et al. [2] used the extended finite element method (XFEM) with cohesive segments to capture the dependency of fracture behavior on build/raster orientations. The classical laminate theory was also utilized to model orthotropic mechanical properties of 3D printed parts [19,23].

Although research has illuminated the effect of raster orientation through different modeling methods and experimentation, no study to date has studied complex deposition trajectories. Previous studies have almost exclusively focused on modeling unidirectional or bidirectional raster orientations. No numerical study to date has examined complex deposition trajectories.

In this paper, the authors intend to propose a simplified finite element method that surmounts the weld lines and the air-gap problems by approaching the behavior in a global way. The model considers the material as homogenous and reproduces the built trajectories using local references assignment in the mesh element [24]. A transverse isotropic law is therefore defined at each integration point to model the behavior in the elastic region. The material constants were determined from a tensile specimen fabricated using the same parameters as the pre-structured material. Then, the Hill criterion coupled with a plastic flow model was identified to go until the fracture limit. The finite element software ABAQUS has been used to implement the identified elastoplastic behavior. Nevertheless, local phenomena such as filaments buckling and air gap repartition are not modeled. The related problems will be discussed later on where this numerical tool will be tested on the compact tension CT specimen [4] and the Single Edge Notched Bend SENB specimen defined in [7] where both the optimized deposition and the classical deposition trajectories will be modeled and compared.

The structure of the paper is as follows. In Section 2, the modeling strategy adopted for the 3D printed material is presented. First, the local references assignment technique is explained. Then the pre-structured material behavior definition is presented. Next, the experimental setups and results are displayed. In Section 4, the implementation of the material's behavior in ABAQUS is detailed. In Section 5, the model will be used to predict the tensile tests. Afterward, it will be applied to the CT and SENB specimens. Here, its capability of implementing complex deposition trajectories will be checked and the limitations of the proposed method will be resumed in Section 6.

## 2. Modeling Methodology

The proposed modeling method is based on a simplification of the built configuration by considering a homogenous material. The complex structures containing the weld lines and the air gaps will be represented by implementing equivalent mechanical behavior constitutive equations. The developed workflow is presented in Figure 1 and will be detailed in the following subsections.

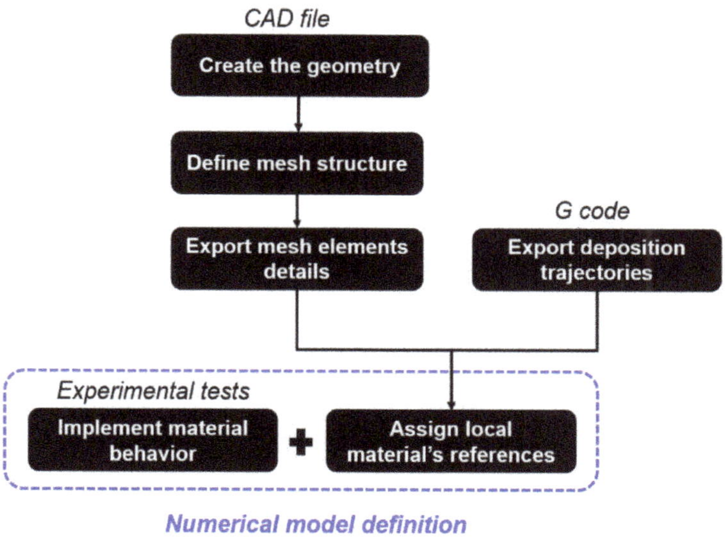


**Figure 1.** Workflow of the modeling methodology.

### 2.1. Local References Assignment in Mesh Elements

The purpose is to reproduce the printing trajectories inside a finite element model. The finite element code ABAQUS is used in this study. From the CAD file representing the part, the geometry is generated. Then an adaptive mesh is generated using an eight-node brick element with reduced integration C3D8R (i.e., one integration point per element). The use of C3D8R elements is motivated by the need to reproduce the layer by layer built in the model. This choice allows a simplification of the problem. The elements heights are fixed and must be equal to the layer thickness in order to ensure the reproduction of the built trajectories in a layer by layer process and facilitate the localization of elements in a certain layer (see Figure 2).

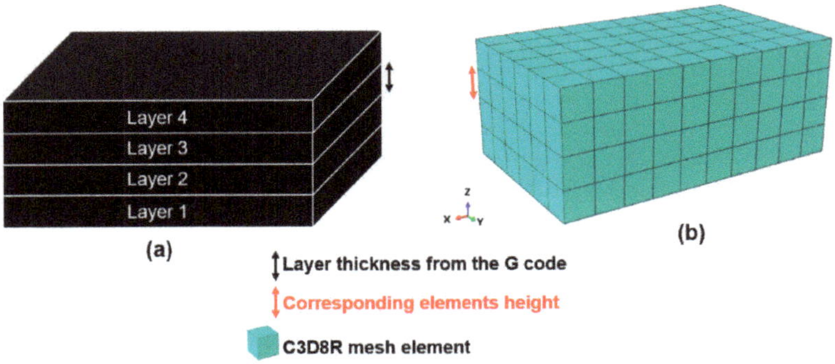


**Figure 2.** Correspondence between the layers heights (**a**) and the elements heights in ABAQUS mesh (**b**). Every layer from the G code contains different deposition trajectories.

The procedure starts by treating the G-code and defining the trajectories that only correspond to the real material deposition. This means that the trajectories where the extrusion nozzle is off are eliminated. Assigning the material's references goes through associating the material deposition trajectory with the elements of the numerical model. In each element, the vector tangent to a representative trajectory that intercepts the element faces is identified. This step requires distinguishing several cases according to the configuration of interception between the deposition trajectories and the element faces. Figure 3 demonstrates the possible cases. (i) If numerous trajectories belong to the element, the nearest to its Gauss's point is chosen. (ii) If a discontinuity between the two trajectories is the closest to its Gauss's point, the first trajectory analyzed is chosen with respect to the printing history. (iii) In the case where no trajectory intercepts the elements, the closest deposition trajectories are again selected.

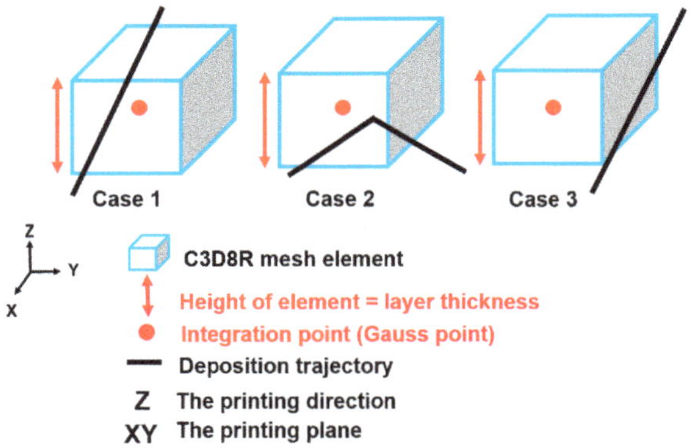


**Figure 3.** Possible interception cases between one mesh element C3D8R and the deposition trajectories.

The representative trajectory selected previously is considered as the first vector of the local reference. The third vector is always the same since it is the built direction (Z) perpendicular to the building plate (XY). A complete coordinate system is defined in each element by computing the cross-product of the first and the third vectors already known. This method was successfully applied on a compact tension CT specimen where complex trajectories were reproduced inside a finite element model after assigning the adequate materials' references [24].

## 2.2. Mechanical Behavior Definition

After assigning local references inside the mesh elements, a mechanical behavior must be defined at the integration points. An orthotropic behavior is commonly used to model the mechanical behavior of FFF parts in the elastic domain. For the sake of simplification, both directions 2 and 3 defined in Figure 4 are considered equivalent since the built-in of both orientations have similar deposition patterns and weld lines structure. A transverse isotropic is therefore chosen. It requires the identification of five independent elastic constants; (i) $E_L$: The Young's modulus in the longitudinal direction of the filament, (ii) $E_T$: The Young's modulus in the transverse direction of the filament, (iii) $\nu_{LT}$: the in-plane Poisson's ratio, (iv) $\nu_{TT}$ : the out of the plane Poisson's ratio, and (v) $G_{LT}$ : the in-plane shear modulus. Tensile tests with digital image correlation (DIC) are used to identify them. Thus, three orientations for the tensile specimens were chosen 0°, 90°, and 45°. The tensile specimens must be printed using the exact same printing parameters as the studied pre-structured material. This will ensure the reproduction of the mesostructure of the pre-structured material. As described previously, the authors decided to approach the material as homogenous. This means that the air gap portion is not subtracted from the geometrical cross-section. Therefore, the effect of the air gap volume fraction is directly incorporated in the identified material's constants.

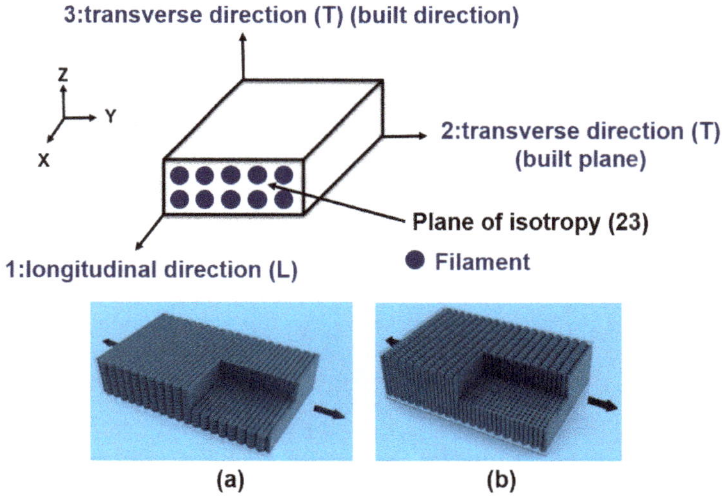


**Figure 4.** Transverse isotropic behavior for 3D printed parts; 1 is the longitudinal direction of the filaments, 2 is the transverse direction in the built plane illustrated in (**a**) and 3 is the second transverse direction that represents the built direction (**b**) [25].

# 3. Experimental Setups

## 3.1. Fabrication of the Specimen

The specimens were designed according to the ASTM D638-03 norm "Standard Test Method for Tensile Properties of Plastics" [26]. The fused deposition modeling fabrication process was conducted on a Makerbot replicator ×2 with a flat printing configuration. The filament is made of white ABS (Acrylonitrile Butadiene Styrene) material so that a better contrast can be achieved while performing the DIC analysis. The printing parameters described in Table 1 are the same for all the specimens studied in this paper.

**Table 1.** Printing settings.

| Printing Configuration | Built Flat |
|---|---|
| Layer thickness | 0.25 mm (0.35 for the first one) |
| Extrusion temperature | 235 °C |
| Built plate temperature | 120 °C |
| Thread diameter | 1.7 mm |
| Infill density | 100% |
| Material | ABS (white) |

The slicing software Slic3r [27] was used to generate the G codes. This software was chosen since it allows a unique infill angle for all layers via the modifier feature. Three infill angles are used 0° representing the longitudinal direction of the filament, 90° for the transverse direction of the filament, and a 45° angle to study the shear behavior. Six specimens were fabricated for each orientation.

### 3.2. Tensile Tests with Digital Image Correlation

The tensile test machine Instron 4411 was used to perform the tensile tests. The loading rate was fixed at 1 mm/min. A camera with a resolution of 1280 pixels × 960 pixels was used to record a snapshot every 0.6 s while performing the test. This frequency gives an accurate reproduction of the tensile curves. The speckle was applied manually through a black paint spray. After collecting the snapshots, the DIC software NCORR [28] allows a quantitative analysis of the strain. It tracks the displacement of the speckles on each snapshot during the loading phase and uses it to interpret the kinematic field maps on a specific region of interest (ROI). A subset size of 15 pixels and a spacing of 2 pixels were used for data treatment. The images were treated as described in [29]. The method uses specific dummy sensors placed next to the jaw's location in order to calculate the true value of displacement.

### 3.3. Tensile Tests Results

For every performed tensile test, the calculated Young's modulus value will indirectly take into account the existing air gap by computing the engineering stress via the geometrical section. The calculations are made with respect to the D638 ASTM norm [26]: Young's modulus is calculated as the slope of the tangent to the true stress-strain curve at low stress. To get such a curve, the assumption of volume conservation was made since the volume variation data is not available. The DIC analysis gives an estimation of the average value of the transversal and the longitudinal strain, which are used to calculate the Poisson's ratio values. The yield strength is calculated via the 0.2% strain offset method. The results are displayed in Table 2.

**Table 2.** Results of tensile tests.

| Orientation | 0° | | 90° | | 45° | |
|---|---|---|---|---|---|---|
| | Mean | Std * | Mean | Std | Mean | Std |
| Young's modulus (MPa) | 1680 | 71 | 1414 | 133 | 1484 | 53 |
| Offset yield strength 0.2% (MPa) | 23.54 | 1.59 | 13.3 | 0.78 | 17.53 | 0.95 |
| Tensile strength (MPa) | 28.67 | 0.79 | 13.86 | 0.65 | 20.25 | 1.16 |
| Elongation at fracture (%) | 4.90 | 1.31 | 1.12 | 0.18 | 3.65 | 0.46 |
| Poisson's ratio | 0.371 | 0.032 | 0.312 | 0.032 | 0.342 | 0.028 |

* Std: standard deviation.

The experimental tensile curves show that the material exhibits different behaviors according to the filament orientation (see Figure 5). For the 90° orientation, the elongation at break is almost equal to the elongation at yield strength. Thus, the behavior can be assumed perfectly elastic with a brittle fracture. On the other hand, the 0° specimens display the highest tensile strength of 23.54 MPa since the loading is directly applied on the longitudinal direction of the filament considered as the strongest direction. Here the behavior has a pronounced plastic behavior directly linked to a ductile fracture. In addition, a softening was observed in this direction. This softening behavior can be explained by referring to the work of Dundar et al. [30]. In fact, ABS material exhibits a strain-softening behavior at a low strain rate, which was around $10^{-4}$ $s^{-1}$ for the tensile tests. Finally, the 45° specimen displays an intermediary behavior between both 0° and 90° with an almost elastic perfectly plastic behavior. The yield strength is found equal to 17.53 MPa. The fracture is also ductile and occurs at a 45° angle at the welding line direction.

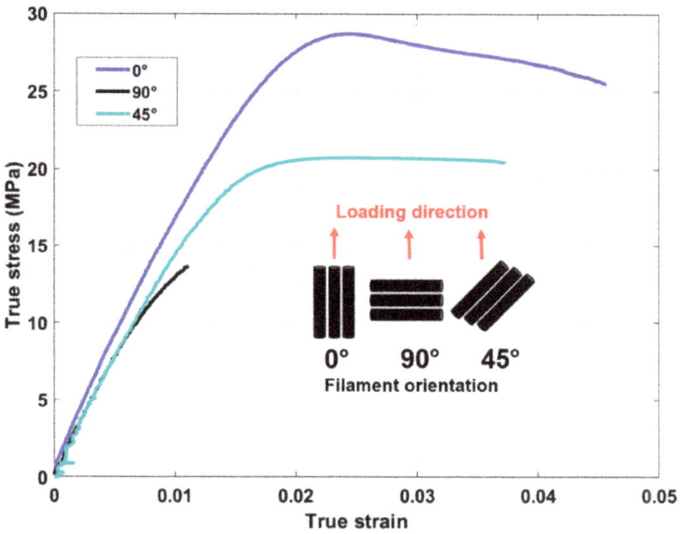


**Figure 5.** Tensile curves of three samples with different filament orientations (0°, 45°, and 90°).

These findings are coherent with the work of Alaimo et al. [31] which showcased the same behavior for the studied orientations. Young's modulus is slightly lower but this can be explained by the difference in the extrusion temperature. The extrusion temperature used to fabricate the specimen is 235 °C while the temperature of extrusion used by Alaimo was around 255 °C. A higher extrusion temperature enhances the bonding between the filament and thus, the mechanical stiffness increases [32]. For the Poisson's ratio of the 0° oriented sample, the same value was found by Rodriguez et al. [33]. For the yield strength, Alaimo [31] found a yield strength equal to 26.3 MPa, 14.6 MPa, and 18.54 MPa while the authors found 23.54 MPa, 13.3 MPa, and 17.53 MPa for 0°, 90°, and 45° respectively. It should be noted that all comparisons with the bibliographic references must be done with a similar printing configuration to avoid any perturbation from one of the printing parameters.

Conclusively, the material exhibits an overall quasi-isotropic stiffness. The difference between the 0° Young's modulus and the 90° Young's modulus does not exceed 16%. This means that welding lines are not affecting the material stiffness in the elastic domain. The anisotropy was mainly observed at the yielding and fracture points. Therefore, it was decided to employ an anisotropic yielding criterion to model the transition between elastic and plastic domains.

## 4. Implementation of the Pre-Structured Material Behavior

### 4.1. Transverse Isotropic Law Identification

As described previously, a transverse isotropic law is chosen to model the 3D printed material behavior. Thus, five independent elastic constants must be identified. Both Young's modulus $E_L$ and $E_T$ and the in-plane Poisson's ratio $\nu_{LT}$ are directly identified from the tensile test results. The longitudinal Young's modulus and the in-plane Poisson's ratio are the ones calculated on the 0° specimen. The transverse Young's modulus is the one calculated with the 90° specimen.

#### 4.1.1. The In-Plane Shear Modulus $G_{LT}$

By establishing the relationship between the global reference and the local reference defined in Figure 6, where the plane stress assumption is presumed, Young's modulus in the loading direction (global reference 1) can be expressed as a function of the transverse isotropic elastic constants and the printing angle. The same relation (see Equation (1)) [34] is used to study the stress analysis of fiber-reinforced composite materials.

$$E_1 = \frac{E_L}{\cos^4\theta + \left(\frac{E_L}{G_{LT}} - 2\nu_{LT}\right)\sin^2\theta\cos^2\theta + \frac{E_L}{E_T}\sin^4\theta} \quad (1)$$

where:

- $\theta$: Printing angle
- $E_L$: Young's modulus in the longitudinal direction
- $E_T$: Young's modulus in the transverse direction
- $\nu_{LT}$: The in-plane Poisson's ratio
- $G_{LT}$: The in-plane shear modulus
- $E_1$: Young's modulus in the loading direction (global reference)

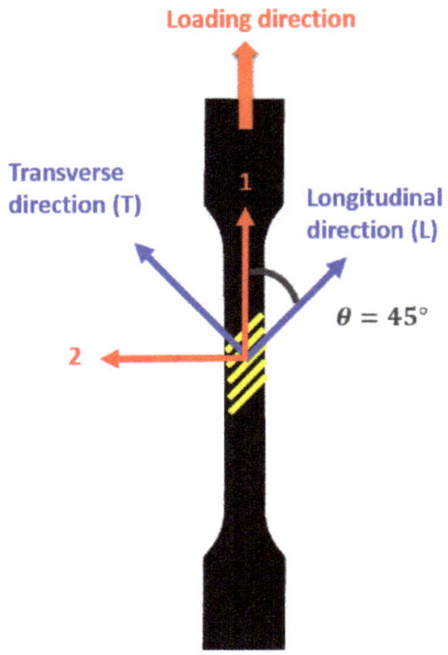


**Figure 6.** Local and global references definition for the 45° specimen.

By using the 45° specimen, the following equation is deducted:

$$G_{LT} = \left(\frac{4}{E_{45°}} - \frac{1}{E_L} - \frac{1}{E_T} + \frac{2\nu_{LT}}{E_L}\right)^{-1} \quad (2)$$

where: $E_{45^\circ}$ is the experimental Young's modulus from the 45° specimen. A value of 545 MPa of the in-plane shear modulus was calculated. A higher value was given by Alaimo et al. [31] equal to 641 MPa, which was already explained by the difference of the extrusion temperature.

4.1.2. The Out-of-Plane Poison's Ratio $\nu_{TT}$

The compliance matrix $\underline{\underline{C}}^{-1}$ of the transverse isotropic material shows that the out-of-plane Poisson's ratio $\nu_{TT}$ is not activated in the case of the plane stress assumption unless the out-of-plane displacement is calculated. However, the DIC analysis employed here provides only the displacement field in 2D ROI. Therefore, the out-of-plane Poisson's ratio $\nu_{TT}$ cannot be identified using the available data. Otherwise, it was proven that its value would not influence the 2D material behavior because it only affects the out-of-plane strain component. For the developed model, $\nu_{TT}$ was taken as the average value between $\nu_{TL}$ and $\nu_{LT}$ . A value of $\nu_{TT}$ =0.3416 will be employed.

4.1.3. Stiffness Matrix $\underline{\underline{C}}$

The transverse isotropic behavior is implemented in ABAQUS by computing all the components of the stiffness matrix $\underline{\underline{C}}$ as shown by Equation (3).

$$\underline{\underline{C}} = \begin{pmatrix} 2592.3 & 1229.5 & 1229.5 & & & \\ 1229.5 & 2184.1 & 1130.1 & & 0 & \\ 1229.5 & 1130.1 & 2184.1 & & & \\ & & & 545 & & \\ & 0 & & & 545 & \\ & & & & & 527 \end{pmatrix} \quad (3)$$

This stiffness matrix $\underline{\underline{C}}$ is defined at each C3D8R element with respect to the local reference defined in its integration point. This allows a local description of the material behavior based on the deposition trajectories at each point of the structure.

*4.2. Anisotropic Yielding Criterion: Hill Criterion*

The anisotropy was mainly observed at the yielding point. Indeed the 0° sample has the highest yield strength while the 90° sample has the lowest one. Where an anisotropic yielding criterion must be used, Hill's criterion is used for this type of problem. It generalizes the Von Mises criterion by introducing anisotropy coefficients [35]. It has been used to describe the elastoplastic behavior of composites reinforced with short carbon fibers from 3D printing [36]. Likewise, it was efficient in underlining the anisotropy of the tensile strength of parts printed in PLA material.

The material has four different yield strengths according to the loading state, as described in Table 3, which must be identified to implement the Hill criterion. $\sigma_L^0$ is the yield strength obtained from the 0° tensile tests and $\sigma_T^0$ is obtained from the 90° tensile tests. They are equal to 23.54 MPa and 13.3 MPa respectively. The in-plane shear yield strength $\sigma_{LT}^0$ is calculated by applying the laminate theory on the 45° specimen. The stress analysis described in Figure 6 is utilized. The tensile stress in the loading direction (global reference) is related to the stress components in the local reference and the printing angle through Equation (4).

$$\sigma_{11}(\theta) = \left[\frac{\cos^4\theta}{\sigma_L^2} + \frac{\sin^4\theta}{\sigma_T^2} - \frac{\cos^2\theta\sin^2\theta}{\sigma_L^2} + \frac{\cos^2\theta\sin^2\theta}{\sigma_{LT}^2}\right]^{-\frac{1}{2}} \quad (4)$$

where

- $\sigma_{11}(\theta)$: The tensile strength in loading direction (global reference)
- $\theta$: Printing angle $\sigma_L$: The stress in the longitudinal direction of the filament
- $\sigma_T$: The stress in the transversal direction

- $\sigma_{LT}$: The in-plane shear stress

**Table 3.** Yield strengths designation for the transverse isotropic material.

| Yield Strength Designation | Loading State on the Filaments |
|---|---|
| Tensile yield strength in the longitudinal direction: $\sigma_L^0$ |  Longitudinal loading |
| Tensile yield strength in the transversal direction: $\sigma_T^0$ |  Transversal loading |
| In-plane shear yield strength: $\sigma_{LT}^0$ |  Delamination on the Weld lines |
| Out-of-plane shear yield strength: $\sigma_{TT}^0$ |  Shear of fibers |

To access the in-plane shear yield strength, results from the 45° specimen are used. The following relation (see Equation (5)) is then established by taking $\theta = 45°$ and replacing $\sigma_{11}(\theta)$ by the experimental yield strength of the 45°spceicmen.

$$\sigma_{LT}^0 = \left( \frac{4}{\sigma_{45°}^2} - \frac{1}{\sigma_T^2} \right)^{-\frac{1}{2}} \tag{5}$$

A value of 11.65 MPa was found similar to the value found by [31] 12 ± 0.7 MPa.

For the out-of-plane shear yield strength, the torsion test is commonly used to identify its value. The experimental results reported in [37] by Balderrama-Armendariz et al. show that the out-of-plane shear yield is about 25 MPa. However, this value cannot be utilized herein since the printing parameters are substantially different. Therefore, both in-plane and out-of-plane shear yield strengths were assumed equal $(\sigma_{TT}^0 = \sigma_{LT}^0)$.

The Hill criterion was implemented in ABAQUS software using the user-defined Hill's potential function in ABAQUS [38]. It implies the definition of six different anisotropic yield stress ratios. As $\sigma_{TT}^0 = \sigma_{LT}^0$, three independent ratios must be introduced as inputs. Those ratios (see Table 4) are expressed using a reference tensile yield stress $\sigma^0$, which was chosen as the tensile yield strength in the longitudinal direction $\sigma_L^0$, and a reference shear yield stress $\tau^0$ where $\tau^0 = \frac{\sigma^0}{\sqrt{3}}$.

**Table 4.** Implementation of Hill's potential function in ABAQUS using the yield stress ratios for the transverse isotropic material.

| Ratio | $R_L$ | $R_T$ | $R_{LT}$=$R_{LT}$ |
|---|---|---|---|
| Expression | $\frac{\sigma_L^0}{\sigma^0}$ | $\frac{\sigma_T^0}{\sigma^0}$ | $\frac{\sigma_{LT}^0}{\tau^0} = \frac{\sigma_{TT}^0}{\tau^0}$ |
| Value | 1 | 0.565 | 0.857 |

*4.3. Plastic Flow Curve of the Filament (0° Specimen)*

In order to study the material behavior in the plastic domain, it was decided to couple the Hill criterion with the curve flow for the 0° orientation. For the 0° specimen as shown in Figure 5, when the stress is below the yield stress, the material behaves elastically. After the yielding point, an isotropic hardening is observed and will be modeled via an exponential law defined by Equation (6) where: $\sigma_L^0$ is the yield stress and $\varepsilon_p$ is the plastic strain and $\sigma_{limit}$ is the hardening limit for the material and $m$ is a positive dimensionless exponent.

$$\sigma = \sigma_L^0 + \sigma_{limit}\left(1 - e^{-m\varepsilon_p}\right) \tag{6}$$

Using the experimental data, a least-square fit was used to identify those constants (see Table 5).

**Table 5.** Exponential law fitting results.

| $\sigma_{limit}$(MPa) | | $m$ | |
|---|---|---|---|
| Mean value | Std | Mean value | Std |
| 6.58 | 3.41 | 221.6 | 96.19 |

This model is only valid until reaching the tensile strength. The observed softening was discarded for the sake of simplicity. A stress saturation is assumed up to the fracture point with only an increase of the plastic strain. This model was implemented in ABAQUS using a tabular law that describes the evolution of the plastic strain from the yielding point up to the breaking point.

## 5. Results and Discussions

All the simulations in this study were carried out using ABAQUS implicit solver. A quasi-static time-independent loading step is defined to model the various tests with an applied displacement as a loading condition. All models utilized C3D8R mesh elements with adequate mesh sizes according to the geometry and the deposition trajectories requirement. The implementation of the mechanical behavior as defined in the previous section is used for all simulations.

*5.1. Tensile Tests Prediction*

The finite element model for the tensile specimen geometry was created in ABAQUS software. Three models were later defined by assigning local material references with respect to the printing trajectories. Three different simulations have been carried out to simulate the behavior of each of the specimens according to the suggested model. The part of the specimen pinched between the fixed jaws is encastered and a displacement is applied on a reference point coupled with the part pinched between the mobile jaws.

Figure 7 compares the stress-strain curves obtained experimentally with those obtained by finite element simulations. The results show that for all three cases, the elastic behavior is well reproduced with a slight difference because of the simplifications described before. Thus, the model predicts well the stress evolution until the yielding point. By comparing the model response for the three different orientations, it is clear that the anisotropic yielding criterion taken in the model captures the yield point in the experimental curves. The transition between the elastic and the plastic domain for the 0° specimen occurs around the value of $\sigma_L^0$. In addition, for this orientation, the saturation of the stress at the average tensile strength is noticed (see Figure 7a). The onset of plastic behavior observed for the 45° tensile sample is well simulated (see Figure 7b) by coupling Hill's criterion with the 0° flow curve. Therefore, the numerical approach is able to highlight the anisotropic strengths according to the filament orientation. For the sample with 90° (Figure 7c), orientation exhibits a brittle fracture. However, a plastic behavior appears in the numerical response. This behavior was intentionally discarded since it does not appear in the experimental tests. The elongation at fracture was managed by applying the appropriate corresponding displacement value without implementing a fracture initiation criterion. The average values reported in Table 2 of the fracture elongation were respected.

The model allows good prediction of the elastic domain without any estimation of the air gap. The plastic component was defined in order to demonstrate the ability of Hill's criterion to predict the anisotropy with the material yield strength. By only implementing the plastic flow curve for the 0° orientation, it has been shown that it is possible to predict the mechanical response of the material in different orientations.

### 5.2. Compact-Tension Specimen

In this section, the model's ability to predict the global response of a compact tension CT specimen, defined in [6], will be checked. After treating both the classical and the optimized G codes of the corresponding geometries, two finite element models were created. The numerical model definition with the boundary conditions is described in [24]. The part is sliced along the Z direction to match the layers from the G code. The local material's references are assigned for both kinds of samples. Figure 8 shows the vectors that correspond to the deposition trajectory in each element. For the optimized specimen, since two principal directions were reproduced, only two material's references definition per layer can be observed. The model's responses will be compared with the experimental data collected from the work of Gardan et al. [6]. The boundaries conditions defined in [24] are maintained the same for this study. Two rigid bodies are used. One to model the fixed pin and the other represents the mobile one.

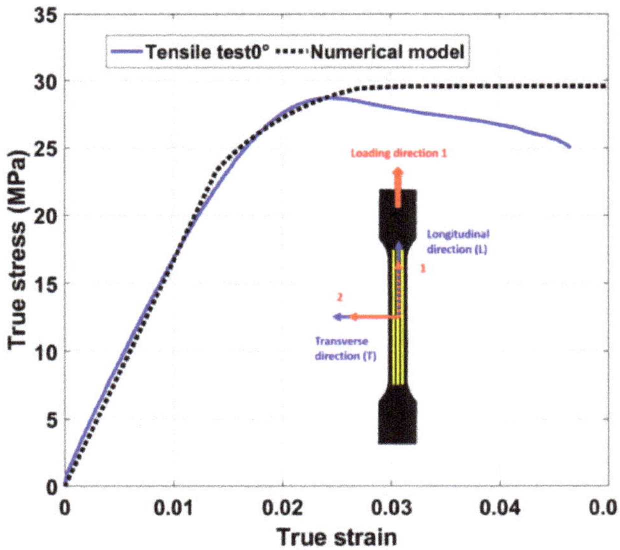


(a)

**Figure 7.** *Cont.*

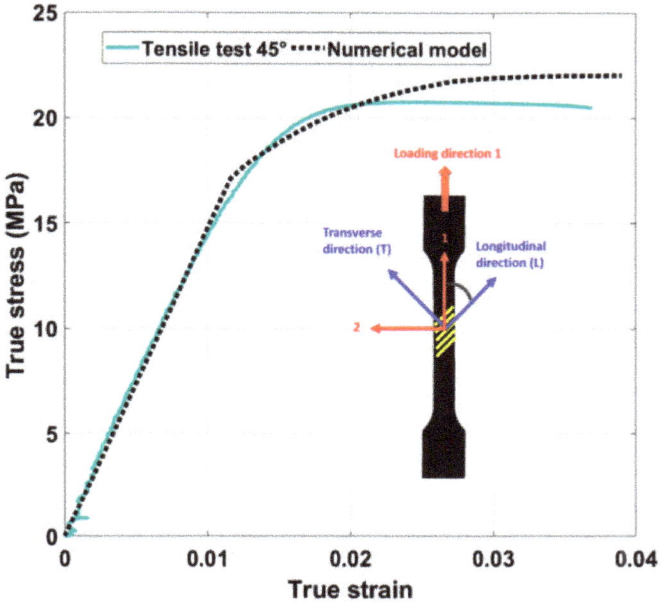


(b)

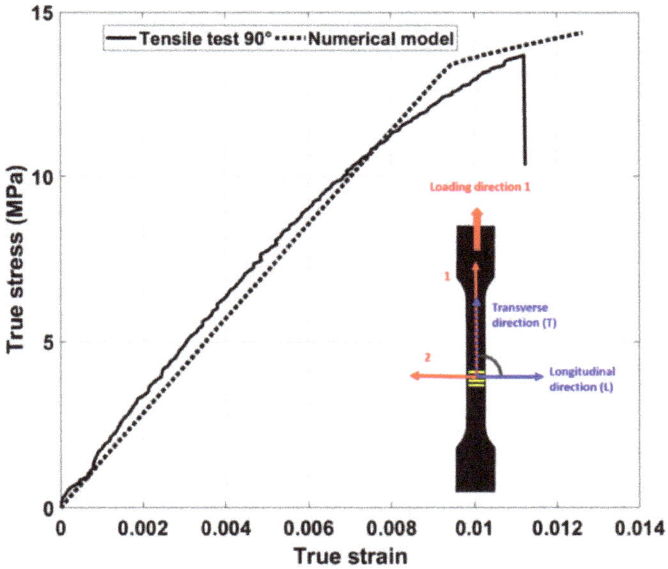


(c)

**Figure 7.** Model response of the tensile test specimens compared to a representative experimental curve. (**a**): 0° raster orientation and (**b**): 45° raster orientation and (**c**): 90° raster orientation.

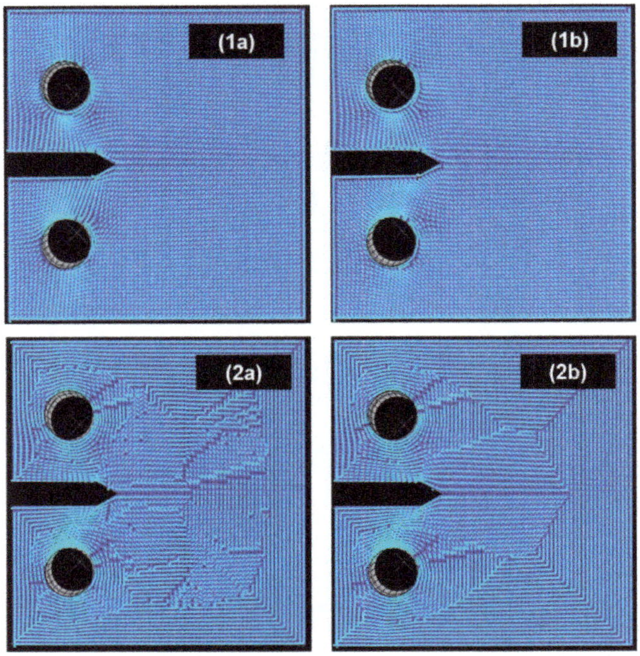


**Figure 8.** Numerical models of the CT specimen. (**1**) Corresponds to the classical deposition with ±**45**° layers alternation. (**2**) For the optimized deposition; (**a**) denotes the first principal direction while (**b**) is for the second principal direction.

Figure 9 compares both experimental and numerical results. The drop of the stress observed in the experimental curves results from the crack initiation and growth from the notch tip. Therefore, the simulation cannot reproduce this behavior because the material damage and fracture are not taken into account by the model. Regarding the linear behavior of the samples, the model gives a good prediction of the stiffness for the optimized specimen and slightly overestimates it for the classical one. For the optimized specimen, where a ductile-like curve is obtained by the test, the model underestimates the load values (see Figure 9a). This is mainly related to a numerical plastic behavior along the transversal direction that is not omitted in the model.

### 5.3. Beam Specimen

The work reported in Lanzilotti et al. [7] focused on applying the optimized deposition method on the Single Edge Notched Bending SENB specimen. It reproduces the principal stress direction in the V notch tip vicinity and keeps a classical deposition ±45° elsewhere. The printing trajectories from the G code are shown in Figure 10.

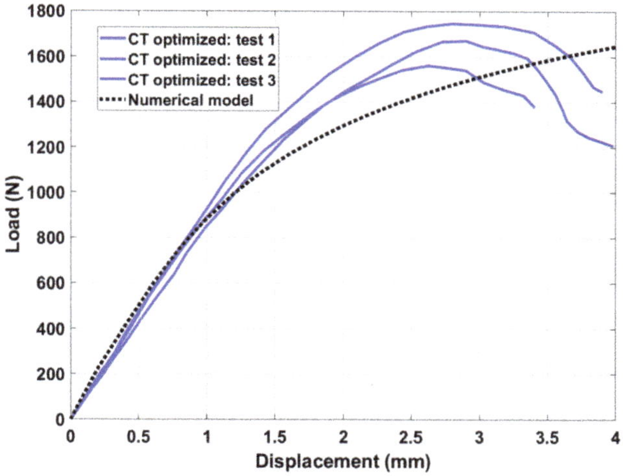


(a)

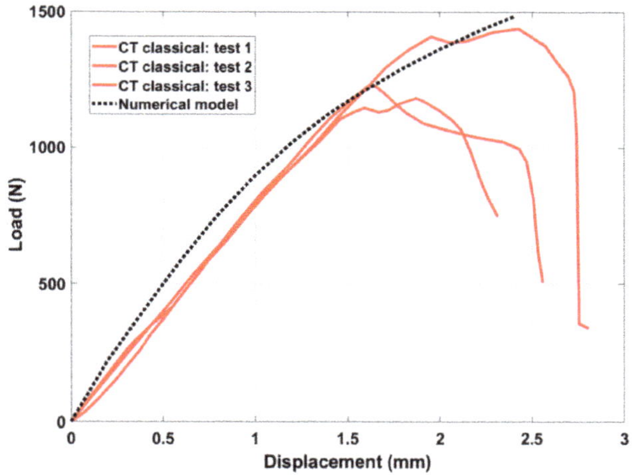


(b)

**Figure 9.** Load-displacement curves of CT test. Numerical model response is compared to experimental results. (**a**) Classical samples are printed with $\pm$ 45° and (**b**) Optimized samples; the filament is oriented according to the principal direction of the stress.

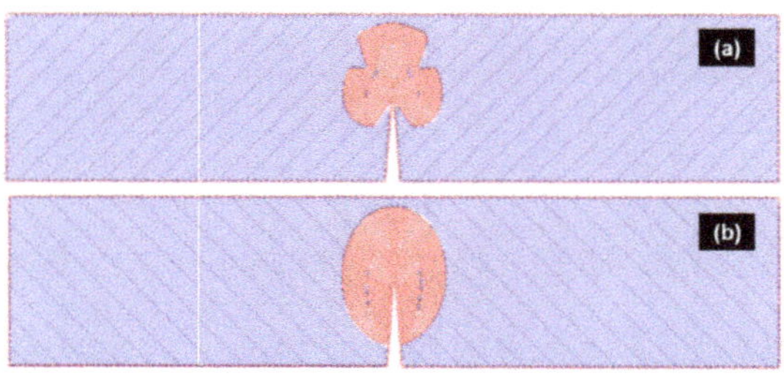


**Figure 10.** Deposition trajectories from the G code of the SENB specimen after applying the optimization method on the V notch vicinity. (**a**) First principal direction and (**b**) second principal direction.

The SENB specimen dimensions and the boundary conditions used for the simulations are shown in Figure 11. A mesh size of 0.56 mm was used in the V notch area to better capture the deposition trajectories. This value is equal to the thread width used during the printing. A bigger mesh size can still capture the trajectories elsewhere since they have unidirectional orientations. This will also allow a reduction in computational time. The first vectors representing the deposition trajectories of all the local references are showcased in all the integration points in Figure 12.

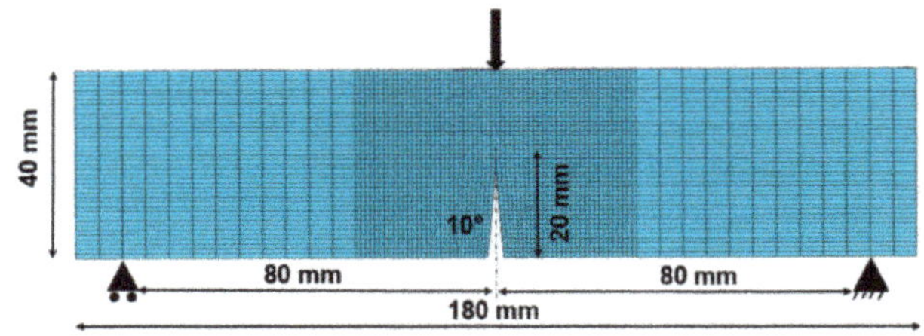


**Figure 11.** SENB specimen geometry's definition and the boundary conditions for the numerical model.

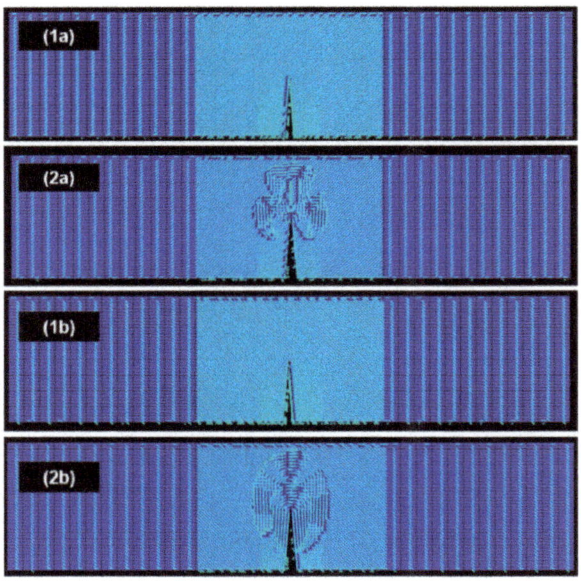


**Figure 12.** Numerical models of the SENB specimen. (**1**) Corresponds to the classical deposition with ±45° layers alternation. (**2**) For the optimized deposition with principal stress reproduction in the V notch tip vicinity; (**a**) denotes the first principal direction while (**b**) is for the second principal direction.

Interestingly, the numerical response seems to be practically the same for both models (classical and optimized). The mesh elements' orientations have almost no effect on the global numerical response. A numerical stiffness equal to 253.15 N/mm was calculated for both models. Although, the classical specimen has an average stiffness of 174.3 ± 11.8 N/mm, the stiffness of the optimized sample is 224.3 ± 12.7 N/mm. This is related to the quasi-isotropic stiffness of the material since no large difference in the elastic modulus was found between the longitudinal and the transversal directions (16%).

Moreover, the modification of the raster orientation is only carried out locally in the notch tip vicinity. The remaining domain is built by $\pm45°$ stacked layers. Although this local modification leads to the improvement of the sample fracture toughness, however, no substantial difference was found in the elastic response of the model. The model assumes the material continuum. This assumption omits any potential effect of the air gaps, weld lines, and even the anisotropy of the filament itself. This partially interprets the difference found between the sample stiffness measured experimentally and numerically. On the other hand, the simulation results show that the influence of the local modification of filament orientation in the notch tip vicinity is vanishingly small so that the model cannot capture it.

The model uses a stiffness matrix and materials constants from tensile tests printed so that the filaments are stacked in the same direction. This is supposed to maximize the bonding conditions between layers, and can therefore explain why the model overestimates the experimental stiffness. The biggest area of the specimen is printed using stacking layers at $\pm45°$. A $90°$ crossing angle between filaments in two different layers would accordingly decrease the bonding conditions.

## 6. Conclusions

A simplified numerical model to predict the mechanical behavior of FFF 3D printing parts has been proposed. The model tackles successfully the anisotropic built configuration by assigning local material references in mesh elements. It approaches the material as homogenous and uses specific materials constants to introduce the anisotropy related to complex deposition trajectories. The present findings confirm that the material has a quasi-isotropic elastic behavior since both longitudinal and transversal Young's modulus are roughly equal. Hill's criterion coupled with the plastic flow curve of the filament has been used to identify the strength anisotropy in tensile specimens with different raster angles. The longitudinal direction has the highest yield strength and a ductile behavior while the transversal one is the weakest and exhibits a fragile behavior at the yielding point.

The model is not able to capture the behavior where the optimized deposition method is applied to a small zone compared to the area of the specimen. Future work could fruitfully explore this issue further by taking into account the bonding conditions between the filaments. Axiomatically the model can be upgraded by assigning local materials' properties according to local bonding conditions.

Despite its limitations in reproducing local phenomena, the model is a key component in future attempts to understand the relation between the local deposition orientation and the material performance. The possibility of utilizing this model in the simulation of the mechanical behavior of 3D printed parts by FDM warrants further investigation and development. Finally, the model can be enriched with another simulation method (e.g., continuum damage mechanics or extended finite elements) in order to study the fracture behavior after crack initiation.

**Author Contributions:** Conceptualization, C.L.; Data curation, M.Z. and J.G.; Formal analysis, A.M. and N.R.; Funding acquisition, M.Z.; Investigation, M.Z., J.G., P.L. and A.M.; Methodology, J.G., P.L., A.M. and C.L.; Project administration, J.G. and P.L.; Software, M.Z. and A.M.; Supervision, J.G. and P.L.; Validation, N.R.; Writing—original draft, M.Z.; Writing—review & editing, J.G., P.L., A.M. and N.R. All authors have read and agreed to the published version of the manuscript.

**Funding:** This research was funded by the French Government, the European Regional Development Fund (ERDF), University of Technology of Troyes (UTT) and EPF Engineering School.

**Institutional Review Board Statement:** Not applicable.

**Informed Consent Statement:** Not applicable.

**Data Availability Statement:** Not applicable.

**Acknowledgments:** The authors gratefully acknowledge the French Government, the European Regional Development Fund (ERDF), University of Technology of Troyes (UTT) and EPF Engineering School for their financial support.

**Conflicts of Interest:** The authors declare that they have no known competing financial interests or personal relationships that could have appeared to influence the work reported in this paper.

## References

1. Gardan, J. Smart materials in additive manufacturing: State of the art and trends. *Virtual Phys. Prototyp.* **2019**, *14*, 1–18. [CrossRef]
2. Li, J.; Yang, S.; Li, D.; Chalivendra, V. Numerical and experimental studies of additively manufactured polymers for enhanced fracture properties. *Eng. Fract. Mech.* **2018**, *204*, 557–569. [CrossRef]
3. Madugula, S.; Giraud-Moreau, L. Renforcement de la structure interne de pièces imprimées par une technique de remaillage adaptée à l'extrusion en continu. In Proceedings of the CSMA 2019, Giens, France, 13–17 May 2019; p. 9.
4. Gardan, J.; Makke, A.; Recho, N. A Method to Improve the Fracture Toughness Using 3D Printing by Extrusion Deposition. *Procedia Struct. Integr.* **2016**, *2*, 144–151. [CrossRef]
5. Gardan, J.; Makke, A.; Recho, N. Fracture Improvement by Reinforcing the Structure of Acrylonitrile Butadiene Styrene Parts Manufactured by Fused Deposition Modeling. *3D Print. Addit. Manuf.* **2019**, *6*, 113–117. [CrossRef]
6. Gardan, J.; Makke, A.; Recho, N. Improving the fracture toughness of 3D printed thermoplastic polymers by fused dep-osition modeling. *Int. J. Fract.* **2018**, *210*, 1–15. [CrossRef]
7. Lanzillotti, P.; Gardan, J.; Makke, A.; Recho, N. Enhancement of fracture toughness under mixed mode loading of ABS specimens produced by 3D printing. *Rapid Prototyp. J.* **2019**, *25*, 679–689. [CrossRef]
8. Kim, J.; Kang, B.S. Optimization of Design Process of Fused Filament Fabrication (FFF) 3D Printing. In *Volume 2: Advanced Manufacturing*; American Society of Mechanical Engineers: Pittsburgh, PA, USA, 2018; p. V002T02A060.
9. Ahn, S.; Montero, M.; Odell, D.; Roundy, S.; Wright, P.K. Anisotropic material properties of fused deposition modeling ABS. *Rapid Prototyp. J.* **2002**, *8*, 248–257. [CrossRef]
10. Rankouhi, B.; Javadpour, S.; Delfanian, F.; Letcher, T. Failure Analysis and Mechanical Characterization of 3D Printed ABS with Respect to Layer Thickness and Orientation. *J. Fail. Anal. Prev.* **2016**, *16*, 467–481. [CrossRef]
11. Xu, L.R.; Leguillon, D. Dual-Notch Void Model to Explain the Anisotropic Strengths of 3D Printed Polymers. *J. Eng. Mater. Technol.* **2019**, *142*. [CrossRef]
12. Allum, J.; Moetazedian, A.; Gleadall, A.; Silberschmidt, V.V. Interlayer bonding has bulk-material strength in extrusion additive manufacturing: New understanding of anisotropy. *Addit. Manuf.* **2020**, *34*, 101297. [CrossRef]
13. Tymrak, B.M.; Kreiger, M.; Pearce, J.M. Mechanical properties of components fabricated with open-source 3-D printers under realistic environmental conditions. *Mater. Des.* **2014**, *58*, 242–246. [CrossRef]
14. Riddick, J.C.; Haile, M.A.; Von Wahlde, R.; Cole, D.P.; Bamiduro, O.; Johnson, T.E. Fractographic analysis of tensile failure of acrylonitrile-butadiene-styrene fabricated by fused deposition modeling. *Addit. Manuf.* **2016**, *11*, 49–59. [CrossRef]
15. Ahn, S.H.; Baek, C.; Lee, S.; Ahn, I.S. Anisotropic Tensile Failure Model of Rapid Prototyping Parts—Fused Deposition Modeling (FDM). *Int. J. Mod. Phys. B* **2003**, *17*, 1510–1516. [CrossRef]
16. Croccolo, D.; De Agostinis, M.; Olmi, G. Experimental characterization and analytical modelling of the mechanical be-haviour of fused deposition processed parts made of ABS-M30. *Comput. Mater. Sci.* **2013**, *79*, 506–518. [CrossRef]
17. Cantrell, J.T.; Rohde, S.; Damiani, D.; Gurnani, R.; DiSandro, L.; Anton, J.; Young, A.; Jerez, A.; Steinbach, D.; Kroese, C.; et al. Experimental characterization of the mechanical properties of 3D-printed ABS and polycarbonate parts. *Rapid Prototyp. J.* **2017**, *23*, 811–824. [CrossRef]
18. Rohde, S.; Cantrell, J.; Jerez, A.; Kroese, C.; Damiani, D.; Gurnani, R.; DiSandro, L.; Anton, J.; Young, A.; Steinbach, D.; et al. Experimental Characterization of the Shear Properties of 3D–Printed ABS and Polycarbonate Parts. *Exp. Mech.* **2018**, *58*, 871–884. [CrossRef]
19. Casavola, C.; Cazzato, A.; Moramarco, V.; Pappalettere, C. Orthotropic mechanical properties of fused deposition mod-elling parts described by classical laminate theory. *Mater. Des.* **2016**, *90*, 453–458. [CrossRef]
20. Biswas, P.; Guessasma, S.; Li, J. Numerical prediction of orthotropic elastic properties of 3D-printed materials using mi-cro-CT and representative volume element. *Acta Mech.* **2020**, *231*, 503–516. [CrossRef]
21. Nasirov, A.; Fidan, I. Prediction of mechanical properties of fused filament fabricated structures via asymptotic homog-enization. *Mech. Mater.* **2020**, *145*, 103372. [CrossRef]
22. Ghandriz, R.; Hart, K.; Li, J. Extended finite element method (XFEM) modeling of fracture in additively manufactured polymers. *Addit. Manuf.* **2020**, *31*, 100945. [CrossRef]
23. Zhao, Y.; Chen, Y.; Zhou, Y. Novel mechanical models of tensile strength and elastic property of FDM AM PLA materials: Experimental and theoretical analyses. *Mater. Des.* **2019**, *181*, 108089. [CrossRef]
24. Zouaoui, M.; Labergere, C.; Gardan, J.; Makke, A.; Recho, N.; Alexandre, Q.; Lafon, P. Numerical Prediction of 3D Printed Specimens Based on a Strengthening Method of Fracture Toughness. *Procedia CIRP* **2019**, *81*, 40–44. [CrossRef]
25. Cole, D.P.; Riddick, J.C.; Jaim, H.M.I.; Strawhecker, K.E.; Zander, N.E. Interfacial mechanical behavior of 3D printed ABS. *J. Appl. Polym. Sci.* **2016**, *133*. [CrossRef]
26. D20 Committee. *Test Method for Tensile Properties of Plastics*; ASTM International: West Conshohocken, PA, USA, 2010.

27. Slic3r Manual—Welcome to the Slic3r Manual. Available online: https://manual.slic3r.org/ (accessed on 24 June 2020).
28. Blaber, J.; Adair, B.S.; Antoniou, A. Ncorr: Open-Source 2D Digital Image Correlation Matlab Software. *Exp. Mech.* **2015**, *55*, 1105–1122. [CrossRef]
29. Zouaoui, M.; Gardan, J.; Lafon, P.; Labergere, C.; Makke, A.; Recho, N. Transverse Isotropic Behavior Identification using Digital Image Correlation of a Pre-structured Material Manufactured by 3D Printing. *Procedia Struct. Integr.* **2020**, *28*, 978–985. [CrossRef]
30. Dundar, M.A.; Dhaliwal, G.S.; Ayorinde, E.; Al-Zubi, M. Tensile, compression, and flexural characteristics of acrylonitrile–butadiene–styrene at low strain rates: Experimental and numerical investigation. *Polym. Polym. Compos.* **2020**, 0967391120916619. [CrossRef]
31. Alaimo, G.; Marconi, S.; Costato, L.; Auricchio, F. Influence of meso-structure and chemical composition on FDM 3D-printed parts. *Compos. Part B Eng.* **2017**, *113*, 371–380. [CrossRef]
32. Attolico, M.A.; Casavola, C.; Cazzato, A.; Moramarco, V.; Renna, G. Effect of extrusion temperature on fused filament fabrication parts orthotropic behaviour. *Rapid Prototyp. J.* **2019**, *26*, 639–647. [CrossRef]
33. Rodríguez, J.F.; Thomas, J.P.; Renaud, J.E. Mechanical behavior of acrylonitrile butadiene styrene (ABS) fused deposition materials. Experimental investigation. *Rapid Prototyp. J.* **2001**, *7*, 148–158. [CrossRef]
34. Hyer, M.W. *Stress Analysis of Fiber-Reinforced Composite Materials*; Updated edition; DEStech Publications: Lancaster, PA, USA, 2009.
35. Hill, R. A theory of the yielding and plastic flow of anisotropic metals. *Proc. R. Soc. Lond. A* **1948**, *193*, 281–297.
36. Bhandari, S.; Lopez-Anido, R.A.; Wang, L.; Gardner, D.J. Elasto-Plastic Finite Element Modeling of Short Carbon Fiber Reinforced 3D Printed Acrylonitrile Butadiene Styrene Composites. *JOM* **2020**, *72*, 475–484. [CrossRef]
37. Balderrama-Armendariz, C.O.; Macdonald, E.; Espalin, D.; Cortes, D.; Wicker, R.; Maldonado-Macías, A. Torsion analysis of the anisotropic behavior of FDM technology. *Int. J. Adv. Manuf. Technol.* **2018**, *96*, 307–317. [CrossRef]
38. Anisotropic Yield/Creep. Available online: https://abaqus-docs.mit.edu/2017/English/SIMACAEMATRefMap/simamat-c-anisoyield.htm (accessed on 15 September 2020).

*applied sciences*

MDPI

*Article*

# Enhancing Structural Performance of Short Fiber Reinforced Objects through Customized Tool-Path

**Jaeyoon Kim and Bruce S. Kang ***

Department of Mechanical and Aerospace Engineering, West Virginia University, Morgantown, WV 26505, USA; jekim@mix.wvu.edu

* Correspondence: bruce.kang@mail.wvu.edu; Tel.: +1-304-293-3232

Received: 24 August 2020; Accepted: 24 September 2020; Published: 18 November 2020

**Abstract:** Fused deposition modeling (FDM) is one of the most common additive manufacturing (AM) technologies for thermoplastic materials. With the development of carbon fiber-reinforced polymer (CFRP) filament for FDM, AM parts with improved strength and functionality can be realized. CFRP is anisotropic material and its mechanical properties have been well studied, however, AM printing strategy for CFRP parts has not been developed. This paper proposes a systematic optimization of the FDM 3D printing process for CFRP. Starting with standard coupon specimen tests to obtain mechanical properties of CFRP, finite element analyses (FEA) were conducted to find principal directions of the AM part and utilized to determine fiber orientations. A specific tool-path algorithm has been developed to distribute fibers with the desired orientations. To predict/assess the mechanical behavior of the AM part, the 3D printing process was simulated to obtain the anisotropic mechanical behavior induced by the customized tool-path printing. Bolt hole plate and spur gear were selected as case studies. FE simulations and associated experiments were conducted to assess their performance. CFRP parts printed by the optimized tool-path shows about 8% higher stiffness than those printed at regular infill patterns. In summary, assisted by FEA, a customized 3D printing tool-path for CFRP has been developed with case studies to verify the proposed AM design optimization methodology for FDM.

**Keywords:** Fused Deposition Modeling (FDM); tool-path; FEA; CFRP

---

## 1. Introduction

Additive manufacturing (AM) technologies have been rapidly advancing and widening its applicability to complex geometries and range of material choice [1–6]. Fused deposition modeling (FDM) is one of the widely used in AM technologies for the thermoplastic material. In contrast to traditional subtractive manufacturing, FDM parts are built by adding materials layer by layer. This layered nature of FDM causes some defects of the printed objects, such as staircase effect, coarse surface, and anisotropic mechanical properties [3,4]. To address these challenges, improvement of the quality of FDM parts has been an active research area. For example, printing process techniques, such as heat treatment [7], machining [8], and chemical treatment [9–12], were investigated to have a better surface quality of the parts. Printing parameters, including the effect of raster angle [13] and building direction [14], were optimized to obtain better bonding strength. However, there is a lack of design strategy to address the anisotropic characteristics of FDM parts. Moreover, the development of carbon fiber-reinforced polymer (CFRP) for FDM requires the optimization of material anisotropy for its best performance. For CFRP, extensive research has been carried out to investigate the anisotropic mechanical properties of CFRP, including ABS, PLA, and nylon. Generally, the anisotropic structural property of FDM parts is highly dependent on the building direction. Short carbon fiber provides 3D printing flexibility to improve the mechanical properties of parts than those printed by continuous fiber-reinforced filaments [15]. The effect of build orientation has been studied to enhance structural performance [16–18]. For FDM 3D printing, commercial slicer programs provide several limited infill patterns to choose a building direction. Once a pattern is selected, it is not allowed to edit its route. This is because the tool-path for FDM was originally developed to control the movement of the CNC machine cutter. When the tool-path algorithm was developed for FDM, it was for printing process improvement, structural strength enhancement of final products was not recognized. In this research, a novel design methodology for short carbon fiber composite built by FDM 3D printing is proposed. It starts with FEA stress analysis to determine principal stress directions of the FDM part with which customized tool-paths are developed. Structural performances of FDM parts built by the proposed tool-path method are verified by both FEA analyses and experiments. Tensile tests and related microstructural analyses using scanning electron microscope (SEM) for CFRP-Nylon were performed to investigate mechanical properties and fiber orientations. The framework for this methodology is shown in Figure 1. The primary contributions of this research are as follows:

(1) A novel design methodology for FDM parts assisted by finite element analysis;
(2) A customized tool-path algorithm for FDM that maximizes the effect of fiber reinforcement under the given loading and boundary conditions.

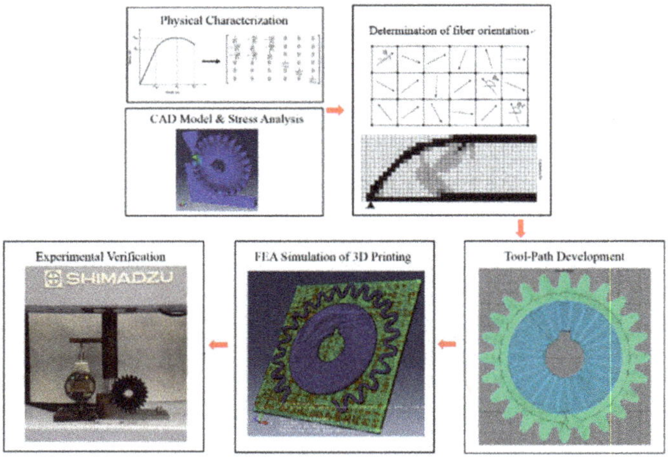


**Figure 1.** The workflow of the design approach.

## 2. Background

### 2.1. Anisotropy of FDM Parts

Many experimental studies have shown that FDM printed parts exhibit anisotropic mechanical properties. Ahn et al. investigated the anisotropic mechanical properties of ABS parts built by FDM [17]. The mechanical properties of PEEK and PC fabricated by FDM were studied by El-Gizawy et al. [19] and Hill and Haghi [20], respectively. They proved that FDM induces a significant structural anisotropy for thermoplastic materials.

There have been several computational methods to address this issue. Hildebrand et al. proposed a computational design methodology to apply different build orientations to each section of an FDM part based on its stress map [20]. Zhou et al. developed a worst-case analysis method assuming the material is orthotropic to find the structurally weak sections [21]. Farbman et al. reported the effect of infill pattern on the mechanical properties of FDM parts by FEA simulation [22].

Besides, build orientation selection methods to control the anisotropy has been proposed. Domingo-Espin et al. investigated the effect of building direction by FEA modeling of FDM parts with orthotropic materials [23]. Richard and Crawford developed a build orientation selection algorithm using the Tsai–Wu failure criterion [24]. However, their works did not try to find the build orientation to minimize internal stress. Similarly, Umetani and Schmidt proposed a method to find the best build orientation for FDM parts [25]. This method accounted for structural anisotropy to stacking layers of extrusion, assuming those are laminate composites. They found that different level of structural performances was captured depending on build direction.

### 2.2. Carbon Fiber Reinforced Polymer (CFRP)

In 3D printing, various fiber-reinforced polymers have been studied to improve mechanical properties using Jute fibers [21], metal [22], glass fibers [23], vapor-grown carbon fiber [24], and continuous fibers [25]. For FDM 3D printing, short fiber reinforced filaments are most commonly used in manufacturing high strength AM part. It has been reported that short carbon fibers reinforced thermoplastic filaments significantly improve the strength of AM parts. These filaments are now commercially available from several manufacturers, such as CarbonX$^{TM}$, Matterhackers$^{TM}$, and ColorFabb$^{TM}$. Figure 2 illustrates a printing extrusion of a short fiber reinforced filament.

**Figure 2.** Short fiber alignment during the extrusion process [26].

### 2.3. Tool Path for FDM

It has been demonstrated that tool-path or build orientation affects the quality and strength of objects. Concerted efforts have been made to optimize tool-path planning for various applications [27,28]. Currenly, contour-parallel and direction-parallel methods are the most widely used for FDM in practice. The contour-parallel tool-path is to move the extruder parallel to the boundaries of cross-sections [29]. Fabrication is precise, but it is computationally expensive. By contrast, the direction-parallel path is to move the extruder in zig-zag motion along a fixed direction within the boundary in the interior region. This method is simple and fast. The algorithm using both methods has been developed and is widely adopted in major slicer programs, such as Simplify3D [30]. Figure 3 illustrates the difference between the contour-parallel and directional-parallel tool-path.

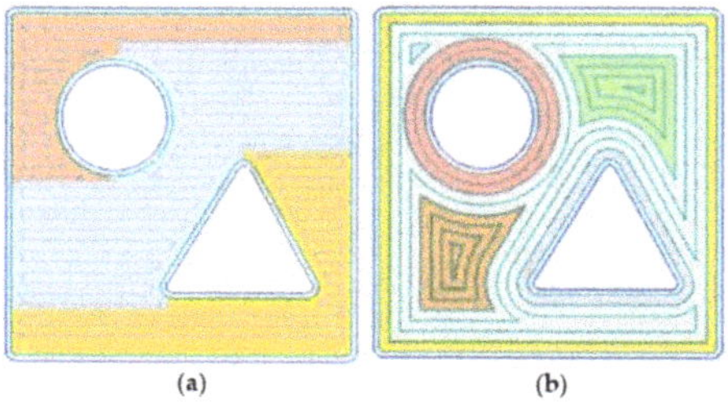

**Figure 3.** Comparison of different tool-path generation strategies; (**a**) direction parallel path, (**b**) contour parallel path, [31].

## 3. Material Characterization

### *3.1. Mechanical Property Measurement*

As various materials for FDM have been developed, associated test methods have been reported. Nevertheless, American Society for Testing and Materials (ASTM) test protocols for additive manufactured objects officially has not been developed [32,33]. In this study, CFRP is treated as a laminated plastic polymer. CFRP-Nylon (20% fiber volume fraction) coupon specimens with printing orientation of 0°, 45°, and 90° were fabricated, and relevant mechanical properties were measured. As shown, direction 1 (red color) is the FDM line of deposition, and direction 2 is perpendicular to this line of deposition. From the printed specimens oriented at 0°, tensile modulus in the deposition direction $E_1$, Poisson ratio $\nu_{12}$ and tensile strength $S_1$ were determined. From the specimens oriented at 90°, tensile modulus perpendicular to the deposition direction $E_2$, Poisson ratio $\nu_{21}$ and tensile strength $S_2$ were determined. From the specimens oriented at ±45°, shear modulus at the 1–2 plane $G_{12}$ and shear strength $S_{12}$ were determined. Three specimens per sample were tested for each one of the three orientation cases, and each printing material, totalizing 9 test runs.

For the determination of stiffness and strength properties at material directions 1 and 2, the ASTM D638 test standard was applied. As shown in Figure 4, I-type (dog-bone) specimen was used, with length and width of 165 mm and 19 mm, respectively. Letter-size sheets (11′ × 8.5″) were printed by the designated printing orientations and cut by a CNC machine to obtain specimens. The nominal specimen thickness was 3.3 mm, with 11 printed layers. ASTM D3518 test method was conducted to obtain the 1–2 plane shear stiffness and strength properties. Rectangular specimens (25 mm × 200 mm) were printed. The nominal thickness is 4.8 mm with 16 printed layers, and the stacking sequence was [±45°]4s (symmetric).

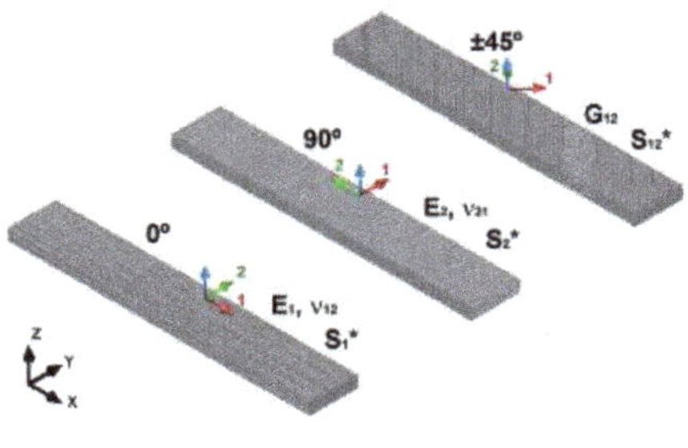


**Figure 4.** Illustration of printing orientations (0°, 90°, and ±45°) [34].

Ultimaker 2+ was used to produce test specimens. It was one of the popular 3D printers at the time of study, allowing users to upgrade it to apply CFRP filaments. The printing envelope of 215 mm × 210 mm × 180 mm was used. For the size of the nozzle, it has been reported that a 0.4 mm nozzle is frequently clogged with short fibers. On the other hand, 0.8 mm is too big to obtain precise fabrication. Manufacturer of Ruby nozzle, which can endure weariness from short fibers, recommended 0.6 mm. Therefore 0.6 mm was chosen in this study. The printing parameters employed were; nozzle extrusion temperature: 260 °C, heat bed temperature: 100 °C, deposition line (layer) height; 0.15 mm, and printing speed: 20 mm/sec. Carbon-fiber nylon filament (NylonX$^{TM}$ from Matterhackers, filament diameter: 2.85 mm) made of 4043D resin reinforced with chopped short carbon fibers (20% weight fraction) was used. The experiments were performed using SHIMAZU$^{TM}$ AGS-X HC universal testing machine (Figure 5b). A fully calibrated extensometer was utilized to precisely capture displacement change. All the specimens were loaded up to material failure at a displacement rate of 1 mm/min. The data acquisition rate was 10 Hz.

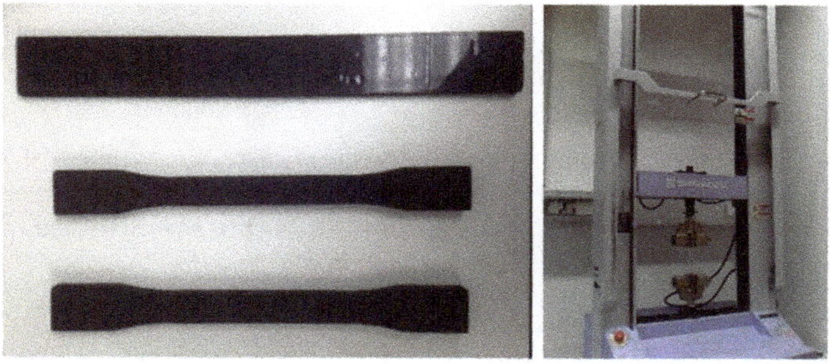

(**a**) (**b**)

**Figure 5.** Test specimens (**a**) tensile ("dog bone", lower) and shear (rectangular, upper) and (**b**) SHIMAZU tensile machine.

Figure 6 shows the stress-strain curves for the specimens printed in the three different directions. Captured data from the machine was 10 points-average smoothed. As shown, the directional dependency of material properties is noted. Table 1 shows the mechanical properties of CFRP-Nylon (NylonX$^{TM}$) which were extracted from the stress-strain curves and used in the FE modeling analysis of FDM parts. Young's moduli and tensile yield strengths, as well as Poisson's ratios, were obtained for the printing directions of 0° and 90°. Moreover, shear strength was experimentally determined per ASTM D3518.

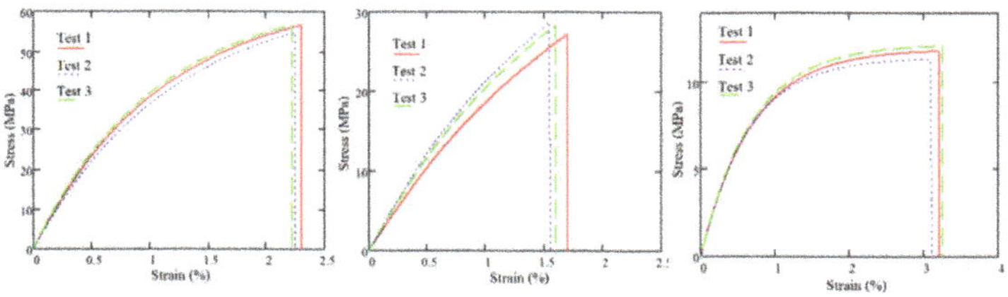


(**a**) (**b**) (**c**)

**Figure 6.** Stress vs. strain data for CFRP-Nylon printed at 0° (**a**), 90° (**b**), and ±45° (**c**).

**Table 1.** Mechanical properties of carbon fiber-reinforced polymer (CFRP)-Nylon.

| Property | Direction | CFRP-Nylon | ASTM |
|---|---|---|---|
| Young's Modulus | 0° | 4.14 GPa | D638 |
| | 90° | 2.15 Gpa | |
| Shear Modulus | ±45° | 1.12 Gpa | D3518 |
| Tensile Strength | 0° | 56.6 Mpa | D638 |
| | 90° | 28.3 Mpa | |
| Shear Strength | ±45° | 11.9 Mpa | D3518 |
| Elongation | 0° | 2.30% | D638 |
| | 90° | 1.59% | |
| Poisson's ratio | $\nu_{12}$ | 0.391 | D638 |
| | $\nu_{21}$ | 0.203 | |

### 3.2. Fiber Orientation

The major assumption underlying the proposed method is that short fibers are fully aligned with the direction of extrusion. Although the fiber orientation of parts was indirectly verified through lab tests, visual verification is required to understand the potential and limitation of the proposed method caused by the inherent drawback of short carbon fibers. The orientation of short fibers embedded in printed parts was investigated by an optical microscope. For this task, the specific method developed by Bay and Tucker [35] was followed to characterize the short fiber orientation of the printed samples. Figure 7. Shows an SEM image of CFRP-Nylon fractured surface, as shown, most of the fibers are aligned in a single direction. Figure 8. shows an optical microscope image of a printed CFRP part. The methodology for fiber alignment analysis is based on the simple geometrical assumption that, ideally, cylindrical fiber should appear on the section as either circular or elliptical shape. However, as shown in Figure 9a, due to variations of fiber tip geometry, such as covering of the matrix material on the fiber tips surface or breakage of fiber tip surface, some of the cross-sectional shapes of fibers are not suitable for fiber alignment analysis—thus, those were excluded.

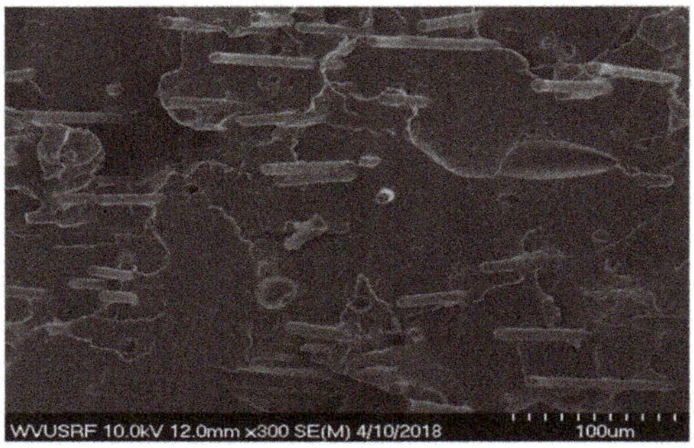


**Figure 7.** SEM image of CFRP specimen printed at 45° direction.

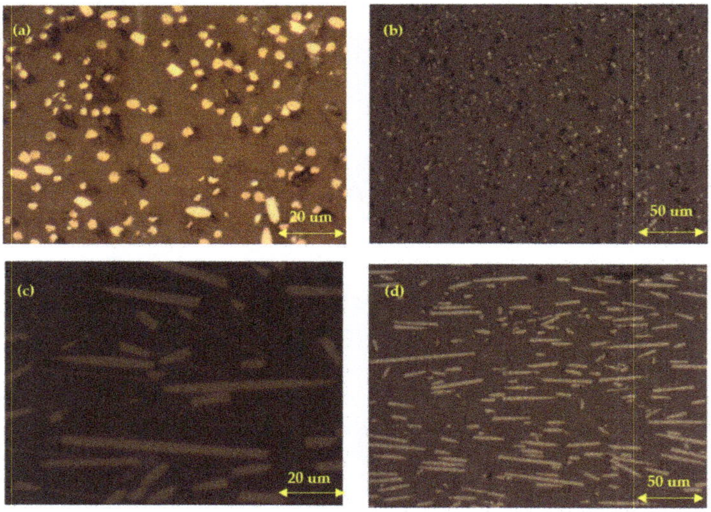


**Figure 8.** Sample micrographs of a polished specimen obtained from optical microscope: (**a**) Cross-section at 0° (printing direction), magnitude 50×; (**b**) cross-section at 0°, magnitude 20×; (**c**) cross-section at 90°, magnitude 50×; (**d**) cross-section at 90°, 20×.

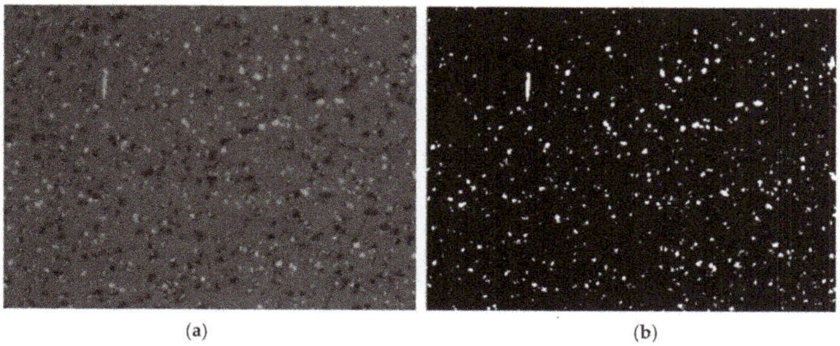

Figure 9. Pixel conversion to white/black. (a) original, (b) converted.

Assuming the length and diameter of the short carbon fibers are uniform as approximately 100 μm and 5 μm, the measurement was processed to find only the ends of the major axis. Each image was broken into pixels; each pixel has a value corresponding to the intensity of light at its Cartesian location. These digital images are first subjected to a threshold operation, making each pixel either black or white, as shown in Figure 9b.

The next step is to identify a group of pixels representing each fiber and determine the relevant dimensions. It measured the chord length in several directions and then took the maximum and minimum values as the major and minor axes, respectively. Using the cross-sectional area and major/minor diameters, each cross-section is checked to confirm they are roughly either circular or elliptical. If those fibers whose cross-sections are not circular or elliptical, they would not be used for fiber orientation analysis, and these fibers are filtered out, as shown in Figure 10b.

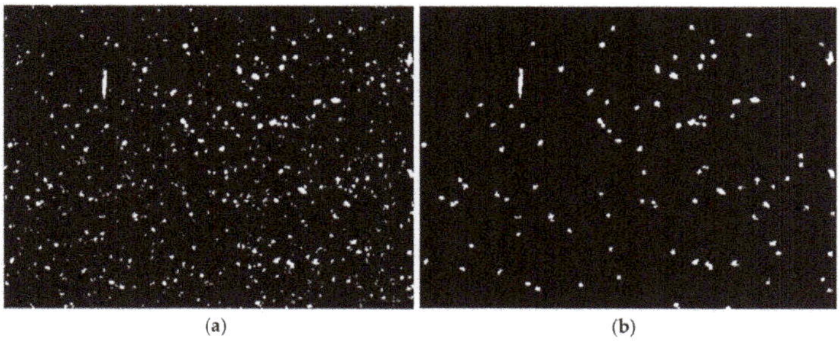

Figure 10. Filtering fibers with circular and elliptical cross-sections. (a) before, (b) after filtering.

Using the processed fiber images, such as Figure 10b, second-order tensors of fiber orientations in the directions of 1, 2, and 3, were computed. Table 2 shows the values of each component, as shown, the majority of the orientation tensor for the FFT sample is a11, an indication that practically most of the short carbon fibers are oriented in the 1-direction, which is the desirable load-bearing direction. This fiber orientation analysis indicated the inherent characteristic of high fiber orientation by the FDM process.

Table 2. Fiber orientation distribution of printed CFRP-Nylon sample.

| Orientation | a11 | a12 | a13 | a22 | a23 | a33 |
|---|---|---|---|---|---|---|
| % | 0.825 | 0.123 | 0.082 | 0.03 | 0.004 | 0.0056 |

## 4. Customized Tool-Path Development

FEA stress field is computed to obtain principal directions of elements, Figure 11 illustrates an example showing element-based principal directions of a cantilever beam under uniformly distributed load. Centroids of elements are connected to create a printing path. The created printing path are aligned with the principal directions of individual elements. Square shell elements are utilized to reduce computational time by creating straight path lines. The size of the shell element is determined based on the printer extruder diameter. The small size of elements guaranteed a more precise printing path; however, if it is smaller than the extruder diameter, the extruding width is set to be invalid and linked to neighboring elements. Figure 12 shows a framework of the optimized tool-path development; more details are explained below.

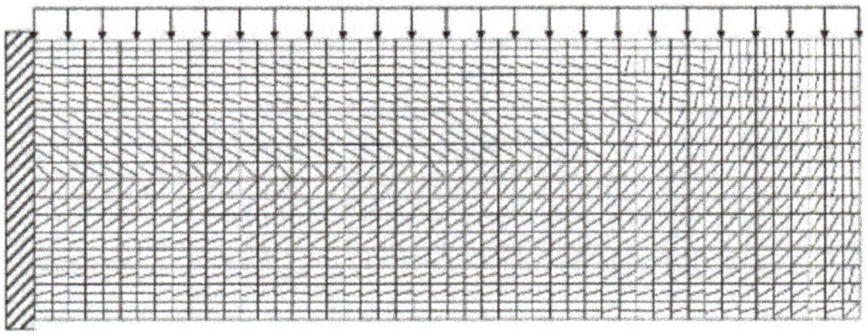

**Figure 11.** Principal directions of elements [36].

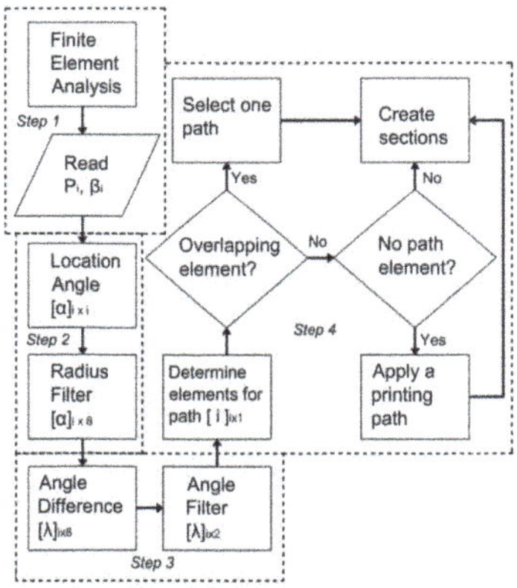


**Figure 12.** Optimized tool-path framework.

### 4.1. FE Analysis

In this step, principal stress $[P_i]$ and direction $[\beta_i]$ matrices of elements are generated from the FEA analysis output. Those are adopted as a basis to create a tool-path to assign fibers aligning with the principal direction of the structure under a given loading condition.

### 4.2. Location Angle and Radius Filter

The curvature of the printing path may affect the quality of the product. This means a sharp curve of the path usually produces more voids, which could lead to poor product quality. To address this issue, an effort to make the printing path straight has been made. Location angle matrix $[\alpha]_{ixi}$ is defined to set the geometrical relationship between two elements on the Cartesian coordinates system. As illustrated in Figure 13b, the angle between connecting lines of adjacent element centroids is computed. The center element (element 23) in Figure 13b is denoted as 'starting element' in this framework, as illustrated in Figure 13a. Then, a radius from the centroid of the starting element is defined to cover neighborhood elements only around the starting element. Neighborhood elements which are filtered by the radius are denoted as 'candidate elements'. Figure 13a describes the concept of elemental printing path connection.

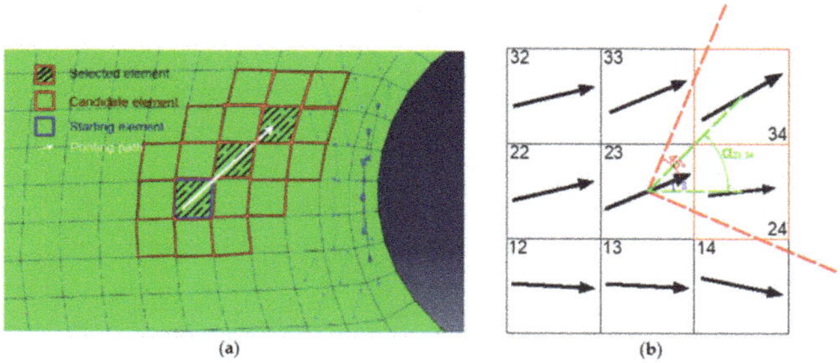

**Figure 13.** (**a**) Concept of elemental printing path connection; (**b**) concept of angle difference filtering.

### 4.3. Angle Difference Filter

Since the printing path must align with the principal directions of elements, the next connecting elements are selected by the principal direction of the starting element. In this step, the angle difference filter is defined as follows.

$$[\lambda]_{ix8} \rightarrow [\lambda]_{ix2} \text{ if } [\lambda]_{ix8} < 45^\circ \text{ Where, } [\lambda]_{ix8} = [\alpha]_{ix8} - [\beta]_i$$

To keep only two elements as candidate elements, ±45° is added to the principal direction of the starting element. As shown in Figure 13b, the red boundary line indicates the final two elements for the printing path connection.

### 4.4. Creation of Printing Path and Sections

From the selected two candidate elements (e.g., Figure 13b), the element with larger principal stress is finally selected to be connected. If there are no candidate elements whose angle differences are less than 45 degrees, the printing path stops the connection. Through the procedure, a set of connecting lines is then created and referred to as a printing path section. Next, the contour-parallel or direct-parallel printing method is determined based on the computed path line and curvature of the path. If a path line moves to align with the product profile, the contour-parallel method is applied; otherwise, the direct-parallel method is applied. For the direct-parallel method, the angle of the infill pattern is determined by the average principal direction of elements occupied the section. To avoid overlapping of printing paths, if an element is selected multiple times for the next connecting element, the principal stresses of starting elements on each path are checked. Then, a printing path with larger principal stress is chosen to continue. Elements covered by similar patterns create a printing section. Printing path of elements having the end of the path, thus, creates section boundaries. Figure 14 shows the workflow of section generation and corresponding printing path.

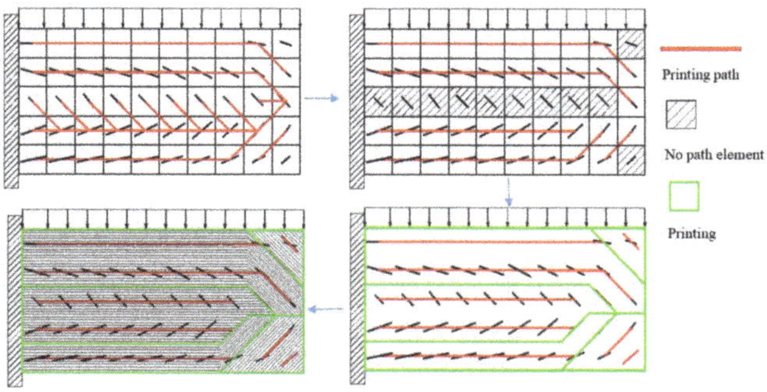


**Figure 14.** Principle of printing path and section development.

In contrast, if an element is not chosen by any path, groups of those elements are merged into neighborhood sections. For those elements, the stress level is usually low, and thus, any printing pattern is allowed, since it has a low impact on the strength of the final product. However, to achieve better product quality a uniform single printing path is applied to the merging section rather than generating printing section boundaries, which could lead to producing voids during the extruding. Lastly, 5% of overlapping is applied to guarantee adequate material bonding. Figure 15 shows an example of an optimized tool-path for a plate with a circular hole under tensile loading.

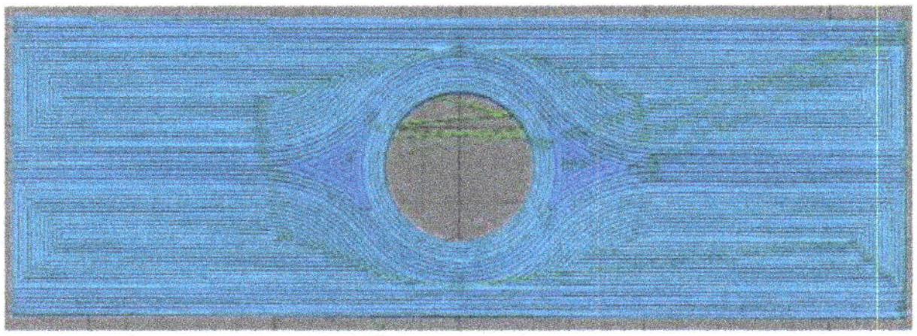

**Figure 15.** Optimized tool-path.

### *4.5. FE Modeling of Printed Object*

Generally, slicer programs, such as Simplify3D, generate G-codes for FDM printing, including controlling of extruder movement. In this research, G-codes are then converted into ABAQUS$^{TM}$ input to define outer boundary lines, material orientations of elements, and mesh size. Specific Matlab$^{TM}$ codes were created to process this step, which provides angle values in the Cartesian coordinates of the FE elements. Material orientations were directly computed from G-codes, as illustrated in Figure 16. Moreover, the orthotropic material properties matrix was applied to elements from the experimental data in Section 3.1.

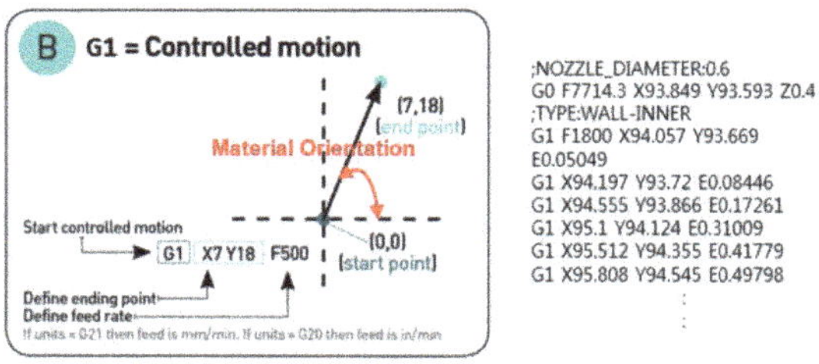


**Figure 16.** Determination of material orientation.

## 5. Case Study-Hole Plate

### *5.1. Problem Statement—Hole Plate*

Stress concentration in a plate with a hole was chosen to demonstrate the advantage of the customized tool-path printing method. Figure 17 shows the maximum principal stress flow and a printed CFRP sample.

### *5.2. Tool Path Development—Hole Plate*

To compute principal directions of each element, FEA stress analyses were carried out using ABAQUS$^{TM}$. For simplicity, 2D shell elements were applied. As described earlier, square elements were employed to connect elements to generate a custom tool-path for printing paths and sections. The height, width, and radius of the hole are 120 mm, 40 mm, and 10 mm, respectively.

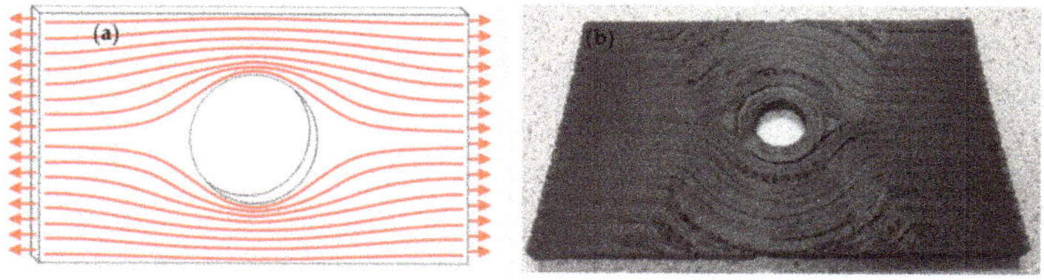


**Figure 17.** (**a**) Stress flow and (**b**) an FDM built CFRP sample of a plate with a circular hole.

Figure 18 shows the maximum principal stress field and the corresponding principal directions of individual elements. As expected, high stresses occurred around the center hole, and their principal directions are aligned with the hole (shown as short length arrows in Figure 18b).

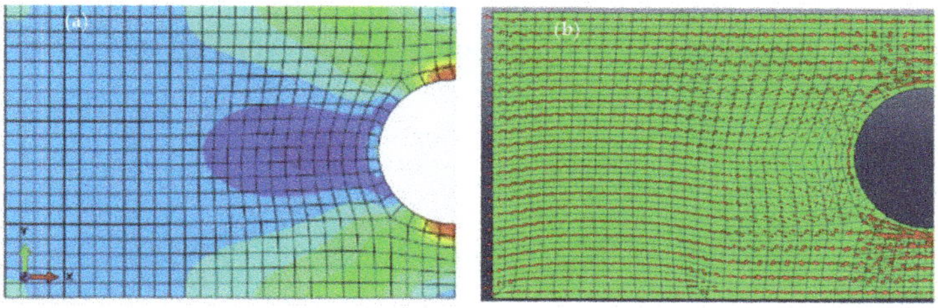

**Figure 18.** (**a**) maximum principal stress field and (**b**) principal directions of the plate with the circular hole under tensile loading.

Figure 19 shows the section division of the plate for tool-path based on the proposed method. Each element is connected to create sections. A customized tool-path is created for each section based on principal directions of elements. For example, the tool-path around the center hole is aligned with the circle as principal directions are parallel to the circle. Moreover, the low-stress region in Figure 19a has a 0° uniform tool-path, which is the same as the tool-path in other major regions. Figure 19b shows the complete tool-path development of the whole plate.

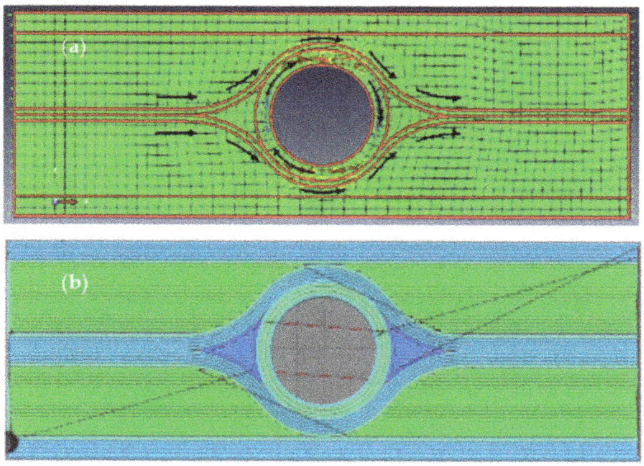


**Figure 19.** (**a**) Section division for tool-path and (**b**) complete optimized tool-path of stress concentration plate.

### 5.3. FE Modeling—Hole Plate

Figure 20 shows the maximum principal stress distribution of the plate with the updated orthotropic material properties induced by the customized tool-path. Non-linear explicit analysis with eight brick elements was performed using ABAQUS$^{TM}$. The thickness of the model was reduced by applying only two layers to reduce the running time. As shown, high stresses (with 8% less magnitude) still occurred around the hole, but were much confined compared to the stress distribution of plate with isotropic material (Figure 18a). Moreover, as an illustration, to show the benefit of tool-path optimization assisted by FEA, Figure 21a is a uniform 0° printing path (i.e., no optimization), and Figure 21b is the optimized printing path. Figure 21c,d show its corresponding shear strain distributions. As shown, for the case of no optimization (Figure 21c), high shear strains at the edge of the hole will be the likely failure initiation site. As for the optimized case (Figure 21d), the magnitude of the highest shear strain is 16% less than the non-optimized case; moreover, the high shear strain regions are away from the hole edge, due to the optimized printing paths.

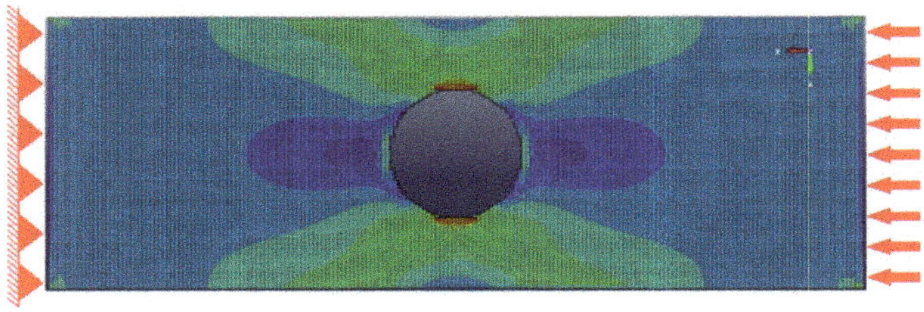

**Figure 20.** Stress distribution of the plate printed based on optimized tool-path.

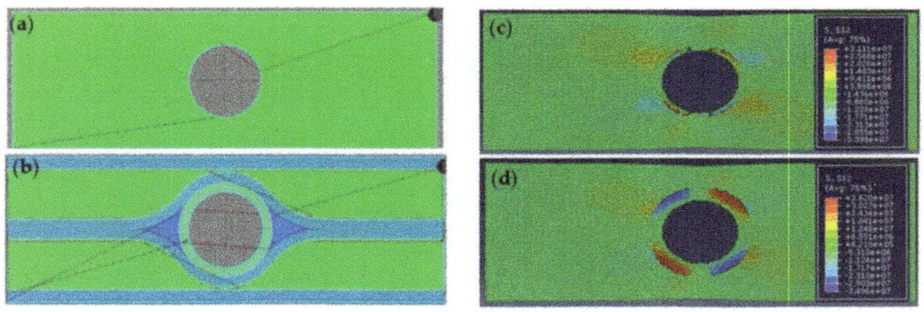


**Figure 21.** (**a**) 0° uniform; (**b**) customized tool-path; (**c**) shear strain distribution of samples printed by 0° uniform; (**d**) customized tool-path.

### 5.4. Tensile Test—Hole Plate

As shown in Figure 22, relevant tensile tests for the printed plates were conducted [37] to validate FEA simulation results. The loading rate was 1 mm/min, with data recorded at every 0.01 sec. Using the force and displacement data, the stiffness response was computed. Figure 22b shows load vs. displacement curves for each case. Averaged test results of each case are shown in Table 3.

The plate printed by the optimized tool-path shows the highest stiffness (6% higher than the direct-parallel (0°) case). However, there is no significant improvement in overall failure strength. Figure 23 represents the comparison of stiffness response results between FEA simulations and tensile tests. Discrepancies between analysis and test for 0°, contour, ±45°, and optimized tool-path show 9.4%, 10.1%, 10.2%, and 11%, respectively, which suggests the optimized tool-path generated slightly more voids than other conventional tool-path methods.

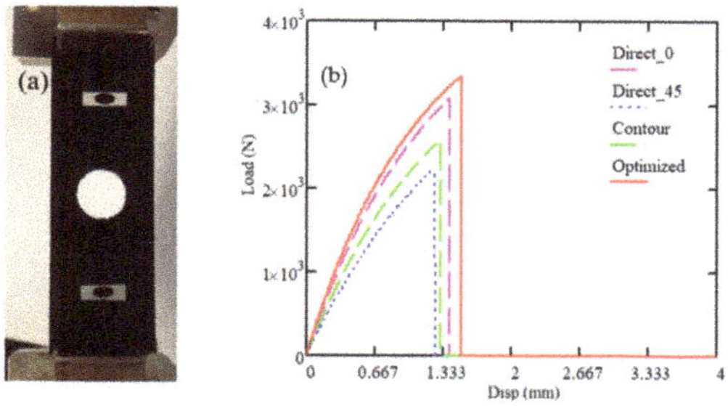


**Figure 22.** (**a**) Tensile test and (**b**) test result of stress concentration plate.

**Table 3.** Tensile test results of stress concentration plate printed by various tool-paths.

| | Stiffness Response (N/mm) | Failure Strength (N) | Max. Displacement (mm) |
|---|---|---|---|
| Direct-parallel (0°) | 2544.0 | 3087 | 1.39 |
| Contour-parallel | 2161.7 | 2592 | 1.30 |
| Direct-parallel (±45°) | 1983.4 | 2262 | 1.25 |
| Optimized | 2687.0 | 3349 | 1.50 |

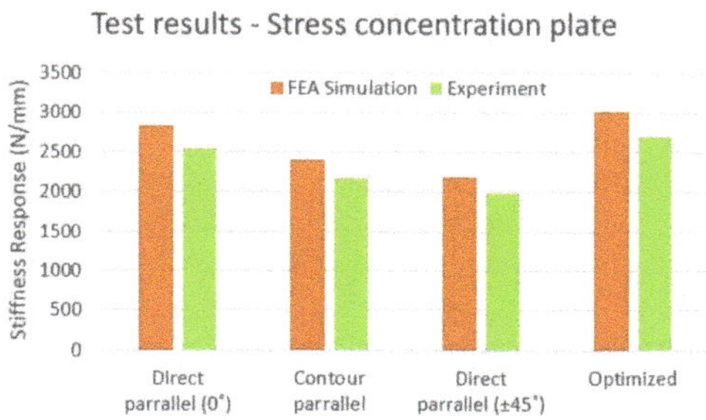


**Figure 23.** Comparison of stiffness response between finite element analyses (FEA) simulations and experiments.

## 6. Case Study—Spur Gear

### *6.1. Problem Statement—Spur Gear*

Gears are mechanical components used for transmitting motion and torque from one shaft to another. A spur gear is the most common type of gear, and the tooth regions are prone to failure, due to high contacting stresses. Figure 24 shows an example of a failed Nylon spur gear in a ball-milling machine.

**Figure 24.** Spur gear failure in a ball-milling machine.

### 6.2. Determination of Fiber Orientations—Spur Gear

Using ABAQUS$^{TM}$, FEA analyses were carried out for the pre-described spur gear geometry under the prescribed loading and boundary conditions. The central main steel gear (Figure 24) was considered as rigid in the FE analysis (shown as the white section in Figure 25a). Figure 25 shows the maximum principal stress field (Figure 25a) and principal directions (Figure 25b) of elements at the spur gear tooth region.

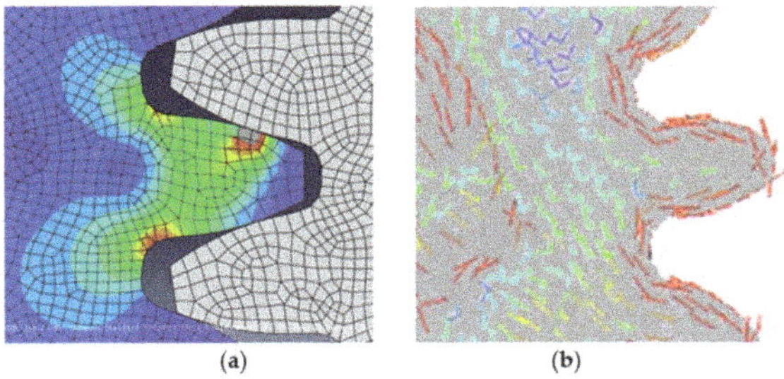

**Figure 25.** (**a**) Maximum principal stresses; (**b**) principle directions at spur gear tooth region.

### 6.3. Tool Path Development—Spur Gear

Figure 26 shows principal directions and section divisions for FDM printing. Different printing patterns were applied to individual sections based on output from the steps described in Section 4. Figure 27 shows the optimized tool-path for the gear. Noted that principal directions of elements in the tooth region are oriented along with the gear tooth profile such that the ideal way to reinforce the gear is to align fibers with the tooth profile in the high-stress region.

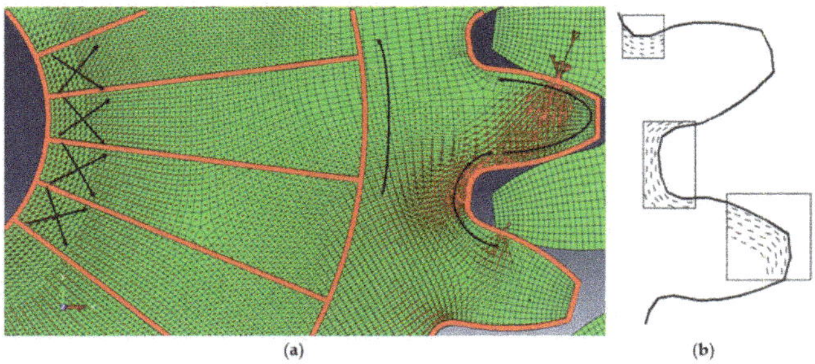

**Figure 26.** (**a**) Section division of spur gear for tool-path; (**b**) principal directions of teeth section.

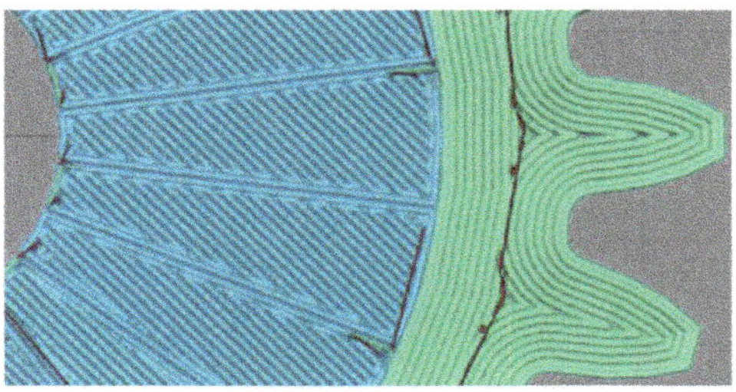

**Figure 27.** Optimized tool-path for spur gear.

### 6.4. FE Modeling—Spur Gear

Coordinate information was exported from G-codes and converted into material orientations of individual elements. Only the initial eight layers were applied to reduce the running time. A total of 180,000 Hexahedron elements having an extruder width (0.6 mm) was generated. Figure 28a is a captured image from the process of mesh generation. Figure 28b shows a FEA simulation of the experiment test (Figure 29). The pushing hammer was treated as rigid, vertical displacement of 1.0 mm was applied.

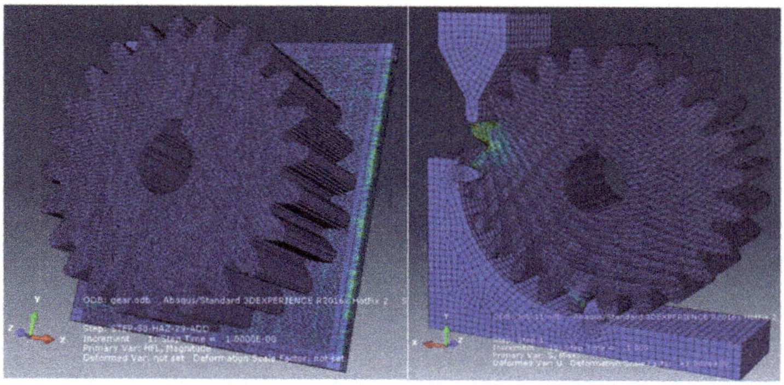

(**a**) (**b**)

**Figure 28.** (**a**) FE modeling of spur gear; (**b**) FE simulation of the experimental test in Figure 29.

**Figure 29.** Spur gear stiffness test for CFRP-Nylon.

Stiffness responses from several different build orientations, including optimal fiber orientations, unidirectional orientation with [0°], [±45°]s, and contour direction were obtained to compare structural performance.

### 6.5. Stiffness Test—Spur Gear

To validate the FEA simulation results, relevant compression tests for spur gear were conducted, as shown in Figure 29. As previously discussed, high stress was generated in the teeth region. The corresponding displacement was measured by a digital dial indicator with a resolution of 0.001 mm. The loading rate is 1 mm/min, and the data was recorded every 10 s.

Figure 30 shows load vs. displacement curves for each case. Averaged test results at three different locations are shown in Table 4. For repeatability of the test results, three different locations were tested, and variation was less than 1%.

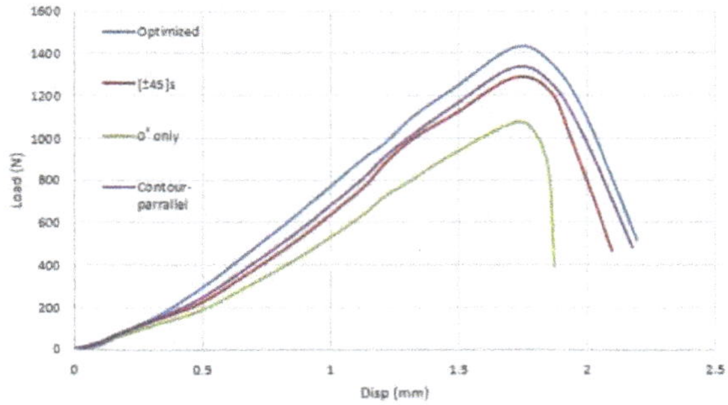


**Figure 30.** Load vs. displacement curves of CFRP spur printed by various tool-path.

**Table 4.** Test results of stiffness response of CFRP-Nylon.

| | Stiffness Response (N/mm) | Failure Strength (N) | Max Disp (mm) |
|---|---|---|---|
| Contour | 778.6 | 1330 | 2.18 |
| Optimized | 834.6 | 1430 | 2.19 |
| [±45°]s | 754.6 | 1287 | 2.10 |
| 0° only | 607.0 | 1073 | 1.87 |

As shown in Figure 30, gear printed by the optimized tool-path shows the highest stiffness response and failure strength. It shows approximately 7% higher in stiffness response than gear printed by contour-parallel. For the failure strength, optimized gear shows 8% higher than contour-parallel.

For the result of FEA simulations, the principal stresses at the critical location of each case were measured, and the stiffness response was computed. The gear with the optimal fiber orientation shows the highest stiffness response of 981 N/mm. For other cases, contour-parallel, [±45°]s, and [0°] showed 894 N/mm, 847 N/mm, and 682 N/mm, respectively. Approximately 9% improvement of stiffness was observed when compared with the contour-parallel tool-path. Figure 31 shows a comparison of stiffness response results between FEA simulation and laboratory experiments. The discrepancy between FEA modeling analyses and experiments is 12% for the optimized tool-path and 9% for other conventional tool-path.

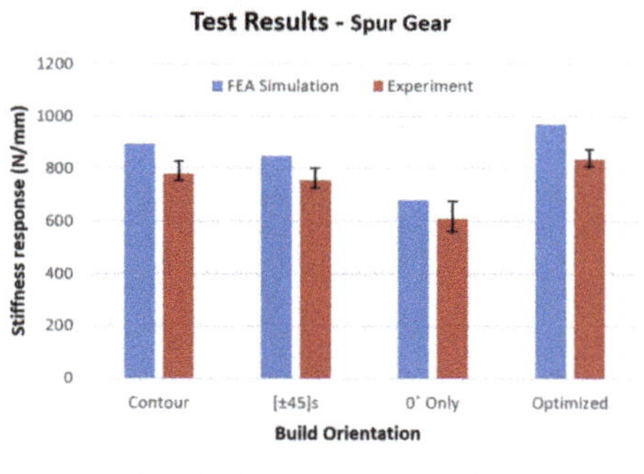


**Figure 31.** Comparison of results between FEA simulation and experimental tests.

## 7. Summary

A 3D printing methodology for fused deposition modeling (FDM) integrating fiber placement and tool-path development has been proposed. It starts with FEA to obtain principal stress fields and principal directions. Optimal fiber orientations of individual elements were determined. Using the stress output data from ABAQUS$^{TM}$, a tool-path optimization algorithm to maximize the effect of fiber reinforcement of CFRP was developed. To predict the mechanical behavior of the printed parts, G-codes from the tool-path were used to model printed objects.

The proposed methodology demonstrates that the optimized tool-path can be applied to the 3D printer to extrude fibers aligned with principal directions. The flow distribution of printed fibers was verified by image analysis, which showed that approximately 83% of fibers were oriented as intended. Both FEA and preliminary experimental case study results show that CFRP-Nylon parts printed by the optimized tool-path achieved approximately 8% improvement in structural performance over parts printed at regular uniform printing direction. Associated experimental test results represent 15% lower stiffness responses than those from FEA predictions. Although 8% improvement is insignificant, nevertheless, with further AM printing process optimization to better control the fiber orientation, as well as advanced material development, the proposed customized tool-path method presented in this paper can be utilized as a design and printing methodology for CFRP structural parts by FDM technology.

The proposed methodology can be extended to other materials. Moreover, with the development of a dual extruder system, dual material optimization would be another topic to overcome the brittle nature of CFRP. To predict their durability more accurately, fatigue tests are required. Due to the limitation of the current FDM printing system, only in-plane tool-path optimization has been allowed in this research. However, if AM with a tilted bed is fully developed, it may provide research opportunities for three-dimensional tool-path optimization.

**Author Contributions:** Conceptualization, B.S.K.; methodology, J.K.; software, J.K.; validation, J.K.; formal analysis, J.K.; investigation, J.K.; resources, J.K.; data curation, J.K.; writing—original draft preparation, J.K.; writing—review and editing, B.S.K. and J.K.; visualization, J.K.; supervision, B.S.K.; project administration, B.S.K.; funding acquisition, B.S.K. All authors have read and agreed to the published version of the manuscript.

**Funding:** This research received no external funding.

**Conflicts of Interest:** The authors declare no conflict of interest.

## References


1. Campbell, T.; Williams, C.; Ivanova, O.; Garrett, B. *Could 3D Printing Change the World?* Atlantic Council: Washington, DC, USA, 2011; p. 16.
2. Vaezi, M.; Seitz, H.; Yang, S. A review on 3D micro-additive manufacturing technologies. *Int. J. Adv. Manuf. Technol.* **2013**, *67*, 1721–1754. [CrossRef]
3. Scott, J.; Gupta, N.; Weber, C.; Newsome, S. Additive manufacturing: Status and opportunities. *Sci. Technol. Policy Inst.* **2012**, 1–19. [CrossRef]
4. Gibson, I.; Rosen, D.W.; Stucker, B. Additive manufacturing technologies: Rapid prototyping to direct digital manufacturing. *Assem. Autom.* **2010**. [CrossRef]
5. Seepersad, C.C. Challenges and opportunities in design for additive manufacturing, 3D print. *Addit. Manuf.* **2014**, *1*, 10–13.
6. Lipson, H.M.K. *Fabricated the New World of 3D Printing*, 1st ed.; John Wiley & Sons Inc.: Hoboken, NJ, USA, 2013; pp. 1–5.
7. Rodríguez, J.F.; Thomas, J.P.; Renaud, J.E. Mechanical behavior of acrylonitrile butadiene styrene (ABS) fused deposition materials. Experimental investigation. *Rapid Prototyp. J.* **2001**, *7*, 48–158. [CrossRef]
8. Pandey, M.P.; Venkata, R.N.; Dhande, S.G. Slicing procedures in layered manufacturing: A review. *Rapid Prototyp. J.* **2003**, *9*, 274–288. [CrossRef]
9. Galantucci, L.M.; Lavecchia, F.; Percoco, G. Experimental study aiming to enhance the surface finish of fused deposition modeled parts. *CIRP Ann. Manuf. Technol.* **2009**, *58*, 189–192. [CrossRef]
10. Galantucci, L.M.; Lavecchia, F.; Percoco, G. Quantitative analysis of a chemical treatment to reduce roughness of parts fabricated using fused deposition modeling. *CIRP Ann. Manuf. Technol.* **2010**, *59*, 247–250. [CrossRef]
11. Garg, A.; Bhattacharya, A.; Batish, A. Chemical vapor treatment of ABS parts built by FDM: Analysis of surface finish and mechanical strength. *Int. J. Adv. Manuf. Technol.* **2017**, *89*, 2175–2191. [CrossRef]

12. Tiwary, V.; Arunkumar, P.; Deshpande, A.S.; Khorate, V. Studying the effect of chemical treatment and fused deposition modelling process parameters on surface roughness to make acrylonitrile butadiene styrene patterns for investment casting process. *Int. J. Rapid Manuf.* **2015**, *5*, 276–288. [CrossRef]
13. Singhal, S.K.; Pandey, A.P.; Pandey, P.M.; Nagpal, A.K. Optimum part deposition orientation in stereolithography. *Comput. Aided. Des. Appl.* **2005**, *2*, 319–328. [CrossRef]
14. Panda, S.K. Optimization of fused deposition modelling (FDM) process parameters using bacterial foraging technique. *Intell. Inf. Manag.* **2009**, *1*, 89–97. [CrossRef]
15. Blok, L.G.; Longana, M.L.; Yu, H.; Woods, B.K.S. An investigation into 3D printing of fi bre reinforced thermoplastic composites. *Addit. Manuf.* **2018**, *22*, 176–186.
16. Ulu, E.; Korkmaz, E.; Yay, K.; Ozdoganlar, B.O.; Burak Kara, B.L. Enhancing the structural performance of additively manufactured objects through build orientation optimization. *J. Mech. Des.* **2015**, *137*, 111410. [CrossRef]
17. Zaldivar, R.J.; Witkin, D.B.; McLouth, T.; Patel, D.N.; Schmitt, K.; Nokes, J.P. Influence of processing and orientation print effects on the mechanical and thermal behavior of 3D-printed ULTEM® 9085 material. *Addit. Manuf.* **2017**, *13*, 71–80. [CrossRef]
18. Buj-Corral, I.; Domínguez-Fernández, A.; Durán-Llucià, R. Influence of print orientation on surface roughness in fused deposition modeling (FDM) processes. *Materials* **2019**, *12*, 3834. [CrossRef]
19. El-Gizawy, A.S.; Corl, S.; Graybill, B. Process-induced properties of FDM products. In Proceedings of the ICMET International Conference on Mechanical Engineerings and Technology Congress & Exposition, Paris, France, 27–29 July 2011; p. 7.
20. Hill, N.; Haghi, M. Deposition direction-dependent failure criteria for fused deposition modeling polycarbonate. *Rapid Prototyp. J.* **2014**, *20*, 221–227. [CrossRef]
21. Torrado Perez, A.R.; Roberson, D.A.; Wicker, R.B. Fracture surface analysis of 3D-printed tensile specimens of novel ABS-based materials. *J. Fail. Anal. Prev.* **2014**, *14*, 343–353. [CrossRef]
22. Hwang, S.; Reyes, E.I.; Moon, S.K.; Rumpf, R.C.; Kim, N.S. Thermo-mechanical characterization of metal/polymer composite filaments and printing parameter study for fused deposition modeling in the 3D printing process. *J. Electron. Mater.* **2015**, *44*, 771–777. [CrossRef]
23. Zhong, W.; Li, F.; Zhang, Z.; Song, L.; Li, Z. Short fiber reinforced composites for fused deposition modeling. *Mater. Sci. Eng. A* **2001**, *301*, 125–130. [CrossRef]
24. Shofner, M.L.; Lozano, K.; Rodríguez-Macías, F.J.; Barrera, E.V. Nanofiber-reinforced polymers prepared by fused deposition modeling. *J. Appl. Polym. Sci.* **2003**, *89*, 3081–3090. [CrossRef]
25. Matsuzaki, R.; Ueda, M.; Namiki, M.; Jeong, T.-K.; Asahara, H.; Horiguchi, K.; Nakamura, T.; Todoroki, A.; Hirano, Y. Three-dimensional printing of continuous-fiber composites by in-nozzle impregnation. *Sci. Rep.* **2016**, *6*, 23058. [CrossRef]
26. Compton, B.G.; Lewis, J.A. 3D-printing of lightweight cellular composites. *Adv. Mater.* **2014**, *26*, 5930–5935. [CrossRef] [PubMed]
27. Jiang, J.; Xu, X.; Stringer, J. Optimization of process planning for reducing material waste in extrusion based additive manufacturing. *Robot. Comput. Integr. Manuf.* **2019**, *59*, 317–325. [CrossRef]
28. Jiang, J.; Xu, X.; Stringer, J. Optimisation of multi-part production in additive manufacturing for reducing support waste. *Virtual Phys. Prototyp.* **2019**, *14*, 219–228. [CrossRef]
29. El-Midany, T.T.; Elkeran, A.; Tawfik, H. Toolpath pattern comparison contour-parallel with direction-parallel. In Proceedings of the Geometric Modeling and Imaging-New Trends (GMAI'06), London, UK, 5–7 July 2006; pp. 77–82.
30. Jin, G.Q.; Li, W.D.; Gao, L. An adaptive process planning approach of rapid prototyping and manufacturing. *Robot. Comput. Integr. Manuf.* **2013**, *29*, 23–38. [CrossRef]
31. Jin, Y.; He, Y.; Fu, G.; Zhang, A.; Du, J. A non-retraction path planning approach for extrusion-based additive manufacturing. *Robot. Comput. Integr. Manuf.* **2017**, *48*, 132–144. [CrossRef]
32. Liao, G.; Li, Z.; Cheng, Y.; Xu, D.; Zhu, D.; Jiang, S.; Guo, J.; Chen, X.; Xu, G.; Zhu, Y. Properties of oriented carbon fiber/polyamide 12 composite parts fabricated by fused deposition modeling. *Mater. Des.* **2018**, *139*, 283–292. [CrossRef]
33. Goh, G.D.; Dikshit, V.; Nagalingam, A.P.; Goh, G.L.; Agarwala, S.; Sing, S.L.; Wei, J.; Yeong, W.Y. Characterization of mechanical properties and fracture mode of additively manufactured carbon fiber and glass fiber reinforced thermoplastics. *Mater. Des.* **2018**, *137*, 79–89. [CrossRef]

34. Ferreira, R.T.L.; Amatte, I.C.; Dutra, T.A.; Bürger, D. Experimental characterization and micrography of 3D printed PLA and PLA reinforced with short carbon fibers. *Compos. Part B* **2017**, *124*, 88–100. [CrossRef]
35. Bay, R.S.; Tucker, C.L. Stereological measurement and error estimates for three-dimensional fiber orientation. *Polym. Eng. Sci.* **1992**, *32*, 240–253. [CrossRef]
36. Stegmann, J.; Lund, E. Discrete material optimization of general composite shell structures. *Int. J. Numerical Methods Eng.* **2005**, *62*, 2009–2027. [CrossRef]
37. Kim, J. Optimization of Design and Manufacturing Process of Fusion Filament Fabrication (fff) 3D Printing. Ph.D. Thesis, West Virginia University, Morgantown, WV, USA, 2018.

applied sciences

MDPI

*Article*

# Novel Resistive Sensor Design Utilizing the Geometric Freedom of Additive Manufacturing

**Hagen Watschke [1,*], Marijn Goutier [1], Julius Heubach [1], Thomas Vietor [1], Kay Leichsenring [2] and Markus Böl [2]**

[1] Institute for Engineering Design, Technische Universität Braunschweig, 38106 Brunswick, Germany; m.goutier@tu-braunschweig.de (M.G.); j.heubach@tu-braunschweig.de (J.H.); t.vietor@tu-braunschweig.de (T.V.)

[2] Institute of Mechanics and Adaptronics, Technische Universität Braunschweig, 38106 Brunswick, Germany; k.leichsenring@tu-braunschweig.de (K.L.); m.boel@tu-braunschweig.de (M.B.)

* Correspondence: h.watschke@tu-braunschweig.de; Tel.: +49-531-391-3356

**Abstract:** Direct additive manufacturing (AM) of sensors has in recent years become possible, but still remains a largely unexplored area. This work proposes a novel resistive sensor design that utilizes the geometric freedom offered by AM, especially by material extrusion, to enable a customizable and amplified response to force and deformation. This is achieved by using a multi-material design made of an elastomer and an electrically conductive polymer that enables a physical shortening of the conductive path under compressive load through a specific definition of shape. A number of different variants of this novel sensor design are tested, measuring their mechanical and electrical behavior under compression. The results of these tests confirm a strong resistive response to mechanical loading. Furthermore, the results provide insight into the influencing factors of the design, i.e., the gap size between the conductive pathing and the stiffness of the sense element support structure are found to be primary influencing factors governing sensor behavior.

**Keywords:** 3D printing; additive manufacturing; design for additive manufacturing; resistive sensors; electrical conductive filament; material extrusion; fused deposition modeling

**Citation:** Watschke, H.; Goutier, M.; Heubach, J.; Vietor, T.; Leichsenring, K.; Böl, M. Novel Resistive Sensor Design Utilizing the Geometric Freedom of Additive Manufacturing. *Appl. Sci.* **2021**, *11*, 113. https://dx.doi.org/10.3390/app11010113

Received: 13 November 2020
Accepted: 22 December 2020
Published: 24 December 2020



## 1. Introduction

Additive manufacturing (AM), also known as 3D printing, is a family of manufacturing processes in which parts are generated based on a 3D model, building them in a layer-by-layer fashion. Some of the most commonly applied AM processes are vat photopolymerization, powder bed fusion for polymers or metals using sintering or melting, and material extrusion (MEX) [1,2]. Of the aforementioned, MEX has a distinct advantage when it comes to producing multi-material parts and integrated electronics, as it can easily employ multiple extruders with distinct materials. By comparison, the other commonly applied processes operate using either a fluid bath or powder bed, in which multi-material parts are difficult to realize. While it is technically possible to manufacture multi-material parts with functional electronics using AM processes other than MEX [3], the present work will focus on MEX due to its advantages in process simplicity and productivity when producing multi-material parts [4] and due to the large variety of technical and functional polymers, e.g., elastomers and electrical conductive polymers [5,6].

Through recent advancements, it has become possible to use MEX to manufacture components from electrically conductive polymers [1,7,8]. Previously, indirect inclusion of conductors and sensors into AM parts had already been demonstrated, e.g., by interrupting the build to embed strain gauges [9], silver paste [10], copper wire, or copper mesh [11]. By applying conductive polymers, it has now become possible to directly include conductive traces, resistors [12], capacitors, inductors, filters [13], capacitive sensors [14,15], piezoresistive sensors [16], thermal sensors [17], and tactile sensors [18], all within the same production process. Conductivity in the plastics is realized through the inclusion of a conductive filler within the polymer matrix. Examples of such fillers include single-wall or multi-wall carbon nanotubes (SWCNT/MWCNT), carbon back (CB), graphene, carbon fibers [19,20], copper nanowires [21], silver nanoparticles, or combinations of different particles [22].

Table 1 provides a literature overview regarding directly produced MEX force and pressure sensors, sorting the sources by sensor working principle and the mechanical loads used to demonstrate its behavior.

**Table 1.** Overview of working principles and tested loads in existing literature on additive manufacturing (AM) of sensors.

| Working Principle | Tensile Load | Compressive Load | Flexural Load |
|---|---|---|---|
| Capacitive | | [7,14,23] | |
| Piezoresistive | [16,22,24–27] | | [7,23,28–30] |
| Resistive path adjustment | [31] | | [31] |

Leigh et al. [23] were amongst the first to demonstrate AM of electrically conductive polymers, using a polycaprolactone (PCL) polymer matrix with a CB conductive filler in a MEX process. The material was applied in piezoresistive flexural load sensing, capacitive detection of touch, and the detection of a fluid within a container. Although there are limitations to the work, such as a low gauge factor and a lack of in-depth consideration of the influences of sensor geometry and process parameters, it provides a proof of concept for directly manufactured AM sensors. Zapciu et al. [7] demonstrated both a capacitive touch sensor and a piezoresistive bending sensor using acrylonitrile butadiene styrene (ABS) polymer with CB conductive particles. Xiang et al. [22] developed a mixture of carbon nanotubes (CNTs) and silver nanoparticles within a thermoplastic polyurethane (TPU) matrix, demonstrating its piezoresistivity under tensile load and the effects of different mixtures of the conductive fillers on the sensor behavior. Dijkshoorn et al. [27] demonstrated the use of polylactide (PLA) with CB to build piezoresistive sensors, and qualitatively validated their response under tensile loading. Maurizi et al. [25] applied the same polylactide (PLA) with CB polymer to build piezoresistive sensors applied in a dynamic strain loading scenario, and their work also gave consideration to the geometric design of the strain gauges and the possibilities of improving output linearity by using two gauges. Dul et al. [26] developed an ABS polymer with CNT conductive filler, and demonstrated its behavior with 0° and 45° infill and strain loading in ramp, creep, and dynamic scenarios. A further development regarding the geometry of a piezoresistive flexural load sensor was carried out by Kim et al. [29], demonstrating a sensor that utilized three sense elements produced from a TPU matrix with MWCNT conductive filler. When combined, the three sense elements are intended to allow for independent detection of forces along three axes, though as noted by the authors there remains a non-negligible cross talk between the sense element signals (i.e., unwanted responses from an element on an unloaded axis). Hohimer et al. [30] also investigated TPU with MWCNT conductive filler, exploring in depth the influences of filler concentration, build orientation, layer height, infill angle, and temperature settings during the build. Their work included the creation of a pneumatic actuator built entirely out of the conductive polymer, which showed piezoresistive sensing properties.

Schouten et al. [14] demonstrated the use of MEX to directly manufacture capacitive force sensors from flexible TPU material with a conductive CB filler. The functionality of these sensors was demonstrated in quasi-static and sinusoidal compressive loading. Using the same TPU with CB material, a piezoresistive sensor in a bending load scenario was also demonstrated, using two mirrored sense elements to allow for improved output linearization [28], though imperfections in the mirrored set-up were reported to negatively affect the result of the linearization attempt. Christ et al. [24] also investigated flexible sensors, applying a TPU matrix with MWCNT. Notably, the work includes experiments performed on various strain gauge geometries to evaluate their effectiveness.

Mousavi et al. [31] demonstrated a PLA matrix with conductive MWCNTs, but used a geometry designed to improve upon the low gauge factors often exhibited in AM piezoresistive sensors. This design resulted in a considerably higher gauge factor, accomplished by a resistive path that is physically lengthened under tensile or flexural loads, thereby providing an amplified response compared to the piezoresistive behavior of the material alone.

Though the feasibility of AM sensors has been demonstrated, and the need to combine electronic design and printing technology acknowledged [32], only limited attempts to use AM's geometric freedom as a means of improving sensor performance have been reported. The use of a pair of AM sensors in a symmetric set-up [28] or with a gauge on the zero strain axis [25] as means of output linearization was reported, but the sense elements themselves retained a geometry similar to traditional strain gauges. In addition, the effect of different trace patterns was tested [24], but this was done largely within the confines of an essentially 2D design. Finally, the work of Mousavi et al. [31] used the geometric freedom of AM to enable a working principle based on resistive path adjustment, but its geometry would not readily translate to compressive loads due to buckling.

In summary, there is limited consideration of AM-enabled freedoms in existing sensor design, which presents an opportunity for further research. For compressive loads, the commonly used solution, as can be seen in Table 1, is a capacitive measurement. However, capacitive measurement requires more complicated circuitry than resistive measurement. This work presents a novel design for a resistive sensor for compressive loads—using the working principle of resistive path adjustment—which utilizes the multi-material properties and the geometric freedom provided by AM. When compared to capacitive compression sensors, this provides an advantage both in circuit simplicity and in a sensor response that is much larger and therefore easily distinguished from noise. This is especially the case when considering AM capacitive sensors, where the capacity deltas reported under load are low compared to the initial values. For example, Schouten et al. [14] reported a 160 fF response, with an unloaded initial value of 105.22 pF, making for a change of only 0.15% from the initial value. Thus, the presented sensor design will provide an alternative to additively manufactured capacitive sensors to measure contact forces or deformation, for example, in gripping systems or robotic hands.

## 2. Materials and Methods

### 2.1. Design Methodology

AM's new design freedom is often considered too late in the design development process, and therefore, the design potentials are only used selectively [33]. Based on limitations of previous works (see Table 1) a novel sensor design (see Figure 1a) for measuring compressive loads using resistive path adjustment is developed by considering AM's unique design possibilities. The systematic development of the novel resistive sensor with customizable force-resistance-behavior is supported by different design tools in order to overcome thinking barriers and ensure a goal-oriented utilization of the design possibilities provided by AM. In the conceptual design phase, the semantic network of AM design potentials by Kumke et al. [34], the design heuristics by Blösch-Paidosh and Shea [35], and the design principles for multi-material AM by Watschke et al. [36] were used. Thus, different design principles and features were identified to realize a novel sensor design with tailored electrical and mechanical properties. Based on these design possibilities and the different sensor principles shown in Table 1, the following design potentials were identified to develop a novel sensor design for adjusting the force-resistance-behavior at specific deformation under compressive load:

- Incorporation of electrical functionalities by using different materials in one part [34–36]
- Tailoring the change in electrical resistance under compressive load by using the principle of shortening the conductor length [31,36], freeform surfaces, and undercuts [34]
- Tailoring the mechanical properties of the substructure by freeform surfaces, wall thickness combinations, and hierarchical structures (e.g., internal structures) [34,36]

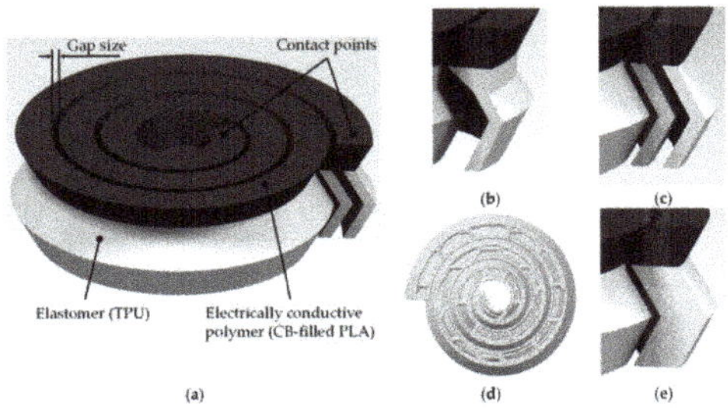


**Figure 1.** Render showing the sensor design (**a**). The black top is electrically conductive polymer, the white substructure is flexible elastomer to control the mechanical behavior of the sensor. Details (**b–e**) show substructure variants 1 to 4 respectively, each structure with increasing stiffness. For variant 3, the toolpath from the underside is depicted, as the CAD (computer-aided design) models for variants 3 and 4 are identical. The difference in stiffness is created through lower infill during printing for variant 3, whereas 4 is solid.

The novel sensor design, as shown in Figure 1, utilizes a conductive spiral-shaped path that shortens under a compressive load, enabling an increased resistive response compared to material piezoresistivity alone. The non-conductive flexible structure underneath the conductor allows for precise control over the sensor's stiffness, and thereby its response under a specific load. The multi-material approach allows the tuning of the mechanical behavior, and thus the force range of the sensor, independent of its resistive range. If, by comparison, a mono-material approach was used, increasing the stiffness would require additional material, which would decrease the electrical resistance. This demonstrates the advantages of both a multi-material solution and of utilizing AM's freedom in design regarding shape complexity. Eight different sensors were designed and manufactured in order to realize specific force-resistance-behaviors. These consisted of four different substructure variants of increasing stiffness, as shown in Figure 1b–e. These substructures were covered by a conductive spiral that had either a 0.3 mm or a 0.5 mm gap between its windings. In all samples, the conductive spiral had a width of 2.4 mm, a thickness of 2.0 mm, and made three windings. The substructure had a height of 5.0 mm and a thickness of 0.8 mm for variant 1 (Figure 1b), a thickness of 0.8 mm—each bar—for variant 2 (Figure 1c), and thickness of 2.4 mm for variants 3 (Figure 1d) and 4 (Figure 1e). However, the outer shapes of variant 3 (Figure 1d) and 4 (Figure 1e) were equal, and a variation of the stiffness of variant 3 (Figure 1d) was achieved by using a honeycomb infill structure with an infill of 20% (see Table 2).

**Table 2.** Process settings for sample production.

| Parameter | Conductive PLA | TPU Variants 1,2, & 4 | TPU Variant 3 |
|---|---|---|---|
| Extrusion width (mm) | 0.4 | 0.4 | 0.4 |
| Nozzle temperature (°C) | 230 | 235 | 235 |
| Bed temperature (°C) | 40 | 40 | 40 |
| Layer height (mm) | 0.2 | 0.2 | 0.2 |
| Infill (%, pattern) | 100, only shells | 100, only shells | 20, honeycomb |
| Extrusion speed (mm/s) | 30 | 35 | 35 |
| Perimeter shells (-) | 3 | 3 | 2 |

PLA: polylactide; TPU: thermoplastic polyurethane.

### *2.2. The Additive Manufacturing Process and Design Variants*

The sensor designs were manufactured by material extrusion, because of the multi-material capability and the variety of commercial available technical polymers, especially regarding elastomers and electrical conductive polymers [6]. In order to ensure a reliable multi-material design, the compatibility of the material combination had to be taken into account. The compact design allowed no additional measures for improving the interface strength between the substructure and the conductive spiral, e.g., interlocking features [37], without a negative impact on the sensor function. In accordance with Freund et al. [37], the material combination of TPU and PLA was chosen because of the good interface strength. The substructure was made using NinjaTek® Ninjaflex TPU [38], and the conductive spiral was manufactured using Protopasta Conductive PLA [39], which is a PLA with approximately 21.5 wt% CB conductive particles. Aside from favorable multi-material behavior, the conductive PLA was selected based on its easy processability, low geometrical deviations, and relatively high electrical resistance. As demonstrated in Watschke et al. [6], resistivity of the extruded material lies in a range of 0.07–0.11 Ωm. Though materials with lower resistivity exist [40], a relatively high initial electrical resistance was considered to be a desirable property in this case, as it enables a larger change in resistance when the resistive path is shortened during sensor operation.

Parts were sliced using Simplify3D® (4.1.2, Simplify3D, LLC, Cincinnati, OH, USA, 2020), and built on the pro-consumer machine X400 by German RepRap GmbH (Feldkirchen, Germany), that comes with a dual extrusion system. The process settings were used as listed in Table 2 (Section 2.1). The build orientation was such that the conductive spiral faces downwards on the build platform. This resulted in only a single tool change per build. Build times ranged from 10 to 15 min per sensor, depending on the density of the substructure.

### 2.3. Experimental Set-Up

Mechanical forces were applied using a uniaxial material test machine (Zwick Z0.5, Zwick GmbH & Co., Ulm, Germany). Force was measured using a 500 N Xforce P load cell (accuracy class 0.5). A custom actuator was used to compress the parts without touching the electrical contact points. The sensor was placed on a flat surface with double-sided tape to prevent unwanted slipping. To enable electrical measurements with low contact resistance, copper wires were attached to both ends of the conductive plastic using MG Chemicals 8331-14G silver epoxy adhesive. The electric signal was passed to an Arduino via a voltage divider and ADC 1115 (Adafruit Industries, LLC, New York City, NY, USA), allowing the logging of a voltage measurement from which resistance was easily derived. Figure 2a,b show the test set-up, and Figure 2c shows a close-up of a sensor with substructure variant 2.

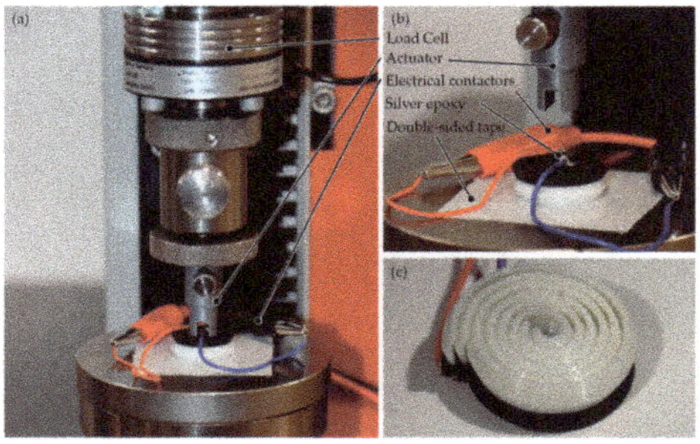


**Figure 2.** (**a**) Overview picture of the test set-up prior to measurement. (**b**) Close-up of the measurement set-up. (**c**) Detail image of substructure variant 2.

During measurement, sensor behavior was determined while compressing each sample by 2.0 mm, then retracting and waiting for 2 s, and repeating this five times. Compression and retraction cycles were performed at constant velocities of 5, 10, and 20 mm/min for each sample.

## 3. Results

### 3.1. Mechanical Behavior

Table 3 and Figure 3 show the mean peak force and its standard deviation for each sensor measurement cycle, which consisted of five actuations. The substructure variants achieved the intended goal of increasing the stiffness of the structure, as shown by the increase in peak force for each variant. A larger gap size between the windings was shown to lead to a lower peak force. This was to be expected, as the increased gap size lead to reduced contact between the windings of the sense element at the same compression. Velocity did not have a pronounced effect on the peak force value in variant 1. In variants 2 to 4, there was a minor increase in peak force with an increase in actuation velocity. Reproducibility of the peak force values with a measurement cycle was generally good, as shown by low standard deviations which ranged from 0.48% to 3.93% of peak force. In measurements where the standard deviations were higher, this was due to a gradual decrease in peak force with repeated actuation, an example of which can be seen in Figure 4. This effect was more pronounced in the stiffer variants 3 and 4, where higher stresses cause creep. The relative standard deviations were also higher in these samples. Lastly, a graph comparing the relationship of force and distance for variants 1–4 (see Figure 1b–e) is shown in Figure 5; the influence of the different substructure variants is clearly visible, as is a degree of hysteresis.

**Table 3.** Mean peak force and its standard deviation in newtons, for each cycle of five actuations.

| Gap Size (mm) | Velocity (mm/min) | Force (N) (Variant 1) | Force (N) (Variant 2) | Force (N) (Variant 3) | Force (N) (Variant 4) |
|---|---|---|---|---|---|
| 0.3 | 5 | 52.05 ± 0.35 | 110.89 ± 0.56 | 316.46 ± 3.51 | 350.99 ± 12.75 |
| | 10 | 52.68 ± 0.35 | 112.93 ± 0.54 | 326.28 ± 7.98 | 435.64 ± 17.14 |
| | 20 | 51.58 ± 0.54 | 114.64 ± 0.61 | 332.75 ± 10.67 | 427.37 ± 7.69 |
| 0.5 | 5 | 40.28 ± 0.25 | 92.05 ± 0.69 | 203.54 ± 2.27 | 317.16 ± 10.02 |
| | 10 | 40.65 ± 0.30 | 99.87 ± 1.60 | 207.25 ± 2.76 | 328.47 ± 7.54 |
| | 20 | 39.24 ± 0.34 | 100.74 ± 0.66 | 206.29 ± 5.88 | 328.65 ± 4.99 |

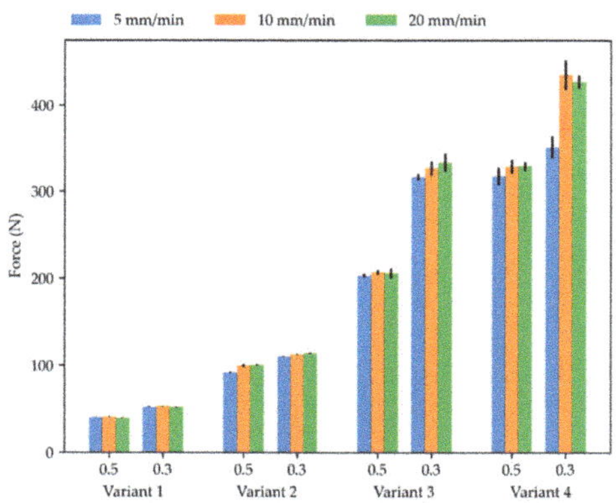


**Figure 3.** Bar chart of mean peak force and its standard deviation for each cycle of five actuations of the eight sensor variants at different speeds and for the gap sizes of 0.5 mm and 0.3 mm (see Figure 1b–e).

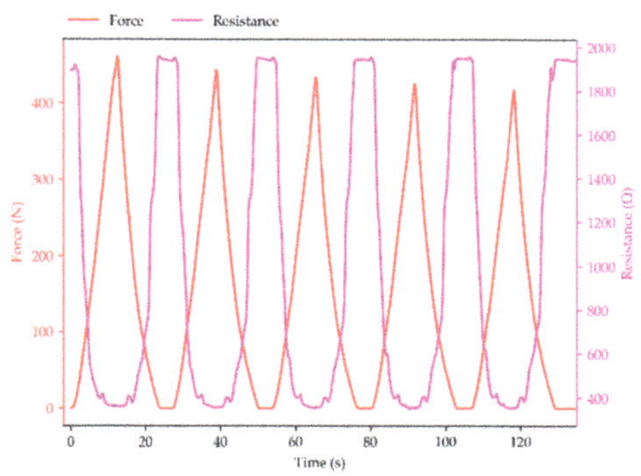


**Figure 4.** Force and resistance plots for the measurement with the highest standard deviation: variant 4 (Figure 1e), gap of 0.3 mm at 10 mm/min velocity. A decrease in peak force is visible with repeated actuation.

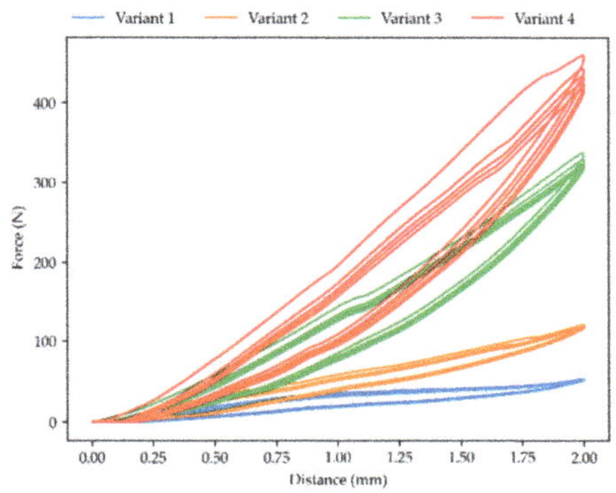


**Figure 5.** Force–distance relationship for variants 1 to 4 (Figure 1b–e), all with gaps of 0.3 mm and velocities of 10 mm/min. All variants show hysteresis, and the stiffer geometries show a decrease in peak force with repeated actuations.

### 3.2. Resistive Behavior

Mean values and standard deviations of the extremes in resistance for each sensor measurement cycle are presented in Table 4. In this case, the values shown are the lowest values reached, as the resistive path shortening under load leads to a reduction in resistance. A decrease in gap size led to stronger resistive response, i.e., lower extreme values. This was to be expected, as with a reduced gap size more windings will contact under identical compression.

The effect of velocity in the 0.3 mm gap size group was not immediately apparent, as different geometric variants appear to exhibit different trends. Closer inspection revealed that the trend observed aligned with the order in which measurements were performed, which is indicated by the arrows in Table 4. Measurements were not performed in identical order on each sensor, specifically to allow for the detection of any unwanted behaviour related to the test order rather than the variable being tested. In this case, sequential measurements showed a reduction in resistance. This reduction in resistance is possibly the result of insufficient wait time between the measurements, rather than an influence of velocity. The effect of this is minor relative to the total resistance change, never exceeding 1% of the total resistance change, and thus was not considered to be problematic. A similar reduction in resistance in the sequential order was not present for the samples with gap size of 0.5 mm.

**Table 4.** Mean values and standard deviation of resistance in ohms, calculated from the lowest resistance value reached in each of the five actuations for each measurement. Note that due to resistive path shortening, a high force and deformation coincide with a low, rather than a high extreme in resistance. The arrows indicate the order in which measurements were performed.

| Gap Size (mm) | Velocity (mm/min) | Resistance (Ω) of Variant 1 | Resistance (Ω) of Variant 2 | Resistance (Ω) of Variant 3 | Resistance (Ω) of Variant 4 |
|---|---|---|---|---|---|
| 0.3 | 5 | 379.01 ± 1.81 | 349.51 ± 2.85 | 339.64 ± 2.91 | 369.54 ± 7.01 |
| | 10 | 383.87 ± 3.18 | 345.60 ± 2.50 | 341.41 ± 2.49 | 353.37 ± 3.11 |
| | 20 | 406.44 ± 3.54 | 342.56 ± 3.73 | 347.43 ± 5.47 | 350.44 ± 1.43 |
| 0.5 | 5 | 560.32 ± 4.03 | 469.12 ± 9.24 | 633.45 ± 17.11 | 622.88 ± 5.93 |
| | 10 | 553.52 ± 0.69 | 455.27 ± 9.58 | 652.59 ± 4.69 | 642.16 ± 14.99 |
| | 20 | 559.84 ± 3.53 | 461.97 ± 3.59 | 631.50 ± 15.98 | 653.15 ± 2.68 |

Despite the minor variances observed in peak resistance, the actuation velocity did not influence the sensor behavior, as can be seen in Figure 6. To ensure this was not related to the limited range of 5 to 20 mm/min, an additional measurement at 100 mm/min was performed which elicited no further change in the sensor response.

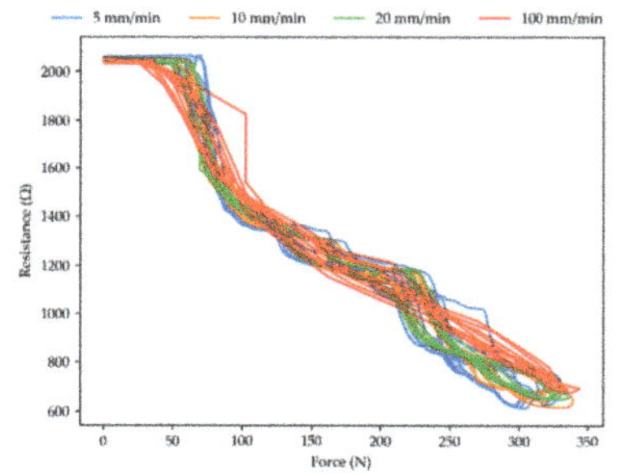


**Figure 6.** Effect of different velocities on the force-resistance behavior of the sensor. Depicted for variant 4, with a gap of 0.5 mm, at velocities 5, 10, 20, and 100 mm/min.

Table 5 demonstrates the high sensitivity of these sensors under the applied load, showing the percentage of the initial resistance R0 that remains at the highest compression levels. Values range from 17.43% to 32.08%, inversely, this means R0 is in the range of 3.1 to 5.7 times higher than the sensors' extreme output value.

**Table 5.** Relative extreme values in resistance, shown as percentage of the initial resistance value R0.

| Gap Size (mm) | Velocity (mm/min) | Relative Peak Resistance (%) (Variant 1) | Relative Peak Resistance (%) (Variant 2) | Relative Peak Resistance (%) (Variant 3) | Relative Peak Resistance (%) (Variant 4) |
|---|---|---|---|---|---|
| 0.3 | 5 | 19.48 | 29.47 | 17.55 | 21.98 |
| | 10 | 19.72 | 28.56 | 17.43 | 18.68 |
| | 20 | 23.79 | 27.23 | 27.08 | 18.14 |
| 0.5 | 5 | 27.64 | 22.9 | 30.99 | 30.62 |
| | 10 | 27.24 | 22.37 | 31.81 | 31.57 |
| | 20 | 27.41 | 22.73 | 30.68 | 32.08 |

The relationship between the distance the part is compressed and its resistive response, for parts with a gap size of 0.5 mm, is displayed in Figure 7. The initial part of the deformation caused no resistive response, which can be attributed to the gap that must first be bridged before the resistive path shortening mechanism has an effect. Once the gap was closed, the resistive response was similar and near linear with compression for variants 1, 3, and 4, with variant 2 showing a non-linear response. Based on the test variables, this effect cannot be readily explained, and further investigation will be performed to find the cause of the observed behavior. Figure 8 displays the relationship between the applied force and resistive response for the same 0.5 mm gap size parts. Like in the distance graph, there is an initial section of force being applied that does not lead to a resistive response. Once the gap was closed, the relationship between force and resistance was initially linear and the resistive response covered the majority of the force range. Only in variants 2 and 3 was there a flattening of the curve towards the end of its deformation.

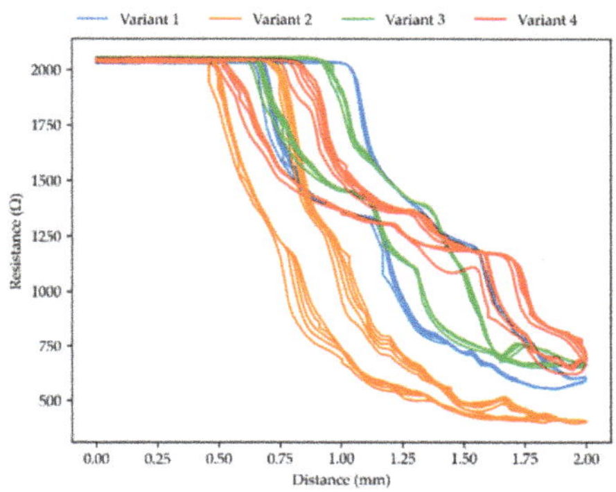


**Figure 7.** Resistance relation to compressed distance for measurements with a gap of 0.5 mm at 10 mm/min.

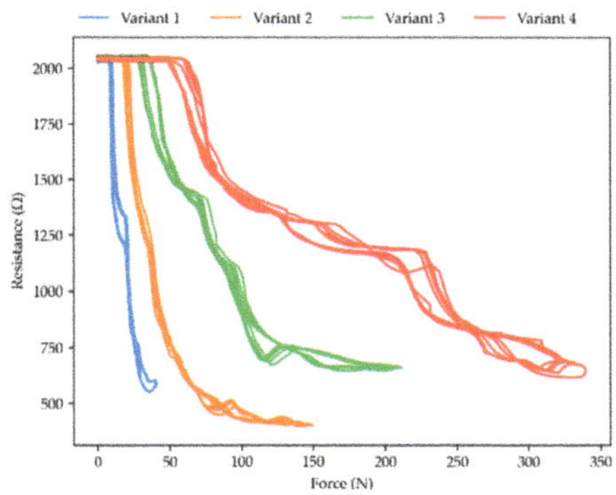


**Figure 8.** Resistance response upon force application for measurements with a gap of 0.5 mm at 10 mm/min.

Figure 9 displays the relationship between the distance the part is compressed and its resistive response, but for parts with a 0.3 mm gap size. The parts initially responded more rapidly to deformation than those with a 0.5 mm gap. The variants displayed a comparable response to a certain distance of deformation, though most of the resistive change occurred in the first millimeter of deformation. Figure 10 displays the relationship between applied force and resistive response for the parts with a 0.3 mm gap size. The substructure variant has an effect on the resistive response to force. However, because most of the peak force is reached in the last millimeter of deformation, which does not relate to a substantial further resistance change, the effect of the substructure was less pronounced than might be expected from observing peak force values alone. The lack of resistive change after the first millimeter of deformation showed that in the 0.3 mm variants, the sense element windings were engaged too quickly to correctly measure 2.0 mm deformation. This is also visible in Figure 4 above, where resistive response flattens off before the peak force is reached.

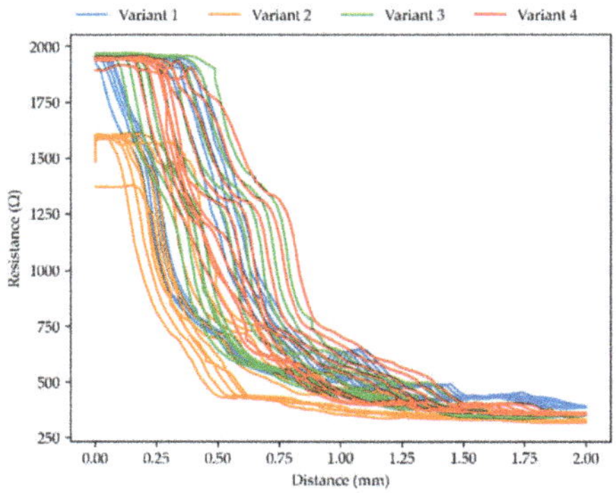


**Figure 9.** Resistance relation to compressed distance for measurements with a gap of 0.3 mm at 10 mm/min.

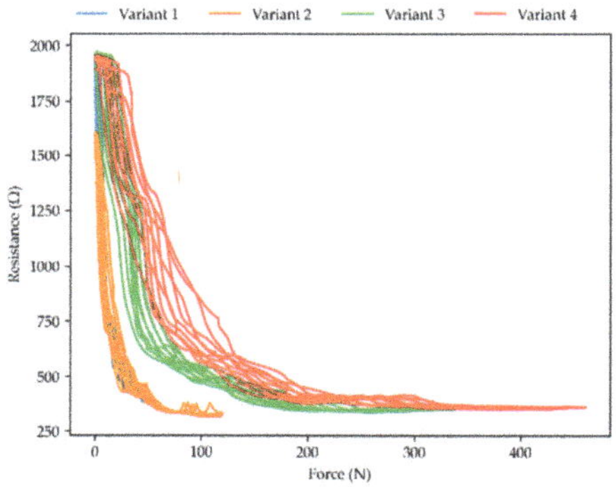


**Figure 10.** Resistance response upon force application for measurements with a gap of 0.3 mm at 10 mm/min.

During the measurements, there were some light audible snaps as the parts deformed. It is believed these were caused by stair-stepping in the angled side walls of the conductive spiral, introduced during the production process. Figure 11 shows a magnified image of this stair-stepping effect, including a tangent line denoting how the part will move along the adjacent surface. These prevent a smooth sliding motion, which could also explain the brief plateaus that are visible in the resistance curves and the slight kinks that are visible in the earlier presented Figure 5. Further investigation is required to confirm if this effect is indeed related to stair-stepping.

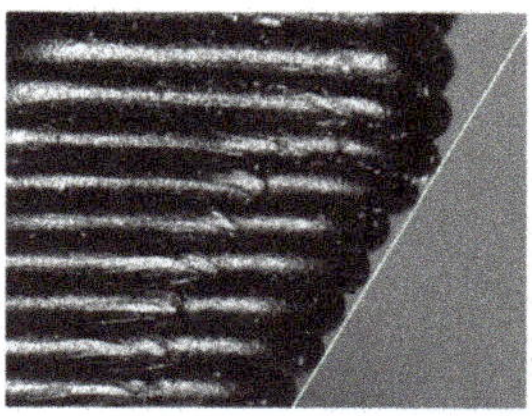

**Figure 11.** Magnified image of the side profile with a tangent line, to highlight the stair-stepping in the shape of the conductive spiral.

## 4. Discussion

The results show that the novel sensor design successfully created a resistive path that is physically shortened under mechanical compression, thereby providing a strong signal response.

A schematic overview of the relationships between the different variables and the sensor resistive behavior is shown in Figure 12. The final resistive response of the sensors is related to the deformation distance, and affected by the gap size. Deformation dictates how much of the shortening resistive path mechanism is engaged. The initial gap size modulates this behavior both through the number of windings that will be engaged at a certain deformation level, and through the distance required to close the first gap and thereby the delay in response.

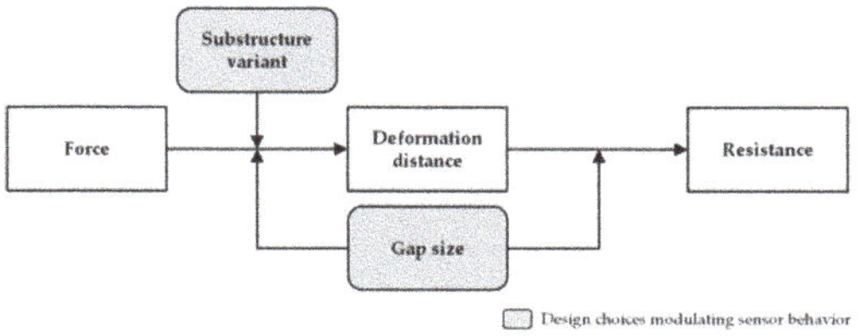


**Figure 12.** Chart showing the relationship between influencing factors in sensor behavior.

The deformation distance in turn is related to the applied force, and affected by the gap size and the stiffness of the substructure variant used. The gap size influences the amount of physical interference, and thereby force, required for deformation, while a stiffer substructure directly resists deformation and affects the force-displacement curve. The actuation velocity has little to no influence on sensor behavior and is omitted from the figure.

For the successful design of a sensor utilizing this working principle, this means that gap size and substructure should be selected in a combination suitable for the force range to be measured. The combination should lead to the forces causing gradual sense element engagement over the whole range.

There are opportunities for further research and improvement on the presented design. The parts with gap size of 0.5 mm allowed a larger measurement range, but had a delayed response to force being applied, whereas the 0.3 mm variants responded more quickly, but nearly all their resistive response occurred in the first millimeter of deformation, limiting the measurement range. A possible improvement might therefore be found in the use of a non-constant gap size that combines both a quick response and a large measurement range. Alternatively, an optimized substructure of different height or graded stiffness could achieve similar results. The application of foaming materials in the substructure could enable use in very low force scenario.

There may also exist more possibilities to optimize the conductive spiral or the substructure, e.g., by using other freeform surfaces or graded lattice structures (see [34]). The spiral sense element can be modified by laying it out on a 3D cone path, rather than as a 2D spiral. This could improve the engagement of the outer windings of the sense element, and allow further design freedom in sensor customization. Another possibility is the use of substructures with auxetic properties, enabling a further amplification of the resistive path shortening mechanism under load. The observed effects of stair-stepping in the side walls of the sense element leading to discontinuous deformation may be improved upon by optimizing the side wall angle, as well as reducing the layer thickness selected during part production.

Further research opportunities also exist in the effects of materials and process parameters. Different polymers and different conductive fillers or combinations thereof may allow for further customization of the sensor performance. In addition, by varying the process parameters, the resistivity, geometry, and mechanical behavior of the sensors could all be influenced and controlled. Lastly, establishing a low-resistance contact to outside equipment currently requires the application of a silver paste, silver ink, or similar medium. Further investigation might eliminate the need for this extra handling after the build.

## 5. Conclusions

The current work presents a novel design for an AM-designed resistive sensor for use under compressive loading. Through the use of a shortening resistive path, this design enables a stronger resistive response than piezoresistive behavior alone. The use of a flexible TPU substructure demonstrates the strength of multi-material AM, allowing both the sense element and the substructure that modulates sensor behavior to be built within the same process. Aside from the substructure, the size of the gap between the sensor windings is shown to be an important design consideration, allowing the designer to control the initial delay in response, as well as the force-distance-resistance relationship. Actuation velocity is shown not to be a major contributing factor. The various interactions in the sensor have been summarized into a model, highlighting the factors that can be readily controlled by the designer. The sensor design can be applied in a wide range of applications in order to measure both contact forces and deformation, such as gripping systems, robotic hands, or prosthesis and, thus, be an alternative to capacitive touch sensors, e.g., [41].

**Supplementary Materials:** Supplementary Materials are available online at https://www.mdpi.com/2076-3417/11/1/113/s1

**Author Contributions:** Conceptualization, H.W., J.H.; design and manufacturing of the test specimens, H.W., J.H.; resources, K.L., J.H.; investigation, M.G.; data curation, M.G.; software, M.G.; formal analysis, M.G.; visualization, M.G.; methodology, M.G.; writing—original draft preparation, M.G.; writing—review and editing, H.W., K.L., T.V., M.B.; supervision and funding acquisition, T.V., M.B. All authors have read and agreed to the published version of the manuscript.

**Funding:** Funded by the Ministry for Science and Culture of Lower Saxony (MWK)–School for Additive Manufacturing SAM (78904-63-3/19).

**Institutional Review Board Statement:** Not applicable.

**Informed Consent Statement:** Not applicable.

**Data Availability Statement:** The data presented in this study are available in supplementary materials.

**Acknowledgments:** We acknowledge the support by the German Research Foundation and the Open Access Publication Funds of the Technische Universität Braunschweig.

**Conflicts of Interest:** The authors declare no conflict of interest.

## References

1. Postiglione, G.; Natale, G.; Griffini, G.; Levi, M.; Turri, S. Conductive 3D microstructures by direct 3D printing of polymer/carbon nanotube nanocomposites via liquid deposition modeling. *Compos. Part A* **2015**, *76*, 110–114. [CrossRef]
2. CEN European Committee for Standardization. *DIN EN ISO/ASTM 52900:2018: Additive Manufacturing-General Principles-Terminology*; CEN European Committee for Standardization: Brussels, Belgium, 2018.
3. Agarwala, S.; Goh, G.L.; Le, T.-S.D.; An, J.; Peh, Z.K.; Yeong, W.Y.; Kim, Y.-J. Wearable bandage-based strain sensor for home healthcare: Combining 3D aerosol jet printing and laser sintering. *ACS Sensors* **2018**, *4*, 218–226. [CrossRef] [PubMed]
4. Lehmhus, D.; Aumund-Kopp, C.; Petzoldt, F.; Godlinski, D.; Haberkorn, A.; Zöllmer, V.; Busse, M. Customized smartness: A survey on links between additive manufacturing and sensor integration. *Procedia Technol.* **2016**, *26*, 284–301. [CrossRef]
5. Vaezi, M.; Chianrabutra, S.; Mellor, B.; Yang, S. Multiple material additive manufacturing–Part 1: A review. *Virtual Phys. Prototyp.* **2013**, *8*, 19–50. [CrossRef]
6. Watschke, H.; Hilbig, K.; Vietor, T. Design and characterization of electrically conductive structures additively manufactured by material extrusion. *Appl. Sci.* **2019**, *9*, 779. [CrossRef]
7. Zapciu, A.; Constantin, G. Additive manufacturing integration of thermoplastic conductive materials in intelligent robotic end effector systems. *Proc. Manuf. Syst.* **2016**, *11*, 201–206.
8. Hampel, B.; Monshausen, S.; Schilling, M. Properties and applications of electrically conductive thermoplastics for additive manufacturing of sensors. *TM-Tech. Mess.* **2017**, *84*, 593–599. [CrossRef]

9. Brink, M.; Ohlendorf, J.-H.; Thoben, K.-D. Development of a handling system with integrated sensors for textile preforms using additive manufacturing. *Procedia Manuf.* **2018**, *24*, 114–119. [CrossRef]
10. Stano, G.; Percoco, G. Design, 3D printing and characterization of a soft actuator with embedded strain sensor. In Proceedings of the 2020 IEEE International Symposium on Medical Measurements and Applications (MeMeA), Bari, Italy, 1 June–1 July 2020; pp. 1–6.
11. Shemelya, C.; Cedillos, F.; Aguilera, E.; Espalin, D.; Muse, D.; Wicker, R.; Macdonald, E. Encapsulated copper wire and copper mesh capacitive sensing for 3-D printing applications. *IEEE Sensors J.* **2014**, *15*, 1280–1286. [CrossRef]
12. Jaksic, N.I.; Desai, P.D. Characterization of resistors created by fused filament fabrication using electrically-conductive filament. *Procedia Manuf.* **2018**, *17*, 37–44. [CrossRef]
13. Flowers, P.F.; Reyes, C.; Ye, S.; Kim, M.J.; Wiley, B.J. 3D printing electronic components and circuits with conductive thermoplastic filament. *Addit. Manuf.* **2017**, *18*, 156–163. [CrossRef]
14. Schouten, M.; Sanders, R.; Krijnen, G. 3D printed flexible capacitive force sensor with a simple micro-controller based readout. In Proceedings of the 2017 IEEE SENSORS, Glasgow, UK, 29 October–1 November 2017; pp. 1–3.
15. Kim, W.S.; Kim, W.S. Toward a smart compliant robotic gripper equipped with 3D-designed cellular fingers. *Adv. Intell. Syst.* **2019**, *1*, 1900019. [CrossRef]
16. Christ, J.F.; Aliheidari, N.; Ameli, A.; Pötschke, P. 3D printed highly elastic strain sensors of multiwalled carbon nanotube/thermoplastic polyurethane nanocomposites. *Mater. Des.* **2017**, *131*, 394–401. [CrossRef]
17. Kwok, S.W.; Goh, K.H.H.; Tan, Z.D.; Tan, S.T.M.; Tjiu, W.W.; Soh, J.Y.; Ng, Z.J.G.; Chan, Y.Z.; Hui, H.K.; Goh, K.E.J. Electrically conductive filament for 3D-printed circuits and sensors. *Appl. Mater. Today* **2017**, *9*, 167–175. [CrossRef]
18. Liu, C.; Huang, N.; Xu, F.; Tong, J.; Chen, Z.; Gui, X.; Fu, Y.; Lao, C. 3D printing technologies for flexible tactile sensors toward wearable electronics and electronic skin. *Polymers (Basel)* **2018**, *10*, 629. [CrossRef]
19. Horst, D.J.; Junior, P.P.A. 3D-printed conductive filaments based on carbon nanostructures embedded in a polymer matrix. *Int. J. Appl. Nanotechnol. Res.* **2019**, *4*, 26–40. [CrossRef]
20. Avilés, F.; Oliva-Avilés, A.I.; Cen-Puc, M. Piezoresistivity, strain, and damage self-sensing of polymer composites filled with carbon nanostructures. *Adv. Eng. Mater.* **2018**, *20*, 1701159. [CrossRef]
21. Palmić, T.B.; Slavič, J.; Boltežar, M. Process parameters for FFF 3D-printed conductors for applications in sensors. *Sensors* **2020**, *20*, 4542. [CrossRef]
22. Xiang, D.; Zhang, X.; Harkin-Jones, E.; Zhu, W.; Zhou, Z.; Shen, Y.; Li, Y.; Zhao, C.; Wang, P. Synergistic effects of hybrid conductive nanofillers on the performance of 3D printed highly elastic strain sensors. *Compos. Part A* **2020**, *129*, 105730. [CrossRef]
23. Leigh, S.J.; Bradley, R.J.; Purssell, C.P.; Billson, D.R.; Hutchins, D.A. A simple, low-cost conductive composite material for 3D printing of electronic sensors. *PLoS ONE* **2012**, *7*, e49365. [CrossRef]
24. Christ, J.F.; Aliheidari, N.; Pötschke, P.; Ameli, A. Bidirectional and stretchable piezoresistive sensors enabled by multimaterial 3D printing of carbon nanotube/thermoplastic polyurethane nanocomposites. *Polymers (Basel)* **2018**, *11*, 11. [CrossRef] [PubMed]
25. Maurizi, M.; Slavič, J.; Cianetti, F.; Jerman, M.; Valentinčič, J.; Lebar, A.; Boltežar, M. Dynamic measurements using FDM 3D-printed embedded strain sensors. *Sensors* **2019**, *19*, 2661. [CrossRef] [PubMed]
26. Dul, S.; Pegoretti, A.; Fambri, L. Fused filament fabrication of piezoresistive carbon nanotubes nanocomposites for strain monitoring. *Front. Mater.* **2020**, *7*, 7. [CrossRef]
27. Dijkshoorn, A.; Werkman, P.; Welleweerd, M.; Wolterink, G.; Eijking, B.; Delamare, J.; Sanders, R.; Krijnen, G.J.M. Embedded sensing: Integrating sensors in 3-D printed structures. *J. Sensors Sens. Syst.* **2018**, *7*, 169–181. [CrossRef]
28. Schouten, M.; Prakken, B.; Sanders, R.; Krijnen, G. Linearisation of a 3D printed flexible tactile sensor based on piezoresistive sensing. In Proceedings of the 2019 IEEE Sensors Applications Symposium (SAS), Montreal, QC, Canada, 27–30 October 2019; pp. 1–4.
29. Kim, K.; Park, J.; Suh, J.-h.; Kim, M.; Jeong, Y.; Park, I. 3D printing of multiaxial force sensors using carbon nanotube (CNT)/thermoplastic polyurethane (TPU) filaments. *Sens. Actuators A Phys.* **2017**, *263*, 493–500. [CrossRef]
30. Hohimer, C.J.; Petrossian, G.; Ameli, A.; Mo, C.; Potschke, P. 3D printed conductive thermoplastic polyurethane/carbon nanotube composites for capacitive and piezoresistive sensing in soft pneumatic actuators. *Addit. Manuf.* **2020**, *34*, 101281. [CrossRef]
31. Mousavi, S.; Howard, D.; Zhang, F.; Leng, J.; Wang, C.H. Direct 3D printing of highly anisotropic, flexible, constriction-resistive sensors for multidirectional proprioception in soft robots. *ACS Appl. Mater. Interfaces* **2020**, *12*, 15631–15643. [CrossRef]
32. Khosravani, M.R.; Reinicke, T. 3D-printed sensors: Current progress and future challenges. *Sens. Actuators A Phys.* **2020**, *305*, 111916. [CrossRef]
33. Pradel, P.; Zhu, Z.; Bibb, R.; Moultrie, J. Investigation of design for additive manufacturing in professional design practice. *J. Eng. Des.* **2018**, *29*, 165–200. [CrossRef]
34. Kumke, M.; Watschke, H.; Hartogh, P.; Bavendiek, A.-K.; Vietor, T. Methods and tools for identifying and leveraging additive manufacturing design potentials. *Int. J. Interact. Des. Manuf.* **2017**, *12*, 481–493. [CrossRef]
35. Bloesch-Paidosh, A.; Shea, K.; Blösch-paidosh, a. design heuristics for additive manufacturing validated through a user study1. *J. Mech. Des.* **2019**, *141*, 399. [CrossRef]
36. Watschke, H.; Kuschmitz, S.; Heubach, J.; Lehne, G.; Vietor, T. A methodical approach to support conceptual design for multi-material additive manufacturing. *Proc. Des. Soc. Int. Conf. Eng. Des.* **2019**, *1*, 659–668. [CrossRef]

37. Freund, R.; Watschke, H.; Heubach, J.; Vietor, T. Determination of influencing factors on interface strength of additively manufactured multi-material parts by material extrusion. *Appl. Sci.* **2019**, *9*, 1782. [CrossRef]
38. Technical data sheet-NinjaFlex® 85A by NinjaTek. Available online: https://ninjatek.com/wp-content/uploads/2019/10/NinjaFlex-TDS.pdf (accessed on 3 November 2020).
39. Technical Data Sheet-Conductive pla. Available online: https://www.proto-pasta.com/pages/technical-data-sheets (accessed on 3 November 2020).
40. Tan, L.J.; Zhu, W.; Zhou, K. Recent progress on polymer materials for additive manufacturing. *Adv. Funct. Mater.* **2020**, *30*, 2003062. [CrossRef]
41. Ntagios, M.; Escobedo, P.; Dahiya, R. 3D printed robotic hand with embedded touch sensors. In Proceedings of the 2020 IEEE International Conference on Flexible and Printable Sensors and Systems (FLEPS), Manchester, UK, 16–19 August 2020; pp. 1–4.

*applied sciences*

MDPI

*Article*

# Additively Manufactured Parametric Universal Clip-System: An Open Source Approach for Aiding Personal Exposure Measurement in the Breathing Zone

**Kirsi Kukko [1,*], Jan Sher Akmal [1], Anneli Kangas [2], Mika Salmi [1], Roy Björkstrand [1], Anna-Kaisa Viitanen [2], Jouni Partanen [1] and Joshua M. Pearce [3,4]**

1 Department of Mechanical Engineering, School of Engineering, Aalto University, 02150 Espoo, Finland; jan.akmal@aalto.fi (J.S.A.); mika.salmi@aalto.fi (M.S.); roy.bjorkstrand@aalto.fi (R.B.); jouni.partanen@aalto.fi (J.P.)

2 Finnish Institute of Occupational Health, FI-00032 Työterveyslaitos, Finland; anneli.kangas@ttl.fi (A.K.); anna-kaisa.viitanen@ttl.fi (A.-K.V.)

3 Department of Electronics and Nanoengineering, School of Electrical Engineering, Aalto University, 02150 Espoo, Finland; joshua.pearce@aalto.fi

4 Department of Materials Science & Engineering and Department of Electrical & Computer Engineering, Michigan Technological University, Houghton, MI 49931, USA

* Correspondence: Kirsi.kukko@aalto.fi; Tel.: +358-50-344-7248

Received: 24 August 2020; Accepted: 22 September 2020; Published: 24 September 2020

**Featured Application: The customizable clip system opens new possibilities for occupational health professionals since the basic design can be altered to hold different kinds of samplers and tools. The solution is shared using an open source methodology to allow for distributed manufacturing of the free designs.**

**Abstract:** Design for additive manufacturing is adopted to help solve problems inherent to attaching active personal sampler systems to workers for monitoring their breathing zone. A novel and parametric 3D printable clip system was designed with an open source Computer-aided design (CAD) system and was additively manufactured. The concept was first tested with a simple clip design, and when it was found to be functional, the ability of the innovative and open source design to be extended to other applications was demonstrated by designing another tooling system. The clip system was tested for mechanical stress test to establish a minimum lifetime of 5000 openings, a cleaning test, and a supply chain test. The designs were also tested three times in field conditions. The design cost and functionalities of the clip system were compared to commercial systems. This study presents an innovative custom-designed clip system that can aid in attaching different tools for personal exposure measurement to a worker's harness without hindering the operation of the worker. The customizable clip system opens new possibilities for occupational health professionals since the basic design can be altered to hold different kinds of samplers and tools. The solution is shared using an open source methodology.

**Keywords:** design for AM; 3D printing; open source; personal exposure; parametric universal clip

## 1. Introduction

The field of occupational health varies with the monitoring of working environment air quality being important in traditional fields ranging from mining to pig farming and new emerging areas such as 3D printing [1–3]. One of the important ways to assess worker's personal exposure is to sample the breathing zone of the workers, which is conventionally defined as the area immediately surrounding a worker's nose and mouth within 30 cm [4]. The purpose of air sampling in a worker's breathing zone in the workplace is to identify and especially, quantify levels of impurities that worker is exposed during working hours. This is to ensure that workers are not exposed to dangerous levels of airborne contaminants such as dust, biological agents, toxic chemicals, or nanoparticles [5–7] and that the protective measures are adequate. With the onset of the COVID-19 pandemic, it has become increasingly important to maintain sufficient working and social distances to avoid the transmission of aerosol SARS-CoV-2 by inhalation [8]. Depending on the compounds, there are several commercial samplers available. For example, cyclone samplers can be utilized to measure respirable dust [4]. For such work, an active personal sampler system is a combination of sampler, which collects emissions from the air, an air pump, which pulls in air at a calibrated speed, and hose, which connects them.

The need to sample the air that workers breathe creates a challenge for attaching samplers, tubes, and air pumps to the workers without interfering with their work. In the worst case, poorly attached samplers can change workers maneuvers which can have an impact to the sample results. Often instructions for samplers include presuppositions of worker's clothing/equipment such as "attach to the label of the worker's coat", "Clip pump to the worker's belt", and "put sampler to the pocket" [9]. This can lead to problems when worker's clothing does not include these aforementioned items. The current state-of-the-art of sampler placement is a clamp with serrated edges, which can usually be attached to the clothing, but slippery fabrics of the work clothes and safety gear may cause problems with it. To enable definite sampler system placement, quite often some kind of harness or back bag system is used [4]. Worker's differing sizes (heights, weights, etc.) and different types of worn safety equipment pose challenges for designing harness systems as the same equipment is routinely reused. Furthermore, additional attachment points are needed to ensure that the sampler system stays in the correct position. For example, the hose leading from the air pump needs to be tightly attached so it will not get caught and cause dangerous situations. Some samplers, such as cyclone samplers, have strict guidelines for staying in the correct position to function properly. The solutions for these problems are either custom fabricated components of high costs or on-the-spot solutions such as the use of duct tape, safety pins, and zip-ties, which are subject to appear as unprofessional; and in the worst case scenario, it can lead to poor data collection e.g., if the sampler becomes loose or poses a contamination risk for some industries e.g., food preparation.

Additive manufacturing (AM) has enabled the production of parts with unparalleled freedom of design, and it is widely adopted in the medical [10–12], dental [13], automotive [14], aerospace [15, 16], and industrial machines [14,17,18] industries. It is mainly used to manufacture functional prototypes [14], rapid tooling [19,20], and end-use parts [17]. AM is also known as 3D printing, and in this study, the term "3D printing" is used for low cost desktop printers. Recently the widespread growth and accessibility of desktop 3D printers [13,21] brought on by the open source development of material extrusion 3D printing [22,23] offers a potential solution to this challenge. The principle of material extrusion is that a filament is extruded through a heated nozzle to create objects layer by layer [24]. 3D printing has been successfully used to make bespoke scientific tools in a number of disciplines for substantially lower costs than commercially available systems [25–28].

Designing for AM offers more geometric freedom, possibility to create easily customizable designs, new creative ways to manufacture instruments, decentralized manufacturing, and a range of material selection [12,18,29–31]. During the COVID-19 outbreak, AM has shown a great potential in mitigating the disruption in global medical and nonmedical supply chains [21].

In this study, the viability of using desktop 3D printing is analyzed to solve problems associated with attachment of sampler systems to workers for monitoring the breathing zone. A parametric and novel 3D printable clip system is designed, manufactured, and tested. The tests included a durability mechanical stress test to establish a minimum lifetime, a chemical solution test to ensure that the system can be appropriately cleaned, and a supply chain test to ensure occupational health professionals can use the digital files to customize and fabricate a custom system for their applications.

The solution is shared using an open source methodology to allow for distributed manufacturing of the free designs, and the costs and functionalities of this system are compared to commercial systems. The open source methodology is a way to easily share designs and research [26]. The best practices of Open-source Hardware (OSHWA) contains guidelines for correctly sharing and properly documenting the studies [32]. The results are analyzed and discussed along with a description of future work.

## 2. Materials and Methods

User needs based on the problems of the current sampler systems and wishes for the improvements were determined in consultation with occupational specialists who use samplers on a daily and weekly basis. A universal clip system for personal exposure measurement in the breathing zone of workers was then designed as free and open source hardware [33] following a general procedure for designing Free and Open-source Scientific Hardware (FOSH) equipment [34].

The first challenge designing the universal clip system was to have a design, which is not reliant on the harness's features and can be attached to any location on the harness. This would eliminate the need to alter the existing harness (e.g., no added hooks or holes). The second challenge was to find a design, which firmly stays in place. The third challenge was to make the design parametric, so that it can be easily altered to fit any harness on the market or can be further developed in the future. Fourth, the design should feature built in instructions so that users do not need to refer to external documentation to use it in the field. Finally, the design should also act as a universal base for different tooling options.

The design concept consists of two parts: (1) the sheath that holds the attachment features and contains the necessary instructions for correct orientation using arrows (Figure 1, middle/bottom left) and (2) the slider that locks the clip in place and also contains necessary tooling for example to hold the hose/sampler in place (Figure 1, middle/bottom right). The mechanism for the clip system works in the way that the harness is inserted inside the sheath by folding it. Then, the slider is slid to the sheath, which provides the necessary friction forces to hold the harness in between the two components in a single location and prevents sliding. The stopping mechanism inside the clip allows for the slider to level with the clip by clicking in place and stopping the sliding. This is done by installing click stoppers at the end of the clip. The release mechanism aids in unlocking the clip system by providing a sufficient amount of friction when a force is applied to it to uncouple the slider from the sheath by the user's hand. This was accomplished by installing grippers to the slider.

The clip concept design was first developed for a simple hose holder clip, which can be used to keep hoses in place and avoid potentially dangerous loops. Previously, this was often prevented by taping the hose in place that resulted in sticky tape residue, which is inappropriate for some work environments (e.g., high ISO class cleanrooms [35]). After the hose holder clip was printed and the concept was found to be functional, the ability of the open source design to be adaptable to additional applications was demonstrated by extending the clip concept to another tooling system—the cyclone sampler holder to the slider.

The principles of fastening and releasing mechanisms, CAD design of the clip with hose holder tooling, and redesigned cyclone sampler tooling are shown in Figure 1.

Two parametric models were created using FreeCAD version 0.16, which is a free, open source software that is readily available over the internet [36]. The parametric nature of the models allows for making changes to the model with ease according to the used harness. The parameters associated with the clip system that is equipped with a hose holder are shown in Figure 2a,b. Parameters that can be modified are harness width, harness thickness, clip thickness, clip height, and assembly print clearance, which is dependent on the used 3D printer. Three variants of the clip system with varying parameters are presented in Figure 2c, however, the most probable changed parameters would be harness width and thickness. In the hose holder clip FreeCAD design, the hose diameter including the clearance is also one parameter that can be changed according to the used hose. The full designs are available on the Open Science Framework (OSF) for downloading [37]. The file formats available are

STL (the standard file format for AM) and FCStd (FreeCADs file format, which enables parameter changing). Printing instructions can be found in Appendix A.

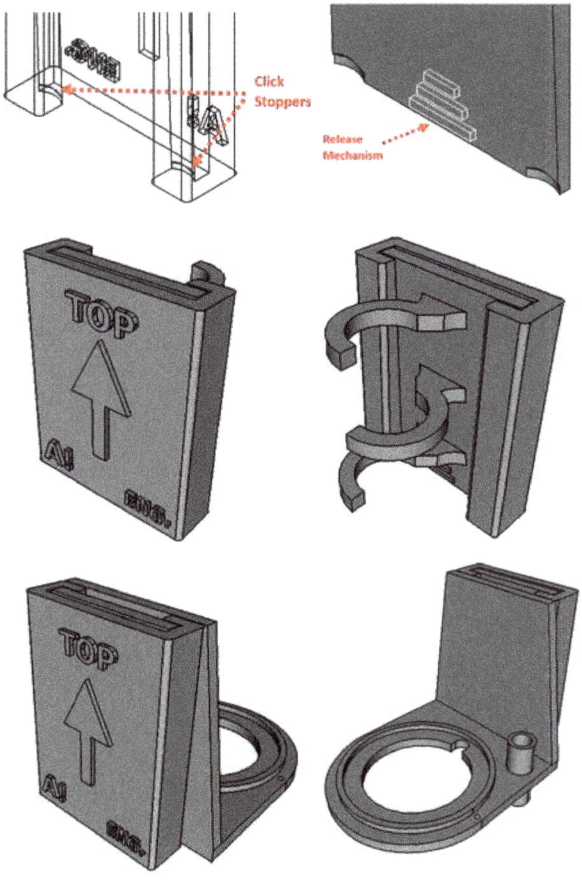


**Figure 1.** FreeCAD renderings of the fastening mechanism on the sheath (**top left**), releasing mechanism on the slider (**top right**), clip system from the back (**middle left**), clip system with hose holder tool from the front (**middle right**), redesigned clip system for the cyclone sampler from the back (**bottom left**), and redesigned clip system for the cyclone sampler from the front (**bottom right**).

The design of the clip system was modelled according to the Fused Deposition Modeling (FDM) Design Guidelines [38] that ensured the ultimate tensile strength of at least 33 MPa according to the datasheet of ABS (acrylonitrile butadiene styrene) material that was used for printing [39]. It should be noted that FDM is a trademark registered by Stratasys and thus, a subset of material extrusion (ME). Any material extrusion-based 3D printing system capable of appropriate polymers (e.g., ABS, PC, etc.) can be used to fabricate the design. The uPrint SE Plus ME printer from Stratasys was used to print the clips with ivory ABS filament. Minimum layer thickness and assembly clearance was set as 0.254 mm. The parts were printed in "upright position" for maximum strength in the X-Y planes following guidelines [38]. The printing times were 2 h 44 min for the hose holder clip and 4 h 13 min for the redesigned cyclone sampler holder clip. For the purpose of testing the clip systems, a prototype harness was fabricated and used for testing.

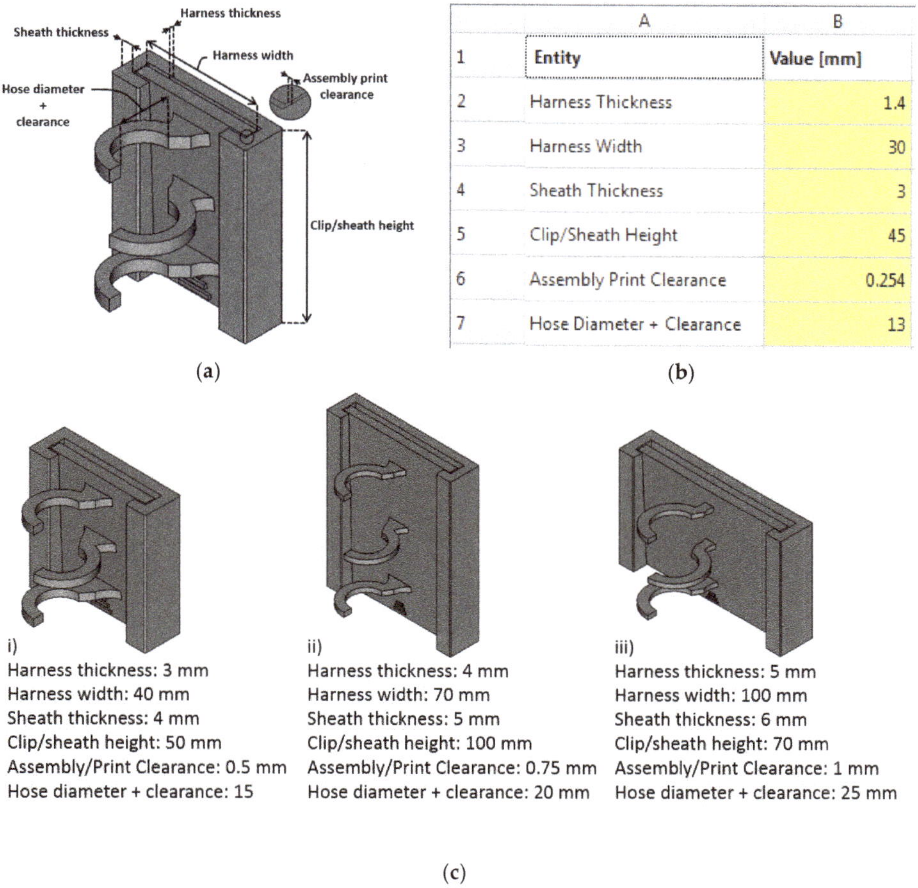


**Figure 2.** (**a**) The 5 + 1 input parameters of the parametric clip system equipped with a hose holder; (**b**) the 5+1 input parameters spreadsheet of the parametric clip system equipped with a hose holder; (**c**) three variants of the parametric clip system equipped with a hose holder.

Due to the nature of the measuring done in different environments from food industry to mining industry, cleaning the clip systems between different environments is important to avoid contamination. The hose holder clip was cleaned by wiping and immersing in an 80% ethanol–water solution. One cleaning cycle lasted for approximately 55 s, which was reiterated 50 times.

The mechanical properties of additively manufactured ABS are well established that range from 21 to 39 MPa in tensile and 38 to 41 MPa in compressive strength [40–46]. Low-cost desktop printers have been used to produce parts with these proprieties reliably for mechanically taxing applications from bicycled parts [47] to surgical tools [48]. The known issue of additively manufactured components is the lower tensile stress values in the Z-direction (upwards). The designs of the tensile and/or compressive specimens in the aforementioned literature ensured that no stress concentration was built up around the fillet of the specimens due to raster termination. This was accomplished either by completely removing the fillet and adapting a flat rectangular specimen according to ASTM D3039 [49] and/or improvising the specimens of ASTM D638 [50]. Likewise, the current design of the clip system ensured that no rasters were terminated at edges of the clip system that were loaded with the most stress during usage. Furthermore, fatigue strength of ABS is recorded as 17.8 MPa with 1.9 × 104 cycles [51] and 40% of the ultimate tensile strength in the order of thousands of cycles [52]. The stresses induced in the clip system during fastening or releasing can be considered as negligible due to the nature of light weight tooling that is used for this study.

Nevertheless, after the cleaning test, the clip system was then put through a durability test similar to those on medical diagnostic equipment [53] consisting of mechanical testing of the clip system through sliding the slider to sheath on and off repeatedly when it was attached to a harness. This procedure was executed 5000 times. This would correlate to 500 clicks annually and is an estimated use case for a 10-year lifecycle if two measurements per week are made. This durability testing consisted of an expected failure analysis at the system designs weakest point. In this failure testing, the clip system was subjected to elastic deformation similar to what would be expected in the application of attaching the system to the harness, but to a greater extent as in this case it was repeated immediately allowing for heat to be generated by the friction of repetitions. The heat generated in the testing was not near notable values. In addition of mechanical friction testing, the clip system was tested by putting it on and off a harness for 500 times.

Before the field testing, both the hose holder clip and cyclone sampler clip were subjected to lead user testing. This test covered ease of use and aesthetics of these clips compared to the traditional systems. During the field testing, test subjects were interviewed to provide feedback concerning the functionality and ergonomics of the clips. Two occupational specialists placed the harness on and used the previous systems of attaching the hose and cyclone sampler. These previous solutions were a custom made metal hose holder and original cyclone sampler holder with a serrated edge clamp. After testing the previous solutions, they tested prototype harness and placed the cyclone sampler with the redesigned clip into it and used the hose holder clip to attach the hose. The clips were also tested if they could be used with commercial harnesses as well as the prototype harness.

Field testing of the hose holder was carried out in three different measurements and workplaces in May, June, and October 2018. Duration of the first measurement was 20 min, and similarly, it was 6 h for the second, and 5 h for the third measurement. During the first measurement, the clip as well as prototype harness was attached to an operator in an industrial workplace. Volatile chemical compounds were measured. The work was partly physical but also included driving a forklift. The hose holder kept the sampling hose attached to a harness and an air pump, although plastic clamps of the prototype harness released during the measurements. A second measurement was also done in an industrial workplace where maintenance worker's exposure to inhalable dust was measured. The worker had hose holder clip(s) with original harnesses. In the third measurement, the cyclone sampler attached to the original harness was used in a machine workshop during welding. A respirable dust was measured from the breathing zone for manganese analysis to evaluate worker's exposure. A hose holder clip was also used. After every measurement, the clip holders were detached from the harness and cleaned according to a cleaning test.

To simulate the situation where environmental experts would decide on their own clip needs, generate correct files, and send the files to be printed in (e.g., using 3D printing services like 3D Hubs, 3DSD, etc.) or print their own systems using low-cost 3D printers, a test was designed. The test occupational specialist/possible end user with no CAD experience was given a link to the files, the FreeCAD software, and instructions for printing. They downloaded the hose holder clip files (FCStd) and opened them in the downloaded FreeCAD. Then, they decided which harness width they wished to use and changed it in the parametric model to generate correct 3D data. Then, they sent the STL data to be printed. There were three cases, in the first case, Aalto University acted as a service provider and printed both clip systems with an industrial level printer (uPrint SE Plus) and printed the hose holder clip with a low-cost printer (Lulzbot Mini). In the second case, a test person's organization, Finnish Institute of Occupational Health (FIOH), had their own in-house printer (Uprint SE Plus) and they tested the printing of their own clips as well. FIOH acted as a pilot organization for this study. Thirdly, an STL file was generated and uploaded to four internet 3D printing service providers to get cost and lead time quotations. These service providers were 3D Hubs [54–57].

The systems were massed with a digital scale (Mettler AE 200 +/− 0.0001 g). The material cost of each systems was calculated by multiplying the aggregate of the component weight by the cost of ABS per gram ($19.99/kg Amazon) for Lulzbot Mini and by the cost of ABSplus + SR30 soluble support per gram (317.25 €/kg + 288.84 €/kg +VAT) for uPrint SE Plus, as previous work has shown that the electrical consumption costs are not necessary [58]. Both the hose holder clip and cyclone sampler holder clip were printed with harness width of 30 mm.

Operating costs per hour were calculated with an assumption of price of the machine, which is amortized in certain amount of years during the expected usage. Price of the machine was 16,000 €, amortizing time was 3 years with an expected usage of 1500 h/year for uPrint SE Plus. Similarly, the price was 1100 € assuming an amortization time of 3 years with 500 h/year of usage for Lulzbot Mini.

## 3. Results

The hose holder parametric clip systems were additively manufactured successfully through uPrint SE Plus (Stratasys, Eden Prairie, MN, USA) and Lulzbot Mini (Aleph Objects Inc., Loveland, CO, USA) in Aalto University and with uPrint SE Plus in pilot organization. Further, cyclone sampler holder clips were printed with uPrint SE Plus in both locations. The hose holder clip was printed with two different harness widths (25 and 30 mm) to prove that the parametric design functioned as planned.

Figure 3 showcases both old traditional and new components along with assembled systems on a prototype harness. Firstly, Figure 3 shows four printed hose holder assemblies in the printing bed with support structures that display recommended printing orientation (a), and instructions on the way in which the slider should slide in the sheath when in the harness (b). Figure 3 also shows the assembled cyclone sampler holder with additively manufactured holder (d), and comparison of new ABS hose holder and traditional metal version, which was attached with green tape (c). Further, Figure 3 shows the blue hose holder clip, which was printed with low-cost 3D printer (f), and also the black hose holder and cyclone sampler holder clip, which were printed as part of supply chain testing in FIOH (e). Finally, Figure 3 demonstrates both the cyclone sampler and hose holder in practice, with the new prototype harness from the front (g) and back (h).

Cleaning test had no significant effect on the test clip, as was expected based on previous work that has shown that additively manufactured ABS is chemically compatible with ethanol–water solutions as well as hydrochloric acid (HCl, 37%), ammonia ($NH_3$, 25%), hydrogen peroxide ($H_2O_2$, 30%), phosphoric acid ($H_3PO_4$, 85%), and nitric acid ($HNO_3$, 69%) [59].

In the expected mechanical failure analysis, the parametric clip system passed a series of altogether 5000 openings and closings without any significant plastic deformation that would inhibit the proper use of the device over a 10-year lifetime. The deformation was too minor to have an impact on the functional integrity of the clip system and was barely noticeable with a naked eye.

Lead user feedback gave the clips good results. Clips could be used with commercially purchased harnesses as easily as with a prototype harness as long as they were the correct width and thickness. Testing subjects gave feedback for both the hose holder clip and cyclone sampler holder clip. They found ergonomics of the new clips to be the same as the traditional design or better. In functionality, they preferred the new clips. The new clip design did not require tape to secure certain placement, which was found to be excellent. Aesthetically, both testers found the new clips to be more professional looking than the tape set-up. In usability, one of them preferred the old attachment of the cyclone sampler, because it had involved less steps for placing it on. They then admitted that the new cyclone sampler clip system gave more secure grip and would less likely be dislocated than serrated clamps. Otherwise in usability, new clips were preferred but difficult to attach in field circumstances when there is usually a hurry.

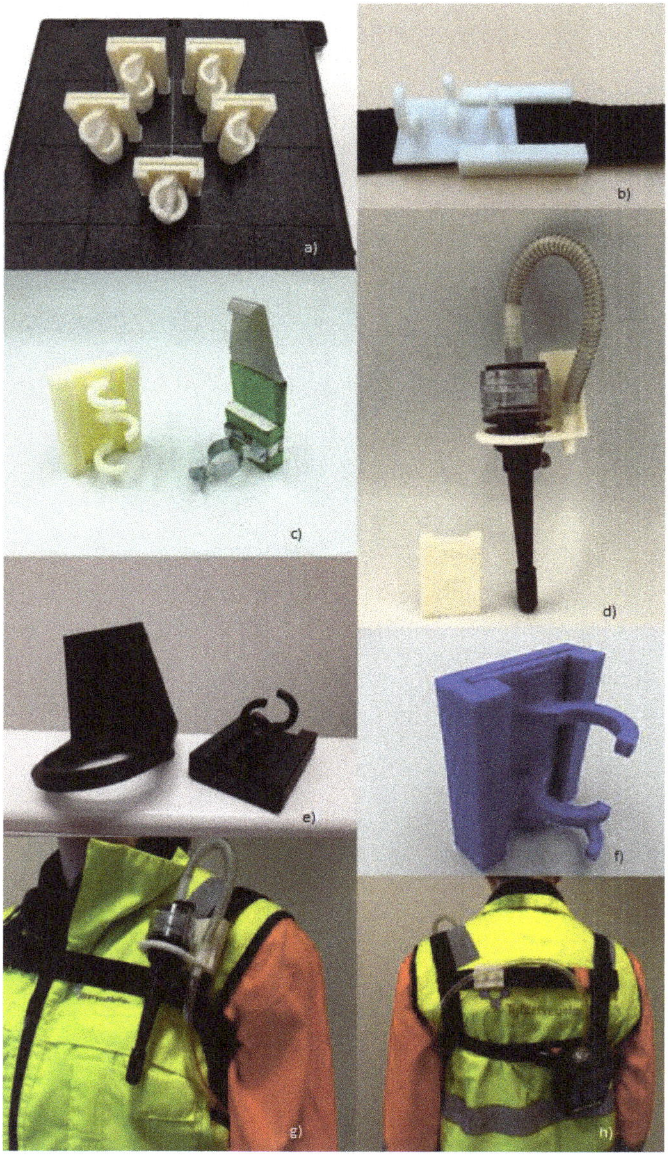


**Figure 3.** (**a**) Five hose holders with support structures in the printing bed; (**b**) sliding the slider into the sheath; (**c**) new (white) and old hose holder (green and metal original cyclone sampler assembly); (**d**) cyclone sampler assembly with the redesigned holder; (**e**) black hose holder clip and cyclone sampler clip printed as part of supply chain testing; (**f**) blue hose holder clip printed with Lulzbot Mini; (**g**) sampler system with cyclone sampler clip from the front; (**h**) harness with sampler system from the back.

The hose holder clips performed adequately in field tests. Zip-ties were used to ensure that no accidental dislocation would happen as the field tests were done in real life sampling situations where it is imperative that sampling is done correctly in one go. These zip-ties did not interfere with working of clips. The occupational specialist was satisfied with performance of the clips and speculated that when they get used to using clips, the zip-ties might be unnecessary. All the clips performed well in the active measuring situation. After the first field test, one of the hose holder clip's prongs were broken during transportation in a bag, presumably because something heavy was placed on top of it. After the third test, one of the hose holder clip's "prongs" broke during removal of the hose. All other clips were a whole and no visible cracks were observed.

The cyclone sampler holder clip test was also successful. The occupational specialist noted that the cyclone sampler clip's hold had been very secure. After the field tests, the occupational specialist's comments about the clips were enthusiastic and several suggestions of future work were mentioned such as longer hose holder, double hose holder, and parametric markings to the clips to express which width and thickness they are meant to have. Reinforcing of the hose holder clip's "prongs" was also suggested including changing the material to a more flexible one.

The supply chain tests showed that a person with no background in AM or experience in CAD designs can change an open source parametric model and generate the needed model with only links and basic written instructions. The occupational specialist had slight problems with FreeCAD, which were already known and are described in Appendix A. Upon following the instructions described in Appendix A, the user was able conveniently solve the Kernel related problems of the open source software. Subsequently, the user was able to generate the necessary models for AM and was able to send it to Aalto University and a pilot organization for production.

The experimental average mass of the hose holder clip system was 13.00 g, which resulted in a range of costs for the components detailed in Table 1. Clip variants 1 and 3 were printed with uPrint Plus SE, and clip variant 2 with Lulzbot mini. Clip variants 1 and 2 were hose holder clips and variant 3 is a cyclone sampler holder clip. All clip variants were designed for 30 mm harness width. The operating cost was calculated, and total price of the clip was obtained by adding together the material cost and operating cost.

**Table 1.** Cost of different variants of the clip system according to experimental average mass.

| | | | Clip Properties | | | |
|---|---|---|---|---|---|---|
| **Clip Variant** | **ABS Mass [g]** | **Support Mass [g]** | **Print Time (Lulzbot Mini)** | **Print Time (uPrint)** | **Printer Operating Cost (Lulzbot Mini) [€/h]** | **Printer Operating Cost (Uprint) [€/h]** |
| (1) hose | 12.2946 | 5.1068 | | 2 h 44 min | | 3.56 |
| (2) hose | 13.7139 | 2.6949 | 1h 47 min | | 0.73 | |
| (3) cyclone | 25.1874 | 11.1652 | | 4 h 13 min | | 3.56 |
| | | | **In House** | | | |
| **Clip Variant** | **Generic Filament Cost [€/clip]** | **Proprietary Filament Cost [€/clip]** | **Printer Operating Cost (Lulzbot mini) [€/clip]** | **Printer Operating Cost (uPrint) [€/clip]** | **Cost [€/clip] (Lulzbot + generic filament)** | **Cost [€/clip] (Uprint + proprietary filament)** |
| (1) hose | | 5.38 | | 9.72 | 15.10 € | |
| (2) hose | 0.33 | | 1.31 | | | 1.64 € |
| (3) cyclone | | 11.20 | | 14.99 | 26.19 € | |
| | | | **Service Providers** | | | |
| **Clip Variant** | **SD3D Cost [€/clip] (Sd3d.com)** | **Craftcloud [€/clip] (craftcloud.all3dp.com)** | **iMaterialise Cost [€/clip] (i.Materialise.com)** | **3Dhubs Cost [€/clip] (3dhubs.com)** | | |
| (1) hose | 21.94 | 21.67 | 73.5 | 100 | | |
| (3) cyclone | 32.38 | 27.05 | 86.8 | 100 | | |

This amounts to savings per replica of 23.36 € for Amazon generic material printed with Lulzbot Mini and 9.9 € for proprietary material printed with uPrint SE Plus, when the price of the original hose holder clip purchased was estimated to be 25 €. These are the cases for in-house production.

The hose holder clip was also uploaded to four different 3D printing service providers via internet and quotes from 21.94 € upwards were received. This is most probable use case for acquiring the clips, because the AM production knowledge and printers are not yet common enough.

## 4. Discussion

Sampling in the worker's breathing zone is one of the most applicable way to assess worker's personal exposure. To get correct positioning in the breathing zone, both commercial and custom-made harnesses may be used. Clamps, tapes, and zip-ties are currently used to attach sampler systems to the harness, creating issues with secureness of the hold, unprofessional looks, and hygienic problems. Designing specifically for AM and 3D printing, the novel design of a universal clip system will help create an aesthetically professional and securely attachable sampler. In this study, such a clip system was designed and tested, first with the hose holder clip and then, the cyclone sampler clip. Both clip systems performed well, although strengthening of hose holder's "prongs" is suggested. The field test showed that especially the cyclone sampler clip is ready for real life sampling situations. For the occupational specialists, a short learning curve is needed before use of clips is instinctual. However, this can be compared with the initial introduction of conventionally used attachment gear and can be circumvented through the supporting documents provided in Appendix A. Field tests also showed that it is important to have back-up clips in case of surprises.

Parametric design of the clip can ensure that the fit of the clip can be made to fit any harness as long as width and thickness of the harness is known. Depending on the 3D printer, some adjustments may be needed to find the correct process parameters which is supported by the parametric settings, in particular the clearance.

The open source files of the design and availability of 3D printing mean that the clip is obtainable to any person, who needs these kind of clip systems. 3D printing also enables the selection of a range of materials and colors.

Using affordable 3D printers means that the costs of the clip systems stay low and the open source nature of the clip system will be able to accommodate customization with minimal additional costs. Therefore, the cost comparison made only on a material level is justified; purchased traditionally manufactured components are compared to the direct material and operating costs of 3D printing.

This customization of the clip system could easily be extended to other tooling options as well since the basic design can be altered to hold other kinds of samplers and tools. To which areas the clip system design can be extended, and what are the limitations, is work for the future studies.

The cost savings shown in Table 1 are the materials and 3D printer operating costs only and they do account for any labor costs. This is a valid assumption as the time needed to customize the system for specific users is comparable (and perhaps shorter) to the time needed to shop for the correct size of the system online. 3D printers do not need to be monitored during use. Thus, the time to print on any of the systems, although hours, is irrelevant as only the time to set up a print is taken into consideration, which is approximately equivalent to ordering a commercial product online. The time to remove the supports, which is less than 10 min for the devices, can be compared to the time it would take to accept delivery of a package and unpack it. Lastly, once all of the components are gathered, they must be assembled. Assembling the complete system takes a similar amount of time to commercial systems. Finally, it is important to point out the life cycle cost advantages of this system. As all the files are shared, regardless of the source of failure of any of the components, the system is easily repaired from readily available components (e.g., reprint a broken plastic component). This ease of repair and upgrading is simply not available for all the commercial systems, which would demand the purchasing of replacement devices. Thus, the value of the open source additively manufactured universal clip system for personal exposure measurement in the breathing zone of workers can be considered higher than the commercial functional equivalent, even though the open source tool costs less money to build upfront.

## 5. Conclusions

This study successfully described the development, design, and testing of a novel 3D printable open source clip system for attaching personal exposure samplers to harnesses. This clip system was extended from a simple hose holder clip to a redesigned cyclone sampler holder. The hose holder clip system proved to be 3D printable with affordable desktop 3D printers, which gives considerable savings compared to custom solutions used before and allows for distributed manufacturing through free designs. The clip system also improved functionality and security of sampler placement. The testing of the clip system showed minimum of 10 years of lifetime.

Secure and user-friendly research instrumentation is essential to ensure consistency and reliability of personal exposure measurements. The open source clip system described here could serve as a reference design for future improvements. The design allows easy customization and modification for other types of instruments. Modifications to the reference design are expected to be faster and more cost efficient than completely new untested clip designs.

**Author Contributions:** Conceptualization, K.K. and J.S.A.; methodology, K.K. and J.S.A.; validation, K.K., J.S.A. and A.K.; investigation, K.K., J.S.A. and A.K.; resources, M.S., R.B., A.-K.V. and J.P.; data curation, K.K. and J.S.A.; writing—original draft preparation, K.K., J.S.A. and J.M.P.; writing—review and editing, K.K., J.S.A., A.K., M.S., R.B., A.-K.V., J.P. and J.M.P.; visualization, K.K. and J.S.A.; supervision, A.-K.V., J.P. and J.M.P.; All authors have read and agreed to the published version of the manuscript.

**Funding:** This research received no external funding.

**Acknowledgments:** This work was supported by Aalto University, Finnish Institute of Occupational Health and Fulbright Finland. The authors would like to acknowledge helpful discussions with Marika Loikala.

**Conflicts of Interest:** The authors declare no conflict of interest.

## Appendix A

Printing instructions

Link to the clip system files can be found osf.io/6u3zr.

Hose holder clip and cyclone sampler holder clip can be printed with the STL files provided in the above link using any ISO/ASTM Standardized Additive Manufacturing Method. These STL files are for harness, which width is 30 mm and thickness is 1.4 mm. The printing materials are recommended to be polymers.

To create new STL files from parametric models, use the following instructions:

1. Download the open source FreeCAD software from https://www.freecadweb.org/.
2. Download the model files provided in the link above.
3. Open the necessary FreeCAD [.FCstd] file using open source FreeCad Software.
4. To access parameters, double click the spreadsheet (left, under Model and labels and Attributes page). Depending of the printer, you may need to adjust assembly clearance and/or harness thickness.
5. Change the required parameters in the Spreadsheet. Please note that regeneration may take some time depending on your hardware specs.'*'
6. Export model as STL to print the model. (Files-> Export). Remember to click the model to activate it before exporting.
7. Use STL file to print the model with (material extrusion) 3D printer. Depending on the printer, you may need to use the brim command to avoid bending. You may also consider printing the hose holder clip arrow downward to ensure good detail of the stopper mechanism.
8. Remove the support structures. They can be removed by hand or with a sharp knife.

'*'Please note that Boolean operations can fail upon regeneration because of the CAD kernel of the open source software. Always use the latest version of the software and try different values of the parametric parameters or delete the Boolean operations to overcome this limitation. Likewise, the fillet operation(s) may fail upon regeneration as well. Remove fillets and redeploy the necessary operation(s) after regeneration to overcome these limitations.

Using the clip with the harness

1. Remove slider from sheath.
2. Place the harness in the slot of the sheath through its open side. The open side will be the front of the harness and the back (arrow) side will be against the person.
3. Slide the sheath on the harness to the needed position. Please note that arrow at the back of the sheath should be pointing up. When using in horizontal harness placement, that arrow points to the direction where slider will be inserted from.
4. Insert the slider with the release mechanism first to the sheath.
5. Push the slider until it clicks. It should be same level as the sheath.
6. Push the release mechanism (segmented arrowhead) upwards to unlock the clip and to separate the components. Please do not push the hose holder or any other parts of tooling.
7. Remove sheath from the harness.

## References


1. Vinson, R.; Volkwein, J.; McWilliams, L. Determining the Spatial Variability of Personal Sampler Inlet Locations. *J. Occup. Environ. Hyg.* **2007**, *4*, 708–714. [CrossRef]
2. Wang, C.-H.; Chen, B.T.; Han, B.-C.; Liu, A.C.-Y.; Hung, P.-C.; Chen, C.-Y.; Chao, H.J. Field Evaluation of Personal Sampling Methods for Multiple Bioaerosols. *PLoS ONE* **2015**, *10*, e0120308. [CrossRef] [PubMed]
3. Stephens, B.; Azimi, P.; El Orch, Z.; Ramos, T. Ultrafine particle emissions from desktop 3D printers. *Atmos. Environ.* **2013**, *79*, 334–339. [CrossRef]
4. Ruzer, L.; Harley, N. *Aerosols Handbook*; CRC Press: Boca Raton, FL, USA, 2013; ISBN 9781439855195.
5. Lidén, G. Evaluation of the SKC Personal Respirable Dust Sampling Cyclone. *Appl. Occup. Environ. Hyg.* **1993**, *8*, 178–190. [CrossRef]
6. Haig, C.W.; Mackay, W.G.; Walker, J.T.; Williams, C. Bioaerosol sampling: Sampling mechanisms, bioefficiency and field studies. *J. Hosp. Infect.* **2016**, *93*, 242–255. [CrossRef] [PubMed]
7. Asbach, C.; Alexander, C.; Clavaguera, S.; Dahmann, D.; Dozol, H.; Faure, B.; Fierz, M.; Fontana, L.; Iavicoli, I.; Kaminski, H.; et al. Review of measurement techniques and methods for assessing personal exposure to airborne nanomaterials in workplaces. *Sci. Total Environ.* **2017**, *603–604*, 793–806. [CrossRef]
8. Vuorinen, V.; Aarnio, M.; Alava, M.; Alopaeus, V.; Atanasova, N.; Auvinen, M.; Balasubramanian, N.; Bordbar, H.; Erästö, P.; Grande, R.; et al. Modelling aerosol transport and virus exposure with numerical simulations in relation to SARS-CoV-2 transmission by inhalation indoors. *Saf. Sci.* **2020**, *130*, 104866. [CrossRef]
9. National Institute for Occupational Safety Health Division of Physical Sciences Engineering. *Manual of Analytical Methods*; DHHS (NIOSH) Publication: Washington, DC, USA, 1994.
10. Akmal, J.S.; Salmi, M.; Hemming, B.; Teir, L.; Suomalainen, A.; Kortesniemi, M.; Partanen, J.; Lassila, A. Cumulative Inaccuracies in Implementation of Additive Manufacturing Through Medical Imaging, 3D Thresholding, and 3D Modeling: A Case Study for an End-Use Implant. *Appl. Sci.* **2020**, *10*, 2968. [CrossRef]
11. Tuomi, J.; Paloheimo, K.; Björkstrand, R.; Salmi, M.; Paloheimo, M.; Mäkitie, A.A. Medical applications of rapid prototyping—From applications to classification. *Innov. Dev. Des. Manuf.* **2010**, 701–704. [CrossRef]
12. Akmal, J.; Salmi, M.; Mäkitie, A.; Björkstrand, R.; Partanen, J. Implementation of Industrial Additive Manufacturing: Intelligent Implants and Drug Delivery Systems. *J. Funct. Biomater.* **2018**, *9*, 41. [CrossRef]
13. Salmi, M.; Tuomi, J.; Sirkkanen, R.; Ingman, T.; Makitie, A. Rapid tooling method for soft customized removable oral appliances. *Open Dent. J.* **2012**, *6*, 85–89. [CrossRef] [PubMed]
14. Wohlers, T. *Wohlers Report 2020*; Wohlers Associates, Inc.: Fort Collins, CO, USA, 2020.
15. Khajavi, S.H.; Partanen, J.; Holmström, J. Additive manufacturing in the spare parts supply chain. *Comput. Ind.* **2014**, *65*, 50–63. [CrossRef]
16. Totin, A.; Macdonald, E.; Conner, B. Additive manufacturing for aerospace maintenance and sustainment. *DSIAC J.* **2019**, *6*, 4–11.
17. Salmi, M.; Partanen, J.; Tuomi, J.; Chekurov, S.; Björkstrand, R.; Huotilainen, E.; Kukko, K.; Kretzschmar, N.; Akmal, J.; Jalava, K.; et al. Digital Spare Parts. Aalto University. 2018. Available online: http://urn.fi/URN:ISBN:978-952-60-3746-2 (accessed on 22 September 2020).

18. Ullah, R.; Akmal, J.S.; Laakso SV, A.; Niemi, E. Anisotropy of additively manufactured 18Ni-300 maraging steel: Threads and surface characteristics. *Procedia CIRP* **2020**, *93C*, 66–76. [CrossRef]
19. Jalava, K.; Salmi, M.; Kukko, K.; Orkas, J. Multi-scale topologically optimized components made by casting and additive manufacturing. In Proceedings of the 73rd World Foundry Congress, Congress Proceedings, Krakow, Poland, 23–27 September 2018; pp. 141–142.
20. Akmal, J. Digital Unique Component Manufacturing Through Direct and Indirect Additive Manufacturing. Aalto University 2017. Available online: http://urn.fi/URN:NBN:fi:aalto-201710307355 (accessed on 22 September 2020).
21. Salmi, M.; Akmal, J.S.; Pei, E.; Wolff, J.; Jaribion, A.; Khajavi, S.H. 3D Printing in COVID-19: Productivity Estimation of the Most Promising Open Source Solutions in Emergency Situations. *Appl. Sci.* **2020**, *10*, 4004. [CrossRef]
22. Jones, R.; Haufe, P.; Sells, E.; Iravani, P.; Olliver, V.; Palmer, C.; Bowyer, A. RepRap: The replicating rapid prototyper. *Robotica* **2011**, *29*, 177–191. [CrossRef]
23. Bowyer, A. 3D printing and humanity's first imperfect replicator. *3d Print Addit. Manuf.* **2014**, *1*, 4–5. [CrossRef]
24. ISO/TC 261 ISO/ASTM 52900:2015(E). *Additive Manufacturing-General Principles-Terminology*, 2nd ed.; ISO/ASTM International: Vernier, Switzerland, 2015.
25. Pearce, J.M. Building Research Equipment with Free, Open-Source Hardware. *Science* **2012**, *337*, 1303–1304. [CrossRef]
26. Pearce, J.M. *Open-Source Lab.: How to Build. Your Own Hardware and Reduce Research Costs*; Elsevier: Amsterdam, The Netherlands, 2014.
27. Baden, T.; Chagas, A.M.; Gage, G.; Marzullo, T.; Prieto-Godino, L.L.; Euler, T. Open labware: 3-D printing your own lab equipment. *PloS Biol.* **2015**, *13*, e1002086. [CrossRef]
28. Coakley, M.; Hurt, D.E. 3D printing in the laboratory: Maximize time and funds with customized and open-source labware. *J. Lab. Autom.* **2016**, *21*, 489–495. [CrossRef]
29. Khajavi, S.; Holmström, J.; Partanen, J. Additive manufacturing in the spare parts supply chain: Hub configuration and technology maturity. *Rapid Prototype J.* **2018**, *65*, 50–63. [CrossRef]
30. Chekurov, S.; Kajaste, J.; Saari, K.; Kauranne, H.; Pietola, M.; Partanen, J. Additively manufactured high-performance counterflow heat exchanger. *Prog. Addit. Manuf.* **2019**, *4*, 55–61. [CrossRef]
31. Thompson, M.K.; Moroni, G.; Vaneker, T.; Fadel, G.; Campbell, R.I.; Gibson, I.; Bernard, A.; Schulz, J.; Graf, P.; Ahuja, B.; et al. Design for Additive Manufacturing: Trends, opportunities, considerations, and constraints. *Cirp. Ann.* **2016**, *65*, 737–760. [CrossRef]
32. Best Practices for Open-Source Hardware 1.0. Available online: https://www.oshwa.org/sharing-bestpractices/ (accessed on 10 September 2020).
33. Gibb, A. *Building Open Source Hardware: DIY Manufacturing for Hackers and Makers*; Pearson Education: London, UK, 2014.
34. Oberloier, S.; Pearce, J.M. General Design Procedure for Free and Open-Source Hardware for Scientific Equipment. *Designs* **2017**, *2*, 2. [CrossRef]
35. ISO. ISO 14644-1:2015; *Cleanrooms and Associated Controlled Environments. Part. 1: Classification of Air Cleanliness By Particle Concentration*; ISO: Geneva, Switzerland, 2015.
36. FreeCAD: An Open-Source Parametric 3D CAD Modeler. Available online: https://www.freecadweb.org/ (accessed on 24 May 2018).
37. Akmal, J.S.; Kukko, K.; Pearce, J.M. Additively Manufactured Parametric Universal-Clip-System. 2019. Available online: https://osf.io/6u3zr/ (accessed on 22 September 2020).
38. Stratasys Direct, Inc. Fused Deposition Modeling (FDM) Design Guidelines. 2015. Available online: https://www.stratasysdirect.com/resources/design-guidelines/fused-deposition-modeling (accessed on 22 September 2020).
39. Stratasys Direct, Inc. ABSplus-P430: Production-Grade Thermoplastic for 3D Printers. 2017. Available online: https://www.stratasys.com/materials/search/absplus (accessed on 22 September 2020).
40. Ahn, S.-H.; Montero, M.; Odell, D.; Roundy, S.; Wright, P.K. Anisotropic material properties of fused deposition modeling ABS. *Rapid Prototype J.* **2002**, *8*, 248–257. [CrossRef]

41. Croccolo, D.; De Agostinis, M.; Olmi, G. Experimental characterization and analytical modelling of the mechanical behaviour of fused deposition processed parts made of ABS-M30. *Comput. Mater. Sci.* **2013**, *79*, 506–518. [CrossRef]
42. Tymrak, B.M.; Kreiger, M.; Pearce, J.M. Mechanical properties of components fabricated with open-source 3-D printers under realistic environmental conditions. *Mater. Des.* **2014**, *58*, 242–246. [CrossRef]
43. Letcher, T.; Rankouhi, B.; Javadpour, S. Experimental study of mechanical properties of additively manufactured ABS plastic as a function of layer parameters. In Proceedings of the ASME International Mechanical Engineering Congress and Exposition, Houston, TX, USA, 13–19 November 2015; Volume 57359, p. V02AT02A018.
44. Rankouhi, B.; Javadpour, S.; Delfanian, F.; Letcher, T. Failure analysis and mechanical characterization of 3D printed ABS with respect to layer thickness and orientation. *J. Fail. Anal. Prev.* **2016**, *16*, 467–481. [CrossRef]
45. Tanikella, N.G.; Wittbrodt, B.; Pearce, J.M. Tensile strength of commercial polymer materials for fused filament fabrication 3D printing. *Addit. Manuf.* **2017**, *15*, 40–47. [CrossRef]
46. Lee, C.S.; Kim, S.G.; Kim, H.J.; Ahn, S.H. Measurement of anisotropic compressive strength of rapid prototyping parts. *J. Mater. Process. Technol.* **2007**, *187–188*, 627–630. [CrossRef]
47. Tanikella, N.G.; Savonen, B.; Gershenson, J.; Pearce, J.M. Viability of distributed manufacturing of bicycle components with 3-D printing: CEN standardized polylactic acid pedal testing. *J. Hum. Eng.* **2017**, *5*, 1.
48. Rankin, T.M.; Giovinco, N.A.; Cucher, D.J.; Watts, G.; Hurwitz, B.; Armstrong, D.G. Three-dimensional printing surgical instruments: Are we there yet? *J. Surg. Res.* **2014**, *189*, 193–197. [CrossRef] [PubMed]
49. ASTM. *ASTM D3039-76, Test. Method for Tensile Properties of Polymer Matrix Composite Materials*; ASTM: West Conshohocken, PA, USA, 1976.
50. ASTM. *ASTM D638-97, Test. Method for Tensile Properties of Plastics*; ASTM: West Conshohocken, PA, USA, 1997.
51. Padzi, M.M.; Bazin, M.M.; Muhamad, W.M.W. Fatigue Characteristics of 3D Printed Acrylonitrile Butadiene Styrene (ABS). *IOP Seri. Mater. Sci. Eng.* **2017**, *269*, 012060. [CrossRef]
52. Lee, J.; Huang, A. Fatigue analysis of FDM materials. *Rapid Prototype J.* **2013**, *19*, 291–299. [CrossRef]
53. Michaels, R.E.; Pearce, J.M. 3-D printing open-source click-MUAC bands for identification of malnutrition. *Public Health Nutr.* **2017**, *20*, 2063–2066. [CrossRef]
54. 3D HUBS: Network of Manufacturing Hubs. Available online: https://www.3dhubs.com/ (accessed on 15 March 2019).
55. CraftCloud: Service Provider. Available online: https://craftcloud.all3dp.com/ (accessed on 15 March 2019).
56. SD3D: Service Provider. Available online: https://www.sd3d.com/ (accessed on 15 March 2019).
57. iMaterialise: Service Provider. Available online: https://i.materialise.com/ (accessed on 8 March 2019).
58. Wittbrodt, B.T.; Glover, A.G.; Laureto, J.; Anzalone, G.C.; Oppliger, D.; Irwin, J.L.; Pearce, J.M. Life-cycle economic analysis of distributed manufacturing with open-source 3-D printers. *Mechatronics* **2013**, *23*, 713–726. [CrossRef]
59. Heikkinen IT, S.; Kauppinen, C.; Liu, Z.; Asikainen, S.M.; Spoljaric, S.; Seppälä, J.V.; Savin, H.; Pearce, J.M. Chemical Compatibility of Fused Filament Fabrication -based 3-D Printed Components with Solutions Commonly Used in Semiconductor Wet Processing. *Addit. Manuf.* **2018**, *23*, 99–107. [CrossRef]

*applied sciences*

MDPI

*Article*

# Design and Fabrication of Complex-Shaped Ceramic Bone Implants via 3D Printing Based on Laser Stereolithography

**Alexander Safonov *, Evgenii Maltsev, Svyatoslav Chugunov, Andrey Tikhonov, Stepan Konev, Stanislav Evlashin, Dmitry Popov, Alexander Pasko and Iskander Akhatov**

Skolkovo Institute of Science and Technology, Center for Design, Manufacturing and Materials, 30/1 Bolshoi Boulevard, Moscow 121205, Russia; e.maltsev@skoltech.ru (E.M.); s.chugunov@skoltech.ru (S.C.); a.tikhonov@skoltech.ru (A.T.); s.konev@skoltech.ru (S.K.); s.evlashin@skoltech.ru (S.E.); dmitry.popov@skoltech.ru (D.P.); a.pasko@skoltech.ru (A.P.); i.akhatov@skoltech.ru (I.A.)

* Correspondence: a.safonov@skoltech.ru

Received: 5 August 2020; Accepted: 12 October 2020; Published: 14 October 2020

**Abstract:** 3D printing allows the fabrication of ceramic implants, making a personalized approach to patients' treatment a reality. In this work, we have tested the applicability of the Function Representation (FRep) method for geometric simulation of implants with complex cellular microstructure. For this study, we have built several parametric 3D models of 4 mm diameter cylindrical bone implant specimens of four different types of cellular structure. The 9.5 mm long implants are designed to fill hole defects in the trabecular bone. Specimens of designed ceramic implants were fabricated at a Ceramaker 900 stereolithographic 3D printer, using a commercial 3D Mix alumina ($Al_2O_3$) ceramic paste. Then, a single-axis compression test was performed on fabricated specimens. According to the test results, the maximum load for tested specimens constituted from 93.0 to 817.5 N, depending on the size of the unit cell and the thickness of the ribs. This demonstrates the possibility of fabricating implants for a wide range of loads, making the choice of the right structure for each patient much easier.

**Keywords:** bones implants; stereolithography; 3D printing; aluminum oxide; Function Representation (FRep) method

## 1. Introduction

Ceramic materials demonstrate excellent mechanical performance, high dimensional stability, high wear and corrosion resistance, and exceptional chemical stability. Current advances in 3D printing can potentially solve the problem of efficient production of custom made complex ceramic parts of arbitrary geometry [1–4]. Today, 3D printing is used to manufacture complex shaped ceramic parts for various applications, such as integrated ceramic casting molds [5,6], cutting tools [7], sensors [8–10], structural parts [11–13], photonic crystals [14,15], and dental components [16,17]. Besides, 3D printing is used to manufacture ceramic implants for various applications [18–23], mostly for the replacement of lost bones or restoration of damaged bone tissues [24].

Various porous structures (scaffolds) can be used to ensure successful proliferation of cells, demonstrating good biocompatibility and adhesion of live cultures. Scaffolds can have an irregular [25] or regular structure formed by way of 3D printing processes [26]. 3D printing allows the fabrication of scaffold structures with dimensions varying from several nanometers to several meters [27]. As was already shown in the previous studies, synthetic scaffolds can be designed based on various unit cell models, such as cube, gyroid, diamond, dispheroid, and others [26]. These models make it possible to obtain materials with different mechanical properties and biological activity. Such scaffolds can be used to splice both bone [28] and soft tissues [29] such as nerves. Pores in such scaffolds should have a certain size to ensure effective splicing of tissues. The effect of the pore size on bone tissue splicing efficiency is studied in [30], using 3D printed scaffolds with various pore sizes, fabricated from $Al_2O_3$-based materials. The optimal pore size for bone tissue proliferation was shown to constitute 390 to 590 μm.

Alumina-based ceramics are biologically inert and resistant materials [31]. High compression strength and wear resistance make these materials an ideal choice for use in stomatology and hip joint surgery. Ceramic parts are produced by sintering at high temperatures (1600–1700 °C) and have residual porosity of about 1–3%. The size of the pores depends on thermal treatment conditions and on the micromorphology of feedstock powders, and usually does not exceed dozens of microns [32], which impedes the growth of tissues, blood vessels, and other nutrient ducts. Alumina-based ceramics can perform a supporting function in the defected bone or in joining two broken bone parts, but they cannot be used for full bone remodeling during treatment. A macroporous architecture is suggested to stimulate bone tissue ingrowth in the region of implant placement. There are several methods of producing highly porous ceramic materials, allowing a trade-off between ceramics porosity and the strength of implants. These are, in particular, the use of foaming [33] and burnt-out additives, and the method of replicas [34]. However, these methods are difficult to control. 3D printing allows the fabrication of parts of any shape [35], implementing a personalized approach to patients' treatment [36] that involves the use of computerized tomography, while controlling porosity and the size of pores (from 400 μm and higher) [37], necessary for better implant acceptance within the damaged area. Considering this, the design and fabrication of 3D printed ceramic implants with complex tailored porous structure becomes an urgent issue.

In this work, we use a promising stereolithographic (SLA) method of 3D printing of ceramic parts, based on the layer-by-layer polymerization of photocurable ceramic suspension or paste [38], consisting of photocurable multicomponent binder and ceramic powder. Each layer (slice) of a 3D model is solidified by laser hatching in a working area and then, a new layer of paste is applied by spreading blades. Figure 1 shows an illustration of the stereolithographic printing process. The process allows fabrication of materials with high spatial resolution (about 40 μm in the printing plane, and 10 μm in the direction perpendicular to the printing plane). Printed green parts are then heated for several hours to remove polymer binder. Further on, the parts are sintered to impart the final shape and desired mechanical properties. Both these processes are traditional for ceramics manufacturing.

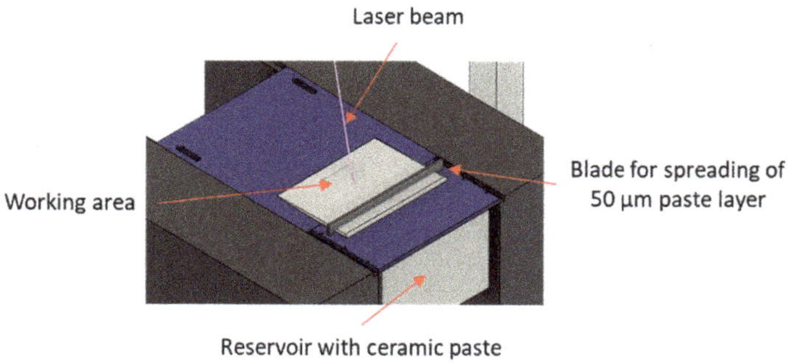


**Figure 1.** Stereolithographic printing process.

There are multiple applications of SLA and a large variety of commercially available 3D printers and ceramic pastes. This method is used for 3D printing of ceramics for at least 20 years. SLA (and its modification, a Digital Light Processing (DLP) method is considered one of the best methods for ceramics manufacturing due to the high precision, low defects, and high density and strength of the produced parts. Other methods, such as Fused Deposition Modeling (FDM) or Selective Laser Sintering (SLS), produce inferior results in terms of resolution and microstructural properties of the material. A video demonstrating the 3D printing process of ceramic parts based on the SLA process can be found at the 3D Ceram company website [39].

In practice, all fabricated ceramic implants would be customized, i.e., tailored to individual patient needs. This raises the question of the quick design of parts with complex porous structure. Traditional methods of geometric modeling based on surface representation of a model cannot be applied here due to a very large amount of computations required to build models with complex internal structure [40]. Considering the above, we used the Function Representation (FRep) approach for geometric modeling of complex shapes with porous, cellular, and irregular microstructure [41]. This approach defines any geometrical object by non-negative values of a real continuous function procedurally evaluated at any given point of coordinate space. Whether a point belongs to an object depends on the sign of the defining function evaluated at that point. Typically, points with positive and zero function values are included in a set, while a zero-level subset is considered a boundary. Basically, FRep modeling is the "extended" implicit modeling [42] with explicitly introduced elementary objects (primitives).

Compared to boundary representation-based Computer-aided design (CAD) systems, the FRep approach offers significant advantages for microstructure modeling. First, the FRep format always ensures the accuracy of the obtained model, as opposed to traditional polygonal representation used in CAD systems where the probability of ending up with cracks or nonmatching edges in a model is rather high, even in simple operations such as offsetting. Second, the FRep has the advantage of complete parametrization of obtained microstructures, giving high flexibility in fast generation of variable 3D models. Third, the FRep offers a wide variety of tools to model various meshed structures, including periodic functions, Triply Periodic Minimal Surfaces (TMPS) [43,44], and replication of a unit cell, based on the method of skeleton-based surfaces [45]. In addition, FRep-based software makes it possible to use controlled blending versions of set-theoretic operations on geometric primitives. It is important for most applications to be able to combine a microstructure with the given external shape of an object or with a rigid platform.

In this study, we investigated the applicability of 3D printing for fabrication of easily modified cellular ceramic structures. While FRep is not a novel method, there are no publications on such application of this approach. Here, we used the FRep method to design 4 mm diameter cylindrical implants. The designed 9.5 mm long implants are intended to fill hole-like defects in the trabecular bone. The implants reproduce the structure of the trabecular bone, having a dense cylindrical base similar to the structure of cortical bone tissue, and porous main body similar to the structure of cancellous bone tissue. Besides, the dense base improves the adhesion of an implant to the building platform during the 3D printing process. The implants were designed on the request of a local medical research institute for the testing of bone substitute materials on rabbits under laboratory conditions. The primary loading mechanism in the implants was specified as compression. Only small tensile or flexural loads were expected for the implants. The cubic diamond-type cell was used as a basis for modeling the cellular microstructure. In the design, we varied the diameter of ribs (from 500 to 700 μm, with 100 μm increment) and the distance between ribs (from 500 to 1000 μm, with 250 μm increment). The size of cells was selected so as to ensure optimal conditions for bone tissues proliferation after sintering and shrinkage. Ceramic implant specimens were fabricated by a stereolithographic 3D printing process at a Ceramaker 900 printer, using a commercial 3DMix alumina feedstock paste. Then, implant specimens were tested in axial compression.

## 2. Materials and Methods

### 2.1. Design

In the FRep approach, the geometry of an object can be represented by the real function, and a general model can be represented as a certain ensemble (M, Φ, W), where M—the set of geometric objects; Φ—the set of geometric operations; W—the set of relationships over the set of objects [46]. Generally, geometric objects of a given model represent closed subsets of the N-dimensional space, $E^n$.

In our case, a 3D Euclidean space, $E^3$, will be considered. The geometry of an object is specified by the function of 3 variables; by using various rules of points classification, we can obtain surfaces or geometric solids. Thus, expression (1) specifies the boundary surface and expression (2) specifies a geometric solid:

$$f(x, y, z) = 0, \tag{1}$$

$$f(x, y, z) \geq 0. \tag{2}$$

In each point in space $P = (x, y, z) \in E^3$, the classification is done by the value of function $f(P)$, such that $f(P) > 0$, if the point P is located within the object; $f(P) = 0$, if the point P is located at the object's boundary; $f(P) < 0$, if the point P is located outside the object.

FRep uses R-functions as set-theoretic operations [46]. The controlled blending versions of set-theoretic operations are applicable to FRep objects. The bounded blending operation [47] is useful for combining a microstructure with the other primitives. This operation allows for the addition or removal of material to/from specified regions of a 3D object, using expression (3) for the bounded blending operation.

$$F_{bb}\ (f_1, f_2, f_3) = R(f_1, f_2) + a_0 disp_{bb}\ (r), \tag{3}$$

$$disp_{bb}\ (r) = \begin{cases} \frac{(1-r^2)^3}{1+r^2}, r < 1 \\ 0, r \geq 1 \end{cases},\ r^2 = \begin{cases} \frac{r_1^2}{r_1^2 - r_2^2}, r_2 > 0 \\ 1, r_2 = 0 \end{cases},\ r_1^2(f_1, f_2) = \left(\frac{f_1}{a_1}\right)^2 + \left(\frac{f_2}{a_2}\right)^2,\ r_2^2(f_3) = \begin{cases} \left(\frac{f_3}{a_3}\right)^2, f_3 > 0 \\ 0, f_3 \leq 0 \end{cases},$$

where $f_1, f_2$—the implicit functions of 2 geometric primitives; $f_3$—the implicit function of a primitive restricting the region of blending operations; R—the R-function; $a_0, a_1, a_2, a_3$—blending operation parameters controlling the symmetry and a resulting shape.

In the course of 3D modeling, we used Function Representation of 3D objects and, particularly, the skeleton-based surfaces approach in combination with the definition of a scalar field and a distance function [48]. As the unit cell, we take a tetrahedral structure built as a crystal lattice of a diamond, where each atom (red points in Figure 2) forms bonds with its 4 nearest neighbors (blue points in Figure 2), with angles between the bonds constituting 109.47°.

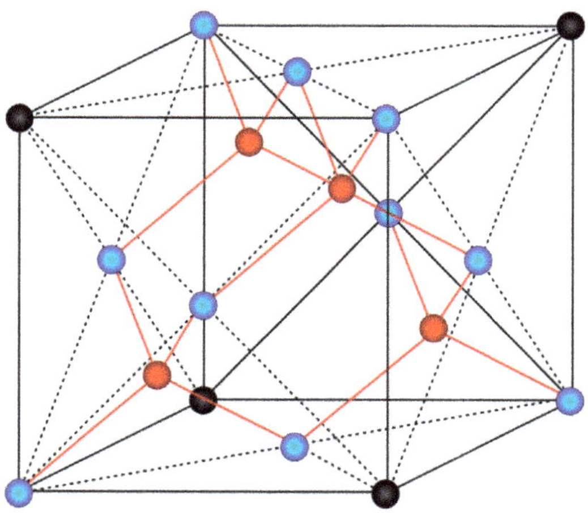

**Figure 2.** Tetrahedral structure of a diamond.

Implant models were built in Rhinoceros suite (version 5), using the Symvol FRep plugin for Rhino version 1.5. Figure 3 shows the workflow of the modeling process consisting of 7 general steps: 1. Construction of the wireframe of the unit cell; 2. Replication of the unit cell; 3. Trimming the frame to its external form (cylinder); 4. Removing unnecessary parts; 5. Application of vTube operation; 6. Adding a cylindrical base plate; 7. Applying the bounded blending operation.

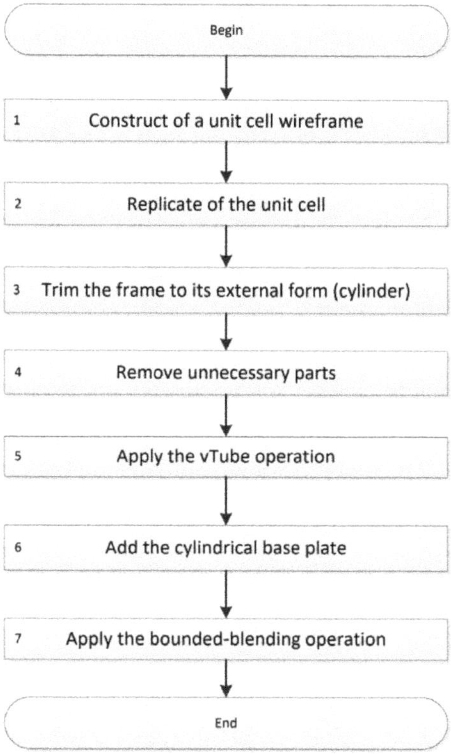


**Figure 3.** Workflow of the modeling process with Symvol FRep plugin for Rhino version 1.5.

## 2.2. Fabrication

To analyze the possibility of fabricating $Al_2O_3$ ceramic implants by 3D printing, we conducted several experiments to fabricate cylindrical specimens. As a ceramic feedstock, we used 3DMix alumina paste produced by 3DCeram (Limoges, France). The weight fraction of the alumina powder in the paste constitutes 84%. We used a Ceramaker 900 (3DCeram, Limoges, France) stereolithographic printer to print green body specimens. The printing scene was prepared in the Netfabb suite (Autodesk, San Rafael, USA). Slicing of the model was done with a slicer tool of DMC software (Direct Machining Control, Vilnius, Lithuania). Processing parameters for the 3D printer were provided by the 3DCeram company in the form of an encrypted formulation file, together with the feedstock material. The precise parameters of the laser system are proprietary information not disclosed by the manufacturer.

After printing, we cleaned all specimens from the excess uncured ceramic feedstock at the CeraKleaner cleaning station (3DCeram, Limoges, France), using the pulverized proprietary solvent provided by 3DCeram. Then, printed specimens were debinded in the Kittec CLL-15 laboratory kiln (Kittec, Rosenheim, Germany). Specimens were gradually heated to 500 °C in the course of 6 h to remove organic binder, constituting over 30% of the volume of the green body. To improve the mechanical strength of the material for further processing, specimens were additionally heated for an hour at a temperature of 1100 °C in the same kiln without prior cooling, for presintering.

Then, specimens were further sintered in the ThermConcept HTL 20/17 furnace (ThermConcept, Bremen, Germany), using the procedure proposed by the furnace manufacturer. After placing the specimens in the kiln, a dome of ceramic bricks was constructed around specimens in order to mitigate spatial temperature gradients. The specimens were heated to a temperature of 1700 °C and held at this temperature for 1.5 h and then, cooled down. The heat up and cooldown rate constituted 3 °C/min.

### 2.3. Scanning Electron Microscopy

After sintering, we analyzed the shape and microstructure of implants using the Apreo scanning electron microscope (SEM) from Thermo Fisher Scientific (Waltham, MA, USA). No additional surface treatments were applied to the specimens before the SEM study. For the SEM study, we used the following parameters: working distance of 13 mm, accelerating voltage of 5 kV, and beam current of 50 pA. Secondary electrons were collected with the in-lens detector.

### 2.4. Mechanical Testing

Fabricated implant specimens were tested under axial compression. Such type of implants can be used for compression tests because the top and the bottom contact faces of the implants are parallel. The bottom face has a tight contact with the compression plate of the mechanical press. The top surface contacts the compression plate at a number of points.

We tested specimens at the Instron 5969 electromechanical testing machine equipped with a 50 kN load cell. We used two 150 mm diameter hardened, polished compression plates, with an adjustable upper plate. Before the test, the plates were brought together with a small load and the upper plate was adjusted so as to ensure parallelism. To avoid premature failure of the ceramic specimens at spots of contact with the metal, thin cardboard pads were glued onto both compression plates with double-sided adhesive tape. For full-field strain measurements, we applied speckles of black acrylic paint onto the specimens, using a spray bottle.

The upper plate was brought to the position of approx. 5 mm above a specimen, then the load was balanced, and the test started with the constant traverse speed of 0.2 mm/min. Tests were stopped manually, on visible signs of specimen fracture. We used a Digital Image Correlation (DIC) system from Correlated Solutions, taking one photograph a second from a 5 MP camera with a Tokina 100 mm F2.8 Macro D lens with simultaneous recording of force via a National Instruments Data acquisition (DAQ). Speckle correlation was performed in the VIC-2D-2009 program. After obtaining the strain field, we applied a virtual extensometer incorporated in the program to extract the tabulated data of engineering strain vs. force relationship.

In this study, we do not consider the testing of specimens with cylindrical base plates at both faces of a cylinder due to the following reasons. The proposed design with a single base plate represents a real implant to be used in further clinical testing. As the mechanical performance of such items is a very important part of the engineering design for biomedical applications, the real implants need to be tested "as is", without artificial parallelization of surfaces.

### 2.5. Mechanical Simulation

For mechanical simulation of implant specimens under compression testing, we used the Abaqus finite element analysis (FEA) suite (version 6.14) [49]. To build the finite element model of the implant, we used its geometrical model in STL format. The model was built with quadratic tetrahedral elements of C3D10-type. To model the printed ceramic material, we used a linear elastic material with elastic modulus of E = 350 GPa and Poisson's ratio of $\nu = 0.21$.

## 3. Results

### 3.1. Design

Four different types of parametric models of bioimplants were built, using four different types of unit cells and filling. Implant models were built by replicating unit cells along three coordinate axes (X,Y,Z) to obtain a wire frame with the addition of a cylindrical base plate. The following unit cells were constructed (see Figure 4): UC001—the cell built on the basis of tetrahedral diamond structure, with rib thickness of 0.4 mm, cell length (L) of 2.3 mm, cell width (W) of 2.3 mm, and cell height (H) of 1.7 mm; UC002—the cell built on the basis of the UC001 unit cell by removing dangling ribs, with rib thickness of 0.7 mm, L = 1.6 mm, W = 1.6 mm, H = 2.3 mm; UC003—the cell built by duplicating the UC002 unit cell along the X and Y axes, with the offset along the Z-axis by H/2, and along the X-axis by W/2, with rib thickness of 0.7 mm, L = 3.5 mm, W = 3.5 mm, H = 3.8 mm; UC004—the cell built by compressing the UC003 unit cell along X and Y axes by a factor of 2, with rib thickness of 0.3 mm, L = 2.0 mm, W = 2.0 mm, H = 4.1 mm. Based on the unit cells data, we modeled the bioimplants with the following wireframe (mesh) types: BI001—the wireframe built by replicating the UC001 unit cell along three coordinate axes—by a factor of 3 along X and Y axes, and by a factor of 5 along Z axis; BI002—the UC002 unit cell is replicated by a factor of 2 along X and Y axes, and by a factor of 4 along Z axis; BI003—the UC003 unit cell is replicated by a factor of 1 along X and Y axes, and by a factor of 3 along Z axis; BI004—the UC004 unit cell is replicated by a factor of 3 along X, Y, and Z axes. The above structures were limited by a 5 mm thick cylinder of 3.8 mm in diameter. These structures were then trimmed by using the SelBoundary operation. Dangling ribs in the structures were also deleted. A 2.5 mm thick cylinder of 4 mm in diameter was modeled as the implant base plate, allowing: (a) better adhesion of a part to a modeling base for 3D printing; (b) tight closure of the inlet hole of a bone defect; (c) machining of a part to the shape of the bone for better implantation into the defect. Modeling and unit cell replication parameters are given in Table 1. Implant models are shown in Figure 5.

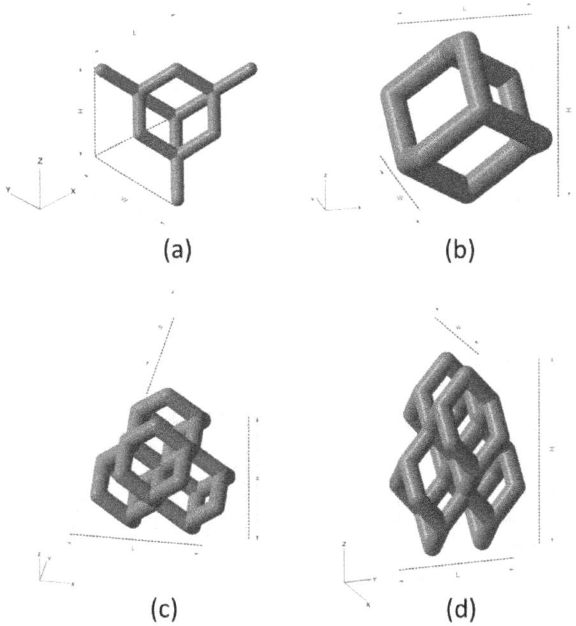

**Figure 4.** Unit cells used for implant modeling: (**a**) UC001—the unit cell built on the basis of tetrahedral diamond structure (rib thickness = 0.4 mm, cell length (L) = 2.3 mm, cell width (W) = 2.3 mm, cell height (H) = 1.7 mm); (**b**) UC00202—the cell built on the basis of the UC001 unit cell by removing open ribs (rib thickness = 0.7 mm, L = 1.6 mm, W = 1.6 mm, H = 2.3 mm); (**c**) UC003—the cell built by replicating the UC002 unit cell along the X and Y axes by a factor of 2, with the offset along Z-axis by H/2, and along X-axis by W/2 (rib thickness of 0.7 mm, L = 3.5 mm, W = 3.5 mm, H = 3.8 mm); (**d**) UC004—the cell built by compressing the UC003 unit cell along X and Y axes by the factor of 2 (rib thickness =0.3 mm, L = 2.0 mm, W = 2.0 mm, H = 4.1 mm).

**Table 1.** Implants modeling parameters.

| Model | Unit Cell | Rib Thickness T (mm) | Cell Length L; Cell Width W; Cell Height H (mm) | Comments | Unit Cell Replication Factor along Coordinate Axes | | |
|---|---|---|---|---|---|---|---|
| | | | | | X | Y | Z |
| BI001 Figure 5a | UC001 Figure 4a | 0.4 | L = 2.3, W = 2.3, H = 1.7 | Unit cell based on the tetrahedral diamond structure. | 3 | 3 | 5 |
| BI002 Figure 5b | UC002 Figure 4b | 0.7 | L = 1.6, W = 1.6, H = 2.3 | Built of UC001 unit cell by removing dangling ribs. | 2 | 2 | 4 |
| BI003 Figure 5c | UC003 Figure 4c | 0.7 | L = 3.5, W = 3.5, H = 3.8 | Built by replicating the UC002 unit cell along the X and Y axes by a factor of 2, with the offset along Z-axis by H/2, and along X axis by W/2. | 1 | 1 | 3 |
| BI004 Figure 5d | UC004 Figure 4d | 0.3 | L = 2.0, W = 2.0, H = 4.1 | Compression of the unit cell along X and Y axes by the factor of 2 | 3 | 3 | 3 |

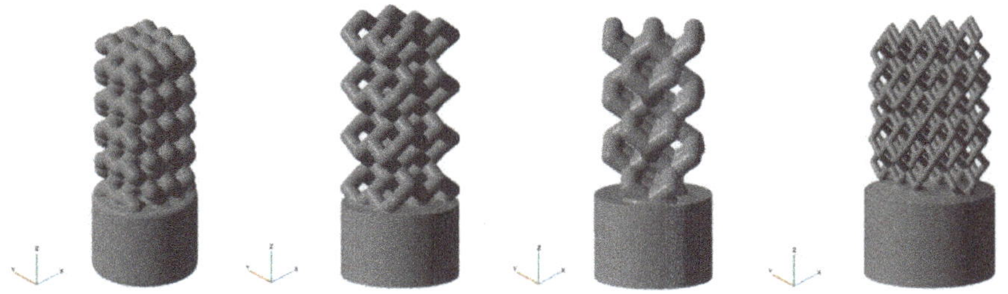

(a) (b) (c) (d)

**Figure 5.** The models of implants. Combination of a wireframe with a cylindrical base plate: (**a**) BI001—the wireframe built by replicating the UC001 unit cell along three coordinate axes, by a factor of 3 along X and Y axes, and by a factor of 5 along Z-axis; (**b**) BI002—the UC002 unit cell is replicated by a factor of 2 along X and Y axes, and by the factor of 4 along Z axis; (**c**) BI003—the UC003 unit cell is replicated by a factor of 1 along X and Y axes, and by a factor of 3 along Z-axis; (**d**) BI004—the UC004 unit cell is replicated by a factor of 3 along X,Y, and Z axes.

To improve stiffness of the structure, we added an extra material between the cylindrical base plate and the wireframe, using the bounded blending operation. Figure 6 shows the result of the bounded blending operation applied to assign stiffness to the implant structure. In our case, the bounded blending operation includes the implicit functions of two geometric primitives: the cylindrical base plate and the wireframe; the R-function is union.

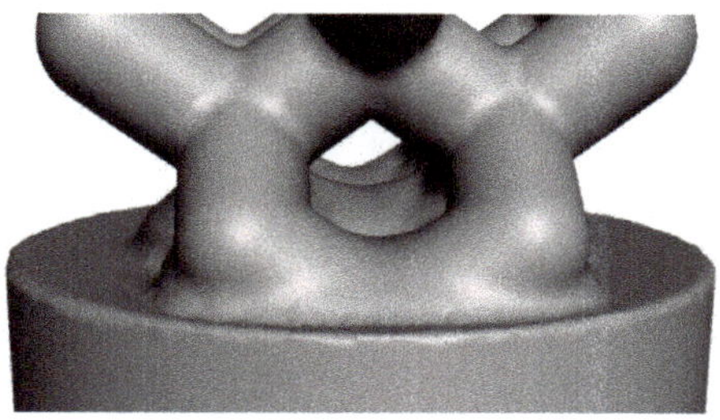

**Figure 6.** Application of bounded blending operation to assign stiffness to the implant structure.

The files of the BI001–BI004 models designed in Rhino software with the Symvol plugin can be found in the Supplementary Materials. The screenshots of the executed operations are also shown in the Supplementary Materials. File sizes of the models in various formats are given in Table 2.

**Table 2.** File sizes of the models in various formats are given in MB.

| Model | Symvol for Rhino | STL |
|---|---|---|
| BI001 | 0.193 | 24.347 |
| BI002 | 0.128 | 39.920 |
| BI003 | 0.110 | 11.694 |
| BI004 | 0.141 | 24.180 |

Table 3 shows porosity values for the models built both with and without account for the cylindrical base plate.

**Table 3.** Porosity of the models with and without account for the cylindrical base plate (%).

| Model | With Account for the Cylindrical Base Plate (%) | Without Account for the Cylindrical Base Plate (%) |
|---|---|---|
| BI001 | 52.7 | 68.3 |
| BI002 | 53.0 | 70.5 |
| BI003 | 53.7 | 70.6 |
| BI004 | 63.9 | 84.4 |

To demonstrate the capabilities of the method described here, we have built an array of 20 × 20 BI004 models (see Figure 7), consisting of 10,800 unit cells. The file size of the array model built in Symvol format constitutes 0.201 Mb (see the Supplementary Materials).

**Figure 7.** The array of 20 × 20 BI004 models.

### 3.2. Fabrication

Four cylindrical implant specimens with dimensions of d = 4 mm, h = 9.5 mm were fabricated for the study. The structure of the implants is described in Table 1. Implant models were additionally scaled to account for shrinkage, using standard shrinkage coefficients for 3DMix alumina paste ($k_X = 1.159$, $k_Y = 1.146$, $k_Z = 1.187$). To prepare the models for printing and to set up printing parameters (laser power, slicing, etc.), we used the Ceramaker software (3DCeram, Limoges, France). With a slice thickness of 50 μm, the printing of the specimens took approximately 5 h. Specimens were printed at 71% of the power of the laser system, which is equivalent to 120 mW/cm$^2$. To remove excess uncured ceramic feedstock, we used a CeraKleaner cleaning station. Due to the complex macroporous geometry of the implants, an ultrasonic cleaning bath was used to speed up the cleaning process. Figure 8 shows photographs of the obtained specimens after printing.

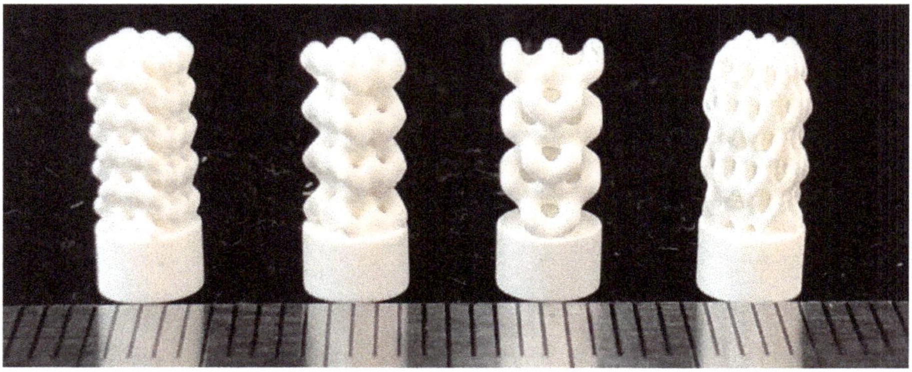

**Figure 8.** Photographs of printed implants fabricated from models developed earlier and shown in Figure 5.

Parts were debinded in the Kittec CLL-15 laboratory kiln at a temperature of 550 °C, using the multistep procedure, and presintered at 1100 °C. Presintered specimens were then placed in the Thermoconcept HTL 20/17 high temperature furnace and heated for 1.5 h at a temperature of 1700 °C for final sintering. Figure 9 shows BI002 model specimens after printing (left) and after sintering (right).

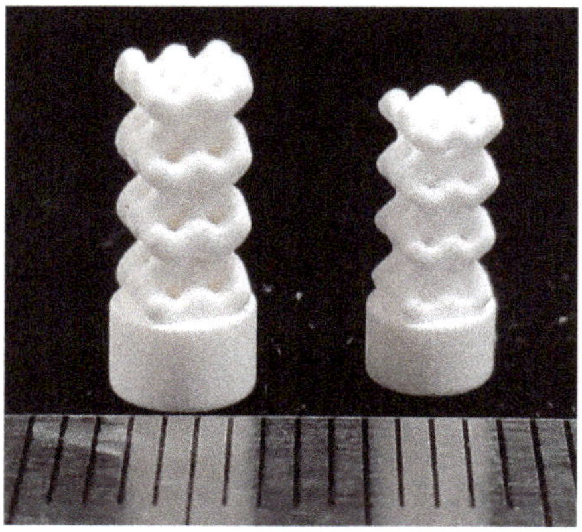

**Figure 9.** BI002 model specimens after printing (**left**) and after sintering (**right**).

Thus, the full manufacturing cycle for the porous $Al_2O_3$ implant (see Figure 10) constituted approximately 7 days.

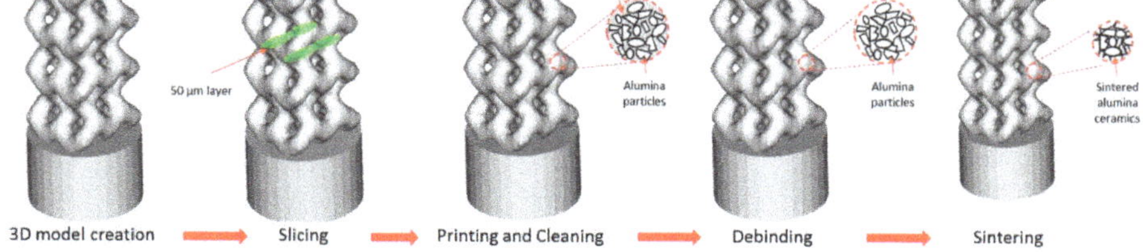


**Figure 10.** Fabrication of $Al_2O_3$ ceramic implant at the Ceramaker 900 stereolithographic 3D printer.

We manufactured 11 specimens of each type. Ten specimens were given to a local medical institute for bone substitute testing on rabbits. The remaining specimens were subjected to mechanical testing. The authors also intend to conduct biomedical testing including cytotoxicity and osteogenic studies. The results of those studies will be reported in future publications.

### 3.3. Scanning Electron Microscopy

According to the SEM images, fabricated structures feature pore sizes of 440 to 700 μm, which is in good agreement with the results obtained earlier [6]. Such structures feature the optimal morphology for bone tissue proliferation. Figure 11 shows the typical example of BI004 structure: (a) scaffold; (b) and (c) pores and ribs after sintering, enlarged view; (d) ceramic microstructure. The average size of grains constitutes several micrometers and the surface of the structure has no pronounced pores and defects. The density of the obtained structures was studied in our previous work and constitutes approximately 3.8 g/cm$^3$ [50]. The porosity of the sintered ceramic material does not exceed 5%.

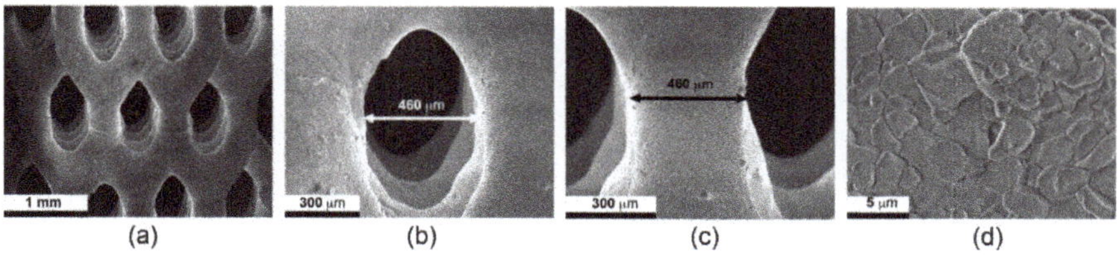


**Figure 11.** SEM images of the BI004 implant after sintering: (**a**) scaffold; (**b**) enlarged view of a pore; (**c**) enlarged view of a rib; (**d**) sintered ceramics.

### 3.4. Mechanical Testing

Figure 12 shows loading curves for implants tested under axial compression. The testing machine crosshead travel (mm) is plotted on a horizontal axis; the testing load (N) is plotted on the vertical axis. Shown also are the photographs of implant specimens after fracture. Table 2 shows maximum load values for specimens under axial compression. The results show that the BI001 specimen is almost three times stronger, compared to the other three specimens. This can be explained by a denser packing of a wireframe, resulting from the use of the UC001 unit cell with rib thickness of 0.4 mm and dimensions of L = 2.3 mm, W = 2.3 mm, H = 1.7 mm. Moreover, the BI001 specimen features three rows of contact surfaces in the upper part of the model, thus forming a larger area of contact with the loading plate and providing more uniform and less intense redistribution of load from the testing machine. At the same time, the BI004 specimen also features three rows of contact surfaces at the upper part of the model; however, the thinner and longer ribs of the UC004 unit cell (thickness = 0.3 mm, L = 2.0 mm, W = 2.0 mm, H = 4.1 mm) make the structure susceptible to premature local buckling and result in a significantly lower strength of the implant. The BI002 specimen has two rows of contact surfaces in the upper part of the model and features higher strength compared to that of the BI004 specimen, due to the higher thickness of the ribs of the UC002 unit cell (rib thickness = 0.7 mm, L = 1.6 mm, W = 1.6 mm, H = 2.3 mm). The lowest strength was demonstrated by the BI003 specimen, having only one row of contact surfaces. The increase in unit cell dimensions (from UC001 to UC004) results in lower maximum load the specimen can carry and in significantly lower physical permeability of an implant, which is necessary for successful integration of the implant into the patient organism. Reducing unit cell dimensions while keeping the thickness of the ribs the same or lower is currently impossible with the SLA printing process. Due to the low strength of the printed layers, they tend to be destroyed by a printer spreading blade when applying new layers of ceramic paste (see Figure 1). Smaller unit cell dimensions in combination with an increase in rib thickness result in a collapse of gaps between neighboring ribs, thus making the removal of excess feedstock from printed parts almost impossible (even in an ultrasound bath).

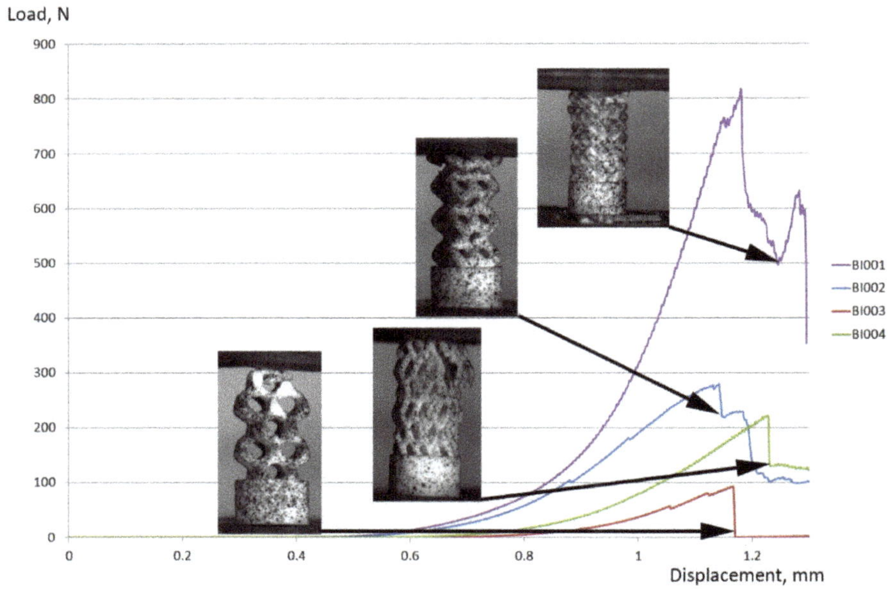


**Figure 12.** Loading curves for implant specimens tested under axial compression. The crosshead travel (mm) during testing is plotted on the horizontal axis; the load (N) is plotted on the vertical axis.

Table 4 also shows the effective compressive strength of implants, calculated by division of the maximum load by the area of a cylindrical base plate of 4 mm in diameter. According to the available data, the ultimate compression strength in the longitudinal direction constitutes 193 MPa for a cortical bone and ~50 MPa for a trabecular bone [51], which is comparable to the effective compressive strength of the BI001 implant (65.1 MPa) shown in Table 4.

**Table 4.** Implants modeling parameters.

| | BI001 | BI002 | BI003 | BI004 |
|---|---|---|---|---|
| Maximum load at a specimen, N | 817.5 | 279.9 | 93.0 | 221.6 |
| Effective compressive strength of the implant [1], MPa | 65.1 | 22.3 | 7.4 | 17.6 |

[1] The cross-section area of the implant is taken to be the area of a circle of 4 mm in diameter.

Figure 13 shows the distribution of vertical strains at the surface of the BI002 specimen during testing at the 185.7 N load, obtained by digital image correlation. We also measured strains between reference points A and B, located at the cylindrical base plate and in the lower part of the wireframe, respectively (see Figure 13). Before loading, the distance between points constituted 7.26 mm. At the load of 185.7 N, the distance between reference points A and B decreased by $\Delta U_{ABexp} = -0.00356$ mm; thus, the resulting strain constituted $\varepsilon_{ABexp} = -0.000492$.

**Figure 13.** The distribution of vertical strains at the surface of the BI002 specimen during testing at the 187.5 N load, obtained by digital image correlation (DIC).

### *3.5. Mechanical Simulation*

Figure 14 shows the compression test simulation results for the BI002 specimen. The computational model includes the implant model and the discrete rigid plane (see Figure 14a) located at the distance of 0.45 mm from the edge of the wireframe part of the implant. The number of C3D10-type finite elements in the model constitutes 432265. General contact conditions are imposed between the implant model and the discrete rigid plane. The implant model is fixed to the foundation of the cylindrical base plate. The loading of the implant is given by the displacement of the discrete rigid plane.

Figure 14b shows the distribution of vertical displacements at the load of 185.7 N. Calculated displacement of the reference points B relative to the point A constituted $\Delta U_{ABcal} = -0.00342$ mm. Thus, the difference between predicted and measured changes in distance between reference points A and B does not exceed 5%. Figure 14c shows the distribution of maximum principal stresses at the load of 279.9 N (see Table 4). The region with maximum principal stress is located in the bridged area between lateral and central rows of pillars. Stresses in this area constitute 400 MPa, exceeding the tensile strength of $Al_2O_3$-based ceramics. During tests, the initial fracture was also observed in this area (see Figure 14d).

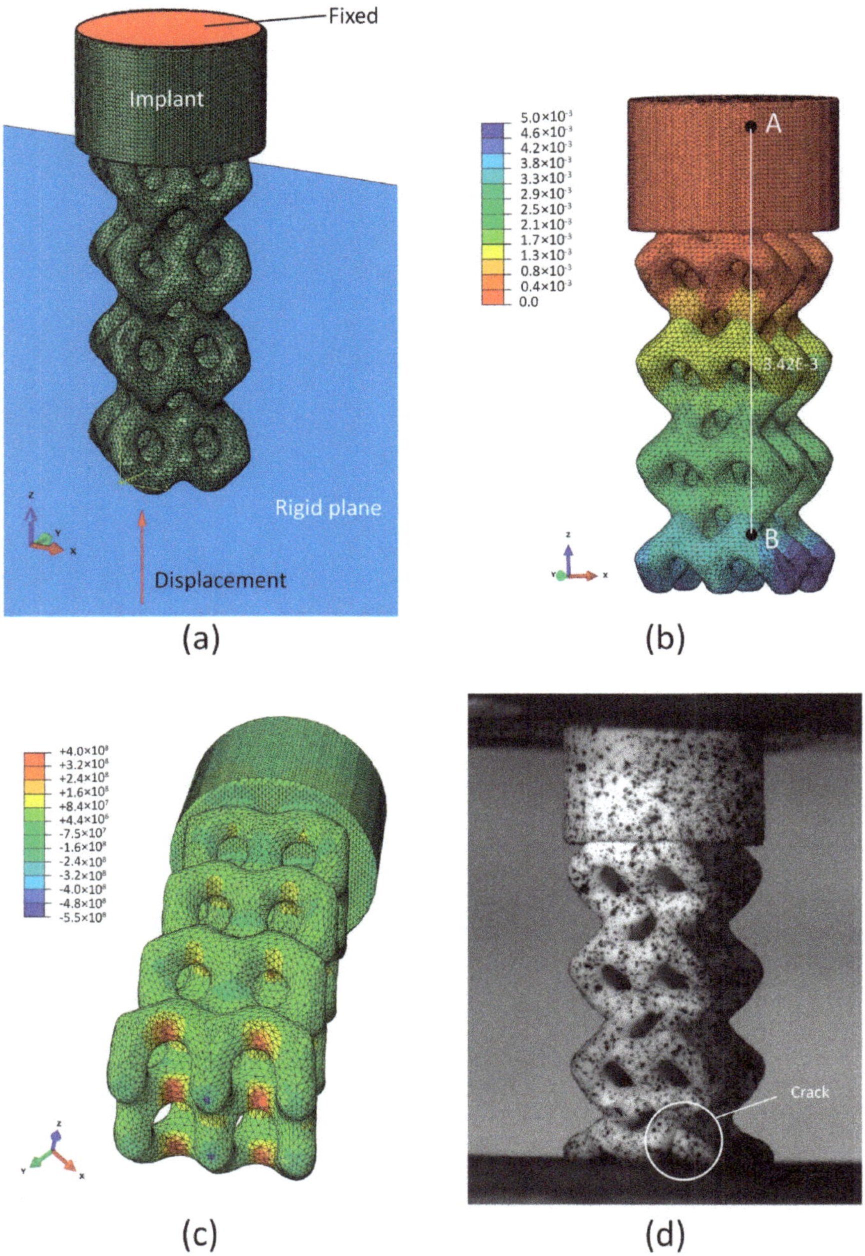


**Figure 14.** Compression test simulation results for the BI002 specimen: (**a**) Computational model; (**b**) distribution of vertical displacements (mm) at the load of 185.7 N; (**c**) distribution of maximal principal stress (Pa) at the load of 279.9 N; (**d**) initial fracture of the specimen during compression test.

## 4. Discussion

The FRep approach allows an engineer to build parametric 3D models of implants, while providing flexibility in the generation of variable 3D objects for further fabrication. The FRep provides an analytical description of models, reducing the volume of data necessary to describe a single model to several hundred kilobytes of a source file, which is several dozen times less compared to the traditional definition of a model that uses boundary representation and a polygonal format (such as STL, for example).

After modeling, the models should be exported to the polygonal STL format suitable for 3D printers for further slicing and 3D printing. The size of the obtained STL files constitutes approximately 20 Mbytes. Polygonization of a model is a time-consuming operation and is often the cause of errors in generated surfaces of 3D objects. However, Jamieson and Hacker in [52] have proposed a method of direct 3D printing, which was further modified and tested for FRep models in [53,54]. The method makes it possible to slice an FRep-model and prepare the job for a 3D printer directly, without the intermediate stage of polygonization, thus reducing the total time required for 3D processing and eliminating the errors related to polygonization. Therefore, the further developments in this field may include the application of a direct 3D printing method.

The ultimate compressive stress of implants studied here corresponds to the strength of porous ceramics and lies between the compressive strength of trabecular (0.1–50 MPa) and cortical (150–200 MPa) bone tissues. Therefore, by varying the size of a unit cell and the thickness of ribs, we can obtain implants with effective strength in the range of 7.4 to 65.1 MPa, making it easier to choose the right material for each patient.

Advantages of studied structures are their high porosity and sufficient distance between ribs, optimal for the ingrowth of bone tissue during treatment. The combination of chemical composition (alumina) and macroporous 3D structure makes it possible to obtain highly porous, permeable, strong bone implants.

In future research, we intend to conduct biomedical testing of fabricated implants, including cytotoxicity and osteogenic studies. The results of these studies will be reported in further publications.

The authors also intend to conduct in-depth finite element analysis of the experimental data in order to obtain a validated computational model. This model will be used for the topological optimization [55] of bioimplants with account for geometric requirements, loading conditions, and anisotropy of mechanical properties of printed ceramics.

## 5. Conclusions

In this study, four types of macroporous ceramic implants based on $Al_2O_3$ have been designed and printed. The BI001 type implant has demonstrated effective compressive strength comparable to that of the trabecular bone, making it a promising bone substitute able to withstand high operating loads.

**Supplementary Materials:** The following are available online at http://www.mdpi.com/2076-3417/10/20/7138/s1, Models.zip: the files of the models designed in Rhino software with the Symvol plugin, Description.docx: the description of the modeling process in Rhino software with the Symvol plugin.

**Author Contributions:** Conceptualization, A.S. and E.M.; methodology, A.P. and E.M.; software, D.P.; validation, A.P., E.M. and S.C.; formal analysis, S.C.; investigation, A.T. and E.M.; resources, S.C.; data curation, S.K. and S.E.; writing—original draft preparation, A.S., A.T. and E.M.; writing—review and editing, E.M., S.E. and A.T.; visualization, E.M. and A.T.; supervision, A.S.; project administration, A.S.; funding acquisition, I.A. All authors have read and agreed to the published version of the manuscript.

**Funding:** S.C. and I.A. acknowledge financial support by the Ministry of Science and Higher Education of the Russian Federation in the framework of the project with a unique identifier RFMEFI61820X0006.

**Acknowledgments:** The authors would like to express their sincere gratitude to the handling editor Joy Sun for the invaluable help in preparation of the manuscript. The authors also would like to thank the anonymous reviewers for their thoughtful and detailed comments on our paper.

**Conflicts of Interest:** The authors declare no conflict of interest.

## References


1. Travitzky, N.; Bonet, A.; Dermeik, B.; Fey, T.; Filbert-Demut, I.; Schlier, L.; Schlordt, T.; Greil, P. Additive Manufacturing of Ceramic-Based Materials. *Adv. Eng. Mater.* **2014**, *16*, 729–754. [CrossRef]
2. Zocca, A.; Colombo, P.; Gomes, C.M.; Günster, J. Additive Manufacturing of Ceramics: Issues, Potentialities, and Opportunities. *J. Am. Ceram. Soc.* **2015**, *98*, 1983–2001. [CrossRef]
3. Deckers, J.; Vleugels, J.; Kruth, J.-P. Additive manufacturing of ceramics: A review. *J. Ceram. Sci. Technol.* **2014**, *5*, 245–260. [CrossRef]
4. Chen, Z.; Li, Z.; Li, J.; Liu, C.; Lao, C.; Fu, Y.; Liu, C.; Li, Y.; Wang, P.; He, Y. 3D printing of ceramics: A review. *J. Eur. Ceram. Soc.* **2019**, *39*, 661–687. [CrossRef]
5. Zhou, W.Z.; Li, D.; Chen, Z.W.; Chen, S. Direct fabrication of an integral ceramic mould by stereolithography. *Proc. Inst. Mech. Eng. Part B J. Eng. Manuf.* **2010**, *224*, 237–243. [CrossRef]
6. Chen, Z.; Li, D.; Zhou, W. Process parameters appraisal of fabricating ceramic parts based on stereolithography using the Taguchi method. *Proc. Inst. Mech. Eng. Part B J. Eng. Manuf.* **2012**, *226*, 1249–1258. [CrossRef]
7. Zhou, M.; Liu, W.; Wu, H.; Song, X.; Chen, Y.; Cheng, L.; He, F.; Chen, S.; Wu, S. Preparation of a defect-free alumina cutting tool via additive manufacturing based on stereolithography—Optimization of the drying and debinding processes. *Ceram. Int.* **2016**, *42*, 11598–11602. [CrossRef]
8. Nguyen, N.T.; Delhote, N.; Ettorre, M.; Baillargeat, D.; Le Coq, L.; Sauleau, R. Design and Characterization of 60-GHz Integrated Lens Antennas Fabricated Through Ceramic Stereolithography. *IEEE Trans. Antennas Propag.* **2010**, *58*, 2757–2762. [CrossRef]
9. Leigh, S.J.; Purssell, C.P.; Bowen, J.; Hutchins, D.A.; Covington, J.A.; Billson, D.R. A miniature flow sensor fabricated by micro-stereolithography employing a magnetite/acrylic nanocomposite resin. *Sens. Actuators A Phys.* **2011**, *168*, 66–71. [CrossRef]
10. Tian, Z.; Yang, Y.; Wang, Y.; Wu, H.; Liu, W.; Wu, S. Fabrication and properties of a high porosity h-BN–$SiO_2$ ceramics fabricated by stereolithography-based 3D printing. *Mater. Lett.* **2019**, *236*, 144–147. [CrossRef]
11. Wu, H.; Liu, W.; He, R.; Wu, Z.; Jiang, Q.; Song, X.; Chen, Y.; Cheng, L.; Wu, S. Fabrication of dense zirconia-toughened alumina ceramics through a stereolithography-based additive manufacturing. *Ceram. Int.* **2017**, *43*, 968–972. [CrossRef]
12. Liu, W.; Wu, H.; Tian, Z.; Li, Y.; Zhao, Z.; Huang, M.; Deng, X.; Xie, Z.; Wu, S. 3D printing of dense structural ceramic microcomponents with low cost: Tailoring the sintering kinetics and the microstructure evolution. *J. Am. Ceram. Soc.* **2018**, *102*, 2257–2262. [CrossRef]
13. He, R.; Liu, W.; Wu, Z.; An, D.; Huang, M.; Wu, H.; Jiang, Q.; Ji, X.; Wu, S.; Xie, Z. Fabrication of complex-shaped zirconia ceramic parts via a DLP- stereolithography-based 3D printing method. *Ceram. Int.* **2018**, *44*, 3412–3416. [CrossRef]
14. Chen, W.; Kirihara, S.; Miyamoto, Y. Fabrication and Measurement of Micro Three-Dimensional Photonic Crystals of SiO2Ceramic for Terahertz Wave Applications. *J. Am. Ceram. Soc.* **2007**, *90*, 2078–2081. [CrossRef]
15. Kirihara, S.; Niki, T. Three-Dimensional Stereolithography of Alumina Photonic Crystals for Terahertz Wave Localization. *Int. J. Appl. Ceram. Technol.* **2015**, *12*, 32–37. [CrossRef]
16. Sarment, D.P.; Al-Shammari, K.; Kazor, C.E. Stereolithographic surgical templates for placement of dental implants in complex cases. *Int. J. Periodontics Restor. Dent.* **2003**. [CrossRef]
17. Lian, Q.; Sui, W.; Wu, X.; Yang, F.; Yang, S. Additive manufacturing of ZrO2 ceramic dental bridges by stereolithography. *Rapid Prototyp. J.* **2018**, *24*, 114–119. [CrossRef]
18. Scalera, F.; Corcione, C.E.; Montagna, F.; Sannino, A.; Maffezzoli, A. Development and characterization of UV curable epoxy/hydroxyapatite suspensions for stereolithography applied to bone tissue engineering. *Ceram. Int.* **2014**, *40*, 15455–15462. [CrossRef]
19. Du, D.; Asaoka, T.; Ushida, T.; Furukawa, K.S. Fabrication and perfusion culture of anatomically shaped artificial bone using stereolithography. *Biofabrication* **2014**, *6*, 045002. [CrossRef]
20. Ribas, R.G.; Schatkoski, V.M.; Montanheiro, T.L.D.A.; De Menezes, B.R.C.; Stegemann, C.; Leite, D.M.G.; Thim, G.P. Current advances in bone tissue engineering concerning ceramic and bioglass scaffolds: A review. *Ceram. Int.* **2019**, *45*, 21051–21061. [CrossRef]
21. Yelten, A.; Yilmaz, S. A novel approach on the synthesis and characterization of bioceramic composites. *Ceram. Int.* **2019**, *45*, 15375–15384. [CrossRef]

22. Liu, Z.; Liang, H.; Shi, T.; Xie, D.; Chen, R.; Han, X.; Shen, L.; Wang, C.; Tian, Z. Additive manufacturing of hydroxyapatite bone scaffolds via digital light processing and in vitro compatibility. *Ceram. Int.* **2019**, *45*, 11079–11086. [CrossRef]
23. Wang, Z.; Huang, C.; Wang, J.; Zou, B. Development of a novel aqueous hydroxyapatite suspension for stereolithography applied to bone tissue engineering. *Ceram. Int.* **2019**, *45*, 3902–3909. [CrossRef]
24. Samavedi, S.; Whittington, A.R.; Goldstein, A.S. Calcium phosphate ceramics in bone tissue engineering: A review of properties and their influence on cell behavior. *Acta Biomater.* **2013**, *9*, 8037–8045. [CrossRef]
25. Evlashin, S.; Dyakonov, P.; Tarkhov, M.; Dagesyan, S.; Rodionov, S.; Shpichka, A.; Kostenko, M.; Konev, S.; Sergeichev, I.; Timashev, P.; et al. Flexible Polycaprolactone and Polycaprolactone/Graphene Scaffolds for Tissue Engineering. *Materials* **2019**, *12*, 2991. [CrossRef]
26. Yang, Y.; Wang, G.; Liang, H.; Gao, C.; Peng, S.; Shen, L.; Shuai, C. Additive manufacturing of bone scaffolds. *Int. J. Bioprinting* **2019**, *5*, 148. [CrossRef]
27. Gibson, I.; Rosen, D.W.; Stucker, B. *Additive Manufacturing Technologies*; Springer: New York, NY, USA, 2010.
28. Zhou, X.; Nowicki, M.; Cui, H.; Zhu, W.; Fang, X.; Miao, S.; Lee, S.-J.; Keidar, M.; Zhang, L.G. 3D bioprinted graphene oxide-incorporated matrix for promoting chondrogenic differentiation of human bone marrow mesenchymal stem cells. *Carbon* **2017**, *116*, 615–624. [CrossRef]
29. Jakus, A.E.; Secor, E.B.; Rutz, A.L.; Jordan, S.W.; Hersam, M.C.; Shah, R.N. Three-Dimensional Printing of High-Content Graphene Scaffolds for Electronic and Biomedical Applications. *ACS Nano* **2015**, *9*, 4636–4648. [CrossRef]
30. Entezari, A.; Roohani, I.; Li, G.; Dunstan, C.R.; Rognon, P.; Li, Q.; Jiang, X.; Zreiqat, H. Architectural Design of 3D Printed Scaffolds Controls the Volume and Functionality of Newly Formed Bone. *Adv. Healthc. Mater.* **2019**, *8*, 8. [CrossRef]
31. Piconi, C.; Porporati, A.A. Bioinert Ceramics: Zirconia and Alumina. In *Handbook of Bioceramics and Biocomposites*; Springer: Berlin/Germany, Germany, 2016; ISBN 9783319124605.
32. Ben-Nissan, B.; Choi, A.H.; Cordingley, R. Alumina ceramics. In *Bioceramics and Their Clinical Applications*; Woodhead Publishing: Cambridge, UK, 2008; ISBN 9781845692049.
33. Su, B.; He, X.; Dhara, S.; Mansell, J.P. Porous and bioactive alumina ceramics for bone grafts and tissue engineering scaffolds. In *Key Engineering Materials*; Trans Tech Publications Ltd.: Stafa, Switzerland, 2007; Volumes 330–332.
34. Ohji, T.; Fukushima, M. Macro-porous ceramics: Processing and properties. *Int. Mater. Rev.* **2012**, *57*, 115–131. [CrossRef]
35. Galante, R.; Figueiredo-Pina, C.G.; Serro, A.P. Additive manufacturing of ceramics for dental applications: A review. *Dent. Mater.* **2019**, *35*, 825–846. [CrossRef]
36. Jariwala, S.H.; Lewis, G.S.; Bushman, Z.J.; Adair, J.H.; Donahue, H.J. 3D Printing of Personalized Artificial Bone Scaffolds. *3D Print. Addit. Manuf.* **2015**, *2*, 56–64. [CrossRef]
37. Gulan, G.; Jurdana, H.; Gulan, L. Personalized Total Knee Arthroplasty: Better Fit for Better Function. In *Personalized Medicine in Healthcare Systems*; Springer: Cham, Switzerland, 2019; pp. 307–314.
38. Safonov, A.; Chugunov, S.; Tikhonov, A.; Gusev, M.; Akhatov, I. Numerical simulation of sintering for 3D-printed ceramics via SOVS model. *Ceram. Int.* **2019**, *45*, 19027–19035. [CrossRef]
39. Ceramic 3D Printing Solutions. Available online: https://3dceram.com/en/ (accessed on 5 August 2020).
40. Pasko, A.; Fryazinov, O.; Vilbrandt, T.; Fayolle, P.-A.; Adzhiev, V. Procedural function-based modelling of volumetric microstructures. *Graph. Model.* **2011**, *73*, 165–181. [CrossRef]
41. Shapiro, V. Real functions for representation of rigid solids. *Comput. Aided Geom. Des.* **1994**, *11*, 153–175. [CrossRef]
42. Bloomenthal, J.; Bajaj, C.; Blinn, J.; Wyvill, B.; Cani, M.P.; Rockwood, A.; Wyvill, G. (Eds.) *Introduction to Implicit Surfaces*; Morgan Kaufmann Publishers Inc.: San Francisco, CA, USA, 1997; ISBN 978-1-55860-233-5.
43. Kapfer, S.C.; Hyde, S.T.; Mecke, K.; Arns, C.H.; Schröder-Turk, G.E. Minimal surface scaffold designs for tissue engineering. *Biomaterials* **2011**, *32*, 6875–6882. [CrossRef]
44. Tikhonov, A.A.; Evdokimov, P.V.; Putlyaev, V.I.; Safronova, T.V.; Filippov, Y.Y. On the Choice of the Architecture of Osteoconductive Bioceramic Implants. *Inorg. Mater. Appl. Res.* **2019**, *10*, 242–247. [CrossRef]
45. Fryazinov, O.; Vilbrandt, T.; Pasko, A. Multi-scale space-variant FRep cellular structures. *CAD Comput. Aided Des.* **2013**, *45*, 26–34. [CrossRef]

46. Pasko, A.; Adzhiev, V.; Sourin, A.; Savchenko, V. Function representation in geometric modeling: Concepts, implementation and applications. *Vis. Comput.* **1995**, *11*, 429–446. [CrossRef]
47. Pasko, G.; Pasko, A.; Ikeda, M.; Kunii, T. Bounded blending operations. In Proceedings of the SMI 2002: Shape Modeling International 2002, Banff, AB, Canada, 17–22 May 2002; pp. 95–103.
48. Sanchez, M.; Fryazinov, O.; Fayolle, P.-A.; Pasko, A. Convolution Filtering of Continuous Signed Distance Fields for Polygonal Meshes. *Comput. Graph. Forum* **2015**, *34*, 277–288. [CrossRef]
49. Abaqus Analysis User Manual, Version 6.14. 2014. Available online: https://www.3ds.com/products-services/ (accessed on 5 August 2020).
50. Chugunov, S.; Adams, N.A.; Akhatov, I.S. Evolution of SLA-Based $Al_2O_3$ Microstructure during Additive Manufacturing Process. *Materials* **2020**, *13*, 3928. [CrossRef] [PubMed]
51. Hart, N.H.; Nimphius, S.; Rantalainen, T.; Ireland, A.; Siafarikas, A.; Newton, R.U. Mechanical basis of bone strength: Influence of bone material, bone structure and muscle action. *J. Musculoskelet. Neuronal Interact.* **2017**, *17*, 114–139. [PubMed]
52. Jamieson, R.; Hacker, H. Direct slicing of CAD models for rapid prototyping. *Rapid Prototyp. J.* **1995**, *1*, 4–12. [CrossRef]
53. Song, Y.; Yang, Z.; Liu, Y.; Deng, J. Function representation based slicer for 3D printing. *Comput. Aided Geom. Des.* **2018**, *62*, 276–293. [CrossRef]
54. Popov, D.; Maltsev, E.; Fryazinov, O.; Pasko, A.; Akhatov, I. Efficient contouring of functionally represented objects for additive manufacturing. *Comput. Des.* **2020**, *129*, 102917. [CrossRef]
55. Safonov, A.A. 3D topology optimization of continuous fiber-reinforced structures via natural evolution method. *Compos. Struct.* **2019**, *215*, 289–297. [CrossRef]

*applied sciences*

MDPI

*Article*

# Leveraging the Advantages of Additive Manufacturing to Produce Advanced Hybrid Composite Structures for Marine Energy Systems

**Paul Murdy [1,*], Jack Dolson [1], David Miller [2], Scott Hughes [1] and Ryan Beach [1]**

[1] National Renewable Energy Laboratory, Golden, CO 80401, USA; jack.dolson@nrel.gov (J.D.); scott.hughes@nrel.gov (S.H.); ryan.beach@nrel.gov (R.B.)
[2] Department of Mechanical and Industrial Engineering, Montana State University, Bozeman, MT 59715, USA; davidmiller@montana.edu
* Correspondence: paul.murdy@nrel.gov; Tel.: +1-303-275-4982

**Featured Application: This work outlines the opportunities for producing novel hybrid structures for marine energy systems by leveraging the benefits of combining additive and composite manufacturing techniques, and the potential this has for rapid prototyping.**

**Abstract:** Many marine energy systems designers and developers are beginning to implement composite materials into the load-bearing structures of their devices, but traditional mold-making costs for composite prototyping are disproportionately high and lead times can be long. Furthermore, established molding techniques for marine energy structures generally require many manufacturing steps, such as secondary bonding and tooling. This research explores the possibilities of additively manufactured internal composite molds and how they can be used to reduce costs and lead times through novel design features and processes for marine energy composite structures. In this approach, not only can the composite mold be additively manufactured but it can also serve as part of the final load-bearing structure. We developed a conceptual design and implemented it to produce a reduced-scale additive/composite tidal turbine blade section to fully demonstrate the manufacturing possibilities. The manufacturing was successful and identified several critical features that could expedite the tidal turbine blade manufacturing process, such as single-piece construction, an integrated shear web, and embedded root fasteners. The hands-on manufacturing also helped identify key areas for continued research to allow for efficient, durable, and low-cost additive/composite-manufactured structures for future marine energy systems.

**Keywords:** marine energy; tidal turbine; design for additive manufacturing; composite manufacturing; hybrid structures; composite structures

**Citation:** Murdy, P.; Dolson, J.; Miller, D.; Hughes, S.; Beach, R. Leveraging the Advantages of Additive Manufacturing to Produce Advanced Hybrid Composite Structures for Marine Energy Systems. *Appl. Sci.* **2021**, *11*, 1336. https://doi.org/10.3390/app11031336

Academic Editor: Marco Mandolini

Received: 4 January 2021
Accepted: 29 January 2021
Published: 2 February 2021



## 1. Introduction

Marine energy systems can take many forms, such as tidal/ocean current turbines, crossflow turbines, wave point absorbers, and oscillating surge flaps and water columns [1]. Because of the nascent nature of the industry, system designers and developers are yet to converge on a single design philosophy and may never do so because of the broad variety of potential operating conditions (see Figure 1). Many marine energy developers are still very much in the prototyping phases of their designs and system deployments. As a result, there is a great opportunity for low-cost, short-lead-time prototype manufacturing techniques to expedite the deployment of devices for testing and validation.

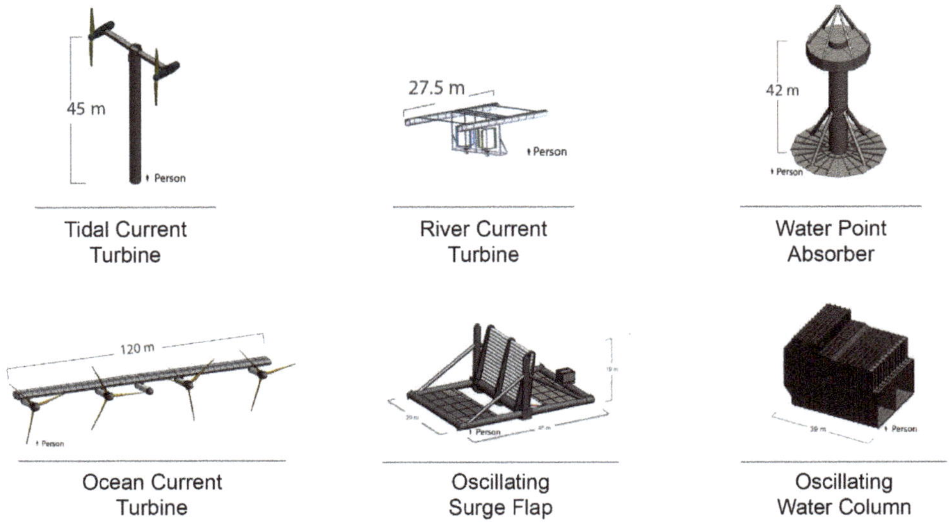


**Figure 1.** All six reference models developed by the Department of Energy Reference Model project [1]. Graphics courtesy of Sandia National Laboratories.

Composite materials, such as fiberglass/epoxy, are an attractive material choice for marine energy systems because of their low cost, high stiffness and strength, and resistance to environmental degradation [2]. Material density and structural mass are of lower importance because the structures are submerged underwater where buoyancy forces offset gravity-induced forces. In fact, marine energy structures are often filled with ballast such as concrete, epoxy slurry, or water to minimize buoyancy loads [3]. Naturally, a large amount of design and manufacturing knowledge has been taken from similar industries, such as wind blade manufacturing and composite boat building. Tidal turbine blades in particular are very similar in design to wind turbine blades, but on a smaller scale and with different loading and stiffness requirements [3]. They also rely on very similar manufacturing techniques and processes; for example, vacuum-assisted resin infusion molding (VARIM) of multiple shells and webs that are subsequently bonded together. The mold making for this process is incredibly expensive and time-consuming [4]. Depending on blade length, molds can cost in the order of hundreds of thousands to millions of dollars, with lead times in excess of 6 months [4], although very little public information is available in this area. This is acceptable on a mass-manufacturing scale in which hundreds of components will be produced from the same set of molds but may be unattainable for small-scale developers in the prototype testing phases of their development, particularly if designs are continuing to evolve. Despite this, large-scale mold manufacturing processes have changed very little in the last 20 years, highlighting a need for faster and cheaper molding practices for large-scale composite structure prototyping and development.

A variety of additive manufacturing processes have the potential to fill these gaps in the marine energy system structural design and validation process. In particular, recent advances in large-scale fused deposition modeling (FDM) have shown exceptional promise for reducing costs and lead times for wind turbine blade molds and components [5–7]. This is particularly exciting for exploring on-site manufacturing and prototyping unique test deployment concepts [8,9]. Another area of interest in FDM style printing is the co-extrusion of continuous fiber polymer composites [10], allowing for the direct printing of fiber-reinforced polymers with no molding required at all. This technology is still relatively new and only available at a small scale, with low-fiber-volume contents (<35%) [11] and a high probability of manufacturing defects like poor consolidation, but may provide an alternative to traditional composite manufacturing processes in the future.

Based on these current advances in additive manufacturing and industry needs, this research hypothesizes that they will allow for novel innovations in marine energy system composite structure prototyping and manufacturing. Instead of using additive manufacturing to produce external/female composite molds, we propose that internal/male composite molds can be additively manufactured and be a permanent part of the final structure, resulting in low-cost, short-lead-time hybrid structures. This approach would allow for novel manufacturing and structural features previously not possible with traditional composite manufacturing techniques. This article serves as a starting point to explore this unique manufacturing approach by identifying several key features and process steps that can be coalesced for potentially more efficient manufacturing and more reliable composite structures. We demonstrated the innovations via a conceptual design and manufacturing of a reduced-scale tidal turbine blade section, which enabled us to identify key features that were particularly beneficial to the manufacturing process and blade structural design. It also allowed us to pinpoint critical areas for future research into the additive/composite hybrid manufacturing process. In this research, the conceptual design was solely a design to demonstrate the manufacturing process, rather than complete holistic approach including the structural and hydrodynamic design. Future research will concentrate on the full structural design to allow for comprehensive structural, techno-economic and process time analyses.

## 2. Design Methods

To show the true potential for using additive manufacturing to produce internal composite molds for tidal turbine blades, we evaluated traditional mold and blade manufacturing practices to identify key process steps and features for innovation. We then used the analysis to develop a conceptual design of a hybrid composite/additively manufactured tidal turbine blade with novel integrated features.

### 2.1. Key Features

Figure 2 shows the general process steps for manufacturing a traditional tidal turbine blade mold set. The mold set generally consists of two halves: one for the low-pressure side of the turbine blade and one for the high-pressure side of the blade. First, a master plug is machined from structural foam, which is then sanded, finished, and prepared to create a splash mold from the low-pressure side of the blade plug. For the splash mold, fiberglass/epoxy or a material similar to the final blade is typically laid over the plug, cured, and removed from the plug. The same is repeated for the high-pressure side. Support structures are then built around the splash molds and their surfaces are sanded and prepared for composite manufacturing. Overall, the mold manufacturing process is slow, expensive, and labor-intensive. Figure 2 also shows how additive manufacturing can significantly simplify this process. In addition, using a permanent internal mold means that only one mold is required instead of two halves. After printing, molds typically require a moderate amount of machining and sanding to smooth the filament lines, which sometimes can be integrated into the printing process [5,7,12]. Finally, the molds are typically sealed to promote vacuum integrity for typical composite manufacturing techniques. Post et al. have demonstrated these techniques at a large scale for female molds and shown that large-scale additive manufacturing processes can significantly reduce the costs and lead times of large composite molds [5,7]. The disadvantage of using an internal mold as a permanent part of a hybrid structure is that one must be printed and prepared for every component to be manufactured, whereas a pair of high-quality traditional female molds can be used to produce hundreds of components, but structural or hydrodynamic design iterations cannot be implemented.

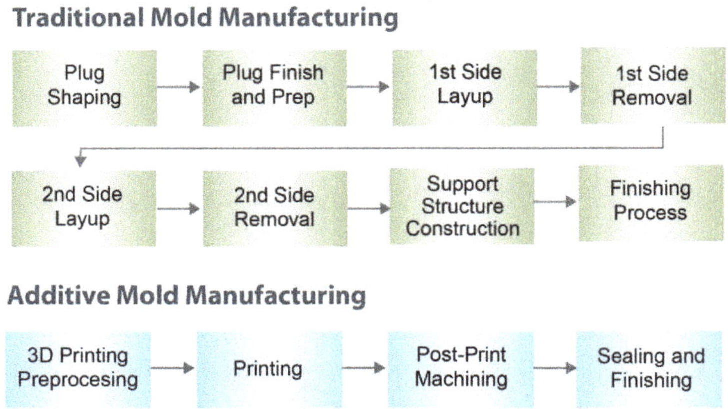


**Figure 2.** A comparison of the general process steps required to produce a traditionally manufactured composite mold vs. an additively manufactured composite mold.

Manufacturing composite tidal turbine blades typically requires several significant process steps (see Figure 3). Blades are made in multiple components: a high-pressure skin, a low-pressure skin, and a shear web or box spar depending on the length of the blade [4,13,14]. These components are often manufactured from fiber-reinforced composite and composite sandwich structures. The components are bonded together using structural adhesives. Several postprocessing steps are also required, such as trimming excess composite from the two bonded skins, and sanding and grinding the outer surfaces to their final hydrodynamic shapes. The root end of the blade must also be prepared for connection to the hub of the tidal turbine, typically in the form of a bolted connection. Bolted connections in composite laminates and composite/metal interfaces can be particularly problematic [15], and a variety of different designs exist in this area. The composite layup in the root section is built particularly thick, so that lengthwise holes can be included or drilled through the laminate and threaded sleeves can be bonded or secured inside. This can be a particularly time-consuming and expensive manufacturing step, as a result of expensive preforms, adhesives, and specialized equipment.

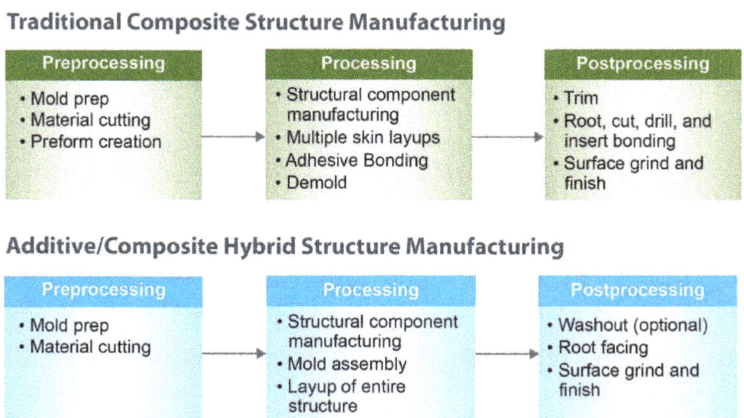


**Figure 3.** Comparison of the generalized traditional process steps required to manufacture composite tidal turbine blades [4,13] vs. those that are proposed using the additively manufactured internal molds with novel integrated features.

Manufacturing a tidal turbine blade with a single internal permanent mold may allow for significant innovations to reduce the manufacturing steps shown at the top of Figure 3 and potentially increase the reliability of the structural design. A major advantage of internal molding is that composite plies can be wrapped around the outside of critical design features, such as leading and trailing edges, to create a continuous construction in failure-prone areas, with continuous fibers. Adhesive bond lines are a common place for manufacturing defects, which have regularly contributed to structural failures of wind and tidal turbine blades [16–19]. Structural adhesives are also significantly more expensive per kg than the low-cost composites they typically bond together [4]. Eliminating these adhesives has generally been hypothesized to lead to significant cost and time savings [4,20].

The same could be achieved with shear webs and box spars. Instead of adhesively bonding premade shear webs during the final blade assembly, an additively manufactured mold could be designed in such a way that it could also serve as a fixture to hold a premade shear web in place to be co-infused with the rest of the composite during the VARIM process. Specifically, printed fixtures to hold mechanical fasteners in place during composite layup have also proven to be beneficial for complex composite component construction [21,22]. This technique could be used for root fasteners and other bolted connections. A well-designed fixture could be printed into the mold to hold the fasteners in place while glass or carbon fabric is wrapped on the top and bottom of the fasteners so they can be co-infused into the structure. More expensive adhesive would be eliminated from the structure, as well as a significant postprocess machining step.

During the layup of the composite for the VARIM process, resin infusion lines are typically placed on top to direct and control the flow of resin through the glass/carbon fabrics. It is often difficult to get good resin flow through thick stacks of fabrics, resulting in porosity and dry areas. Additively manufactured internal molds could allow for the integration of infusion lines directly into the mold for precise control of the resin flow through all layers of fabric. Resin would flow from the mold and into the glass/carbon fabrics, potentially reducing the likelihood of detrimental levels of porosity.

Blades for tidal/ocean current turbines are anticipated to grow to be in excess of 20 m in length, depending on their site location [1]. Blades of this length can be expensive to transport across land, so there is a desire for on-site manufacturing in shipyards and dry docks to minimize these costs. Segmented mold construction would ensure easy on-site manufacturing or transportation to the manufacturing site. Blade length would not be limited by the size of 3D printers and manufacturing sites. Additively manufactured molds could easily and efficiently be printed in sections and assembled on-site before beginning the composite manufacturing process.

### 2.2. Conceptual Design

We selected a conceptual tidal turbine blade section for this manufacturing demonstration with an additively manufactured internal mold, incorporating all of the novel features and unique innovations described earlier. The blade sections' geometry was based on the blades from the Reference Model 1 (RM1) tidal turbine from the U.S. Department of Energy Reference Model Project [3]. RM1 is a turbine with 9-m rotor blades with NACA $63_1$-424 primary airfoil shapes. The blade section was designed to include the root section through to the maximum chord region (~1.5 m outboard). It should be noted that we simplified the overall blade geometry to some degree for this research. The main goal was to focus on the manufacturing aspects of the blade design, so it is not structurally or hydrodynamically optimized in any way. The demonstration is primarily meant to be representative of a tidal turbine blade.

The additively manufactured mold was to be printed in four separated sections (see Figure 4): a root fixture (a), a root section (b), a transition section (c), and a main foil section (d). The four sections were designed to join together with steel dowel pins for easy and fast construction of the mold. Internal resin tubes were designed to be printed directly into the internal cavities of the sections, which were aligned by the dowel pins (see Figure 5b). They were designed so that during the VARIM process, resin could flow from the resin inlet ports in the root section along the length of the blade and then to the high- and low-pressure surfaces into the composite layers.

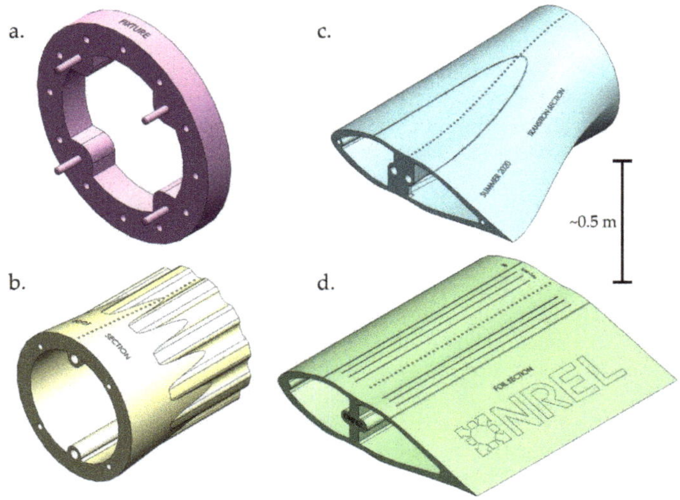


**Figure 4.** Models of the four mold sections to be printed for the conceptual tidal turbine blade section design where (**a**) is the sacrificial root fixture, (**b**) is the root section, (**c**) is the transition section, and (**d**) is the main hydrofoil section.

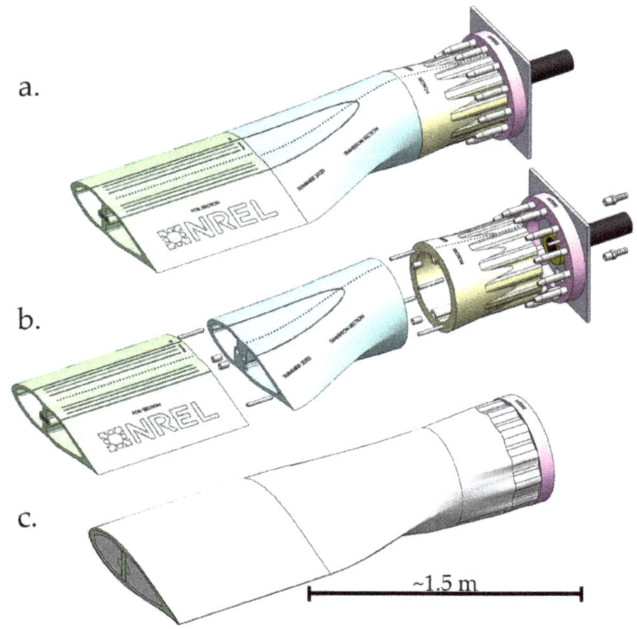


**Figure 5.** Full assembly drawings of the blade section mold, where (**a**) is fully assembled, (**b**) is exploded to show the connections between sections, and (**c**) includes the composite overlay.

A shear web fixture was also integrated into the main foil section. It was designed so that a premade web could slide into the fixture, which was then sealed when the mold sections were assembled. The integrated resin tubes were located so resin could flow into the remaining free volume around the shear web before continuing to the surface of the mold, co-infusing the web in place with the resin.

The root section and fixture are particularly unique. The root fixture is a sacrificial feature, functioning as an alignment and fixturing member for the 12 threaded root inserts around the printed root section. The 3D-printed root mold section is designed to account for the glass/carbon fabric and the cylindrical shape of the inserts. To build up the composite root, half of the root plies are laid down before attaching the root inserts, then the remaining root plies can be laid over the top of the inserts. The inserts themselves are tapered at the end to prevent a sudden change in geometry for the overlaying plies. The root fixture is a sacrificial print to be faced off after the composite manufacturing process is complete, leaving a clean, circular root section with the threaded fasteners in situ.

Overall, the outer additively manufactured mold shape and composite layup must be designed concurrently. The outer shape of the mold accounts for the varying thicknesses of the composite layup to preserve the intended hydrodynamic shapes and profiles. The root of the blade has a thick composite buildup that gradually tapers in thickness toward the transition section, hence the reverse taper of the root mold section to be printed. Thicker spar cap sections are also included in the composite design to taper in thickness along the length of the transition and foil sections. Their influence on the mold shape can be observed at the center chord of the main foil section (Figure 4d). The blade is also designed to have a thin, continuous skin layer covering the entire outer surface of the blade section.

Finally, a thick, reusable, machined steel plate was mounted to the root fixture, so the fully assembled mold could be secured to a stand. This allowed the mold to be easily maneuvered and rotated to expedite the composite layup and infusion process.

## 3. Manufacturing

To properly demonstrate the potential of internal molds for hybrid composite marine energy structures, the conceptual design described above was used to manufacture a reduced-scale model. The purpose of the model was not only to display the capabilities but also to clearly identify specific features that were particularly useful, as well as those that were not. It also helped us recognize key areas for continued research to allow this design and manufacturing approach to be viable for future iterations. The blade section was reduced to a blade approximately 0.75 m in length with a 0.3 m maximum chord length. Manufacturing was conducted at the National Renewable Energy Laboratory's Composites Manufacturing Education and Technology (CoMET) facility.

### 3.1. Additively Manufactured Mold Preparation

The four mold sections were printed from acrylonitrile butadiene styrene (ABS) plastic with a Stratasys Fortus 400MC FDM printer. The sections were printed with a double-dense sparse in-fill to reduce material usage and printing times and ensure geometric integrity when vacuum pressure was applied for the VARIM process. Before assembly, each mold section was sealed to ensure that vacuum integrity could be maintained throughout the resin infusion process. This was done following Stratasys's recommendations for sealing additively manufactured composite molds [12]. The internal and external surfaces were sanded, coated with a fast-curing wet-layup epoxy from West Systems, and sanded again (see Figure 6).

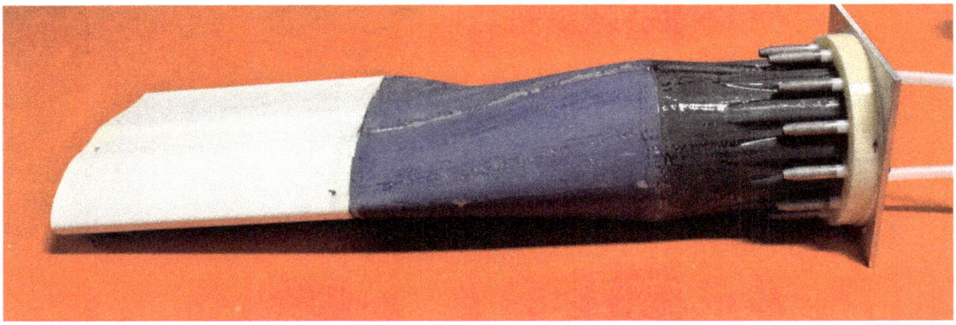

**Figure 6.** The 3D-printed mold sections after being sanded, sealed with epoxy, and fully assembled, ready for the composite manufacturing process. The root inserts were bolted in place for display purposes.

After sealing, the mold sections were joined and bonded using the prescribed dowel pins, nylon bushings, and structural adhesive, although other techniques such as thermal welding could have achieved the same effect without the expensive adhesive [12,20]. The premade shear web was also installed before bonding the main foil and transition section together, so it was secured in place for the VARIM process. The shear web was a flat sandwich panel manufactured from epoxy-resin-infused triaxial fiberglass and a structural foam core. Finally, the mold was vacuum-bagged, and a leak test was performed to confirm adequate vacuum integrity for the VARIM process. Figure 6 shows the final assembled additively manufactured mold ready for the composite layup and infusion process.

### 3.2. Composite Layup and Infusion

As mentioned in Section 2.2, we designed the composite layup concurrently with the additively manufactured mold to achieve the desired outer hydrodynamic profiles. A triaxial fiberglass fabric with an average ply thickness of 0.9 mm was used for all areas of the blade section. The root was designed to incorporate 14 plies tapering in length over the root section of the mold. The spar caps were designed to be four plies thick and tapered in both length and width toward the main foil section of the mold. Finally, a single skin ply was to be wrapped over the full length and chord of the mold. A structural flow fiberglass media was used as the first ply to encourage resin flow across the entire mold. As previously stated, this was not intended to be an optimized structural design, but rather a representative design for manufacturing purposes.

Figure 7 shows the layup process. First, the structural flow media was applied to the mold and held in place with spray adhesive. The first half of the root plies could then be applied, followed by all 12 of the root inserts (see Figure 7a). The remaining 7 root plies were then placed over the top of the inserts (see Figure 7b). The high-pressure and low-pressure spar cap plies were then installed (see Figure 7c), followed by the final skin ply, which was wrapped tightly around the leading and trailing edges to ensure tight contours in those critical areas (see Figure 7d). The blade section was then vacuum-bagged with vacuum lines installed along the leading and trailing edges to ensure resin flow from the center of the mold over the entire surface and to control it during the infusion process.

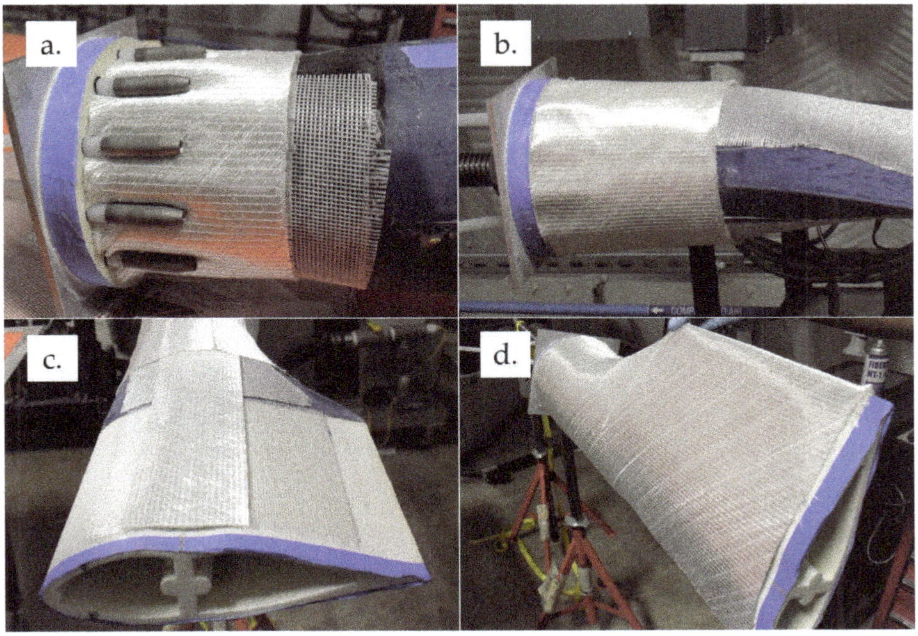


**Figure 7.** Showing the composite layup process with (**a**) the first half of the root plies and root inserts installed, (**b**) the second half of the root plies, (**c**) the spar caps, and (**d**) the final skin ply.

For the resin infusion process, we used Hexion's Epikote RIMR135/Epikure RIMH1366 two-part epoxy resin system, which is commonly used in the manufacturing of wind turbine blades [23]. Figure 8 shows the VARIM process. The transition section wet out first, followed by the thicker root section. The main foil section was last to fully saturate because the epoxy resin had to flow around the shear web before reaching the main surfaces. Overall, the entire blade section became fully saturated with epoxy as intended.

**Figure 8.** Photo taken mid-infusion, showing the epoxy resin saturating the fiberglass around the transition section and beginning to flow through the root and main foil sections.

### 3.3. Final Finishing

Once cured, the vacuum bag and other infusion materials were removed from the blade section. The root was then prepared by removing the fasteners and slicing off the sacrificial root fixture portion of the mold. The surface was then faced to be flush with the threaded root inserts. Figure 8 shows the draping of the triaxial fiberglass fabric around the cylindrical insert and the remaining root section of the mold in place.

The end of the main foil section was also sliced to better observe the resin flow through the fixture surrounding the premade shear web (see Figure 9). The epoxy resin flowed well around the web, with no visible dry spots around the structural foam. This cross-section also shows that the resin flowed into portions of the reduced density cavities in the 3D-printed mold.

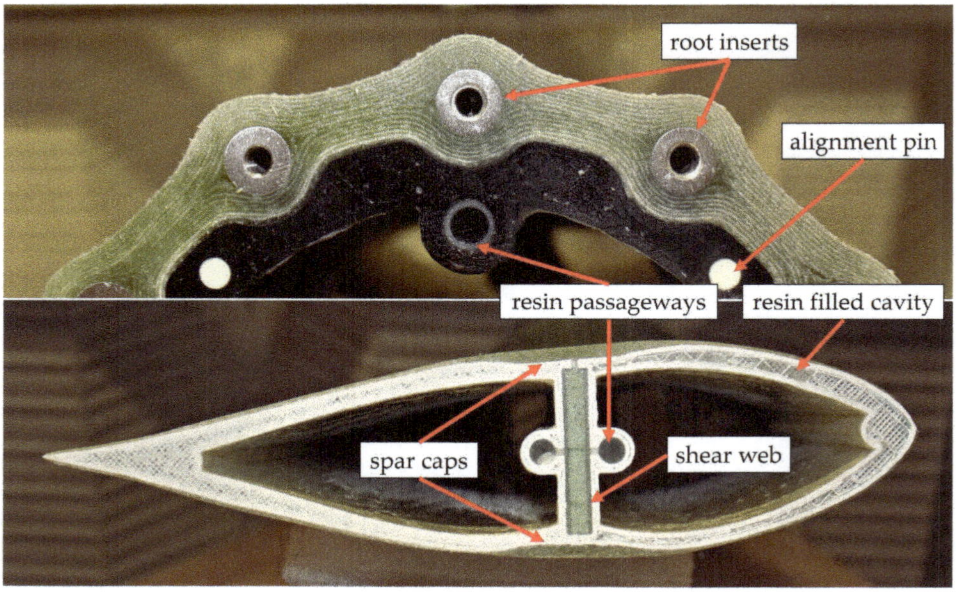


**Figure 9.** The final faced surface of the root end of the blade section (**top**) showing the draping of the fiberglass plies over and under the root inserts, and the main foil section (**bottom**) to show the profiles of the final spar caps and shear web. Epoxy resin in the sparsely filled internal cavities of the 3D print can also be observed.

The surface was ground and sanded smooth using an orbital sander and gradually increasing sandpaper grits, as is standard practice, until the surface was free of visible waves and discontinuities. Figure 10 shows the final, finished, reduced-scale tidal turbine blade section. Overall, the additively manufactured internal mold approach to manufacturing the blade section required significantly less time and effort than applying conventional mold manufacturing techniques to produce two half-blade section molds and then manufacturing a blade section from them. This is in comparison with the extensive experience gained through conventional manufacturing of tidal and wind turbine blades previously in the CoMET facility [13,14], as well as with a variety of industry partners. The facing of the blade root was quick and simple, and there were absolutely no secondary bonding processes required. Surface grinding and finishing was more time-consuming than typical but was more than made up for by the reduction and elimination of other process steps.

**Figure 10.** The final finished blade section with visible out-of-plane waves stemming from the root inserts and the delamination along the leading edge of the transition region.

## 4. Discussion

The goal of this research was to identify the manufacturing potential for using additive manufacturing for internal molding to produce hybrid composite structures for marine energy systems. Through this research, we demonstrated the many potential benefits and key features that permanent additively manufactured tooling could allow for. As a proof of concept, we conceptually designed and manufactured a reduced-scale tidal turbine blade section, which integrated a number of novel features made possible through additive manufacturing. Gaining hands-on experience allowed for a much greater understanding of how to properly implement the technology in a manufacturing environment and helped identify important requirements for future research. This section provides a discussion of the key findings from the design and manufacturing process.

### 4.1. Mold Design

The process of designing the additively manufactured mold itself did not have a significant influence on the design process from a manufacturing perspective. Once the composite ply layout had been designed, the composite thicknesses were subtracted from the intended net external geometry of the blade section, leaving the final outer surface geometry of the mold to be printed. This required some intuition during the CAD drawing process but could be much more streamlined with the use of specialized composite design and ply draping software. The implementation of the integrated shear web, resin infusion lines, root fasteners, and mold segmentation were also straightforward in the design for manufacturing approach, although ensuring that the complex internal structures of the mold could be perfectly sealed required forethought. Nonetheless, there is plenty of room for further optimization of this design in the future, especially from a structural integrity perspective.

Another area for extensive optimization in the additively manufactured mold design is the in-fill and wall thickness parameters for the internal cavities of the 3D print. We chose a sparse, double-dense in-fill to reduce material usage and print times, and also to minimize distortion of the mold as a result of vacuum pressure (see Figure 9). In future designs, this aspect must be carefully considered, especially if the mold is intended to remain as a load-bearing component in the full-scale hybrid additively manufactured/composite structure. Segmented optimization or even full topology optimization of mold designs could give way to highly efficient, advanced hybrid composite structures. In addition, print and lead times, as well as material costs, could be significantly reduced for prototype and large-scale-manufacturing marine energy structures. This will be a specific focus for future research.

Material selection for the blade section design and manufacturing was relatively conservative. The mold was printed with ABS plastic because of its low cost, and the composite was fiberglass infused with a room-temperature-cure epoxy. We chose this specific epoxy because of its low exotherm temperatures during curing, so it was compatible with the ABS mold. ABS has a relatively low heat deformation temperature and may have permanently deformed if a higher temperature cure resin was used. To 3D-print a mold also intended to be a structural element, chopped fiber-filled or even continuous fiber thermoplastic filaments would be a better choice, because of their higher stiffnesses, strengths, and heat deformation temperatures. Their lower coefficients of thermal expansion would also be beneficial. In turn, this would allow for a much broader selection of resin systems for the VARIM process. Novel, infusible thermoplastic resin systems are showing great promise for renewable composite applications [13]. They exhibit similar mechanical properties to epoxy resin systems, with the added benefit of being recycled more easily [24]. The combination of a thermoplastic, 3D-printed hybrid mold with a thermoplastic composite would allow for fully recyclable structures at the end of their service life [25]. This would be an important economic and sustainable benefit for the marine energy industry.

### 4.2. Manufacturing

Once printed, the mold sections were sanded and then sealed with a fast-curing, wet layup epoxy resin system. This was by far the most time-consuming step of the manufacturing process, mainly because of the complex internal shapes of the mold. The inside of the resin infusion passageways and the shear web fixture could not be sealed. This was sufficient to provide a good vacuum seal for the VARIM process but did lead to other issues. The resin was drawn into the internal cavities of the 3D prints (see Figure 9), leading to a significant increase in the resin required for the infusion. Based on estimates from an assumed fiber mass fraction for the VARIM process, the total mass of the fiberglass, and the volumes of the internal mold cavities, the part absorbed almost three times as much epoxy as was calculated to be required. Because of this, better sealing methods are required for 3D-printed molds with hard-to-reach, complex geometries and design features. Techniques such as vapor smoothing with acetone [26] may be very useful for this type of application, but will require stricter controls when applied on a larger scale. More research is required in this area.

Hands-on manufacturing of the conceptual design provided a true understanding of which design features were particularly beneficial for the manufacturing process. The segmented mold construction was simple to implement and would be beneficial for transportation or on-site manufacturing of large components. In fact, it made sealing the harder-to-reach areas of the mold easier while it was still segmented. Integrating the root fasteners also worked as intended, as well as the integrated shear web fixture. The premade shear web was easily inserted and sealed into the mold for infusion. On a larger scale, more considerations would need to be made for the thickness taper and hydrodynamic twist of the tidal turbine blade. Both the integrated root fasteners and shear web features proved to be incredibly useful, removed time-consuming steps from the manufacturing process, and eliminated the need for costly, defect-prone adhesive bonding steps.

The only design features that proved to be a hindrance were the integrated resin passageways. As described earlier, it was not possible to seal the inside of them, which led to excessive resin uptake. In this case, they were not worth the additional design and manufacturing effort, although their presence was useful for providing resin flow around the shear web. Despite the disadvantages, this research has demonstrated that it is possible to implement integrated resin passageways as a 3D-printed design feature. They may still become useful for other more complex composite geometries in the future, wherein accurate resin flow is critical.

Final finishing of the blade section was a critical step in the manufacturing process. It was already understood that this was where the novel internal molding process would be at a disadvantage to traditional female molding techniques. Overall, the finishing process was not much more time-consuming than is typically required for a tidal turbine blade based on past manufacturing experience in the CoMET, but other issues were uncovered. The spar caps needed a substantial amount of material removed to meet the designed hydrodynamic surface geometry. This would be detrimental to the integrity of a full-scale blade structure. Great care also needed to be taken around the trailing edge region to avoid grinding through the skin plies to the mold surface. In light of these issues, the grinding and sanding process could be accounted for in the design process. Low-cost "sacrificial" plies, such as chopped-strand mats could be included in the composite layup design, so that the critical skin and spar cap plies would not get damaged by the finishing process.

Figure 10 shows some visible out-of-plane waves in the composite around the root of the blade section that were not removed during the final finishing process. Although they were far away from the main hydrofoil portion of the blade section, manufacturing defects like these can become damage initiation points under loading of the structure [27]. They were caused by the sharp changes in geometry of the mold surface and root inserts, resulting in the triaxial fabric wrinkling. Further design of the root insert geometry or different glass fabric selection would help alleviate these wrinkles. The triaxial fabric used was relatively stiff and not easily draped over complex contours. The drapability of fabrics should be carefully considered for future designs when sharp contours are present. The scale of the blade section may have contributed to this issue. It is expected that the waves would be much less prominent on a larger-scale component.

The final defect observed on the manufactured blade section was a large delamination along the leading edge of the transition region (see Figure 10). This was caused by the removal of the vacuum bag from the cured part. Ultimately, the 3D-printed mold will be intended to be a load-bearing component of the composite structure, so delaminations such as these would be unacceptable, and good adhesion between the 3D print and the composite overlay is critical. The mold surface was prepared in a comparable way to adhesive bonding surfaces but, clearly, that was insufficient. More research is required to develop surface preparation practices for bonding these dissimilar materials. It appears there is currently very little information available for such requirements.

To summarize, the main goal of this research was to investigate the potential for additive manufacturing to be applied to the creation of marine energy composite structures through internal molding to produce advanced hybrid structures. Our investigation has indicated that additive manufacturing of internal composite molds can allow for a unique design approach which can have the potential to reduce costs and lead times. Future research will focus on a fully optimized structural design, which will then allow for proper in-depth techno-economic analyses for comparison with conventional manufacturing techniques. Ultimately, additive manufacturing of internal hybrid structural molds could help pave the way for a new generation of low-cost, highly reliable, recyclable, and advanced marine energy composite structures.

**Author Contributions:** Conceptualization, P.M. and D.M.; methodology, P.M., J.D., and R.B.; design, P.M., J.D., D.M., S.H., and R.B.; manufacturing and validation, P.M., J.D., and R.B.; writing—original draft preparation, P.M.; writing—review and editing, J.D., D.M., S.H., and R.B.; funding acquisition, P.M., D.M., and S.H. All authors have read and agreed to the published version of the manuscript.

**Funding:** This work was authored by the National Renewable Energy Laboratory, operated by the Alliance for Sustainable Energy, LLC, for the U.S. Department of Energy (DOE) under Contract No. DE-AC36-08GO28308. Funding provided by the U.S. Department of Energy Office of Energy Efficiency and Renewable Energy Water Power Technologies Office. The views expressed in the article do not necessarily represent the views of the DOE or the U.S. Government. The U.S. Government retains and the publisher, by accepting the article for publication, acknowledges that the U.S. Government retains a nonexclusive, paid-up, irrevocable, worldwide license to publish or reproduce the published form of this work, or allow others to do so, for U.S. Government purposes.

**Institutional Review Board Statement:** Not applicable.

**Informed Consent Statement:** Not applicable.

**Data Availability Statement:** The data presented in this study are available on request from the corresponding author.

**Acknowledgments:** The authors would like to thank all the engineers and technicians who provided their knowledge and expertise for this project, particularly David Barnes and Casey Nichols. The authors would also like to thank Bernadette Hernandez-Sanchez from Sandia National Laboratories and Lauren Ruedy from the U.S. Department of Energy's Water Power Technologies Office for their valued input throughout the project. Finally, we would like to thank Paul Ubrich from Hexion Inc. for supplying the epoxy resin system used in this study.

**Conflicts of Interest:** The authors declare no conflict of interest.

## References

1. Neary, V.S.; Previsic, M.; Jepson, R.A.; Lawson, M.J.; Yu, Y.H.; Copping, A.E.; Fontaine, A.A.; Hallet, K.C.; Murray, D.K. *Methodology for Design and Economic Analysis of Marine Energy Conversion (MEC) Technologies*; Sandia National Larboratories: Albuquerque, NM, USA, 2014.
2. Nunemaker, J.D.; Voth, M.M.; Miller, D.A.; Samborsky, D.D.; Murdy, P.; Cairns, D.S. Effects of moisture absorption on damage progression and strength of unidirectional and cross-ply fiberglass-epoxy laminates. *Wind Energy Sci.* **2018**, *3*, 427–438. [CrossRef]
3. Bir, G.; Lawson, M.J.; Li, Y. Structural design of a horizontal-axis tidal current turbine composite blade. In Proceedings of the ASME 30th International Conference on Ocean, Offshore and Arctic Engineering, Rotterdam, The Netherlands, 19–24 June 2011. [CrossRef]
4. Murray, R.E.; Jenne, S.; Snowberg, D.; Berry, D.; Cousins, D.S. Techno-economic analysis of a megawatt-scale thermoplastic resin wind turbine blade. *Renew. Energy* **2019**, *131*, 111–119. [CrossRef]
5. Post, B.; Richardson, B.; Lloyd, P.; Love, L.; Nolet, S.; Hannan, J. *Additive Manufacturing of Wind Turbine Molds*; Oak Ridge National Laboratory: Oak Ridge, TN, USA, 2017. [CrossRef]
6. Post, B.K.; Chesser, P.C.; Roschli, A.C.; Love, L.J.; Gaul, K.T. *Large-Scale Additive Manufacturing for Low Cost Small-Scale Wind Turbine Manufacturing*; Oak Ridge National Laboratory: Oak Ridge, TN, USA, 2018. [CrossRef]
7. Post, B.K.; Richardson, B.; Lind, R.; Love, L.J.; Lloyd, P.; Kunc, V.; Rhyne, B.J.; Roschli, A.; Hannan, J.; Nolet, S.; et al. Big Area Additive Manufacturing Application in Wind Turbine Molds. In Proceedings of the 28th Annual International Solid Freeform Fabrication Symposium—An Additive Manufacturing Conference, Austin, TX, USA, 7–9 August 2017.
8. Bassett, K.; Carriveau, R.; Ting, D.S.K. 3D printed wind turbines part 1: Design considerations and rapid manufacture potential. *Sustain. Energy Technol. Assess.* **2015**, *11*, 186–193. [CrossRef]
9. Li, H.; Taylor, G.; Bheemreddy, V.; Iyibilgin, O.; Leu, M.; Chandrashekhara, K. Modeling and characterization of fused deposition modeling tooling for vacuum assisted resin transfer molding process. *Addit. Manuf.* **2015**, *7*, 64–72. [CrossRef]
10. Wang, X.; Jiang, M.; Zhou, Z.; Gou, J.; Hui, D. 3D printing of polymer matrix composites: A review and prospective. *Compos. Part B Eng.* **2017**, *110*, 442–458. [CrossRef]
11. van der Kilft, F.; Koga, Y.; Todoroki, A.; Ueda, M.; Hirano, Y.; Matsuzaki, R. 3D Printing of Continuous Carbon Fibre Reinforced Thermo-Plastic (CFRTP) Tensile Test Specimens. *Open J. Compos. Mater.* **2016**, *6*, 18–27. [CrossRef]
12. Stratasys. *FDM for Composite Tooling 2.0: Design Guide*; Stratasys, Ltd.: Eden Prairie, MN, USA, 2017.
13. Murray, R.E.; Snowberg, D.; Berry, D.; Beach, R.; Rooney, S.; Swan, D. Manufacturing a 9-meter thermoplastic composite wind turbine blade. In Proceedings of the American Society for Composites 32nd Technical Conference, West Lafayette, IN, USA, 23–25 October 2017. [CrossRef]
14. Murray, R.E.; Beach, R.; Barnes, D.; Snowberg, D.; Berry, D.; Rooney, S.; Jenks, M.; Gage, B.; Boro, T.; Wallen, S.; et al. Structural validation of a thermoplastic composite wind turbine blade with comparison to a thermoset composite blade. *Renew. Energy* **2021**, *164*, 1100–1107. [CrossRef]
15. Camanho, P.P.; Matthews, F.L. Stress analysis and strength prediction of mechanically fastened joints in FRP: A review. *Compos. Part A Appl. Sci. Manuf.* **1997**, *28*, 529–547. [CrossRef]
16. Al-Khudairi, O.; Hadavinia, H.; Little, C.; Gillmore, G.; Greaves, P.; Dyer, K. Full-Scale Fatigue Testing of a Wind Turbine Blade in Flapwise Direction and Examining the Effect of Crack Propagation on the Blade Performance. *Materials* **2017**, *10*, 1152. [CrossRef] [PubMed]
17. Chen, X.; Berring, P.; Madsen, S.H.; Branner, K.; Semenov, S. Understanding progressive failure mechanisms of a wind turbine blade trailing edge section through subcomponent tests and nonlinear FE analysis. *Compos. Struct.* **2019**, *214*, 422–438. [CrossRef]
18. Jensen, F.M.; Falzon, B.G.; Ankersen, J.; Stang, H. Structural testing and numerical simulation of a 34 m composite wind turbine blade. *Compos. Struct.* **2006**, *76*, 52–61. [CrossRef]
19. Lahuerta, F.; Koorn, N.; Smissaert, D. Wind turbine blade trailing edge failure assessment with sub-component test on static and fatigue load conditions. *Compos. Struct.* **2018**, *204*, 755–766. [CrossRef]
20. Murray, R.E.; Roadman, J.; Beach, R. Fusion joining of thermoplastic composite wind turbine blades: Lap-shear bond characterization. *Renew. Energy* **2019**, *140*, 501–512. [CrossRef]
21. Türk, D.-A.; Kussmaul, R.; Zogg, M.; Klahn, C.; Leutenecker-Twelsiek, B.; Meboldt, M. Composites Part Production with Additive Manufacturing Technologies. *Procedia CIRP* **2017**, *66*, 306–311. [CrossRef]

22. Türk, D.-A.; Triebe, L.; Meboldt, M. Combining Additive Manufacturing with Advanced Composites for Highly Integrated Robotic Structures. *Procedia CIRP* **2016**, *50*, 402–407. [CrossRef]
23. Murdy, P. Combining Acoustic Emission and Guided Ultrasonic Waves for Global Property Prediction and Structural Health Monitoring of Glass Fiber Composites. Ph.D. Thesis, Montana State University, Bozeman, MT, USA, 2018.
24. Cousins, D.S.; Suzuki, Y.; Murray, R.E.; Samaniuk, J.R. Recycling glass fiber thermoplastic composites from wind turbine blades. *J. Clean. Prod.* **2019**, *209*, 1252–1263. [CrossRef]
25. Rybnicek, J.; Lach, R.; Lapcikova, M.; Steidl, J.; Krulis, Z.; Grellmann, W.; Slouf, M. Increasing recyclability of PC, ABS and PMMA: Morphology and fracture behavior of binary and ternary blends. *J. Appl. Polym. Sci.* **2008**, *109*, 3210–3223. [CrossRef]
26. Singh, R.; Singh, S.; Singh, I.P.; Fabbrocino, F.; Fraternali, F. Investigation for surface finish improvement of FDM parts by vapor smoothing process. *Compos. Part B Eng.* **2017**, *111*, 228–234. [CrossRef]
27. Murdy, P.; Hughes, S. Investigating Core Gaps and the Development of Subcomponent Validation Methods for Wind Turbine Blades: Preprint. In Proceedings of the AIAA SciTech Forum, Orlando, FL, USA, 6–10 January 2020. [CrossRef]